NEW CONCEPTS OF ANTIVIRAL THERAPY

New Concepts of Antiviral Therapy

Edited by

ELKE BOGNER

*Friedrich-Alexander Universität Erlangen-Nürnberg,
Institut für Klinische und Molekulare Virologie,
Erlangen, Germany*

and

ANDREAS HOLZENBURG

Texas A&M University, Microscopy and Imaging Centre, College Station, U.S.A.

 Springer

A C.I.P. Catalogue record for this book is available from the Library of Congress.

ISBN-10 0-387-31046-0 (HB)
ISBN-13 978-0-387-31046-6 (HB)
ISBN-10 0-387-31047-9 (e-books)
ISBN-13 978-0-387-31047-3 (e-books)

Published by Springer,
P.O. Box 17, 3300 AA Dordrecht, The Netherlands.

www.springer.com

Printed on acid-free paper

Dedication

This Book is dedicated to our parents and our colleagues Dr. Ernst Kuechler & Dr. Susanna Prösch who passed away in 2005

Contents

Part 1. Concepts of therapy for RNA viruses

Part 2. Concepts of therapy for DNA viruses

Part 3. Concepts of therapy for emerging viruses

Part 4. General concepts of therapy

Contributors

Dr. Stephen Barr Department of Microbiology, University of Pennsylvania, 3610 Hamilton Walk, Philadelphia, PA 19104, USA

Dr. Ralf Bartenschlager Department for Molecular Virology, Hygiene Institute, University of Heidelberg, Im Neuenheimer Feld 345, 69120 Heidelberg, Germany

Dr. Karen K. Biron International Clinical Virology, GlaxoSmithKline R&D, Research Triangle Park, NC 27709, USA

Dr. Elke Bogner Institut für Molekulare und Klinische Virologie, Universität Erlangen-Nürnberg, Schloßgarten 4, 91054 Erlangen, Germany

Dr. Peter Borowski Abteilung für Virologie, Bernhard-Nocht-Institut für Tropenmedizin, Bernhard-Nocht-Str. 74, 20359 Hamburg, Germany

Dr. Mark Bower The Chelsea and Westminster Hospital, 369 Fulham Road, London, SW10 9NH, UK

Dr. Mike Bray Biodefense Clinical Research Branch, Office of Clinical Research, National Institute of Allergy and Infectious Diseases, National Institute of Health, Bethesda, Maryland 20892, USA

Dr. Hyeryun Choe Department of Pediatrics, Children's Hospital, Harvard Medical School, Boston, MA 01772-9102, USA

Dr. Erik De Clercq Laboratory of Virology, Rega Institute for Medical Research, Katholieke Universiteit Leuven, Minderbroedersstraat 10, 3000 Leuven, Belgium

Dr. John C. Drach Department of Biologic and Materials Sciences, School of Dentistry and Interdepartmental Graduate Program in Medicinal Chemistry, College of Pharmacy, University of Michigan, Ann Arbor, Michigan 48109, USA

Dr. Armin Ensser Institut für Molekulare und Klinische Virologie, Universität Erlangen-Nürnberg, Schloßgarten 4, 91054 Erlangen, Germany

Dr. Michael Farzan Department of Microbiology and Molecular genetics, Harvard Medical School, New England Primate Research Center, Southborough, MA 01772-9102, USA

Dr. Michael Frese Department for Molecular Virology, Hygiene Institute, University of Heidelberg, Im Neuenheimer Feld 345, 69120 Heidelberg, Germany

Dr. Holger Garn Institut für Klinische Chemie und Molekulare Diagnostik, Philipps-Universität Marburg, 35033 Marburg, Germany

Dr. Brian Gazzard The Chelsea and Westminster Hospital, 369 Fulham Road, London, SW10 9NH, United Kingdom

Dr. Lutz Gissmann Deutsches Krebsforschungszentrum Heidelberg, Schwerpunkt Infektionen und Krebs, In Neuenheimer Feld 280, 69120 Heidelberg, Germany

Dr. Jaap Goudsmit CRUCELL Holland B.V., Archimedesweg 4, 2301 CA Leiden, The Netherlands

Dr. Thomas Herget Merck KGaA Darmstadt, Frankfurter Str. 250/C 10/351, 64293 Darmstadt, Germany

Marion Kaspari   Institute of Virology, Charité – University of Medicine Berlin, Schumannstr. 20/21, 10117 Berlin, Germany

Dr. Hans-Dieter Klenk   Institut für Virologie, Philipps-Universität Marburg, Robert-Koch-Str. 17, 35037 Marburg, Germany

Dr. Klaus Korn   Institut für Molekulare und Klinische Virologie, Universität Erlangen-Nürnberg, Schloßgarten 4, 91054 Erlangen, Germany

Dr. Ernst Kuechler†   Max. F. Perutz Laboratories, University Departments at the Vienna Biocenter, Department of Medical Biochemistry. Medical University Vienna, Dr. Bohrgasse 9/3, 1030 Vienna, Austria; died March 5th 2005

Dr. Jens Kuhn   Department of Microbiology and Molecular genetics, Harvard Medical School, New England Primate Research Center, Southborough, MA 01772-9102, USA and Department of Biology, Chemistry, Pharmacy, Freie Universität Berlin, Berlin, Germany

Dr. Stephan Ludwig   Institute of Molecular Virology (IMV), ZMBE, Westfälische-Wilhelms-Universität, Von-Esmarch Str. 56, D-48149 Münster, Germany

Dr. Wenhui Li   Department of Microbiology and Molecular genetics, Harvard Medical School, New England Primate Research Center, Southborough, MA 01772-9102, USA

Dr. Mengji Lu   Universitätsklinikum Essen, Hufelandstr. 55, 45122 Essen, Germany

Dr. Manfred Marschall   Institut für Molekulare und Klinische Virologie, Universität Erlangen-Nürnberg, Schloßgarten 4, 91054 Erlangen, Germany

Dr. Jan ter Meulen   CRUCELL Holland B.V., Archimedesweg 4, 2301 CA Leiden, The Netherlands

Dr. Johan Neyts Laboratory of Virology, Rega Institute of Medical Research, Katholieke Universiteit Leuven, Minderbroeders straat 10, 3000 Leuven, Belgium

Dr. Oliver Planz Friedrich-Loeffler-Institute (FLI), Paul-Ehrlich Str. 28, D-72076 Tübingen, Germany

Dr. Stephan Pleschka Institut für Virologie, Justus-Liebig-Universität Giessen, Frankfurter Str. 107, 35392 Giessen, Germany

Dr. Stefan Pöhlmann Institut für Molekulare und Klinische Virologie, Universität Erlangen-Nürnberg, Schloßgarten 4, 91054 Erlangen, Germany

Dr. Susanna Prösch† Institute of Virology, Charité –University of Medicine Berlin, Schumannstr. 20/21, 10117 Berlin, Germany; died November 4th 2005

Sheli Radoshitzky Department of Microbiology and Molecular genetics, Harvard Medical School, New England Primate Research Center, Southborough, MA 01772-9102, USA

Dr. Jaqueline D. Reeves Department of Microbiology, University of Pennsylvania, 3610 Hamilton Walk, Philadelphia, PA 19104, USA

Dr. Michael Roggendorf Universitätsklinikum Essen, Hufelandstr. 55, 45122 Essen, Germany

Dr. Barbara Schmidt Institut für Molekulare und Klinische Virologie, Universität Erlangen-Nürnberg, Schloßgarten 4, 91054 Erlangen, Germany

Dr. Joachim Seipelt Max. F. Perutz Laboratories, University Departments at the Vienna Biocenter, Department of Medical Biochemistry. Medical University Vienna, Dr. Bohrgasse 9/3, 1030 Vienna, Austria

Dr. Justin Stebbing The Chelsea and Westminster Hospital, 369 Fulham Road, London, SW10 9NH, UK

Dr. Jürgen Stech Institut für Virologie, Philipps-Universität Marburg, Robert-Koch-Str. 17, 35037 Marburg, Germany

Dr. Leroy B. Townsend College of Pharmacy, University of Michigan, 428 Church Street, Ann Arbor, Michigan 48109, USA

Monika Tschoschner Institut für Molekulare und Klinische Virologie, Universität Erlangen-Nürnberg, Schloßgarten 4, 91054 Erlangen, Germany

Dr. Hauke Walter Institut für Molekulare und Klinische Virologie, Universität Erlangen-Nürnberg, Schloßgarten 4, 91054 Erlangen, Germany

Dr. Thorsten Wolff Robert-Koch-Insitut, Division of Viral Infections, Nordufer 20, 13353 Berlin, Germany

Dr. Swee Kee Wong Department of Microbiology and Molecular genetics, Harvard Medical School, New England Primate Research Center, Southborough, MA 01772-9102, USA

Preface

Tempus fugit irreparabile (after *Publius Vergilius Maro*, 70-19 B.C.)

In an ideal world virus-induced diseases could be entirely eliminated. While elimination is out of the question in many cases as viruses may persist in the host in a latent stage, minimizing their pathogenic effects through therapeutic drugs often remains the only viable approach. The only other alternative on offer is prevention. Unfortunately, effective vaccines are only available for a limited number of viruses and the required area-wide administering schemes are not always supported. To this end, new therapies are in high demand. This book deals with and introduces possible ways forward with the view of providing a state-of-the-art platform for current and future developments in the field.

All therapeutic compounds effective against pathogenic viruses must target some critical features in the replication cycle in the host. One of the first antiviral compounds was Amantadin, which was described in 1964 and had been approved by the Federal Drug Administration in 1966. Amantadin blocks the correct viral maturation of influenza A by targeting the ion channel protein M2. Many ways of antiviral therapies have been developed in the last decades, but numerous problems have arisen. Recent problems occurred because most of the drugs approved for clinical treatment target identical steps in viral replication thus leading to a dramatic increase in drug resistances. A novel preferred drug would be one that targets an entirely different mode of action (e.g. inhibition of viral packaging processes vs. nucleotide inhibitors) and, of course, an action that is unique to the virus. Another challenge that needs to be dealt with is the sudden arrival of emerging viruses, again requiring new concepts for treatment. A typical example of this was the discovery of the severe acute respiratory syndrome (SARS) coronavirus, that caused epidemic outbreaks in China, Taiwan and Canada from November 2002 (Guangdong) until July 2003. In addition to the emerging viruses, there is a possibility of simultaneously occurring outbreaks of highly pathogenic viruses like the avian influenza virus H5N1, which, combined with a large geographic spread, increase the chance of virus transmission across host barriers and may lead to the induction of a global pandemic. The purpose of this book is to provide the reader with a comprehensive overview over novel and already existing concepts for antiviral therapies. The content of this book is broken down into four sections dealing specifically with RNA, DNA, and emerging viruses as well as alternative therapeutic approaches that may be applicable to a wide range of viruses. More specifically, the first section focuses on flaviviridae, influenza virus and HIV. In section two, the selective inhibition of poxviruses, the advantages of the benzimidazole nucleosides D and L enantiomers concerning HCMV therapy, and the therapy based on kinase inhibitors and immune therapy in conjunction with papillomavirus-induced tumors are described. The third section reports on the treatment of emerging viruses including SARS coronavirus and the extremely virulent filoviruses. The last section alludes to proteasome inhibitors, human monoclonal antibodies as well as vectors-based therapies.

We hope that this book will prove particularly useful for researchers, educators, and advanced graduate students in the fields of medicine, microbiology, and human biology whether in an academic or an industrial environment.

Elke Bogner
Andreas Holzenburg

Acknowledgements

Elke Bogner and Andreas Holzenburg would like to thank the Alexander von Humboldt-Foundation (Feodor-Lynen Program) for laying the foundations of our long-term collaborative research activities and enabling us to seed ideas for this book. We are especially grateful to all our colleagues who submitted timely data in a timely fashion. We would also like to thank the German Research Foundation (DFG), the Wilhelm Sander Foundation, Dr. Bernhard Fleckenstein (Erlangen) and the Office of the Vice President for Research (Texas A& M University, College Station) for their support.

1. Concepts of therapy for RNA viruses

Chapter 1.1

THERAPEUTIC VACCINATION IN CHRONIC HEPADNAVIRUS INFECTION

M. ROGGENDORF[1] and M. LU[1]
[1]Institut für Virologie, Universitätsklinikum Essen, Hufelandstraße 55, 45122 Essen, Germany

Abstract: Interferon α and nucleoside analogues are available for treatment of chronic hepatitis B virus (HBV) infection but do not lead to a satisfactory result. New findings about the immunological control of HBV during acute infection suggest the pivotal role of T-cell mediated immune responses. Several preclinical and clinical trials were undertaken to explore the possibility to stimulate specific immune responses in chronically infected animals and patients by vaccination. Vaccinations of patients with commercially available HBV vaccines did not result in effective control of HBV infection, suggesting that new formulations of therapeutic vaccines are needed. Some new approaches including DNA vaccines and combinations with antiviral treatments were tested in the woodchuck model. It could be shown that therapeutic vaccinations are able to stimulate specific B- and T-cell responses and to achieve transient suppression of viral replication. These results suggest the great potential of therapeutic vaccination in combination with antivirals to reach an effective and sustained control of HBV infection.

1. INTRODUCTION

Hepatitis B virus (HBV) is the most common among those hepatitis viruses which cause chronic infections of the liver in humans worldwide, and represents a global public health problem. According to WHO estimates, there were over 5.2 million cases of acute hepatitis B infection in 2000. More than 2 billion people have been infected worldwide and, of these, 360 million suffer from chronic HBV infection. The incidence of HBV infection

E. Bogner and A. Holzenburg (eds.), New Concept of Antiviral Therapy, 3-20.

and patterns of transmission vary greatly throughout the world in different ethnic groups. It is influenced primarily by the age at which infection occurs. Endemicity of infection is considered high in those parts of the world (e.g. Africa and Asia) where at least 8% of the population is HBsAg-positive. In these areas, almost all infections occur during either the perinatal period or early in childhood, a fact that accounts for the high rates of chronic HBV infection in these populations. 70-90% of the population generally have serological evidence of previous HBV infection. Chronic hepatitis B may progress to cirrhosis and death from liver failure and is the major cause of hepatocellular carcinoma (HCC) worldwide. HCC prevalence is known to vary widely among the world population, and those areas with higher prevalence of chronic HBV infections present the highest HCC rates. HBV causes 60-80% of the world's HCCs, one of the three major causes of death in Africa an Asia.

The development of effective strategies to control chronic HBV infections is an area of considerable current interest. While significant efforts focused on developing prophylactic vaccines against HBV, there also is a great need to develop strategies to boost immune responses and viral control in individuals who are already infected and HBV carriers. Effective therapeutic vaccination would offer an attractive method of boosting immune responses and enhancing viral control during chronic viral infections. For example, during human immunodeficiency virus (HIV) infection, potent CD8 T-cell responses are associated with an initial drop in viremia, and in some cases, with long-term control of viral replication. Therapeutic interventions that boost specific T-cell responses and lower the viral load may not only prevent the progression of infection but also reduce the rate of transmission of HBV. Furthermore, the principles of effective therapeutic vaccination may apply not only to chronic viral infections but also to prevent development of HCC. Thus, it is important to evaluate how to elicit the most successful immune response following therapeutic intervention during chronic HBV infection. Several studies have examined the potential benefits of therapeutic vaccination in chronic hepatitis B. Some of reports have demonstrated enhanced immune responses following therapeutic vaccination.

In this review we describe the immune pathogenesis of hepadnaviral infections leading to acute resolving or chronic hepatitis. The defect of cellular immune response observed in chronic hepatitis let to the concept of therapeutic vaccination to stimulate first of all the T cell response in HBV carriers. The efforts and partial success to develop a therapeutic vaccine will presented for the woodchuck model the third part of this review will cover trials in patients with chronic hepatitis B carried out so far.

2.   PATHOGENESIS OF HEPADNAVIRUS INFECTIONS

Acute hepatitis B runs a self-limited course in 95% of adult subjects, with most patients recovering completely. Fulminant hepatitis occurs in 1-2% of acute infections resulting in the death of most of these patients. 5% of adults and 90% of newborns infected with HBV develop chronic hepatitis B. About 15-40% of chronically infected subjects will develop complications earlier or later, leading to an estimated 520 000-1 200 000 deaths each year due to chronic hepatitis, cirrhosis, and HCC. In general the innate and adoptive immune responses play a key role in the defense of viral infections, resulting in the elimination of virus (Fig. 1). In hepadnaviral infection the innate immune response has been poorly investigated.

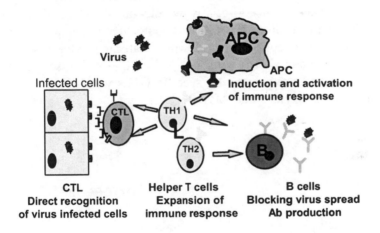

Figure 1. Immune responses to HBV infection. Both cell-mediated and humoral immune responses contribute to the control of HBV infection. Antibodies are needed to neutralize virus and to inhibit virus entry into host's cells. T cells are needed to recognize and eliminate intracellular virus. Liver damage in HBV infection is immune-mediated (designed by A. Bertoletti).

However, the adoptive immune response, especially the function of cytotoxic T cells, has been characterized in detail (Guidotti *et al.*, 1996). In transgenic mice expressing the complete HBV genome and chimpanzees infected with HBV T cell response has been characterized in very elegant experiments. The clearance of intracellular viruses by the immune response require the destruction of infected cells by virus specific CTLs via perforin of Fas dependent pathways. In recent years, however, it became evident that antiviral cytokines like interferon-gamma or TNF-alpha secreted by CTLs can purge viruses from living cells. This process can very efficiently reduce viral replication in living cells without killing them. This direct antiviral potential of cytokines produced by CTLs has clearly been demonstrated in HBV transgenic mice. Chisari and his colleagues have shown that the antiviral activity of interferon-gamma and TNF-alpha was produced by adoptively transferred virus specific CTLs which completely inhibited HBV replication in the liver (Guidotti *et al.*, 1996). In a recent study in acutely infected HBV infected chimpanzees (Guidotti *et al.* 1999) showed that non-cytopathic antiviral mechanisms contribute to HBV clearance since HBV DNA disappeared from the liver of acutely infected chimpanzees largely before the onset of liver disease concomitantly with intrahepatic appearance of interferon-gamma. Furthermore the appearance of this cytokine in the chimp liver preceded the peak of T cell infiltration suggesting that interferon-gamma was produced initially by cells other CTLs perhaps NK or NKT cells. The elimination of HBV from hepatocytes in hepatitis B occurs in two steps. First, the replication is downregulated by cytokines secreted from NK, NKT cells and shortly after by specific CTLs. The second step is the killing of infected hepatocytes by specific CTLs which is characterized by elevation of liver enzymes like ALT (Fig. 2).

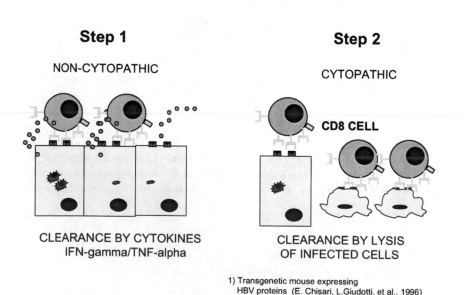

Step 1

NON-CYTOPATHIC

CD8 CELL

CLEARANCE BY CYTOKINES
IFN-gamma/TNF-alpha

Step 2

CYTOPATHIC

CLEARANCE BY LYSIS
OF INFECTED CELLS

1) Transgenetic mouse expressing
HBV proteins (E. Chisari, L.Giudotti, et al., 1996)

2) Chimpanzee L. Giudotti et al. Science (1999) 284, 825

Figure 2. The specific CTLs may inhibit HBV replication by a non-cytolytic mechanism (step 1) or to destroy infected cells directly (step 2).

Several studies in patients with acute hepatitis B demonstrated the presence of specific CTLs to different epitopes of HBV gene products whereas patients with chronic HBV infection are characterized by the absence of these specific CTLs. In addition to CTLs it has been shown that T helper cells may also directly act on replication by secreting cytokines which down regulate HBV replication. Humoral immune response in acute HBV infection results in the production of virus neutralizing antibodies to the surface protein (anti-HBs) which prevent re-infection of non-infected hepatocytes.

Patients with chronic hepatitis B who have been treated with interferon-alpha or nucleoside analogs show a reduction of viral replication and an appearance of virus specific Th cells and CTLs (Rehermann *et al.*, 1996). These data indicate that chronic carriers of HBV per se are not strictly tolerant to HBV. This observation is the basis of the recent development of strategies of therapeutic vaccination. Therapeutic vaccinations are aimed to induce a specific T cell response in chronic carriers which may migrate to the liver and downregulate replication by secretion of IFN-gamma and specific B cells producing anti-HBs.

Additional evidence has shown that tolerance to HBV in patients with chronic hepatitis B can be broken by immune transfer during transplantation. Transfer of HBV immune memory from an immune donor by transplantation of PBMCs to a HBs carrier resulted in the clearance of HBsAg following the engraftment of the HLA identical bone marrow. The resolution of chronic HBV infection is associated with a transfer of CD4+ T lymphocytes reactivity to HBcAg rather than to HBV envelope proteins (Lau *et al.*, 2002).

3. TESTING NEW APPROACHES FOR THERAPEUTIC VACCINATION IN THE WOODCHUCK MODEL

Current therapies like IFN-alpha and nucleoside analogues are effective in only a fraction of patients and have severe side effects or induce resistant viral mutations. Thus, therapeutic vaccines against chronic HBV infection are of general interest, both in the basic science and clinical use. Among the new types of vaccines, genetic vaccines based on purified plasmid DNA provide a series of new features in contrast to classical protein vaccines and seem to be the most promising candidate for future development. Through the progress on the characterization of the woodchuck immune system and the development of specific immunological assays, the woodchuck model became an informative animal model for vaccine development (Lu and Roggendorf, 2001; Roggendorf and Lu, 2005 a; b). The woodchuck is an excellent model to assess prototype prophylactic or therapeutic vaccines as efficacy can be tested by challenge experiments or viral elimination in chronic carriers respectively.

Up to date, clinical trials using conventional HBV vaccines to stimulate the specific immune response to HBV were carried out but did not reach the control over HBV in chronic carriers (see below). It appears that the conventional HBV vaccines consisting of recombinant HBsAg absorbed to alum adjuvant are not suitable for therapeutic vaccination, possibly due their preference to stimulate Th2-biased immune responses. Thus, the recent approaches are mainly directed by the idea that the stimulation of cell-mediated immune responses may be the key to reach the control over HBV. To test this idea, woodchucks with chronic WHV infection provide excellent opportunities to test the effectiveness of new options for therapeutic vaccinations (Tab. 1).

Table 1: Therapeutic vaccinations in the woodchuck model

Vaccines	Applica-tion	Outcome	Reference
WHcAg	i.m.	Viral elimination in 1 of 6 animals	Roggendorf 1995
WHsAg and Th-peptide	i.m.	Transient anti-WHs antibody response Two woodchucks died	Hervas-Stubbs *et al.*, 1997
WHsAg in combi-nation with L-FMAU	i.m.	Stimulation of T-cell responses to WHV proteins, anti-WHs antibody response	Menne *et al.*, 2002 Korba *et al.*, 2004
WHsAg in adjuvans	i.m.	Antibodies to the preS region of WHsAg	Lu *et al.*, 2003
WHsAg-anti-WHs immune complex and DNA vaccines in combi-nation with lamivudine	i.m.	Stimulation of anti-WHs antibody and suppression of WHV titer	Lu *et al.*, 2005 (unpublished results)

Based on the assumption that the specific T-cell responses to hepadnaviral nucleocapsid protein are important for viral control, immunizations of chronic carrier woodchucks with WHcAg in incomplete Freundsche adjuvant were carried out. However, no antiviral effects were achieved in chronically WHV-infected woodchucks with WHcAg only or in combination with famciclovir (Roggendorf and Tolle, 1995). Likely, such immunization procedures were not sufficient to overcome the T-cell unresponsiveness.

In an attempt to circumvent the T-cell unresponsiveness to WHV proteins in chronic carriers, a T helper cell determinant peptide FISEAIIHVLHSR encompassing amino acids 106-118 from sperm whale myoglobin (short named as FIS) was added to the vaccine preparation in subsequent experiments (Hervas-Stubbs *et al.*, 1997; 2001). This peptide, in combination with HBsAg, could induce anti-HBsAg in SJL/J mice, a non-responder strain to immunization with HBsAg alone. Indeed, PBMCs from FIS-immunized woodchucks produced IL-2 upon restimulation with FIS *in vitro* IL-2. Immunizations of seven chronic WHV carrier woodchucks with WHsAg in combination with FIS led to the induction of anti-WHsAg. However, two woodchucks developed high titres of anti-WHsAg and died from severe liver damage. Nevertheless, this experiment demonstrated that the deficiency of specific T-cell responses to viral proteins in chronic carriers could be overcome with foreign T helper cell determinant peptides.

The development of new adjuvants was another approach to improving therapeutic immunization. Monophosphoryl lipid A was one interesting candidate adjuvant that promotes Th1 type responses. A group of chronically WHV-infected woodchucks were immunized with plasma-derived WHV surface antigens (WHsAg) adsorbed to aluminum salt with monophosphoryl lipid A (Lu *et al.*, 2003). Anti-WHsAg antibodies were detected in all immunized woodchucks and persisted for a time period of up to 2 years after immunizations. Despite the induction of anti-WHsAg antibodies, neither WHV DNA nor WHsAg titers in immunized woodchucks changed significantly. Sequence analysis of the WHV pres- and s-genes of WHV isolates from these woodchucks showed that no WHV mutants emerged after the induction of anti-WHsAg/anti-WHpreS antibodies. These results indicate that immunizations with WHsAg could partially induce specific B-cell responses to WHV proteins in chronically WHV-infected woodchucks. However, additional components for the stimulation of T-cell responses are necessary to achieve therapeutic effects against chronic hepatitis B.

The T-cell response to HBV was successfully restored in patients treated with lamivudine or interferon-α, as published by Boni *et al.* (1998). Thus, a reduction of the viral load by antiviral treatments may enhance the effect of therapeutic vaccines. A combination of an antiviral treatment using a potent drug L-FMAU and immunization with WHsAg induces WHV-specific lymphoproliferation in chronic carriers (Menne *et al.*, 2002). Immunizations with WHsAg were able to induce lymphoproliferative responses to WHV proteins in both treated and untreated woodchucks. However, the antiviral treatment led to a broader spectrum of specific T-cell responses to different

WHV proteins core and X-proteins. In accordance with previous experiments, immunizations with WHsAg consistently induced a low level of anti-WHsAg in chronic carrier woodchuck. This finding indicates that a combination of antiviral treatment and immunizations is more effective to stimulate specific T-cell responses in chronic carriers.

A novel prototype therapeutic vaccine based on an antigen-antibody complex was developed by the group of Wen *et al.* (1995) (see below). A triple combination of antiviral treatment with lamivudine and therapeutic vaccination with DNA vaccines and antigen-antibody complexes was carried out in the woodchuck model to evaluate the efficacy (Lu *et al.*, unpublished). Ten woodchucks chronically infected with WHV were treated with 15 mg of lamivudine/day for 4 months. 6 weeks after starting of lamivudine treatment, one group with woodchucks were immunized with pWHsIm, a plasmid expressing WHsAg and a second group with WHsAg-anti-WHsAg complex and pWHsIm. Two woodchucks treated with lamivudine only served as controls. The treatment with lamivudine led to a marginal decrease of WHV DNA concentrations in woodchucks. Interestingly, 3 woodchucks immunized with WHsAg-anti-WHs complexes and pWHsIm developed anti-WHs antibodies and showed a further decrease of serum WHV DNA and WHsAg concentrations. The anti-WHsAg antibodies persisted in two woodchucks for a period of 8 weeks. These results indicated that this triple combination therapy was an effective treatment against chronic hepadnaviral infection. Further efforts should focus on the induction of sustained immune responses to maintain control over viral replication.

The present work done in the woodchuck model proved the feasibility of therapeutic vaccination against chronic HBV infection. It became clear, that the induction of antibodies to WHsAg could be achieved in chronically infected individuals while the control of viral replication needs multiple branches of immune responses, especially T-cell branches. Particularly, immunizations with immune complexes appear to be an effective method to stimulate antibody responses with antiviral action.

DNA immunization is a powerful method to induce protective immune responses to viral infection, particularly with the option to induce cellular immune responses (Donnelly *et al.*, 1997; Ulmer *et al.*, 1996) and may therefore be an excellent tool in chronic infections which lack a specific T cell response. DNA vaccination has been tested in different animal models including mouse, duck, woodchuck, and chimpanzees (Davis *et al.*, 1996; Kuhrober *et al.*, 1997; Lu *et al.*, 1999; Michel *et al.*, 1995; Pancholi *et al.*, 2001; Prince *et al.*, 1997; Schirmbeck *et al.*, 1996; Siegel *et al.*, 2001; Thermet *et al.*, 2003; Triyatni *et al.*, 1998). These experiments clearly demonstrated that DNA vaccines are able to induce specific humoral and cellular responses to HBV proteins. In mice, plasmids expressing HBsAg or

HBcAg are able to induce specific antibody and CTL-responses to HBsAg or HBcAg, respectively (Davis *et al.*, 1996; Kuhrober *et al.*, 1997; Michel *et al.*, 1995; Schirmbeck *et al.*, 1996;). In addition, results of DNA immunization in the duck model have demonstrated that DNA vaccines prime the specific immune responses and lead to control of hepatitis B virus infection (Triyatni *et al.*, 1998).

In woodchucks, DNA vaccines are able to prime the immune response to WHcAg and WHsAg and confer protection against WHV challenge (Lu *et al.*, 1999). The intramuscular route of DNA immunization requires high amounts of plasmids up to 1 mg per injection to achieve a measurable immune response to WHV proteins. Nevertheless, a single, low dose DNA vaccination by codelivery of the expression vectors for WHcAg and woodchuck interferon-gamma by gene gun was sufficient to limit the WHV infection (Siegel *et al.*, 2001). The serological profiles of challenged woodchucks indicate that DNA-primed immune responses to WHcAg did not prevent the infection of hepatocytes by input viruses but limited the virus spread. In a similar experiment IL12, a potent inducer of IFN-γ-production has been shown to enhance the protective immunity induced by DNA vaccine (Garcia-Navarro *et al.*, 2001). An interesting approach is to fuse a bioactive domain like cytotoxic T lymphocyte-associated protein 4 (CTLA-4), a ligand of CD80 and CD86 molecules, to an viral antigen. Such fusion antigens may be expressed *in vivo* and directed to antigen-presenting cells carrying CD80 and CD86 by the specific bioactive domain, and therefore possess a great potential to induce and modulate an antigen-specific immune responses. In a new study this approach was tested for the immuno-modulation against hepadnaviral infection in the woodchuck model (Lu *et al.* 2005). Plasmids expressing the nucleocapsid protein (WHcAg) and e antigen (WHeAg) of woodchuck hepatitis virus (WHV) only or in fusion with the extracellular domain of the woodchuck CTLA-4 and CD28 were constructed. While immunizations with plasmids expressing WHeAg or WHcAg only led to a specific antibody response of Th1 type, fusions of WHcAg to the woodchuck CTLA4 and CD28 induced an antibody response of both Th1 and Th2 at comparable level s. The Th2 type response seems to be greatly accelerated, resulting in a fast appearance of high titers of antibodies after only a single immunization with the CTLA4-WHcAg fusion antigen. Furthermore, woodchucks immunized with plasmids expressing CTLA-4-WHcAg fusion antigen showed a rapid antibody response to WHsAg and cleared WHV early after challenge with WHV.

Therapeutic DNA vaccination in woodchucks has been done recently (Lu *et al.*, unpublished). Chronic carrier woodchucks were treated with lamivudine for 8 weeks and then immunized with a combination of plasmids expressing WHsAg, WHcAg, and woodchuck interferon-gamma. After immunizations, all woodchucks showed a transient decrease of viremia in different extents. Lymphoproliferative responses to WHsAg and WHcAg were detected. These results indicate that DNA vaccines are promising candidates for immunotherapies.

The future investigations will take the major advantages in the woodchuck model as an authentic infection model. No other animal model is available to mimic the chronic course of hepadnaviral infection and present the features in pathogenesis and virus-host interaction in such a satisfactory way as the woodchuck model.

4.   THERAPEUTIC VACCINATION AGAINST HBV IN PATIENTS WITH CHRONIC HEPATITIS B

Parallel to preclinical studies in animal models using therapeutic vaccines several studies have been performed in patients with chronic hepatitis B (Tab. 2). Pilot studies of Pol *et al.* (1993; 1994; 1998) have established that specific vaccine therapy by a standard anti-HBV vaccination may reduce HBV replication in chronic carrier patients. A recent control study showed the efficacy and limitation of standard vaccine therapy in chronic HBV infection (Pol *et al.*, 2001). 118 patients who never received any previous HBV therapy were immunized five times with 20 µg of presS2/S vaccine (Genehevac B) or S vaccine (Recombivax) or placebo as the control. After 12 months follow-up after five vaccine injections there was no difference in the number of HBV DNA molecules between vaccinated and unvaccinated subjects. However, in the first six months following vaccination patients treated with either vaccine were significantly more likely to clear serum HBV DNA and seroconverted to anti HBV than untreated controls. At 12 months these differences lost significance due to increase seroconversion in the placebo group. In another randomized placebo controlled therapeutic vaccination study including 22 chronically infected patients a vaccine containing preS1, preS2 and S antigen components did not induce an HBs specific induction of T helper 1 or HBV specific CD8 T cell response. These studies indicate that the conventional vaccines used for prophylactic immunizations are not suitable as a therapeutic vaccine. Specific vaccine formulation including different adjuvants may be needed to induce a proper T cell response.

Three additional studies of therapeutic vaccination of HBsAg carriers have been published in 2003. Yalcien *et al.* (2003) immunized 31 patients with a preS2/S vaccine. 40 patients served as a non-vaccinated control group. Post-vaccination follow-up (month 12) three of the 31 patients (10%) cleared HBsAg and developed anti-HBs. Whereas none of the 40 controls lost surface antigen. Ren *et al.* (2003) immunized 30 patients with HBsAg. Six patients were used as controls. Serum HBV DNA levels decreased significantly at three months after completion of therapy and were significant lower in vaccinated patients than in controls at 12 and 18 months after completion of the study. Vaccination-induced antigens specific CD4+ T cell proliferative response was found in four patients. No CD8+ T cell response was observed. These results suggest that envelope specific CD4+ T cell may control direct the HBV replication by producing antiviral cytokines. However, no effect on HBV was observed in the study of Dikici *et al.* (2003). 43 children with chronic hepatitis B infection were immunized 3 times with standard HBV vaccines.

Post-vaccination serological and virological evaluation was performed six months after the first injection and at the end of the 12 months. There was no statistically significant difference between the vaccinated and control group with respect to viral load. Taken together, all studies performed so far with conventional vaccines used for prophylactic immunization have very little or no effect to reduce viremia or eliminate HBV patients with chronic hepatitis B. Different immunization protocols should be considered for future investigations in the immune tolerant phase of patients with chronic hepatitis B infection.

Overall, immunizations of chronic HBV-infected patients with standard HBV vaccines did not lead to a satisfactory result. Standard HBV vaccines may potently stimulate Th cells, as repeatedly reported by several studies performed so far. However, standard HBV vaccines are designed mainly to prime humoral responses to HBsAg. They are not suitable for the enhancement of cell-mediated immune responses that are critical for the control of HBV infection. Many research groups aimed to boost cellular responses using different approaches.

Wen *et al.* (1995) demonstrated that immunizing with immune complexes of HBsAg and anti-HBs antibodies reduced viremia in patients with chronic hepatitis B and also reduced viremia in transgenic mice expressing the complete HBV genome (Wen *et al.*, 1995; Zheng *et al.*, 2004). The immune complexes of HBsAg and anti-HBs are supposed to be taken up by antigen presenting cells in a facilitated way and may enhance the priming of specific T-cell responses. The same approach in the woodchuck model led to a reduction of viremia and production of anti-WHs antibodies in chronically infected animals. (see above). The immune complex has now passed the clinical trial phase I and phase IIa in China (Xu *et al.*, 2005). It could effectively induce HBsAg-specific immune responses in healthy persons.

Table 2: Therapeutic vaccine studies in patients with chronic hepatitis B

Protein vaccine	Patients	Application	Reduction of viremia	T-cell responses	Anti-HBs antibody	References
HBsAg / anti-HBs immune complex		i.m.	yes		no	Wen et al., 1995
HBsAg[1]	32		yes	yes		Pol et al., 1998
HBsAg[1,2]	118	5 x i.m.	yes/no			Pol et al. 2001
Peptid vaccine CTL epitop	19	s.c.	no	yes		Heathcote et al., 1999
HBsAg[3]	22	3 x i.m.	yes	yes		Jung et al., 2002
glyco gp26 and preS2	13		yes	yes		Ren et al., 2003
HBsAg/preS1/preS2	42	3 x p.o.	yes	yes		Safadi et al., 2003
PreS2/S[1]	31	3 x i.m.	no			Yalcin et al., 2003
HBsAg[2]	43	3 x i.m.	no		no	Dikici et al., 2003
DNA vaccine						
HBsAg	10	3- 4x i.m.	yes	yes		Mancini-Bourgine et al., 2004
Combination Nucleosid analogs and Vaccine						
HBsAg	72	12 x i.d.	yes			Horiike et al., 2005

[1] presS2/S genHevacB (Pasteur Mérieux)
[2] S vaccine RecombiVax (Merck)
[3] presS1/preS2/Sag
[4] healthy individuals

i.d. = intradermal
p.o. = per os
i.m. = intramuscular
s.c. = subcutaneous

An alternative vaccination strategy focusing on T cell response using peptide epitopes for cytotoxic T cells was developed and used for prophylactic and therapeutic vaccination studies (Vitiello *et al.*, 1995). A lipopeptide vaccine (CY1899) was designed consisting of HBV core antigen peptide 18 to 27 as a CTL epitope, the T helper peptide derived from tetanus toxoid, and two palmitic acid molecules as lipids were included. Immunization trials in 26 healthy subjects showed that this type of vaccine was safe and was able to induce HBV specific CTL responses (Livingstone *et al.*, 1997). Subsequently 19 patients with chronic HBV were immunized four times with this T cell vaccine. This vaccine initiated weak CTL responses in patients, however, the CTL activity was not associated with reduction of viral replication in the liver and viral load in the serum (Heathcote *et al.*, 1999). Obviously the response to a single T-cell epitope was insufficient to recreate enough T cells in the liver to have an effect in HBV replication.

DNA vaccines against HBV as described above have been used in a pilot study (Roy *et al.*, 2000). In a phase I clinical trial healthy volunteers received DNA including the surface antigen of HBV with a dose of 1,2 of 4 µg. The vaccine was safe and well-tolerated. All volunteers developed productive antibody response. In volunteers who were HLA-A2 positive, an antigen specific CD8+ T cell response could be detected. This is the first demonstration of a DNA vaccine inducing productive antibody and cell mediated immune response in humans. Mancini-Bourgin *et al.* (2004) immunized 10 chronically HBV infected patients with DNA vaccines. DNA vaccinations appeared to enhance T-cell responses in these patients. 5 patients had reduced viremia after 3 injections. Further studies on DNA vaccines are needed to improve their achievement in chronically HBV infected patients.

Wherry *et al.* (2005) have shown very elegantly that high viral load limits the effectiveness of therapeutic vaccination in mice chronically infected with LCM virus. Therefore, the reduction of viremia in patients with chronic hepatitis B by nucleoside analogues for several logs may help to overcome unresponsiveness to therapeutic vaccination with either surface or core protein of HBV. Horiike *et al.* (2005) treated 72 patients with chronic hepatitis B with Lamivudine at a dose of 100 µg daily for 12 months. 15 patients received vaccines containing 20 µg of HBsAg intradermaly once every two weeks. 12 months after start of therapy, HBV DNA became negative in nine of nine patients receiving combination therapy and in 15 of 31 patients receiving Lamivudine mono therapy. The rate of seroconversion from HBeAg to anti-HBV was also significantly higher in patients receiving combination therapy 56% versus monotherapy 16%. Breakthrough of HBV

DNA was found in 10 patients, but in none of the patients receiving combination therapy. This study shows for the first time that a combination therapy with nucleoside analogues to reduce viremia and vaccination may be the therapy of choice to eliminate HBV infection in chronic HBV patients. In this study intradermal vaccination was given twelve times which may have induced a strong T cell response in those patients. Unfortunately in these patients T cell response has not investigated to clarify whether CTLs have been induced by this immunization approach.

5.   PERSPECTIVES

The rapid progress of immunology has opened many important aspects of HBV infection for future investigation and has provided new clues for vaccine development. It is generally accepted that the improvement of specific CTL responses is most crucial for therapeutic vaccination. Though there are many approaches suitable to prime specific CTL responses, their ability to break the immune tolerance mechanisms in chronically HBV infected patients needs to be investigated. In this respect, understanding the tolerance mechanisms of chronic HBV infection will further help the design of therapeutic vaccines. It is assumed that a high level of viral replication may overwhelm host immune responses. Thus future approaches of therapeutic vaccination will include a vigorous antiviral treatment to reduce viral replication. Preferably, the expression of viral proteins should be reduced or completely inhibited as viral proteins represents effectors to keep the tolerance mechanisms. The new, potent antivirals could even greatly lower the level of cccDNA that is mainly responsible for the persistence of HBV. The new development of RNA interference may provide suitable tools to contribute to antiviral treatments.

Other components of the immune system beside specific CTLs are needed for a sustained response. It is clear that a CTL response could not be maintained without Th cells. Further, humoral responses, particularly the anti-HBs antibody response, are crucial to terminate HBV infection by clear free virions in periphery. As we understand that a immune response including multiple effectors is necessary to control primary HBV infection, therapeutic vaccines should stimulate all effectors involved. This concept suggests that effective therapeutic vaccines will be a combination of different components like DNA, proteins, cytokines, and adjuvants. It is to be mentioned that HBcAg is a good candidate to promote cell-mediated immune responses. Previous studies showed the potential of the hepadnaviral core antigens to induce protective immune responses (Roos *et al.*, 1989; Schödel *et al.*, 1993). Priming-boosting protocols using different components will provide additional options to target specific branches of the immune system.

REFERENCES

Boni, C., Bertoletti, A., Penna, A., *et al.*, 1998, Lamivudine treatment can restore T cell responsiveness in chronic hepatitis B. *J. Clin. Invest.* **102**:*968-976*

Couillin, I., Pol, S., Mancini, M., Driss, F., Brechot, C., Tiollais, P., and Michel, M.L., 1999, Specific vaccine therapy in chronic hepatitis B: induction of T cell proliferative responses specific for envelope antigens. *J. infect. Dis.* **180**:*15-26*

Davis, H.L., McCluskie, M.J., Gerin, J.L., and Purcell, R.H., 1996, DNA vaccine for hepatitis B: evidence for immunogenicity in chimpanzees and comparison with other vaccines. *Proc. Natl. Acad. Sci. USA* **93**:*7212-7218*

Dikici, B., Kalayci, A.G., Ozgenc, F., Bosnak, M., Davutoglu, M., Ece, A., Ozkan, T., Ozeke, T., Yagci, R.V., and Haspolat, K., 2003, Therapeutic vaccination in the immunotolerant phase of children with chronic hepatitis B infection. *Pediatr. Infect. Dis. J.* **22**:*345-349*

Donnelly, J.J., Ulmer, J.B., Shiver, J.W., and Liu, M.A., 1997, DNA vaccines. *Ann. Rev. Immunol.* **15**:*617-648*

Garcia-Navarro, R., Blanco-Urgoiti, B., Berraondo, P., Sanchez de la Rosa, R., Vales, A., Hervas-Stubbs, S., Lasarte, J.J., Borras, F., Ruiz, J., and Prieto, J., 2001, Protection against woodchuck hepatitis virus (WHV) infection by gene gun coimmunization with WHV core and interleukin-12. *J. Virol.* **75**:*9068-9076*

Guidotti, L.G., Ishikawa, T., Hobbs, M.V., Matzke, B., Schreiber, R., and Chisari, F.V., 1996, Intracellular inactivation of the hepatitis B virus by cytotoxic T lymphocytes. *Immunity* **4**:*25-36*

Guidotti, L.G., Rochford, R., Chung, J., Shapiro, M., Purcell, R., and Chisari, F.V., 1999, Viral clearance without destruction of infected cells during acute HBV infection. *Science* **284**:*825-829*

Heathcote, J., McHutchison, J., Lee, S., *et al.*, 1999, A pilot study of the CY-1899 T-cell vaccine in subjects chronically infected with hepatitis B virus. The CY1899 T Cell Vaccine Study Group. *Hepatology* **30**:*531-536*

Hervas-Stubbs, S., Lasarte, J.J., Sarobe, P., Prieto, J., Cullen, J., Roggendorf, M., *et al.*, 1997, Therapeutic vaccination of woodchucks against chronic woodchuck hepatitis virus infection. *J. Hepatol.* **27**:*726-737*

Hervas-Stubbs, S., Lasarte, J.J., Sarobe, P., Vivas, I., Condreay, L., Cullen, J.M., Prieto, J., and Borras-Cuesta, F., 2001, T-helper cell response to woodchuck hepatitis virus antigens after therapeutic vaccination of chronically-infected animals treated with lamivudine. *J. Hepatol.* **35**:*105-111*

Horiike, N., Fazle Akbar, S.M., Michitaka, K., Joukou, K., Yamamoto, K., Kojima, N., Hiasa, Y., Abe, M., and Onji, M., 2005, *In vivo* immunization by vaccine therapy following virus suppression by lamivudine: a novel approach for treating patients with chronic hepatitis B. *J. Clin. Virol.* **32**:*156-161*

Jung, M.C., Gruner, N., Zachoval, R., Schraut, W., Gerlach, T., Diepolder, H., Schirren, C.A., Page, M., Bailey, J., Birtles, E., Whitehead, E., Trojan, J., Zeuzem, S., and Pape, G.R., 2002, Immunological monitoring during therapeutic vaccination as a prerequisite for the design of new effective therapies: induction of a vaccine-specific CD4+ T-cell proliferation response in chronic hepatitis B carriers. *Vaccine* **20:3598-3612**

Korba, B.E., Cote, P.J., Menne, S., Toshkov, I., Baldwin, B.H., Wells, F.V., Tennant, B.C., and Gerin, J.L., 2004, Clevudine therapy with vaccine inhibits progression of chronic hepatitis and delays onset of hepatocellular carcinoma in chronic woodchuck hepatitis virus infection. *Antivir. Ther.* **9:937-952**

Kuhrober, A., Wild, J., Pudollek, H.P., Chisari, F.V., and Reimann, J., 1997, DNA vaccination with plasmids encoding the intracellular (HBcAg) or secreted (HBeAg) form of the core protein of hepatitis B virus primes T cell responses to two overlapping Kb- and Kd-restricted epitopes. *Int. Immuno.* **9:1203-1212**

Lau, G.K., Suri, D., Liang, R., Rigopoulou, E.I., Thomas, M.G., Mullerova, I., Nanji, A.,Yuen, S.T., Williams, R., and Naoumov, N.V., 2002, Resolution of chronic hepatitis B and anti-HBs seroconversion in humans by adoptive transfer of immunity to hepatitis B core antigen. *Gastroenterology* **122:614-624**

Livingston, B.D., Crimi, C., Grey, H., Ishioka, G., Chisari, F.V., Fikes, J., Grey, H., Chesnut, R.W., and Sette, A., 1997, The hepatitis B virus-specific CTL responses induced in humans by lipopeptide vaccination are comparable to those elicited by acute viral infection. *J. Immunol.* **159:1383-1392**

Lu, M., Hilken, G., Kruppenbacher, J., Kemper, T., Schirmbeck, R., Reimann, J., and Roggendorf, M., 1999, Immunization of woodchucks with plasmids expressing woodchuck hepatitis virus (WHV) core antigen and surface antigen suppresses WHV infection. *J. Virol.* **73:281-289**

Lu, M. and Roggendorf, M., 2001, Evaluation of new approaches to prophylactic and therapeutic vaccinations against hepatitis B viruses in the woodchuck model. *Intervirology* **44:124-131**

Lu, M., Klaes, R., Menne, S., Gerlich, W., Stahl, B., Dienes, H.P., Drebber, U., and Roggendorf, M., 2003, Induction of antibodies to the PreS region of surface antigens of woodchuck hepatitis virus (WHV) in chronic carrier woodchucks by immunizations with WHV surface antigens. *J. Hepatol.* **39:405-413**

Menne, S., Maschke, J., Tolle, T.K., Lu, M., and Roggendorf, M., 1997, Characterization of T-cell response to woodchuck hepatitis virus core protein and protection of wood-chucks from infection by immunization with peptides containing a T-cell epitope. *J. Virol.* **71:65-74**

Menne, S., Roneker, C.A., Korba, B.E., Gerin, J.L., Tennant, B.C., and Cote, P.J., 2002, Immunization with surface antigen vaccine alone and after treatment with 1-(2-fluoro-5-methyl-beta-L-arabinofuranosyl)-uracil (L-FMAU) breaks humoral and cell-mediated immune tolerance in chronic woodchuck hepatitis virus infection. *J. Virol.* **76:5305-5314**

Menne, S., Roneker, C.A., Tennant, B.C., Korba, B.E., Gerin, J.L., and Cote, P.J., 2002, Immunogenic effects of woodchuck hepatitis virus surface antigen vaccine in combination with antiviral therapy: breaking of humoral and cellular immune tolerance in chronic woodchuck hepatitis virus infection. *Intervirology* **45:237-250**

Michel, M.L., Davis, H.L., Schleef, M., Mancini, M., Tiollais, P., and Whalen, R.G., 1995, DNA-mediated immunization to the hepatitis B surface antigen in mice: aspects of the humoral response mimic hepatitis B viral infection in humans. *Proc. Natl. Acad. Sci. USA* **92:5307-5311**

Pancholi, P., Lee, D.H., Liu, Q., Tackney, C., Taylor, P., Perkus, M., Andrus, L., Brotman, B., and Prince, A.M., 2001, DNA prime/canarypox boost-based immunotherapy of chronic hepatitis B virus infection in a chimpanzee. *Hepatology 33:448-454*

Pol, S., Driss, F., Michel, M.L., Nalpas, B., Berthelot, P., and Brechot, C., 1994, Specific vaccine therapy in chronic hepatitis B infection. *Lancet 344:342*

Pol, S., 1995, Immunotherapy of chronic hepatitis B by anti HBV vaccine. *Biomed. & Pharmacother. 49:105-109*

Pol, S., Couillin, I., Michel, M.L., *et al.*, 1998, Immunotherapy of chronic hepatitis B by anti-HBV vaccine. *Acta Gastroenterol. Belg. 61:228-233*

Pol, S., Nalpas, B., Driss, F., *et al.*, 2001, Efficacy and limitations of a specific immuno-therapy in chronic hepatitis B. *J. Hepatol. 34:917-921*

Prince, A.M., Whalen, R., and Brotman, B., 1997, Successful nucleic acid based immunization of newborn chimpanzees against hepatitis B virus. *Vaccine 15:916-919*

Rehermann, B., Lau, D., Hoofnagle, J.H., and Chisari, F.V., 1996, Cytotoxic T lymphocyte responsiveness after resolution of chronic hepatitis B virus infection. *J. Clin. Invest.* 97:1655-1665

Roggendorf, M., and Tolle, T.K., 1995, The woodchuck: an animal model for hepatitis B virus infection in man. *Intervirology 38:100-12*

Rollier, C., Sunyach, C., Barraud, L., Madani, N., Jamard, C., Trepo, C., and Cova, L. 1999, Protective and therapeutic effect of DNA-based immunization against hepadnavirus large envelope protein. *Gastroenterol. 116:658-665*

Roos, S., Fuchs, K., and Roggendorf, M., 1989, Protection of woodchucks from infection with woodchuck hepatitis virus by immunization with recombinant core protein. *J. Gen. Virol. 70:2087-95*

Roy, M.J., Wu, M.S., Barr, L.J., *et al.*, 2000, Induction of antigen-specific CD8+ T cells, T helper cells, and protective levels of antibody in humans by particle-mediated administration of a hepatitis B virus DNA vaccine. *Vaccine 19:764-778*

Safadi, R., Isreali, E., Papo, O., Shibolet, O., Melhem, A., Bloch, A., Rowe, M., Alper, R., Klein, A., Hemed, N., Segol, O., Thalenfeld, B., Engelhardt, D., Rabbani, E., and Ilan, Y., 2003, Treatment of chronic hepatitis B virus infection via oral immune regulatione toward hepatitis B virus proteins. *Am. J. Gastroenterol. 98:2505-2515*

Schirmbeck, R., Bohm, W., Ando, K., Chisari, F.V., and Reimann, J., 1995, Nucleic acid vaccination primes hepatitis B virus surface antigen- specific cytotoxic T lymphocytes in nonresponder mice. *J. Virol. 69:5929-5934*

Schödel, F., Neckermann, G., Peterson, D., Fuchs, K., Fuller, S., Will, H., and Roggendorf, M., 1993, Immunization with recombinant woodchuck hepatitis virus nucleocapsid antigen or hepatitis B virus nucleocapsid antigen protects woodchucks from woodchuck hepatitis virus infection. *Vaccine 11:624-628*

Siegel, F., Lu, M., and Roggendorf, M., 2001, Coadministration of gamma interferon with DNA vaccine expressing woodchuck hepatitis virus (WHV) core antigen enhances the specific immune response and protects against WHV infection. *J. Virol. 75:5036-42*

Thermet, A., Rollier, C., Zoulim, F., Trepo, C., and Cova, L., 2003, Progress in DNA vaccine for prophylaxis and therapy of hepatitis B. *Vaccine 21:659-662*

Triyatni, M., Jilbert, A.R., Qiao, M., Miller, D.S., and Burrell, C.J., 1998, Protective efficacy of DNA vaccines against duck hepatitis B virus infection. *J. Virol. 72:84-94*

Ulmer, J.B., Sadoff, J.C., and Liu, M.A., 1996, DNA vaccines. *Curr. Opin. Immunol. 8:* 531-536

Vitiello, A., Ishioka, G., Grey, H.M., *et al.*, 1995, Development of a lipopeptide-based therapeutic vaccine to treat chronic HBV infection. I. Induction of a primary cytotoxic T lymphocyte response in humans. *J. Clin. Invest.* **95:**341-349

Von Herrath, M.G., Berger, D.P., Homann, D., Tishon, T., Sette, A., and Oldstone, M.B., 2000, Vaccination to treat persistent viral infection. *Virology* **268:**411-419

Wen, Y.M., Wu, X.H., Hu, D.C., Zhang, Q.P., and Guo, S.Q., 1995, Hepatitis B vaccine and anti-HBs complex as approach for vaccine therapy. *Lancet* **345:**1575-1576

Wherry, E.J., Blattman, J.N., and Ahmed, R., 2005, Low CD8 T cell proliferative potential and high viral load limit the effectiveness of therapeutic vaccination. *J. Virol.* **79:**8960-8968

World Health Organization, 1996, The World Health Report 1996. Geneva: World Health Organization

Xu, D.Z., Huang, K.L., Zhao, K., Xu, L.F., Shi, N., Yuan, Z.H., and Wen, Y.M., 2005, Vaccination with recombinant HBsAg-HBIG complex in healthy adults. *Vaccine* **23:**2658-2664

Yalcin, K., Acar, M., and Degertekin, H., 2003, Specific hepatitis B vaccine therapy in inactive HBsAg carriers: a randomized controlled trial. *Infection* **31:**221-225

Zheng, B.J., Ng, M.H., He, L.F., Yao, X., Chan, K.W., Yuen, K.Y., and Wen, Y.M., 2001, Therapeutic efficacy of hepatitis B surface antigen-antibodies-recombinant DNA composite in HBsAg transgenic mice. *Vaccine* **19:**4219-4225

Chapter 1.2

CHARACTERIZATION OF TARGETS FOR ANTIVIRAL THERAPY OF *FLAVIVIRIDAE* INFECTIONS

P. BOROWSKI [1]

[1] *Abteilung für Virologie, Bernhard-Nocht-Institut für Tropenmedizin, Hamburg, Germany*

Abstract: The members of the Flaviviridae family are the cause of the explosively increasing number of emerging infections of humans and economically important animals. Up to date there is no existing effective therapy targeting theses viruses. Recently obtained knowledge about the replication cycle and molecular organization of the Flaviviridae helps to understand the mechanisms by which the propagation of the viruses could be blocked. In this chapter the state of art regarding information about possible targets for antiviral strategies will be summarized.

1.   CHARACTERIZATION OF FLAVIVIRIDAE

This virus family was named after the jaundice occurring in course of Yellow fever virus (YFV) infection, the first identified virus of the Flaviviridae, which causes disease (Monath, 1987; Halstead, 1992). In humans infections with Flaviviridae viruses may lead to fulminant, hemorrhagic diseases [YFV, dengue fever virus (DENV) and omsk hemorrhagic fever virus (OHFV)], viral encephalitis [japanese encephalitis virus (JEV), tick-borne encephalitis virus (TBEV), West Nile virus (WNV), St. Louis encephalitis virus (SLEV)] or chronic hepatitis C, formerly referred to as non-A, non-B hepatitis, [hepatitis C virus (HCV)] (Monath and Heinz, 1996; Rice, 1996). Some members of the Flaviviridae can infect only animals, leading to severe disease of the host, usually followed by death

E. Bogner and A. Holzenburg (eds.), New Concepts of Antiviral Therapy, 21–46.

[bovine viral diarrhea virus (BVDV), classical swine fever virus (CSFV) and border disease virus (BDV)] (Nettelton and Entrican, 1995).

The members of the Flaviviridae can be classified into three genera: hepaciviruses, flaviviruses and pestiviruses (Westaway, 1987; Chambers *et al.*, 1990; Monath and Heinz, 1996). Recently a HCV-related virus was characterized, the hepatitis G virus (HGV), formerly referred to as "GB-agent" (Muerhoff *et al.*, 1995). The phylogenetic classification of the virus is, however, not yet established (Leyssen *et al.*, 2000). The amino acid sequence of the HGV-polyprotein displays 28% homology to HCV and 20% homology to YFV (Muerhoff *et al.*, 1995). Thus, the HGV with its three genotypes, HGV-A, HGV-B and HGV-C may constitute a further separate genus of Flaviviridae (Muerhoff *et al.*, 1995).

Members of the family of Flaviviridae are small (40 to 50 nm), spheric, enveloped RNA viruses of similar structure (Westaway, 1987; Monath and Heinz, 1996; Rice, 1996). The genome of the viruses consists of one single-stranded, positive-sense RNA with a length of 9100 to 11000 bases [e.g. 10862 for YFV (strain 17D), 10477 for Russian spring-summer encephalitis virus (RSSEV), approx. 9500 for HCV and 9143 to 9493 for HGV]. The RNA possesses a single open reading frame (ORF) flanked by 5'- and 3'-terminally located untranslated regions (5'UTR and 3'UTR respectively).

The replicative cycle of all viruses of the Flaviviridae is similar. After binding to the target receptor [i.e. CD81 molecule for HCV (Pileri *et al.*, 1998) or heparan sulfate for DENV (Chen *et al.*, 1997)] (see below) the virus penetrates the cell and its plus-strand RNA is released from the nucleocapsid into the cytoplasm (Leyssen *et al.*, 2000). The released viral RNA is translated into a polyprotein consisting of approximately 3000 to 3500 amino acids (e.g. 3010 for HCV, 3411 for YFV and 3412 for RSSEV). In the course of the infection the polyprotein is cleaved co- and post-translationally by both host cell proteases [signalases (Pryor *et al.*, 1998)] and virus-encoded proteases. The amino terminus of the polyprotein is processed into 3 (hepaciviruses, flaviviruses, HGV) or 5 (pestiviruses) structural proteins. The proteolytic processing of the carboxy terminus of the polyprotein of hepaciviruses and of HGV results in 6 mature proteins (NS2, NS3, NS4A, NS4B, NS5A, and NS5B), the polyprotein of the genus flavivirus is processed into 7 proteins (NS1, NS2A, NS2B, NS3, NS4A, NS4B, NS5) and the polyprotein of pestiviruses into 5 fragments (NS2/NS3, NS4A, NS4B, NS5A and NS5B) (Westaway, 1987; Leyssen *et al.*, 2000; Monath and Heinz, 1996; Rice, 1996; Meyers and Thiel, 1996). In addition, a small product of the proteolytical cleavage of the polyprotein of Flaviviridae, positioned between the structural and nonstructural regions of the polyprotein, has been detected. The function of the strong hydrophobic peptide, named p7, remains unknown (Bartenschlager, 1997; Zhong *et al.*,

1999). Some reports suggest, however, that p7 could act as an anchor for the proteins of the replication complex, or enhance the ion-permeability of the membranes fulfilling the function of a viroporin (Carrasco, 1995). Figure 1 (Fig. 1) presents, schematically, the structure of the *Flaviviridae* polyprotein.

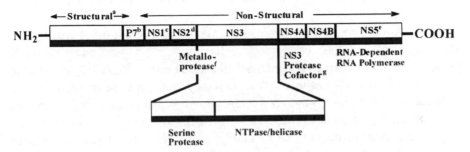

Figure 1: Simplified representation of the structure of Flaviviridae polyprotein. The NS3 region was expanded to better presentation of the protease and NTPase/helicase domains. Enzymatic activities associated with the non-structural proteins are indicated (compare the text). **(a)** The NH_2-terminal part of flaviviruses polyprotein is processed into three structural proteins: a nucleocapsid protein, precursor membrane protein and one envelope protein; the polyprotein of hepaciviruses into nucleocapsid protein and two envelope proteins; that of pestiviruses into five proteins: autoprotease, nucleocapsid protein, envelope protein with RNase activity and two envelope proteins. **(b)** Peptide p7 of putative viroporin function. **(c)** The NS1 protein is encoded exclusively by flaviviruses. **(d)** The NS2 protein of flaviviruses is processed into two proteins: NS2A and NS2B of unknown function. **(e)** The NS5 of hepaciviruses and pestiviruses is cleaved into NS5A and NS5B. The RNA polymerase activity is associated with the NS5B protein. **(f)** The functions of the NS2/NS3 metalloprotease are attributed exclusively to hepaciviruses. **(g)** The function of the NS3 protease cofactor is limited to hepaci- and pestiviruses.

The viral plus-strand RNA serves further as template for the synthesis of several minus-strand RNAs (Westaway, 1987; Westaway *et al.*, 1995; Monath and Heinz, 1996; Rice, 1996). This synthesis is carried out by a membrane associated replicase complex consisting of two essential proteins of the replication cycle: NS3 protein with its nucleoside-triphosphatase and helicase (NTPase/helicase) activities, and NS5B (for hepaci- and pestiviruses) as well as NS5 protein (for flaviviruses) with their RNA-dependent RNA polymerase (RdRp) activity. The minus-strand RNAs are transcribed into respective plus-strand RNAs which, in turn, are assembled into the nucleocapsid (Westaway, 1987; Leyssen *et al.*, 2000; Meyers and Thiel, 1996; Monath and Heinz, 1996; Rice, 1996).

Although the NS proteins are not constituents of the virus particle their intact function, particularly of the components of the replication complex, is essential for virus replication (Leyssen *et al.*, 2000; Bartenschlager, 1997; Ishido *et al.*, 1998; Neddermann *et al.*, 1999; Koch and Bartenschlager,

1999). In this context the NTPase and/or helicase activities of NS3, as well as NS5-associated RdRp, appear to be exceptionally attractive targets for termination of viral replication.

2.   TARGETS FOR POTENTIAL ANTIVIRALS

The presumed replicative cycles of Flaviviridae consist of: I) Adsorption and receptor-mediated endocytosis; II) low pH-dependent fusion in lysosomes and uncoating; III) internal ribosomal entry site (IRES) - (in case of hepaciviruses and pestiviruses) or cap-mediated (in case of flaviviruses) initiation of translation of the viral RNA into viral precursor polyprotein; IV) co- and post-translational proteolytic processing of the viral polyprotein by cellular and viral proteases; V) metabolism of RNA - membrane-associated synthesis of template minus-strand RNA and progeny plus-strand RNA; VI) assembly of the nucleocapside, budding of virions in the endoplasmic reticulum (ER), transport and maturation of virions in the ER and the Golgi complex, vesicle fusion and release of mature virions. Some of these steps could be potential targets for antiviral compounds.

2.1   Adsorption and receptor-mediated endocytosis

To date a broad range of receptors for members of the Flaviviridae have been identified. Predominantly there are oligo- and polysaccharides, particularly heparin, heparan sulfate and glycosoaminoglycans (GAGs) (Lee and Lobigs, 2000; Barth *et al.*, 2003). Nevertheless, in some cases membrane proteins, especially receptor proteins, serve as binding site(s) for viruses of the Flaviviridae family. Interestingly, the majority of flaviviruses and hepaciviruses require cooperation of numerous receptors for successful entry into the cell (Heo *et al.*, 2004). For example, it was demonstrated that HCV bind to CD81, human scavenger receptor class B type 1, dendritic cell-specific intracellular adhesion molecule 3-grabbing nonintegrin (C-type lectin) (DC-SIGN), related L-SIGN, DC-SIGNR and heparan sulfate (Moriishi and Matsuura, 2003; Heo *et al.*, 2004). However, there are further candidates serving as co-receptors or, at least, factors facilitating the virus entry e.g. low density lipoprotein receptor (Agnello *et al.*, 1999) or high density lipoproteins (Voisset *et al.*, 2005; Bressanelli *et al.*, 2004). It was previously suggested that the envelope glycoprotein E2 binds to the multiple receptors at different domains in a noncompetitive manner (Heo *et al.*, 2004). This could explain why the blockade of one of the receptor leads to only incomplete inhibition of infection (Heo *et al.*, 2004; Bartosch *et al.*, 2003). In agreement with this is the observation that neutralizing human antiserum or monoclonal antibodies against glycoprotein envelope E2 protein reduce activity towards the binding of HCV E2 protein to its receptor, but does not abolish it completely (Heo *et al.*, 2004). The possible mechanism of this action might be a sterical or allosterical blockade of the interaction. The interaction of HCV envelope glycoprotein with heparan sulfate strongly suggests that binding is, at least partially, dependent on the charge of the ligand and receptor. Indeed, the binding of the virus to human hepatoma cells was reduced by using heparin, polysulfate PAVAS (a copolymer of acrylic acid and vinyl alcohol sulfate), polysulphone suramin and numerous other sulfated polymers (Garson *et al.*, 1999).

There is only little known about cellular receptors for flaviviruses and pestiviruses. Recently, it was reported that antibodies against a 105 kDa glycoprotein on the plasma membrane of Vero and murine neuroblastoma cells with complex N-linked sugars could block West Nile virus entry efficiently (Chu and Ng, 2003). Pretreatment of the cells with proteases and glycosidases, or gene silencing of the 105 kDa glycoprotein (identified as alpha(v)beta(3) integrin) strongly inhibited entry of the virus. The interaction of the glycoprotein with the WNV occurs via its envelope glycoprotein E2, more exactly by the Domain III of the E2 (WNV-DIII). Recombinant WNV-DIII protein was able to inhibit WNV entry into Vero cells and C6/36 mosquito cells (Yu *et al.*, 2004). Thus, alpha(v)beta(3) integrin could be the putative receptor for WNV. The interaction between alpha(v)beta(3) integrin and WNV-DIII of envelope glycoprotein E2 plays a role – although not prominent - in infections of cells by other flaviviruses like dengue virus serotype 2 (DENV-2) and JEV (Yu *et al.*, 2004; Chu and Ng, 2004).

As for HCV there are indications that DENV demands numerous receptors. In addition to the above-mentioned alpha(v)beta(3) integrin and heparan sulfate, the virus use receptors that are seroptype specific. The 37-kilodalton/67-kilodalton (37/67-kDa) high-affinity laminin receptor was recently identified as a selective binding site for DENV-1, but not for DENV serotypes 2, 3 or 4. The entry of DENV-1 was reduced by antibodies against (37/67-kDa) laminin receptor or by soluble laminin (Thepparit and Smith, 2004). However, further understanding of the mechanisms involved in binding and entry of Flaviviridae to their target cells is needed until candidate substances can be developed.

2.2   Low pH-dependent fusion in lysosomes and uncoating

After the attachment of an enveloped virus to the cell surface receptors (see above), the fusion of the viral envelope with the host membrane follows. This process is mediated by virus-specific fusion proteins that merge the viral and cellular membranes. The fusion proteins (E in case of flaviviruses) contain a striking motif, so-called fusion peptide, that becomes exposed in the course of conformational changes, and is inserted into the target membrane. To date two different classes of fusion proteins have been described. The proteins (class I and II) differ dramatically in their structure and molecular architecture (Weissenhorn et al., 1999; Colman and Lawrence, 2003). The fusion protein of class I is represented by hemagglutinin of influenza virus and of other related viruses (Wilson et al., 1981; Weissenhorn et al., 1999). Class II fusion protein is represented by the E of flaviviruses mentioned above (Modis et al., 2003). The fusion of the viral and host membranes may occur at neutral or alternatively at lowered pH in endocytic vesicles. Flaviviridae and numerous enveloped viruses, like alphaviruses, orthomyxoviruses, rhabdoviruses or rhinoviruses require acidic pH for successful fusion with the cell membrane, followed by uncoating of the virus (Bressanelli et al., 2004). However, although uncoating is dependent on lowered pH, the rate of infection is determined by the location where the process occurs (Gollins and Porterfield, 1986). In contrast to the fusion and uncoating, endocytosis of the cell surface-attached virus is pH-independent (Kimura, et al., 1986).

As demonstrated for DENV-2 and TBEV at neutral pH, the envelope glycoproteins (in the form of homodimers or homotrimers, respectively) are so closely packed that the viral membrane is practically inaccessible and fusion does not occur. The lowered pH leads to changes of the conformation of envelope glycoprotein resulting, in formation of E homotrimer (for DENV-2) or rearrangement of the structure, but without changes of the polymerisation status of the E homotrimer (for TBEV). In the course of these conformational alterations, fragments of the surface area of the viral membrane will be exposed, making possible penetration of the virus particle into the cell. Although, to date, the processes of fusion and uncoating were very well documented only for DENV and TBEV, it appears that all flaviviruses, as well as hepaciviruses and pestiviruses, use the same mechanism for infection (Bressanelli et al., 2004). In this context, it is likely that compounds or short peptides competing with the regions of E protein which mediate the low pH-induced rearrangements of the structure of the virus surface mentioned above, would be potential antivirals. Such inhibitors for class I viral fusion protein of HIV-1 are already developed (Eckert and

Kim, 2001). Another mechanism of action of potential flaviviruses inhibitors of fusion and uncoating could be a reduction of the pH-gradient between acidified and pH-neutral cell compartments. The macrolide antibiotic bafilomycin A1 (Baf-A1), a specific inhibitor of vacuolar-type H(+)-ATPase, is commonly used to demonstrate the requirement of low endosomal pH for viral uncoating (Bayer, *et al.*, 1998; Nawa, 1998; Natale and McCullough, 1998). Treatment of the cell with the compound induced complete disappearance of acidified cell compartments. The effect of Baf-A1 is concentration-dependent. As demonstrated for JEV growth in Vero cells, the rate of infection decreases proportionally to the degree of depletion of the pH-lowered compartments (Andoh, 1998). Nevertheless, there are indications that Baf-A1 acts additionally on further intracellular processes, like blockade of transport from early to late endosomes. Since the early endosomes are suspected to lack components essential for uncoating, this activity of Baf-A1 could result in a further antiviral mechanism of action of the compound (Bayer *et al.*, 1998).

The neutralization of the lysosomal proton gradient alone appears, however, to be sufficient to exert an antiviral effect. Compounds that abolish or reduce the pH-gradient between cell compartments, like ammoniun chloride and chloroquine, inhibit the intracellular transport from the formerly acidified organells (Furuchi *et al.*, 1993). In the case of WNV, infection of Vero cells treated with ammonium chloride inhibited the uncoating of already internalized virus (Gollins and Porterfield, 1986).

2.3 IRES - (in case of hepaciviruses and pestiviruses) or cap-mediated (in case of flaviviruses) initiation of translation of the viral RNA into viral precursor polyprotein

Signals required for replication of plus-strand RNA viruses are usually located in the 5'-terminal regions of the template strands. They act as promoter elements for initiation of minus- and plus-strand RNA synthesis. The 5'UTR of hepaciviruses and pestiviruses contain IRES located at the 3' end of the 5' UTR and is composed of three stem-loop structures. The IRES of both genera initiate translation by directing the first triplet coding for the polyprotein (the initiator AUG of the polyprotein) to the ribosome (Friebe *et al.*, 2001). A 20 Å-resolution 3D model of the mammalian ribosomes complexed with the complete IRES demonstrated that this binding induces significant conformational changes of the 40S ribosomal subunit (Penin *et al.*, 2004) and *vice versa* the ribosomes along with the associated proteins might modify the RNA-RNA interaction within the IRES structure (Martines-Salas *et al.*, 2001). The important interactions between the RNA IRES and proteins of translation apparatus are demonstrated in figure 2A (Fig. 2A). Thus, the HCV IRES may play the role of a matchmaker of the ribosomes (Racker, 1991) and therefore modulate the host translational machinery. These results have supplied important knowledge about direct contact site(s) of the ribosome and RNA, allowing the design of peptides mimicking the IRES-binding domains, and thus abolishing the interaction. On the other hand, using a small RNA (60 nucleotides) that competes for critical IRES cellular polyprotein binding sites, a reduction of virus replication was demonstrated. The results were, however, obtained in a replication system employing a chimeric poliovirus whose IRES was replaced by the HCV IRES (Zhao *et al.*, 1999; Lu and Wimmer; 1996).

   In flaviviruses, the translation is initiated by a process called capping. The cap is a unique structure found at the 5' terminus of viral and cellular eukaryotic mRNA, which is important for mRNA stability and binding to the ribosome during translation. The viral mRNA capping is a cotranscriptional modification resulting from three chemical reactions mediated by viral enzymes: First, 5'-triphosphate of the mRNA is converted to diphosphate by an RNA triphosphatase. In the polyprotein of members of the Flaviviridae family, the RNA triphosphatase activity has been mapped to the carboxy-terminus of the NS3 protein. It is established that the energy resulting from triphosphate hydrolysis is used for the unwinding reaction mediated by the NS3-associated helicase (see below). The second reaction is the transfer of guanosine monophosphate (GMP) from GTP to the 5'-diphosphate RNA. This reaction is mediated by a guanylyltransferase, which, however, has not yet been identified in Flaviviridae. In a third reaction the transferred guanosine moiety is methylated at the N7 position. A second methylation at the first nucleotide 3' to the triphosphate bridge yields 7MeG 5'–ppp 5'-NMe. Sequence analysis revealed the presence of the characteristic motif of S-adenosyl-L-methionine (SAM)-dependent methyltransferases within the NH_2-terminal domain of the NS5 protein of flaviviruses.

   The prevention of capping, e.g. by inhibition of cap synthesis, should result in an antiviral effect by disabling the RNA of the progeny virus. There is evidence indicating that some of the enzymes mediating cap synthesis could be selectively inhibited by small-molecule compounds. Thus, the NS3-associated nucleotide triphosphatase may be inhibited (in a competitive manner) by a broad range of nucleotide analogues either with a modified nucleobase [(e.g. ribavirin-5'-triphosphate, paclitaxel and some ring expanded nucleosides (REN's) triphosphates (Borowski *et al.*, 2000; Borowski *et al.*, 2002a; Zhang *et al.*, 2003a; Zhang *et al.*, 2003b), or by nucleotide derivatives that posses a nonhydrolysable bound between the beta- and gamma-phosphates (Kalitzky, 2003; Borowski *et al.*, unpublished data). Also a noncompetitive inhibitor of NTPase, trifluoperacine (selective towards the HCV enzyme) was described (Borowski *et al.*, 1999). Nevertheless, inhibition was obtained only under selective reaction conditions and the antiviral effect of the compounds, as well as the exact in vivo mechanism of action, remain to be classified.

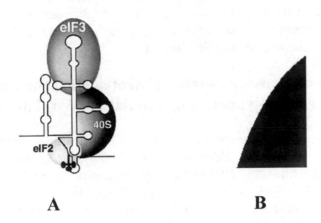

A   B

Figure 2: Schematical presentation of the functional interactions between IRES (A) or cap-structures (B) and target proteins. (A) Along with 40S ribosomal subunit IRES segments bind and modulate a range of proteins amongst others numerous eucaryotic initiation factors (eIF). On the other hand the eIFs modulate the structure of RNA structures of IRES in the sense of matchmakers. (B) The 7-methylated base of the cap-structure lies deep within the cap-binding slot of eIF4E. There are numerous interactions between the cap and the protein (not indicated here). One of the most important, with a regulatory function, is the interaction of Ser209 with phosphates of cap-structure. The phosphorylation of the serine destabilizes the binding and reduces the affinity of eIF4E for the ligand. Data were taken from Martinez-Salas *et al.* (2001) as well as from Sheper and Proud (2002).

The second promising target for potential antivirals is the viral NS5 protein-associated methyltransferase. The majority of the viral and cellular methyltransferases could be inhibited by derivatives of SAM, the S-adenosylhomocysteine (De Clercq, 1993). In this context, specific inhibitors of a cellular SAH hydrolase might inhibit the replication of Flaviviridae RNA, as demonstrated for rhabdoviruses or reoviruses (De Clercq, 1993). Such substances have already been found. For example, Neplanocin A (NPA), a naturally occurring carbocyclic nucleoside, and related abacavir and carbovir, in which the absence of a true glycosidic bond makes the compounds chemically more stable, as they are not susceptible to enzymatic cleavage by SAH hydrolase (Song *et al.*, 2001). A further possibility to inhibit the interactions between cap-structure and its target protein eucaryotic initiation factor 4E (eIF4E) is the reduction of the affinity of the last to the ligand. There are findings suggcsting strongly that this affinity is regulated by intracellular protein phosphorylation taking place in the cap-structure-binding pocket of eIF4E (Scheper and Proud, 2002). In figure 2B (Fig. 2B) is demonstrated the putative mechanism of this affinity regulation. Whether the activation of the protein kinases or inhibition of the cellular phosphatases determining the status of the phosphorylation, influences the infection rate of the cell, should be clarified.

2.4   Co- and post-translational proteolytic processing of the viral polyprotein by cellular and viral proteases

The genome of the Flaviviridae encodes NS3 associated serine protease. The HCV genome encodes additionally for a NS2/NS3 protease. The only known function of the NS2/NS3 protease is the autoproteolytic cleavage of the NS2/NS3 junction. Although no significant amino acid sequence similarities between NS2/NS3 and other proteases have been found, there are sequences within NS2 and amino-terminus of NS3 protein that are characteristic for Zn^{2+}-binding proteins. This strongly suggested that the HCV enzyme is a metalloprotease. Indeed, EDTA inhibits, and Zn^{2+} stimulates, the enzymic activity of the NS2/NS3 protease (Wu *et al.*, 1998). Although the genetic characterization of the protease is far advanced, knowledge about the biochemical nature of the enzyme is rather limited; mainly because of the difficulties with expressing and purification of active protein and, consequently, the lack of a proper *in vitro* assay.

Considerable interest has been devoted to the characterization of the proteolytic serine protease activity associated with NS3 protein. This protein is essentially involved in the formation of the replication complex, and is therefore, an attractive target for potential antivirals. Although there is significant amino acid sequence variability among different HCV isolates, the enzyme displays a conserved pattern of characteristic motifs common to all viral and cellular serine proteases (Bartenschlager *et al.*, 1993; Lohmann *et al.*, 1997). The enzyme comprises the 189 amino-terminal amino acids of NS3 protein. The NS3 is released from the polyprotein by an autoproteolytic process occurring in cis. The intramolecular cleavage at the NS3/NS4A junction is followed by proteolysis at the NS5A/NS5B, NS4A/NS4B and NS4B/NS5A sites that occur in trans and cis (Neddermann *et al.*, 1997). For full proteolytic activity, the NS3 protease demands a cofactor, the NS4A protein, particularly its central part (Lin *et al.*, 1995). The stimulating effect of NS4A on the proteolytic activity of the enzyme can be efficiently mimicked in vitro by synthetic peptides encompassing or reproducing the central region of NS4A (Lin *et al.*, 1995; Tomei *et al.*, 1996, Shimizu *et al.*, 1996).

Analysis of the three-dimensional structures of NS3 protease, and NS3 protease complexed with NS4A, revealed that the HCV enzyme takes the shape of the chymotrypsin fold (Love *et al.*, 1996; Kim *et al.*, 1996). The NS4A protein interacts with the amino-terminus of the NS3 protease, changes significantly the structure of the protein, and participates in formation of the hydrophobic core of the enzyme. Thus, the protein appears to be an integral structural component of the NS3/NS4A complex (Love *et al.*, 1996; Kim *et al.*, 1996).

Since the cleavage sites of the NS3 protease are highly specific, it is obvious that potential inhibitors of the enzyme might constitute substrate- or product-like peptidomimetcs and peptides. A further interesting strategy is based on inhibitors that can block the NS4A binding site(s). The inhibitors represent NS4A analogues alone or linked to peptides mimicking the unique sequences recognized by the protease as substrate(s) mentioned above (Llinas-Brunet *et al.*, 2004; Goudrenau, *et al.*, 2004). Indeed, a broad range of such compounds was designed and synthesized (Gordon and Keller, 2005). Nevertheless, there are some obstacles to bypass: the large size of the compounds and therefore their poor bioavailability; on the other hand the peptide character of the inhibitors often limits their stability in the cell culture medium, serum or in the cells. Due to these difficulties, a high-throughput screening of chemical and natural products was employed. In the course of these studies, some groups of substances have been detected, which inhibited the NS3 protease *in vitro* to 50% (IC_{50}) at concentrations lying in the low micromolar range. Unfortunately, the majority of the compounds exhibit inhibitory effects also against chymotrypsin, trypsin, plasmin or elastase (Gordon and Keller, 2005). Nevertheless, some of the compounds could serve as lead substances for further development.

2.5 Metabolism of RNA - membrane-associated synthesis of templated minus-strand RNA and progeny plus-strand RNA

Formation of a membrane-associated replication complex composed of viral proteins replicating RNA, and altered host-cell membranes is a typical feature of all plus-strand RNA viruses investigated so far (Schwartz *et al.*, 2002). Recently in a tetracycline-regulated cell line inducible expressing the entire HCV polyprotein, a structure was identified, termed membranous web that appears to correspond to the native replication complex previously found in the liver of HCV-infected chimpanzes. The membranous web consisted of all nonstructural proteins (Egger *et al.*, 2002; Gosert *et al.*, 2003). The participation of the majority (or of all) nonstructural proteins on formation of the replication complex is not surprising. There are numerous reports demonstrating multiple protein-protein interactions between the nonstructural proteins and/or between the proteins and host-cell membranes. In many cases the domains of the interacting proteins were mapped (Wolk *et al.*, 2000; Schmidt-Mende *et al.*, 2001). For example, the hydrophobic HCV NS2 intercalates partially into the membrane of ER and the cytoplasmic part of the protein serves as an anchor for NS5A (Santolini *et al.*, 1995). The NS3 protein of the Flaviviridae interacts very strongly with NS5A and NS5B (in the case of hepaciviruses and pestiviruses) as well as with NS5 (in the case of flaviviruses) (Chen *et al.*, 1977; Ishido *et al.*, 1998; Kapoor *et al.*, 1995). This interaction (binding) plays an essential role in the regulation of the status of the hyperphosphorylation and changes the conformation of NS5A or NS5. Moreover, NS3 influences the rate of translocation of the proteins to the nucleus (Kapoor *et al.*, 1995).

Very well characterized is the function of NS4 protein as an anchor for the replication complex to the ER. It was demonstrated that the hydrophilic (cytoplasmic) part of NS4A binds directly to NS3 and NS5B or NS5. NS4B appears, like p7, NS2 and NS4A, to play a role in attachment of the replication complex to ER (Ishido *et al.*, 1998; Kim *et al.*, 1996; Lin *et al.*, 1997).

The negative-stranded RNA of viruses of the Flaviviridae is synthesized with the use of the parental positive-strand RNA as template. The resulting negative-strand RNA is then used as template for the synthesis of the positive-strand progeny RNA, that is then assembled into viral particles. Since the negative and positive oriented RNA strands are complementary, the NS3-associated helicase activity appears to be necessary for strand separation. Among all components of the replication complex, two, mentioned above, enzymatic activities appear to be exceptionally attractive targets for termination of virus replication, e.g. NTPase/helicase associated with the carboxy-terminus of NS3, and RdRp identified in the NS5 (flaviviruses), or NS5B (hepaci- and pestiviruses) proteins. Figure 3 (Fig. 3) presents schematically the collaboration of the enzymes of the Flavivirdae replication complex.

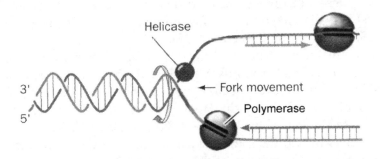

Figure 3: Simplified presentation of the cooperation of enzymes of the *Flaviviridae* replication complex. Two enzymatic activities of the replication complex: the NS3 associated NTPase/helicase along with the NS5 (or NS5B) RdRp are necessary for the replication of the viruses of the *Flaviviridae*. After the penetration of the cell by viral ssRNA a second complementary RNA strain will be synthetised by RdRp. The resulting dsRNA serves than as substrate for the NTPase/helicase. In the course of the unwinding reaction two ssRNA strains will be produced, which again serve as substrates for RdRp.

NTPase/helicases are in general nucleoside triphosphate-dependent ubiquitous proteins, capable of enzymatic unwinding double-stranded DNA or RNA structures by disrupting the hydrogen bonds that keep the two strands together. Approximately 80% of all known plus-strand RNA viruses encode at least for one NTPase/helicase. In many cases, only the detection of short amino acid sequences (motifs) representative for NTPase and helicase, allows us to presume the unwinding activity. The enzymatic evidence, however, is missing. In the case of members of the Flaviviridae, the helicase could be experimentally documented only in a few instances: HCV (Suzich *et al.*, 1993), HGV (Gwack *et al.*, 1999), WNV (Borowski *et al.*, 2001), JEV (Utama *et al.*, 2000), DENV (Li *et al.*, 1999) and BVDV (Tamura *et al.*, 1993).

Insights into the dependency between structure and function have come primarily from X-ray crystallography data of representative family members of the HCV NTPase/helicase. The structure of the HCV enzyme has been solved in the absence and presence of a bound oligonucleotide (Yao *et al.*, 1997; Kim *et al.*, 1998). The protein consists of three equally-sized structural domains separated by series of deep clefts. Domains 1 and 3 share with each other a more extensive interface than either of them shares with domain 2. In consequence, the clefts between domains 1 and 2 and domains 2 and 3 are the largest. The domain 2 is flexibly linked to the other two and could undergo a rigid movement relative to domains 1 and 3.

On the surface of domains 1 and 2 were found seven conserved amino acid sequences (motifs I–VII), characteristic for the majority of known

NTPase/helicases, that determines their affilliation to one of the three superfamilies (SF) of the enzymes. The Flaviviridae enzyme is placed in superfamily II (SFII) (Kadare and Haenni, 1997).

Some of the motifs are attributed to defined function of the enzyme. The motifs I and II, so called Walker motifs A and B, have been described as a key part of NTP binding pocket. Walker motif A binds to the terminal phosphate group of the NTP and the Walker motif B builds a chelate complex with the Mg2+ ion of the Mg2+ − NTP complex (Walker *et al.*, 1983). In the absence of substrate, the residues of the Walker motifs bind one to the other, and to the residues of the conserved T-A-T sequence of motif III. This motif is part of a flexible "switch sequence" connecting domains 1 and 2, which transduces the energy resulting from NTP hydrolysis and participates in the conformational changes induced by NTP binding (Yao *et al.*, 1997; Matson and Kaiser-Rogers, 1990). The role of the highly conserved arginine-rich motif VI, which is located on the surface of domain 2, is controversial. On the basis of crystallographic analyses, it can be assumed that motif VI is important for RNA binding (Yao *et al.*, 1997) or, alternatively, that it is directly involved of its arginine residues in ATP binding (Kim *et al.*, 1998). Similarly controversial is the mechanism by which the enzymatic activities of NTPase/helicases (NTPase and unwinding) are coupled. It is possible to inhibit or stimulate both activities separately (Porter and Preugschat, 2000; Borowski *et al.*, 2001; Borowski *et al.*, 2002).

Analysis of the results obtained from crystallographic, enzymatic and inhibitory studies show that the inhibitors NTPase/helicase could act by the following mechanisms: i) Inhibition of the NTPase activity by competitive blockade of the NTP binding site or by an allosteric mechanism (Borowski *et al.*, 1999; Borowski *et al.*, 2002b); ii) Inhibition of the transmission of energy and conformational changes by immobilisation of the "switch region" (Borowski *et al.*, 1999; Borowski *et al.*, 2002b). ii) Competitive inhibition of RNA binding (Diana *et al.*, 1998; Diana and Bailey, 1996; Phoon *et al.*, 2001); iv) Inhibition of unwinding by sterical blockade of translocation of the NTPase/helicase along the polynucleotide chain (Lun *et al.*, 1998; Bachur *et al.*, 1992).

Another main target for potential antivirals within the replication complex is the RdRp. The enzyme facilitates the synthesis of both the negative-strand RNA intermediate, complementary to the viral genome, and the positive-strand RNA genomes complementary to the negative-strand intermediate. The RNA polymerases of HCV (Lohmann *et al.*, 1997), DENV (Tan *et al.*, 1996), WNV (Steffens *et al.*, 1999), BVDV (Zhong *et al.*, 1998) and CSFV (Steffens *et al.*, 1999) have been cloned and expressed. The structural and kinetic properties of the HCV RdRp have been studied in detail (Lohmann *et al.*, 1997). The catalytic domain of the enzyme was crystallized and its structure resolved. The investigated fragment of NS5B is folded in the characteristic fingers, palm, and thumb subdomains. The finger subdomain contains a region, the "fingertips", that displays the same fold with reverse transcriptases (RT's). Comparison with the known structures of the RT's shows that residues from the palm and fingertips are structurally equivalent (Bressanelli *et al.*, 1999). Conserved, between Flaviviridae, Picornaviridae families and retroviruses, cluster in defined regions of the molecular surface: the RNA and NTP binding groove, the back of the thumb, the NTP tunnel, and acidic path at the top-front of the fingers. The back surface of the thumb could conceivably to be a site of interaction with other components of the replication complex mentioned above or cellular proteins (Bressanelli *et al.*, 1999).

The Flaviviridae RdRp-mediated reaction was found not to be limited to input RNAs containing HCV sequences. The polymerase reaction may be performed also with HCV-unrelated input RNA (Behrens *et al.*, 1996; De Francesco *et al.*, 1996). In comparison with other RNA polymerases, e.g. poliovirus 3D polymerase, the flavivirus enzymes display a relative low turnover rate (Arnold and Cameron, 2000). This finding indicates that other viral and/or cellular proteins must be involved in the reaction as co-factors that determine the specificity of the RNA synthesis and regulate the velocity of the reaction.

As is the case with HIV-1 RT, the majority of currently known HCV polymerase inhibitors fall into two main categories, according to their chemical structure and their mechanism of action. There are nucleoside analogue inhibitors and non-nucleoside inhibitors. Recently a third class of compounds, mimicking the pyrophosphate group and displaying an ability to inhibit HCV RdRp, was separated.

All nucleoside analogues appear to inhibit the polymerase activity in a similar manner. After penetration into the cell, the compounds undergo intracellular phosphorylation to the corresponding triphosphate. Subsequently the nucleotide analogues are incorporated by the viral polymerase into the growing nucleic acid chain. This leads, in turn, to an increased error frequency of the polymerase and, in consequence, to early termination of the elongation reaction. This is probably the mechanism by which ribavirin, a broad-spectrum antiviral nucleoside, exerts its anti-HCV effect. Numerous studies have demonstrated the conversion of ribavirin to ribavirin triphosphate (Miller *et al.*, 1977; Zimmerman and Deeprose, 1978) and incorporation of the nucleotide into the nucleic acid chain by some viral polymerases (Lau *et al.*, 2002; Crotty *et al.*, 2000; Maag, *et al.*, 2001). Nevertheless, only a handful of nucleoside analogues active against HCV and other Flaviviridae polymerases have been identified.

The second category of RdRp inhibitors comprises structurally and chemically heterogenous compounds, not related to the non-nucleosides or nucleotides. The substances are not incorporated into growing DNA or RNA strand. The compounds inhibit the polymerase indirectly by binding to the enzyme in a reversible and non-competitive manner. Cocrystallisation studies of some RdRp inhibitors bound to HCV polymerase reveal an inhibitor binding site located at the base of the thumb subdomain lying in the direct proximity to the polymerase active site (Wang *et al.*, 2003). The studies strongly suggest that binding of the inhibitor prevents the conformational changes of the RdRp necessary for its enzymatic activity.

A third category of polymerase inhibitors consists of the chemically and structurally homogenous pyrophosphate mimics possessing a diketo acid moiety (Altamura *et al.*, 2000). The mechanism by which the compounds exert their inhibitory effect is the blockade of the active site of the enzyme. Thus the binding of the phosphoryl groups of the nucleotide substrate is blocked and formation of complexes $Mg2+$-NTP or $Mn2+$-NTP is abolished.

2.6   Assembly of the nucleocapsid, budding of virions in ER, transport and maturation of virions in the ER and the Golgi complex, vesicle fusion and release of mature virions

The structural protein region of Flaviviridae polyprotein is processed to separate structural proteins by host signal peptidases and intramembrane protease (see above). The processes are best illustrated in the case of HCV. The core protein of the virus is further cleaved and the products then associate with lipid droplets (Lemberg and Martoglio, 2002). There are numerous signal peptidases that cleave the core protein. One of them, the intramembranous protease, was identified as a signal peptide peptidase (SPP), a presenilin-type aspartic protease. The proteolytic activity of SPP is inhibited by (Z-LL)2 ketone (Weihofen *et al.*, 2003). Nevertheless, the SPP cleaves also numerous host proteins, e.g. prolactin or HLA-E (Lemberg *et al.*, 2001). Thus, the development of such more selective compounds could prove useful for the treatment of chronic HCV infection. The E protein (for flaviviruses) and E1, E2 proteins (for hepaciviruses and pestiviruses), resulting from the proteolytical processing, form a homo- or heterodimer respectively, which intercalate into membranes of ER building the prebudding form of the virus (Deleersnyder *et al.*, 1997). The signal peptidases that process the E1, E2 proteins and p7 are not nearly as well characterized, and selective inhibitors of the enzymes do not exist.

Endoplasmic reticulum alpha-glucosidase inhibitors block the trimming stem in the course of N-linked glycosylation, and eliminate the production of several ER-budding viruses. In a recent study, the iminosugar derivative N-nonyl-deoxynojirimycin was found to inhibit the replication of JE and DENV significantly (Wu *et al.*, 2002). This effect was probably mediated by inhibition of secretion of the viral glycoproteins E and NS1. The latter protein is known to be essential for flavivirus replication (Lindenbach and Rice, 1997). The difficulties to obtain therapeutic serum concentrations, and adverse side effects, have limited the clinical usefulness of these compounds.

3. OTHER INHIBITION MODES OF FLAVIVIRIDAE REPLICATION

Ribavirin exerts its antiviral effects by different mechanisms of action: In chronic viral infections like HCV infection, beyond the induction of lethal mutations of the viral genome, mentioned above, the therapeutic effect of the compound seems to depend also on immunomodulating mechanisms. It could be shown that the nucleoside may modulate interleukin-10 expression in mice. In the case of flaviviruses, a curative effect of ribavirin mono-therapy was demonstrated. There are reports of significantly higher survival, and eradication of WNV from brain in mice, after intraperitoneal injection of ribavirin and in vitro studies showing that the nucleoside inhibited WNV replication in human oligodendral cells (Jordan *et al.*, 2000; Morrey *et al.*, 2002). Similarly, acute infections with Lassa and RSV viruses could be treated efficiently by ribavirin monotherapy (McCormick *et al.*, 1986; Taber *et al.*, 1983). However, the need to use very high doses of ribavirin in WNV infections in vivo, proved too toxic to be clinically useful. Multicenter studies showed that ribavirin applied as monotherapy at clinically admissible doses was not more effective than placebo in reducing or eliminating levels. Recent studies with Vero cells showed that interferon alpha-2b inhibited viral cytotoxicity when applied after or before cells were infected with WNV (Zoulim *et al.*, 1998; Lee *et al.*, 1998; Jordan *et al.*, 2000). The effect of combined therapy should be first evaluated in patients infected with flaviviruses, particularly the combination of interferon and ribavirin, which is the standard therapy for chronic HCV infection.

An interesting approach to reduce HCV replication appears to be the use of the so-called small interfering RNA (siRNA; Fire *et al.*, 1998; Paddison and Hannon, 2002). Such small dsRNA fragments are incorporated into RNA-induced silencing complex (RISC), which leads to specific destruction of the target mRNA recognized by the antisense strand of the siRNA

(Hammond *et al.*, 2000). Such a strategy was proposed for HIV (Novina *et al.*, 2002) and for hepatitis B therapy (McCaffrey *et al.*, 2003).

New data regarding the biochemical and biological properties of p7 make the peptide an attractive target for antiviral drugs. This viroporin appears to play an important role in virus particle release and maturation (Carrasco *et al.*, 1995). Although p7 seems to be located mainly in ER membranes, it can be exported to the plasma membrane, and may have a functional role in the secretory pathway (Carrere-Kremer, 2004). As demonstrated in the case of BVDV, p7 is essential for the progeny of the virus (Harada *et al.*, 2000). Nevertheless, the events in which p7 is involved, are only poorly characterized and hence the development of appreciate inhibitor(s) could be a protracted process.

4.   CONCLUSIONS

Combination therapy with interferon alpha-2b and ribavirin leads to remission of disease in only 40% of patients with chronic HCV infection; furthermore this therapy causes significant side effects (Wedemeyer *et al.*, 1998). Moreover, similarly to its HIV counterpart, long-term therapy of HCV infection will be hindered by the emergence of drug-resistant strains. Anti-HCV therapy is further complicated by high genetic diversity and the geographic distribution of different HCV genotypes. Thus, it is very important to develop an alternative combination therapy for chronic hepatitis C with a number of agents acting against numerous, highly conserved essential regions of the HCV functional proteins. Fortunately, there are numerous opportunities for inhibiting HCV replication. There are inhibitors developed for targeting the IRES, NS3 protease and NS5B RdRp, already in phase II clinical trials. Further opportunities are expected from the numerous promising drugs developed to date, and from the increasing knowledge about the virus life cycle, as well as about the targets for these drugs.

REFERENCES

Agnello, V., Abel, G., Elfahal, M., Knight, G. B., Zhang Q. X., 1999, Hepatitis C virus and other Flaviviridae viruses enter cells via low density lipoprotein receptor. *Proc. Natl. Acad. Sci. USA*, **96**: 12766-12771.

Altamura, S., Tomei, L., Koch, U., Neuner, P. J. S., Summa, V., 2000, Diketoacid derivatives as inhibitors of viral polymerases. WO Patent No. 2000006529.

Andoh, T., Kawamata, H., Umatake, M., Terasawa, K., Takegami, T., Ochiai, H., 1998, Effect of bafilomycin A1 on the growth of Japanese encephalitis virus in Vero cells. *J. Neurovirol.* **4**: 627-631.

Arnold, J. J. and Cameron, C.E., 2000, Poliovirus RNA-dependent RNA Polymerase (3Dpol). *J. Biol. Chem.*, **275**: 5329-5336.

Bachur, N., Yu, F., Johnson, R., Hickey, R., Wu, Y., Malkis, L., 1992. Helicase inhibition by anthracycline anticancer agents. *Mol. Pharmacol.*, **41**: 993-998.

Bartenschlager, R., Ahlborn-Laake, L., Mous, J., Jacobsen, H.,1993, Nonstructural protein 3 of the hepatitis C virus encodes a serine-type proteinase required for cleavage at the NS3/4 and NS4/5 junctions. *J. Virol.*, **67**: 3835-3844.

Bartenschlager, R., 1997, Candidate targets for hepatitis C virus-specific antiviral therapy. *Intervirology*, **40**: 378-393.

Barth, H., Schafer, C., Adah, M. I., Zhang, F., Linhardt, R. J., Toyoda, H., Kinoshita-Toyoda, A., Toida, T., van Kuppevelt, T. H., Depla, E., von Weizsacker, F., Blum, H. E., Baumert, T. F., 2003, Cellular binding of hepatitis C virus envelope glycoprotein E2 requires cell surface heparan sulfate. *J. Biol. Chem.*, **278**: 41003-41012.

Bartosch, B., Dubuisson, J., Cosset, F. L., 2003, Infectious hepatitis C virus pseudo-particles containing functional E1-E2 envelope protein complexes. *J. Exp. Med.*, **197**: 633-642.

Bayer, N., Schober, D., Prchla, E., Murphy, R. F., Blaas, D., Fuchs, R., 1998, Effect of bafilomycin A1 and nocodazole on endocytic transport in HeLa cells: implication for viral uncoating and infection. *J. Virol.*, **72**: 9645-9655.

Behrens, S. E., Tomei, L., De Francesco, R., 1996, Identification and properties of the RNA-dependent RNA polymerase of hepatitis C virus. *EMBO J.*, **15**, 12-22.

Borowski, P., Kühl, R., Mueller, O., Hwang, L.-H., Schulze zur Wiesch, J., Schmitz, H., 1999, Biochemical Properties of a minimal functional domain with ATP binding activity of the NTPase/helicase of hepatitis C virus. *Eur. J. Biochem.*, **266**: 715-723.

Borowski, P., Mueller, O., Niebuhr, A., Kalitzky, M., Hwang, L. H., Schmitz, H., Siwecka, M., Kulikowski, T., 2000, ATP-binding domain of NTPase/helicase as target for hepatitis C antiviral therapy. *Acta Biochim. Pol.*, **47**: 173-180.

Borowski, P., Niebuhr, A., Mueller, O., Bretner, M., Felczak, K., Kulikowski, T., Schmitz, H., 2001, Purification and characterization of West Nile virus NTPase/helicase. Evidence for dissociation of the NTPase and helicase activities of the enzyme. *J. Virol.*, **75**: 3220-3229.

Borowski, P., Niebuhr, A., Schmitz, P., Hosmane, R. S., Bretner, M., Siwecka, M., Kulikowski, T., 2002, NTPase/helicase of Flaviviridae: inhibitors and inhibition of the enzyme. *Acta Biochim. Pol.*, **49**: 597-614.

Bressanelli, S., Stiasny, K., Allison, S., Stura, E., Duquerroy, S., Lescar, J., Heinz, F. X., Rey, F., 2004, Structure of flavivirus envelope glycoprotein in its low-pH-induced membrane fusion conformation. *EMBO J.*, **23**: 728-738.

Bressanelli, S., Tomei, L., Roussel, A., Incitti, I., Vitale R. L., Mathieu, M., De Francesco, R., Rey, F. A., 1999, Crystal structure of the RNA-dependent RNA polymerase of the hepatitis c virus. *Proc. Natl. Acad. Sci. USA*, **96**: 13034-13039.

Carrasco, L., 1995, Modification of membrane permeability by animal viruses. *Adv. Virus Res.*, **45**: 61-112.

Carrere-Kremer, S., Montpellier, C., Lorenzo, L., Brulin, B., Cocquerel, L., Belouzard, S., Penin, F., Dubuisson, J., 2004, Regulation of hepatitis C virus polyprotein processing by signal peptidase involves structural determinants at the p7 sequence junctions. *J. Biol. Chem.*, **279**: 41384-41392.

Chambers, T.J., Hahn, C.S., Galler, R., Rice, M.C., 1990, Flavivirus genome organization, expression, and replication.. *Annu. Rev. Microbiol.*, **44**: 649-688.

Chen, C. J., Kuo, M. D., Chien, L. J., Hsu, S. L., Wang, Y. M., Lin, J. H., 1997, RNA-protein interactions: involvement of NS3, NS5, and 3' noncoding regions of Japanese encephalitis virus genomic RNA. *J. Virol.*, **71**: 3466-3473.

Chen, Y., Maguire, T., Hileman, R. E., Fromm, J. R., Esko, J. D., Linhardt, R. J., Marks, R. M., 1997, Dengue virus infectivity depends on envelope protein binding to target cell heparan sulfate. *Nat. Med.*, **3**: 866-871.

Chu, J. J., Ng, M. L., 2004, Interaction of West Nile virus with alpha v beta 3 integrin mediates virus entry into cells. *J. Biol. Chem.*, **279**: 54533-54541.

Chu, J. J.; Ng, M. L., 2003, Characterization of a 105-kDa plasma membrane associated glycoprotein that is involved in West Nile virus binding and infection.. *Virology*, **312**: 458-469.

Colman, P. M., Lawrence, M. C., 2003, The structural biology of type I viral membrane fusion. *Nat. Rev. Mol. Cell Biol.*, **4**: 309-319.

Crotty, S., Cameron, C., Andino, R., 2002, Ribavirin's antiviral mechanism of action: lethal mutagenesis? *J. Mol. Med.*, **80**: 86-95.

Crotty, S., Maag, D., Arnold, J. J., Zhong, W., Lau, J. Y., Hong, Z., Andino, R., Cameron, C. E., 2000, The broad-spectrum antiviral ribonucleoside ribavirin is an RNA virus mutagen. *Nat. Med.*, **6**: 1375-1379.

De Clercq E., 1993, Antiviral agents: characteristic activity spectrum depending on the molecular target with which they interact. *Adv. Virus Res.*, **42**: 1-55.

De Francesco, R., Behrens, S. E., Tomei, L., Altamura, S., Jiricny, J., 1996, RNA-Dependent RNA Polymerase of Hepatitis C Virus. *Meth. Enzymol.*, **275**: 58-67.

Deleersnyder, V., Pillez, A., Wychowski, C., Blight, K., Xu, J., Hahn, Y. S., Rice, C. M., Dubuisson, J., 1997, Formation of native hepatitis C virus glycoprotein complexes. *J. Virol.*, **71**: 697-704.

Deluca, M. R. and Kervin, S. M., 1996, The para-toluenosulfonic acid promoted synthesis of 2-substituted benzoxazoles and benzimidazoles from diacylated precursors. *Tetrahedron*, **53**: 457-464.

Diana, G. D. and Bailey, T. R., 1997, US Patent No. 5633388.

Diana, G. D., Bailey, T. R., Nitz, T. J., 1997, WO Patent No. 9736554.

Eckert, D. M., Kim, P. S., 2001, Design of protein inhibitors of HIV-1 entry from the gp41 N-peptide region.. *Proc Natl. Acad. Sci. USA*, **98**: 11187-11192.

Egger, D., Wolk, B., Gosert, R., Bianchi, L., Blum, H. E., Moradpour, D., Bienz, K., 2002, Expression of hepatitis C virus proteins induces distinct membrane alterations including a candidate viral replication complex. *J. Virol.*, **76**: 5974-5984.

Fire, A., Xu, S., Montgomery, M. K. , Kostas, S. A., Driver, S. E., Mello, C. C., 1998, Potent and specific genetic interference by double-stranded RNA in Caenorhabditis elegans. *Nature*, **391**: 806-811.

Friebe, P., Lohmann, V., Krieger, N., Bartenschlager, R., 2001, Sequences in the 5' Nontranslated Region of Hepatitis C Virus Required for RNA Replication. *J. Virol.*, **75**: 12047-12057.

Furuchi, T., Aikawa, K., Arai, H., Inoue, K., 1993, Bafilomycin A1, a specific inhibitor of vacuolar-type H(+)-ATPase, blocks lysosomal cholesterol trafficking in macrophages. *J. Biol. Chem.*, **268**: 27345-27348.

Garson, J. A., Lubach, D., Passas, J., Whitby, K., Grant, P. R., 1999, Suramin blocks hepatitis C binding to human hepatoma cells in vitro. *J. Med. Virol.*, **57**: 238-242.

Gollins, S. W., Porterfield, J. S., 1986, pH-dependent fusion between the flavivirus West Nile and liposomal model membranes. *J. Gen. Virol.*, **67**: 157-166.

Gordon, C. P., Keller, P. A., 2005, Control of hepatitis C: a medicinal chemistry perspective. *J. Med. Chem.*, **48**: 1-20.

Gosert, R., Egger, D., Lohmann, V., Bartenschlager, R., Blum, H. E., Bienz, K., Moradpour, D., 2003, Identification of the hepatitis C virus RNA replication complex in Huh-7 cells harboring subgenomic replicons. *J. Virol.*, **77**: 5487-5492.

Goudreau, N., Cameron, D. R., Bonneau, P., Gorys, V., Plouffe, C., Poirier, M., Lamarre, D., Llinas-Brunet, M., 2004, NMR structural characterization of peptide inhibitors bound to the Hepatitis C virus NS3 protease: design of a new P2 substituent. *J. Med. Chem.*, **47**: 123-132.

Gwack, Y., Yoo, H., Song, I., Choe, J., Han, J.H., 1999, RNA Stimulated ATPase and RNA helicase activities and RNA binding domain of hepatitis virus G nonstructural protein 3. *J. Virol.*, **73**: 2909-2915.

Halstead, S. B., 1992, The XXth century dengue pandemic: need for survieillance and research. In World Health StatQ, 45, 292-298.

Hammond, S. M., Bernstein, E., Beach, D., Hannon, G. J., 2000, An RNA-directed nuclease mediates post-transcriptional gene silencing in Drosophila cells. *Nature*, **404**: 293-296.

Harada, T., Tautz, N., Thiel, H.J., 2000, E2-p7 region of the bovine viral diarrhea virus polyprotein: processing and functional studies. *J. Virol.*, **74**: 9498-9506.

Heo, T. H., Chang, J. H., Lee, J. W., Foung, S. K., Dubuisson, J., Kang, C. Y., 2004, Incomplete humoral immunity against hepatitis C virus is linked with recognition of putative multiple receptors by E2 envelope glycoprotein. *J. Immunol.*, **173**: 446-455.

Ishido, S., Fujita, T., Hotta, H., 1998, Complex formation of NS5B with NS3 and NS4A proteins of hepatitis C virus. *Biochem. Biophys. Res. Commun.*, **244**: 35-40.

Jordan, I., Briese, T., Fischer, N., Yiu-Nam, J., Lipkin, W. I., 2000, Ribavirin inhibits West Nile virus replication and cytopathic effect in neural cells. *J. Infect. Dis.*, **182**: 1214-1217.

Kadare, G. and Haenni, A., 1997, Virus-encoded RNA helicase. *J. Virol.*, **71**: 2583-2590.

Kalitzky, M., 2003, NTPase/helicase of hepatitis C and West Nile virus: correlation of DNA unwinding activity with NTP hydrolysis. In Faculty of Medicine, Doctoral Thesis, University of Hamburg, Hamburg.

Kapoor, M., Zhang, L., Ramachandra, M., Kusukawa, J., Ebner, K. E., Padmanabhan, R., 1995, Association between NS3 and NS5 proteins of dengue virus type 2 in the putative RNA replicase is linked to differential phosphorylation of NS5. *J. Biol. Chem.*, **270**: 19100-19106.

Kim, J. L., Morgenstern, K. A., Lin, C., Fox, T., Dwyer, M. D., Landro, J. A., Chambers, S. P., Markland, W., Lepre, C. A., O'Malley, E. T., Harbeson, S. L., Rice, C. M., Murcko, M. A., Caron, P. R., Thomson, J. A., 1996, Crystal structure of the hepatitis C virus NS3 protease domain complexed with a synthetic NS4A cofactor peptide. *Cell*, **87**: 343-55.

Kim, J. L., Morgenstern, K. A., Griffith, J. P., Dwyer, M. D., Thomson, J. A., 1998, Hepatitis C virus NS3 RNA helicase domain with a bound oligonucleotide: the crystal structure provides insights into the mode of unwinding. *Structure*, **6**: 89-100.

Kimura, T., Gollins, S. W., Porterfield, J. S., 1986, The effect of pH on the early interaction of West Nile virus with P388D1 cells. *J. Gen. Virol.*, **67**: 2423-2433.

Koch, J. O. und Bartenschlager. R., 1999, Modulation of hepatitis C virus NS5A hyperphosphorylation by nonstructural proteins NS3, NS4A, and NS4B. *J. Virol.*, **73**: 7138-7146.

Lau, J., Tam, RC., Liang, TJ., Hong, Z., 2002, Mechanism of action of ribavirin in the combination treatment of chronic HCV infection. *Hepatol.*, **35**: 1002-1009.

Laxton, C. D., McMillan, D., Sullivan, V., Ackrill, A. M., 1998, Expression and characterization of the hepatitis G virus helicase. *J. Viral Hepat.*, **5**: 21-26.

Lee, E. and Lobigs, M., 2000, Substitutions at the Putative Receptor-Binding Site of an Encephalitic Flavivirus Alter Virulence and Host Cell Tropism and Reveal a Role for Glycosaminoglycans in Entry. *J. Virol.,* **74**: 8867-8875.

Lee, J., von Wagner, M., Reoth, K., Teuber, G., Sarrazin, C, Zeuzem, S., 1998, Effect of ribavirin on virus load and quasispecies distribution in patients infected with hepatitis C virus. *J. Hepatol.,* **29**: 29-35.

Lemberg, M. K. and Martoglio, B., 2002, Requirements for signal peptide peptidase-catalyzed intramembrane proteolysis. *Mol. Cellular Biol.,* **10**: 735-744.

Lemberg, M. K., Bland, F. A., Weihofen, A., Braud, V. M., Martoglio, B., 2001, Intramembrane proteolysis of signal peptides: an essential step in the generation of HLA-E epitopes. *J. Immunol.,* **167**: 6441-6446.

Leyssen, P., De Clercq, E., Neyts, J., 2000, Perspectives for the treatment of infections with Flaviviridae. *Clin. Microbiol. Rev.,* **13**: 67-82.

Li, H., Clum, S., You, S., Ebner, K.E., Padmanabhan, R., 1999, The serine protease and RNA stimulated nucleoside triphosphatase and RNA helicase functional domains of dengue virus type 2 NS3 converge within a region of 20 amino acids. *J. Virol.,* **73**: 3108 3116.

Lin, C., Thomson, J. A., Rice, C. M., 1995, A central region in the hepatitis C virus NS4A protein allows formation of an active NS3-NS4A serine proteinase complex in vivo and in vitro. *J. Virol.,* **69**: 4373-4380.

Lin, C., Wu, J. W., Hsiao, K., Su, M. S., 1997, The hepatitis C virus NS4A protein: interactions with the NS4B and NS5A proteins. *J. Virol.,* **71**: 6465-6471.

Lindenbach, B. D. and Rice, C. M., 1997, Trans-complementation of yellow fever virus NS1 reveal a role in early RNA replication. *J. Virol.,* **71**: 9608-9617.

Liu, Q., Bhat, R. A., Prince, A. M., Zhang, P., 1999, The Hepatitis C Virus NS2 Protein Generated by NS2-3 Autocleavage Is Required for NS5A Phosphorylation. *Biochem. Biophys. Res. Comm.,* **254**: 572–577.

Llinas-Brunet, M., Bailey, M. D., Ghiro, E., Gorys, V., Halmos, T., Poirier, M., Rancourt, J., Goudreau, N. A., 2004, systematic approach to the optimization of substrate-based inhibitors of the hepatitis C virus NS3 protease: discovery of potent and specific tripeptide inhibitors. *J. Med. Chem.,* **47**: 6584-6594.

Lohmann, V., Korner, F., Herian, U., Bartenschlager, R., 1997, Biochemical properties of hepatitis C virus NS5B RNA-dependent RNA polymerase and identification of amino acid sequence motifs essential for enzymatic activity. *J. Virol.,* **71**: 8416-8428.

Love, R. A., Parge, H. E., Wickersham, J. A., Hostomsky, Z., Habuka, N., Moomaw, E. W., Adachi, T., Hostomska, Z., 1996, The crystal structure of hepatitis C virus NS3 proteinase reveals a trypsin-like fold and a structural zinc binding site. *Cell,* **87**: 331-342.

Lu, H. H., Wimmer, E., 1996, Poliovirus chimeras replicating under the translational control of genetic elements of hepatitis C virus reveal unusual properties of the internal ribosomal entry site of hepatitis C virus. *Proc. Natl. Acad. Sci. USA,* **93**: 1412-1417.

Lun, L., Sun, P.-M., Trubey, C., Bachur, N., 1998, Antihelicase action of Cl-958, a new drug for prostate cancer. *Cancer Chemother. Pharmacol.,* **42**: 447-453.

Maag, D., Castro, C., Hong, Z., Cameron, C. E., 2001, Hepatitis C virus RNA-dependent RNA polymerase (NS5B) as a mediator of the antiviral activity of ribavirin. *J. Biol. Chem.,* **276**: 46094-46098.

Martınez-Salas, E., Ramos, R., Lafuente, E., Lopez de Quinto, S., 2001, Functional interactions in internal translation initiation directed by viral and cellular IRES elements. *J. Gen. Virol.,* **82**: 973-984.

Matson, S. W. and Kaiser-Rogers, K. A., 1990, DNA helicases. *Annu. Rev. Biochem.,* **59**: 289-329.

Mc Cormick, J. B., King, I. J., Webb, P. A., Scribner, C. L., Craven, R. B., Johnson, K. M., Elliott, L. H., Williams, R. B., 1986, Lassa fever. Effective therapy with ribavirin. *N. Engl. J. Med.*, **314**: 20-26.

McCaffrey, A. P., Nakai, H., Pandey, K., Huang, Z., Salazar, F. H., Xu, H., Wieland, S. F., Marion, P. L., Kay, M. A., 2003, Inhibition of hepatitis B virus in mice by RNA interference. *Nat. Biotechnol.*, **21**: 639-644.

Meyers, G. and Thiel, H. J., 1996, Molecular characterization of pestiviruses. *Adv. Virus Res.*, **47**: 53-118.

Miller, J. P., Kiowana, L. J., Streeter, D. G., Robins, R. K., Simon, L. N., Roboz, J., 1977, The relationship between metabolism of ribavirin and its proposed mechanism of action. *Ann. N. Y. Acad. Sci. U.S.A.*, **287**: 211-229.

Modis, Y., Ogata, S., Clements, D., Harrison, S. C., 2003, A ligand-binding pocket in the dengue virus envelope glycoprotein. *Proc. Natl. Acad. Sci. USA*, **100**: 6899-6901.

Monath, T. P. and Heinz, F. X., 1996, Flaviviruses. In Fields virology. Fields BN, Knipe, D.M., Howley, P.M. (Eds.) Lippincott-Raven Publishers, Philadelphia, PA, 961-1034.

Monath, T. P., 1987, Yellow fever: a medically neglected disease. Report on a seminar. *Rev. Infect. Dis.*, **9**: 165-175.

Moriishi, K., Matsuura, Y., 2003, Mechanism of hepatitis C virus infection. *Antiviral Chem. Chemother.*, **14**: 285-297.

Morrey, J. D., Smee, D. F., Sidwell, R. W., Tseng, C., 2002, Identification of active antiviral compounds against a New York isolate of West Nile virus. *Antiviral Res.*, **55**: 107-116.

Muerhoff, A. S., Leary, T. P., Simons, J. N., Pilot-Matias, T. J., Dawson, G. J., Erker, J. C., Chalmers, M. L., Schlauder, G. G., Desai, S. M., Mushahwar, I. K., 1995, Genomic organization of GB viruses A and B: two new members of the *Flaviviridae* associated with GB agent hepatitis. *J. Virol.*, **69**: 5621-5630.

Natale, V. A., McCullough, K, C., 1998, Macrophage cytoplasmic vesicle pH gradient and vacuolar H+-ATPase activities relative to virus infection. *J. Leukoc. Biol.*, **64**: 302-310.

Nawa, M., 1998, Effect of bafilomycin A1 on Japanese encephalitis virus in C6/36 mosquito cells. *Arch. Virol.*, **148**: 1555-1568.

Neddermann, P., Clementi, A., De Francesco, R., 1999, Hyperphosphorylation of the hepatitis C virus NS5A protein requires an active NS3 protease, NS4A, NS4B, and NS5A encoded on the same polyprotein. *J. Virol.*, **73**: 9984-9991.

Neddermann, P., Tomei, L., Steinkuhler, C., Gallinari, P., Tramontano, A., De Francesco R., 1997, The nonstructural proteins of the hepatitis C virus: structure and functions. *Biol. Chem.*, **378**: 469-476.

Nettelton, P. E. and Entrican, G., 1995 Ruminant pestiviruses. *Br. Vet. J.*, **151**: 615-642.

Novina, C. D., Murray, M. F., Dykxhoorn, D. M., Beresford, P. J., Riess, J., Lee, S. K., Collman, R. G., Lieberman, J., Shankar, P., Sharp, P. A., 2002, siRNA-directed inhibition of HIV-1 infection. *Nat. Med.*, **8**: 681-686.

Nunes-Correia, I., Nir, S., Pedroso de Lima, M. C., 2003, Kinetics of influenza virus fusion with the endosomal and plasma membranes of cultured cells. Effect of temperature. *J. Membr. Biol.*, **195**: 21-26.

Paddison, P. J., Hannon, G. J., 2003, siRNAs and shRNAs: skeleton keys to the human genome. *Curr. Opin. Mol. Ther.*, **5**: 217-224.

Papa, R. A., Frost, M. J., Mackintosh, S. G., Gu, X., Dixon, R. J., Shannon, A. D., 2002, A single amino acid is critical for the expression of B-cell epitopes on the helicase domain of the pestivirus NS3 protein. *Virus Res.*, **84**: 111-124.

Penin, F., Dubuisson, J., Rey, F. A., Moradpour, D., Pawlotsky, J. M., 2004, Structural biology of hepatitis C virus. *Hepatol.*, **39**: 5-19.

Phoon, C. W., Ng, Y. P., Ting, A. E., Yeo, S. L., Sim, M. M., 2001, Biological evaluation of hepatitis C virus helicase inhibitors. *Bioorg. Med. Chem. Lett.*, **11**: 1647-1650.

Pileri, P., Uematsu, Y., Campagnoli, S., Galli, G., Falugi, F., Petracca, R., Weiner, A. J., Houghton, M., Rosa, D., Grandi, G., Abrignani, S., 1998, Binding of hepatitis C virus to CD81. *Science*, **282**: 938-941.

Porter, D. J. and Preugschat, F., 2000, Strand-separating activity of hepatitis C virus helicase in the absence of ATP. *Biochemistry,* **39**: 5166-5173.

Porter, D., 1998, Inhibition of the hepatitis C virus helicase-associated ATPase activity by the combination of ADP, NaF, MgCl2, and poly(rU). *J. Biol. Chem.*, **273**: 7390-7396.

Poynard, T., Marcellin, P., Lee, S. S., Niederau, C., Minuk, G. S., Ideo, G., Bain, V., Heathcote, J., Zeuzem, S., Trepo, C., Albrecht, J., 1998, Randomised trial of interferon alpha2b plus ribavirin for 48 weeks or for 24 weeks versus interferon alpha2b plus placebo for 48 weeks for treatment of chronic infection with hepatitis C virus. International Hepatitis Interventional Therapy Group (IHIT). *Lancet,* **352**: 1426-1432.

Pryor, M. J., Gualano, R. C., Lin, A. D., Davidson, A. D., Wright, P. J., 1998, Growth restriction of dengue virus type 2 by site-specific mutagenesis of virus-encoded glyco-proteins. *J. Gen. Virol.*, **79**: 2631-2639.

Racker, E., 1991, Chaperones and Matchmakers: Inhibitors and Stimulators of Protein Phosphorylation. *Curr. Top. Cell. Regulation,* **33**: 127-143.

Rice, M. C.,1996, Flaviviridae: the viruses and their replication. In Fields, B.N. Knipe, D.M. Howley, P. M. (Ed) Fields virology 3rd ed., vol. 1. Lippincott-Raven Publishers, Philadelphia, Pa., 931-960.

Santolini, E., Pacini, L., Fipaldini, C., Migliaccio, G., Monica, N., 1995, The NS2 protein of hepatitis C virus is a transmembrane polypeptide. *J. Virol.,* **69**: 7461-7471.

Scheper, G., and Proud, C., 2002, Does phosphorylation of the cap-binding protein eIF4E play a role in translation initiation? *Eur. J. Biochem.,* **269**: 5350-5359.

Schmidt-Mende, J., Bieck, E., Hugle, T., Penin, F., Rice, C. M., Blum, H. E., Moradpour, D., 2001, Determinants for membrane association of the hepatitis C virus RNA-dependent RNA polymerase. *J. Biol. Chem.*, **276**: 44052-44463.

Schwartz, M., Chen, J., Janda, M., Sullivan, M., den Boon, J., Ahlquist, P., 2002, A positive-strand RNA virus replication complex parallels form and function of retrovirus capsids. *Mol. Cell,* **9**: 505-514.

Shimizu, Y., Yamaji, K., Masuho, Y., Yokota, T., Inoue, H., Sudo, K., Satoh, S., Shimotohno, K., 1996, Identification of the sequence on NS4A required for enhanced cleavage of the NS5A/5B site by hepatitis C virus NS3 protease. *J. Virol.,* **70**: 127-132.

Song, G. Y., Paul, V., Choo, H., Morrey, J., Sidwell, R. W., Schinazi, R. F., Chu, C. K., 2001, Enantiomeric synthesis of D- and L-cyclopentyl nucleosides and their antiviral activity against HIV and West Nile virus. *J. Med. Chem.*, **44**: 3985-3993.

Steffens, S., Thiel, H. J., Behrens, S. E., 1999, The RNA-dependent RNA polymerases of different members of the family *Flaviviridae* exhibit similar properties in vitro. *J. Gen. Virol.*, **80**: 2583-2590.

Suzich, J., Tamura, J., Palmer-Hill, F., Warrener, P., Grakoui, A., Rice, C., Feinstone, S., and Collett, M., 1993, Hepatitis C Virus NS3 Protein Polynucleotide stimulated Nucleoside Triphosphatase and Comparison with the related Pestivirus and Flavivirus Enzymes. *J. Virol.*, **67**: 6152-6158.

Taber, L. H., Knight, V., Gilbert, B. E., McClung, H. W., Wilson, S. Z., Norton, H. J., Thurson, J. M., Gordon, W. H., Atmar, R. L., Schlaudt, W. R., 1983, Ribavirin aerosol treatment of bronchiolitis associated with respiratory syncytial virus infection in infants. *Pediatrics,* **72**: 613-618.

Tamura, J. K., Warrener, P., Collett, M. S., 1993, RNA-stimulated NTPase activity associated with the p80 protein of the pestivirus bovine viral diarrhea virus. *Virology*, **193**: 1-10.

Tan, B. H., Fu, J., Sugrue, R. J., Yap, E. H., Chan, Y. C., 1996, Recombinant dengue type 1 virus NS5 protein expressed in Escherichia coli exhibits RNA-dependent RNA polymerase activity. *Virology*, **216**: 317-325.

Thepparit, C., Smith, D. R., 2004, Serotype-specific entry of dengue virus into liver cells: identification of the 37-kilodalton/67-kilodalton high-affinity laminin receptor as a dengue virus serptype 1 receptor. *J. Virol.*, **78**: 12647-12656.

Thurner, C., Witwer, C., Hofacker, I. L., Stadler, P. F., 2004, Conserved RNA secondary structures in Flaviviridae genomes. *J. Gen. Virol.*, **85**: 1113-1124.

Tomei, L., Failla, C., Vitale, R. L., Bianchi, E., De Francesco, R., 1996, A central hydrophobic domain of the hepatitis C virus NS4A protein is necessary and sufficient for the activation of the NS3 protease. *J. Gen. Virol.*, **77**: 1065-1070.

Utama, A., Shimizu, H., Morikawa, S., Hasebe, F., Morita, K., Idarash, A. I., Hatsu, M., Takamizawa, K., Miyamura, T., 2000, Identification and characterization of the RNA helicase activity of Japanese encephalitis virus NS3 protein. *FEBS Lett.*, **465**: 74-78.

Voisset, C., Callens, N., Blanchard, E., op de Beeck, A., Dubuisson, J., Vu-Dac, N., 2005, High density lipoproteins facilitate hepatitis C virus entry through the scavenger receptor class B type I. *J. Biol. Chem.*, **280**: 7793-7799.

Walker, J. E., Saraste, M., Runswick, M. J., Gay, N. J., 1982, Distantly related sequences in the alpha- and beta-subunits of ATP synthase, myosin, kinases and other ATP- requiring enzymes and a common nucleotide binding fold. *EMBO J.*, **1**: 945-951.

Wang, M., Ng, K. S., Cherney, M., Chan, L., Yannopoulus, C. G., Bedard, J., Morin, N., Nguyen-Ba, N., Alaoui-Ismaili, M. H., Bethell, R. C., James, M. N. G., 2003, Non-nucleoside analogue inhibitors bind to an allosteric site on HCV NS5B polymerase. Crystal structures and mechanism of inhibition. *J. Biol. Chem.*, **278**: 9489-9495.

Wedemeyer, H., Caselmann, W. H., Manns, M. P., 1998, Combination therapy of chrnoc hepatits C: an important step but not the final goal! *J. Hepatol.*, **29**: 1010-1014.

Weihofen, A., Lemberg, M. K., Friedmann, E., Rueeger, H., Schmitz, A., Paganetti, P., Rovelli, G., Martoglio, B., 2003 Targeting presenilin-type aspartic protease signal peptide peptidase with gamma-secretase inhibitors. *J. Biol. Chem.*, **278**: 16528-16533.

Weissenhorn, W., Dessen, A., Calder, I. J., Harrison S. C., Skehel, J. J., Wiley, D. C., 1999, Structural basis for membrane fusion by enveloped viruses. *Mol. Membr. Biol.*, **16**: 3-9.

Westaway, E. G., Brinton, M. A., Gaidamovich, Y., Horzinek, M. C., Igarashi, A., Kääriäinen, L., Lvov, D. K., Porterfield, J. S., Russell, P. K., Trent, D. W., 1985, Flaviviridae. *Intervirology*, **24**: 183-192.

Westaway, E. G., 1987, Flavivirus replication strategy. *Adv. Virus Res.*, **33**: 45-90.

Wilson, I. A., Skehel, J. J., Wiley, D. C., 1981, Structure of the hemagglutinin membrane glycoprotein of influenza virus at 3Å resolution. *Nature*, **289**: 366-378.

Wolk, B., Sansonno, D., Krausslich, H. G., Dammacco, F., Rice, C. M., Blum, H. E., Moradpour, D., 2000, Subcellular localization, stability, and trans-cleavage competence of the hepatitis C virus NS3-NS4A complex expressed in tetracycline-regulated cell lines. *J. Virol.*, **74**: 2293-2304.

Wu, S. F., Lee, C. J., Liao, C. L., Dwek, R., Zitzmann, N., 2002, Antiviral effects of an iminosugar derivative on flavivirus infections. *J. Virol.*, **76**: 3596-3604.

Wu, Z., Yao, N., Lee, H. V., Weber P. C., 1998, Mechanism of autoproteolysis at the NS2-NS3 junction of the hepatitis C virus polyprotein. *Trends Biochem. Sci.*, **23**: 92-94.

Yao, N., Hesson, T., Cable, M., Hong, Z., Kwong, A., 1997, Structure of the hepatitis C virus RNA helicase domain. *Nat. Struct. Biol.*, **4**: 463-467.

Yu, S., Wuu, A., Basu, R., Holbrook, M. R., Barrett, A. D., Lee, J. C., 2004, Solution structure and structural dynamics of envelope protein domain III of mosquito- and tick-borne flaviviruses. *Biochemistry,* **43**: 9168-9176.

Zhang, N., Chen, H. M., Koch, V., Schmitz, H., Liao, C. L., Bretner, M., Bhadti, V. S., Fattom, A. I., Naso, R. B., Hosmane, R. H., Borowski, P., 2003a, Ring-expanded ("fat") nucleoside and nucleotide analogues exhibit potent in vitro activity against Flaviviridae NTPase/helicases, including those of West Nile virus, hepatitis C virus and Japanese encephalitis virus. *J. Med. Chem.,* **46**: 4149-4164.

Zhang, N., Chen, H. M., Koch, V., Schmitz, H., Minczuk, M., Stepien, P., Fattom, A. I., Naso, R. B., Kalicharran, K., Borowski, P., Hosmane, R. S., 2003b, Potent inhibition of NTPase/helicase of the West Nile virus by ring-expanded ("fat") nucleoside analogues. *J. Med. Chem.,* **46**: 4776-4789.

Zhao, W. D., Wimmer, E., Lahser, F. C., 1999, Poliovirus/ Hepatitis C virus (internal ribosomal entry site-core) chimeric viruses: improved growth properties through modification of a proteolytic cleavage siteand requirement for core RNA sequences but not for core-related polypeptides. *J. Virol.,* **73**: 1546-1554.

Zhong, W., Gutshall, L. L., Del Vecchio, A. M., 1998, Identification and Characterization of an RNA-Dependent RNA Polymerase Activity within the Nonstructural Protein 5B Region of Bovine Viral Diarrhea Virus. *J. Virol.,* **72**: 9365-9369.

Zhong, W., Ingravallo, P., Wright-Minogue, J., Skelton, A., Uss, A. S., Chase, R., Yao, N., Lau, Y. N., Hong, Z., 1999, Nucleoside triphosphatase and RNA helicase activities associated with GB virus B nonstructural protein 3. *Virology,* **261**: 216-226.

Zimmerman, T. P. and Deeprose, R. D., 1978, Metabolism of 5-amino-1-beta-D-ribofuranosylimidazole-4-carboxamide and related five-membered heterocycles to 5'-triphosphates in human blood and L5178Y cells. *Biochem. Pharmacol.,* **27**: 709-716.

Zoulim, F., Haem, J., Ahmed, S. S., Chossegros, P., Habersetzer, F., Chewallier, M., Bailly, F., Trepo, C., 1998, Ribavirin monotherapy in patients with chronic hepatitis C: a retrospective study of 95 patients. *J. Viral. Hepat.,* **5**: 193-198.

Chapter 1.3

INHIBITION OF HEPATITIS C VIRUS BY NUCLEIC ACID-BASED ANTIVIRAL APPROACHES

M. FRESE and R. BARTENSCHLAGER
Department for Molecular Virology, Hygiene Institute, University of Heidelberg, Im Neuenheimer Feld 345, D-69120 Heidelberg, Germany

Abstract: Persistent infection with the hepatitis C virus (HCV) is a major cause of acute and chronic liver disease and frequently leads to liver cirrhosis and hepatocellular carcinoma. Current treatment is based on a combination therapy with polyethylene glycol-conjugated interferon-alpha and ribavirin, but efficacy is limited and treatment is associated with severe side-effects. More efficient and selective drugs are therefore needed. Apart from small molecule inhibitors targeting the viral key enzymes, especially the NS3 proteinase and the NS5B RNA-dependent RNA polymerase, nucleic acid (NA)-based antiviral intervention is an attractive option. Originally, antisense oligo-nucleotides and ribozymes targeting highly conserved regions in the HCV genome have been developed. More recently, short interfering RNAs (siRNAs) were shown to potently block HCV RNA replication in cell culture. However, the high degree of sequence diversity between different HCV genotypes, the rapid evolution of quasispecies and the delivery of antivirally active NAs are challenging problems of NA-based therapies. Here, we will review the current state of NA-based approaches designed to interfere with HCV replication.

1.   INTRODUCTION

Worldwide, more than 170 million individuals have been infected with hepatitis C virus (HCV) (World Health Organization, 2000). In about 80% of all cases, the virus establishes a persistent infection which frequently leads to chronic liver disease including liver fibrosis, cirrhosis, and eventually

47

E. Bogner and A. Holzenburg (eds.), New Concepts of Antiviral Therapy, 47–86.

hepatocellular carcinoma (Hoofnagle, 1997; Theodore and Fried, 2000). Consequently, HCV infection is in many countries a primary cause for liver transplantation. The standard therapy for chronic hepatitis C patients currently consists of polyethylene glycol (PEG)-conjugated type I interferon (IFN) e.g. IFN-α2, and ribavirin (McHutchison and Fried, 2003).

Type I (alpha and beta) IFNs are key cytokines of the innate immune response to viral infections and induce the expression of numerous effectors of which some may have the potential to interfere directly with HCV replication. It has been demonstrated that both type I and II IFNs inhibit HCV replication in cell culture without the help of professional immune cells such as T cells or natural killer (NK) cells (Blight *et al.*, 2000; Cheney *et al.*, 2002; Frese *et al.*, 2001; 2002; Lanford *et al.*, 2003; Mizutani *et al.*, 1996; Okuse *et al.*, 2005; Robek *et al.*, 2005; Shimizu and Yoshikura, 1994). In addition to their direct antiviral activities, type I IFNs activate professional immune cells. These systemic effects most likely contribute to HCV clearance in IFN-treated patients. However, the molecular mechanisms and the extent to which IFNs cure infected host cells and contribute to the elimination of HCV from infected people are still largely unknown.

Ribavirin is a guanosine analogue bearing antiviral activity against a variety of RNA and DNA viruses (Snell, 2001). Nevertheless, only two human diseases are frequently treated with this drug: severe respiratory syncytial virus infections in certain at-risk patients and chronic hepatitis C. When given alone, ribavirin does not reduce the viral load in chronic hepatitis C patients. In combination with type I IFN, however, it significantly reduces the number of relapse patients i.e. patients in which HCV rebounds after cessation of therapy. Ribavirin may have several mechanisms of action. At high concentrations, it has been shown to increase the mutation rate which would lead to a reduction in the infectivity of progeny viruses due to an accumulation of deleterious mutations (Contreras *et al.*, 2002; Lanford *et al.*, 2003). Furthermore, the drug polarizes the human T cell response towards a type I cytokine profile, thereby strengthening the cellular arm of the immune response and enhancing the chance of virus elimination (Hultgren *et al.*, 1998; Tam *et al.*, 1999).

The combination of type I IFNs and ribavirin has likely saved the lives of many hepatitis C patients, but the treatment has its limitations. Current therapy regimens require a drug administration of at least 6 months, both drugs have serious side effects and, most importantly, certain patients respond only poorly to the therapy. For example, only 50% of patients infected with the HCV genotype 1 mount a sustained viral response, whereas 80 to 90% of those infected with genotype 2 and genotype 3 viruses do so. The correlation between therapy success and the infecting genotype implies an involvement of viral factors, but the underlying molecular mechanisms are not yet understood.

With the establishment of a subgenomic replicon system in 1999 (see below), it became possible to evaluate small molecules for their ability to block HCV RNA replication, and several inhibitors of the NS3/4A proteinase or the NS5B RNA dependent RNA polymerase (RdRp) have been identified. In a proof-of-concept study it has been demonstrated that the NS3/4A inhibitor BILN-2061 can reduce the viral load in persistently infected hepatitis C patients with a few oral doses by more than 1,000-fold, raising hope that HCV eradication by selective drugs is possible (Hinrichsen *et al.,* 2004; Lamarre *et al.,* 2003). Further clinical trials, however, are on hold pending resolution of animal toxicity issues. Nevertheless, these results will provide impetus for the generation of improved molecules. Many other compounds including several nucleoside and non-nucleoside inhibitors of NS5B are currently entering early-phase clinical trials or are under evaluation as reviewed elsewhere (Horscroft *et al.,* 2005; Ni and Wagman, 2004). However, it is unclear whether these approaches will ultimately lead to a cure for chronic hepatitis C and therefore, alternative strategies are being pursued to combat this insidious infection. One attractive approach is nucleic acid (NA)-based therapies such as antisense oligonucleotides (AS-ON), ribozymes and short interfering (si) RNAs. In this chapter, we will summarize the current state of these approaches for HCV-specific therapies.

1.1   HCV genome organization and protein functions

HCV belongs to the genus *Hepacivirus* within the family *Flaviviridae* (Van Regenmortel *et al.,* 2000). At least 6 different HCV genotypes exist which show distinct geographical distributions. For each genotype, a series of more closely-related subtypes have been described that differ from one another by 20 to 25% in their nucleotide sequence. HCV has a ~9.6-kb single-stranded RNA genome of positive polarity carrying one long open reading frame (reviewed in Bartenschlager *et al.,* 2004). The genome termini are formed by highly-structured non-translated regions (NTRs) important for both RNA translation and replication. Within the 5' NTR, an internal ribosome entry site (IRES) has been identified that permits expression of the viral proteins in the absence of a cap structure (Pestova *et al.,* 1998). The HCV genome encodes one large polyprotein of approximately 3,000 amino acids that is co- and post-translationally cleaved by cellular and viral proteinases into 10 polypeptides (core, E1, E2, p7, NS2, NS3, NS4A, NS4B, NS5A, and NS5B). The production of an additional viral protein by ribosomal frameshift and internal translation initiation has also been reported (Varaklioti *et al.,* 2002; Xu *et al.,* 2001), but its function remains to be defined. By contrast, distinct functions within the virus life cycle have been ascribed to almost all other HCV proteins (Fig. 1).

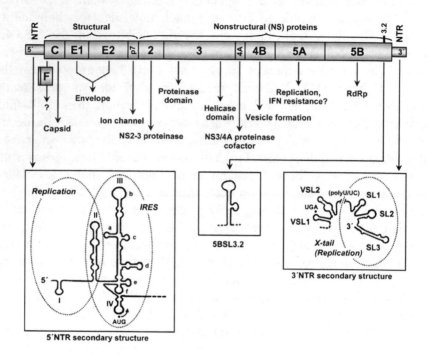

Figure 1. HCV genome organization, presumed functions of viral proteins, and essential *cis*-acting RNA elements. The viral genome contains a large open reading frame (ORF) encoding all major proteins and an alternative ORF that encodes the frame shift protein (F) which has an unknown function. The structural proteins C, E1, E2, and p7 are liberated from the polyprotein by cellular signal peptidases. The junction between NS2 and NS3 is cleaved by the NS2/3 proteinase. All other cleavages are mediated by the NS3/4A proteinase complex. RdRp, RNA-dependent RNA polymerase. The 5' and the 3' NTR structures as well as the stem-loop structure 5BSL3.2 (lower panel) are drawn according to Blight and Rice, 1997; Friebe *et al.*, 2005; Honda *et al.*, 1999; You *et al.*, 2004, respectively. Black dots indicate the position of the start and stop codon of the large ORF. The 5' NTR consists of 4 domains of which the first two are required for RNA replication and domains two to four for RNA translation. The 3' NTR consists of two potential stem-loop structures in the variable regions (VSL1 and VSL2), a polyU/UC tract and three highly conserved stem-loop structures in the 3' terminal region designated the X-tail. Note that the length of the poly U/UC tract can vary dramatically even between different isolates of the same genotype. The minimum regions in the 5' and 3' NTRs required for replication and initiation of translation are encircled with dotted lines. The secondary structure of stem-loop 5BSL3.2 is shown in the middle panel. Note that its upper loop is involved in direct base pairing with the upper loop of SL2 in the 3' NTR.

The core protein resides at the very N terminus of the polyprotein and is released from the precursor by signal peptidase. A further cleavage, performed by signal peptide peptidase, results in the mature protein that most likely makes up the viral capsid. E1 and E2 are highly glycosylated type I transmembrane proteins. Their biogenesis is an intricate process involving ER-resident chaperones and the formation of noncovalently linked heterodimers (Deleersnyder *et al.*, 1997). The relatively small, hydrophobic p7 protein is an integral membrane protein with two transmembrane domains. It has been reported to form hexameric complexes that may function as ion channels (Pavlovic *et al.*, 2003; Sakai *et al.*, 2003).

The nonstructural protein NS2 is liberated from NS3 by an intra-molecular cleavage that is performed by the NS2 protein itself in co-operation with the first 180 amino acids of NS3. The NS2-3 proteinase is most likely a cysteine proteinase with a catalytic triad that requires zinc as a structural component (Pause *et al.*, 2003; Steinkühler *et al.*, 1996). The NS3 protein has multiple functions. Besides its role in NS2/3 cleavage, NS3 cleaves all subsequent junctions in the HCV polyprotein *via* a chymotrypsin-like proteinase domain that needs NS4A for full enzymatic activity (Bartenschlager *et al.*, 1994; Failla *et al.*, 1994). NS3 also contains a helicase domain that appears to bind specifically to HCV promoter sequences in the 5' and 3' NTR and thereby promote the initiation of positive- and negative-strand RNA synthesis (Banerjee and Dasgupta, 2001; Paolini *et al.*, 2000). The transmembranous structure of NS4B, its capacity to induce membranous vesicles, and its subcellular colocalization with newly-synthesized viral RNA, suggest that NS4B plays an important role in RNA replication (Egger *et al.*, 2002; Gosert *et al.*, 2003). It is assumed that NS4B-induced vesicles serve as a scaffold for the HCV replication complex, protecting it from RNA degradation and preventing the induction of a double-stranded RNA-dependent innate antiviral defence. NS5A is a zinc binding highly phosphorylated protein that is indispensable for HCV RNA replication (Tellinghuisen *et al.*, 2004). Based on the recently solved 3D structure of NS5A domain 1, it was suggested that NS5A may bind viral RNA and function as a regulator of viral RNA replication (Tellinghuisen *et al.*, 2005). NS5B is the RdRp forming the catalytic core of the HCV replication machinery (Behrens *et al.*, 1996; Lohmann *et al.*, 1997).

1.2   HCV multiplication cycle

HCV replication takes place in the cytoplasm of the host cell. As shown in Fig. 2, the virus enters the cell *via* a specific interaction between the HCV envelope glycoproteins and an as yet unidentified cellular receptor complex that most likely includes CD81 (Bartosch *et al.*, 2003; Lavillette *et al.*, 2005; Lindenbach *et al.*, 2005; Pileri *et al.*, 1998; Wakita *et al.*, 2005; Zhong *et al.*, 2005). Bound particles are probably internalized by receptor-mediated endocytosis. After the viral genome is liberated from the nucleocapsid, RNA translation is initiated. Newly-synthesized NS4B proteins induce (presumably in conjunction with other viral or cellular factors) the formation of clustered membranous vesicles or membrane invaginations that have been referred to as the membranous web which is most likely derived from the ER (Egger *et al.*, 2002; Moradpour *et al.*, 2003; Mottola *et al.*, 2002) (see Fig. 2). Enzymatically active replication complexes isolated from cells with HCV replicons were found to have a rather closed conformation, because most of their viral RNA content is nuclease resistant (Lai *et al.*, 2003), their enzymatic activity is highly resistant to proteinase K (Miyanari *et al.*, 2003; Quinkert *et al.*, 2005), they are unable to use exogenously added RNA templates (Lai *et al.*, 2003), and with the exception of NS5A, trans-complementation of HCV proteins is not possible (Appel *et al.*, 2005). The intimate interaction of HCV proteins with membranes and with each other may shield existing replication complexes from certain antiviral compounds. In this context, it is interesting to note that several non-nucleoside inhibitors of NS5B failed to block RNA synthesis of native replicase complexes isolated from replicon cells at concentrations 1,000-fold higher than concentrations required for half-maximal inhibition of recombinant NS5B (Ma *et al.*, 2005). The shielded structure of the HCV replication complex may also limit the effectiveness of NA-based antiviral strategies (see below).

Detailed information about the mode of RNA replication is currently not available for HCV, but in analogy to other flaviviruses (Westaway *et al.*, 2002), a model as depicted in Fig. 2 can be proposed: Incoming positive-strand RNA serves as a template for the synthesis of a single, negative-strand RNA molecule that remains base paired with its template. The resulting double-strand RNA, the so-called replicative form (RF), is then copied multiple times into positive-strand RNA *via* a replicative intermediate (RI). In this way, positive-strand RNA progenies are transcribed in 5 to 10-fold molar excess over negative-strand RNA.

The production of authentic HCV particles in cell culture has recently been achieved by several groups (Lindenbach *et al.*, 2005; Wakita *et al.*, 2005; Zhong *et al.*, 2005). However, the site of particle formation and the mechanism by which HCV particles leave the host cell are still unknown.

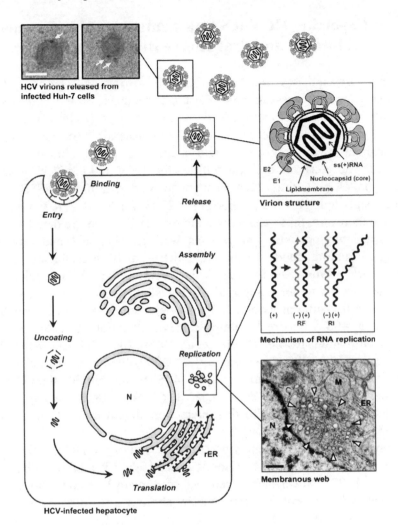

Figure 2. Multiplication cycle of HCV. Virus particles enter the host cell most likely by receptor-mediated endocytosis (binding and entry). The viral genome is liberated from the nucleocapsid (uncoating) and translated at the rough ER. Newly synthesized viral proteins induce the formation of membranous vesicles forming a membranous web, the site of HCV RNA replication (see electron micrograph in the lower right; the arrow heads indicate the position of the web; note the close proximity of the web and ER membranes; bar, 500 nm). After genome amplification and further viral protein accumulation, progeny virions are assembled and released. N, nucleus; rER, rough endoplasmic reticulum. The middle right panel of the figure shows a model for the synthesis of negative (-) and positive-strand (+) RNA *via* a replicative form (RF) and a replicative intermediate (RI). The upper right panel shows a schematic representation of an HCV particle. The envelope proteins E1 and E2 are drawn according to the structure and orientation of the tick-borne encephalitis virus envelope proteins M and E, respectively (Yagnik *et al.*, 2000). The two electron micrographs in the upper left corner of the figure show infectious HCV particles produced in the hepatoma cell line Huh-7 (arrows point to E2-specific immunogold particles; bar, 50 nm).

1.3   *Cis*-acting HCV RNA elements amenable for nucleic acid-based antiviral intervention

Several prerequisites must be fulfilled so that NA-based antiviral treatment can be successful. First, the target sequence in the viral genome must have an essential function in the viral life cycle (at least when using AS-ONs); second, the target sequence must be accessible for the therapeutic compound; third, the target sequence in the viral genome must be highly conserved. These criteria are best fulfilled for the 5' and the 3' NTR as well as for a novel *cis*-acting RNA element located within the NS5B coding region, designated 5BSL3.2 (Friebe *et al.*, 2005; Lee *et al.*, 2004; You *et al.*, 2004) and therefore, these elements will be described in more detail.

By using phylogenetic analyses as well as chemical and enzymatic probing, a structural model of the HCV IRES has been established (Fig. 1; (Honda *et al.*, 1996a; 1996b; 1999; Wang *et al.*, 1994; 1995). It folds into a unique tertiary structure composed of four stem-loop domains (I – IV), a helical structure and a pseudoknot. The key element is domain III capable of recruiting a 40S ribosomal subunit/eukaryotic initiation factor 3 (eIF3) complex with high affinity (Kieft *et al.*, 1999; 2001; Klinck *et al.*, 2000; Kolupaeva *et al.*, 2000; Lukavsky *et al.*, 2000; Spahn *et al.*, 2001). This interaction is determined primarily by the 40S subunit binding the IRES with a 15-fold higher affinity as compared to eIF3 (Kieft *et al.*, 2001). Consequently, interfering with this interaction should prevent the assembly of the 43S subunit-IRES complex.

According to RNA mapping studies, the 5' end of the HCV IRES resides between nucleotides 38 and 46 (Honda *et al.*, 1996b; Rijnbrand *et al.*, 1995; Yoo *et al.*, 1992). Nevertheless, removal of domain I, residing upstream of this region, reduces protein expression by about 3 to 5-fold, arguing that it contributes to RNA translation (Friebe *et al.*, 2001; Kim *et al.*, 2002; Luo *et al.*, 2003). Most studies indicate that sequences located immediately downstream of the start codon of the large open reading frame encoding the HCV polyprotein are required for RNA translation (Honda *et al.*, 1996a; 1996b; Lu and Wimmer, 1996; Reynolds *et al.*, 1995). However, these sequences appear to be primarily required to prevent the formation of stable stem-loop structures in the vicinity of the initiator AUG codon (Rijnbrand *et al.*, 2001). Moreover, the core protein may modulate IRES structure and activity and thereby contribute to a switch from RNA translation to replication and packaging. In agreement with this assumption, it was found that the core protein causes an inhibition of translation from the HCV IRES, arguing for a regulatory function of this protein (Zhang *et al.*, 2002). Apart from the core protein, several cellular factors binding to the HCV IRES have been described, such as ribosomal proteins (Otto *et al.*, 2002), polypyrimidine tract binding protein (PTB) (Ali and Siddiqui, 1995), La autoantigen (Ali and Siddiqui, 1997), the heterogeneous nuclear ribonucleoprotein L (Hahm *et al.*, 1998), the poly(rC)-binding protein (PCBP-2) (Fukushi *et al.*, 2001), and several other proteins that so far have not yet been characterized (Yen *et al.*, 1995). These interactions should be kept in mind when designing NA-based antiviral therapies targeting the 5' NTR because they may affect RNA structure and mask potential target sites in the viral RNA.

In addition to serving as IRES, domains I and II are also indispensable for RNA replication (Friebe *et al.*, 2001). This functional overlap suggests that domain II is involved in regulating a switch from RNA translation to replication. Moreover, we can assume that the complement of positive-strand 5' NTR, corresponding to the 3' end of negative-strand, serves as a recognition site for the viral replication machinery to initiate synthesis of positive-strand RNA. Consequently, structures other than the ones that are formed by the 5' NTR of positive-strand RNA operate as a promoter for the initiation of positive-strand RNA synthesis. Recently, secondary structure models of the 3' end of negative-strand RNA were described and they suggest that this structure is not simply a mirror image of the 5' NTR structure of the positive-strand (Kashiwagi *et al.*, 2002; Schuster *et al.*, 2002; Smith *et al.*, 2002). Importantly, stem-loops I, IV, and parts of stem-loop III (IIIa and IIIb) appear to be conserved in both orientations, whereas stem-loop II and the surrounding interdomain regions are reorganized into two large stem-loops. These models must be kept in mind when evaluating the most adequate target sequence for NA-based antiviral therapy.

Another *cis*-acting RNA element in the HCV genome that fulfills the criteria as a target sequence is a highly-conserved stem-loop in the 3' terminal coding region of NS5B (Fig.1; (Friebe *et al.*, 2005; Lee *et al.*, 2004; You *et al.*, 2004). This element, called 5BSL3.2 forms a kissing-loop interaction with stem-loop II in the X-tail, most likely *via* direct base pairing between the loop regions (Friebe *et al.*, 2005). Mutations disturbing complementarity between the loop regions of 5BSL3.2 and SL2 in the X-tail reduce or block RNA replication that can be rescued when complementarity is restored. This result suggests that the kissing loop interaction is essential for viral RNA replication. Since the nucleotides of the loop regions involved in this RNA-RNA interaction are highly conserved among different HCV genotypes, these RNA elements might represent interesting targets for NA-based antiviral therapies.

2.   EXPERIMENTAL SYSTEMS APPLICABLE TO NA-BASED HCV THERAPY

The development of antiviral drugs as well as novel therapeutic concepts has been slowed down by the lack of adequate culture systems supporting the propagation of the virus in the laboratory. Numerous attempts have been made but all these efforts were of limited success. Although infection of human cell lines and primary human hepatocytes could be achieved, viral RNA replication and virion production was so low that detailed studies of the HCV life cycle have not been possible (for review see Bartenschlager *et al.*, 2004; Kato and Shimotohno, 2000). Efficient virus production in cell culture has only recently been achieved (see below) and thitherto several surrogate systems have been developed of which we will only briefly describe those that are relevant for NA-based antiviral strategy.

2.1   Surrogate systems used to study NA-based HCV-specific therapy

In the absence of systems that support HCV RNA replication, most groups focussed their interest on the IRES in the 5' NTR for which simple surrogate systems could be developed. In several instances, bicistronic vectors were used in which the HCV IRES was inserted between two reporter genes, the first one being translated by a cap-dependent process and the second by the HCV IRES. Inhibition of the latter resulted in a decrease of expression of the second reporter, whereas expression of the cap-dependent reporter was not affected and could be used as an internal standard for normalization. This type of constructs and several other construct designs have been used either in cell-free systems such as rabbit reticulocyte lysates or in cell-based systems by using transient or stable expression (Alt *et al.*, 1995; Hanecak *et al.*, 1996; Mizutani *et al.*, 1995; Vidalin *et al.*, 1996; Wakita and Wands, 1994). Although these are valid approaches, it is unclear whether NAs interfering with HCV functions in these assays are also active on HCV RNAs in an infected cell where virus-induced membranes may prevent access for nucleic acids. In fact it was shown that HCV RNA present in replication complexes isolated from infected cells are refractory to exogenously added proteinase K and nuclease as long as no detergent is added arguing that these complexes are shielded by intracellular membranes (Quinkert *et al.*, 2005). Moreover, in infected cells, viral positive- and negative-strand RNAs are present which may form stable hybrids with blocked target sequences. Finally, RNA structures in full-length genomes may be different from those present in subgenomic constructs due to potential long-range RNA/RNA interactions as well as viral and cellular factors binding to the 5' NTR or other regions.

To overcome some of these limitations, chimeric polioviruses were generated carrying replacements of the poliovirus IRES by that from HCV. The most efficient carries a 5' NTR composed of the poliovirus cloverleaf structure that is essential for RNA replication, the HCV IRES (domain II to IV) and the 5' terminal coding region of the HCV core protein that was fused to the poliovirus polyprotein *via* a 2A proteinase cleavage site (Zhao *et al.,* 1999). Replication of this chimeric virus could be improved further by removal of an unfavourable interaction between the HCV IRES and surrounding poliovirus sequences, resulting in a virus that replicated as efficiently as wild type poliovirus (Zhao *et al.,* 2000). A similar HCV-poliovirus chimera was described by Yanagiya and coworkers (Yanagiya *et al.,* 2003) and shown to replicate in transgenic mice expressing the poliovirus receptor. These model systems are of great value for studies of the HCV IRES in cell culture and *in vivo* as well as for the discovery and evaluation of antiviral compounds targeting the HCV IRES. However, owing to their artificial design, the chimeric polioviruses will probably be replaced by the authentic HCV production system (see below).

To evaluate the efficacy of HCV-specific AS-ONs *in vivo*, a recombinant vaccinia virus was generated carrying an expression cassette composed of the HCV IRES and the core or core-to-E1 coding region fused to the luciferase gene (Zhang *et al.,* 1999). Upon intraperitoneal inoculation of BALB/c mice with this recombinant virus, the luciferase reporter gene was expressed under the control of the HCV IRES. Administration of two different AS-ONs targeting the HCV IRES reduced luciferase expression, providing evidence that these AS-ONs interfered with IRES function *in vivo*.

Thus far, in only one case it was shown that an AS-ON is effective in an HCV-infected cell (Mizutani *et al.,* 1995). By using an experimentally infected T cell line it was shown that an 18-mer phosphorothioate (PO)-ON targeting the sequence surrounding the initiator AUG codon reduced virus replication. When added directly to the cell culture supernatant, positive- and negative-strand HCV RNA as determined by strand-specific RT-PCR was reduced to below the detection limit. However, owing to low virus replication in this cell culture system, no conclusions about the efficacy of this AS-ON could be drawn, and therefore it should be re-evaluated with the novel virus production system.

2.2   HCV replicons

Although the surrogate systems described above allowed a first evaluation of antiviral efficacy, they were limited by their artificial nature. A major step forward was therefore the development of the HCV replicon system. By definition, a replicon is a DNA or RNA molecule capable of self-amplification. The first HCV replicon was derived from a cloned viral consensus genome by deletion of the core to NS2-region and the insertion of two heterologous elements: the selectable marker *neo* encoding for the neomycin phosphotransferase and the IRES from the encephalomyocarditis virus (EMCV; Fig. 3). The resulting replicon was therefore of bicistronic design with the first cistron (*neo*) being translated under the control of the HCV IRES and the second cistron (NS3 to 5B) by the EMCV IRES (Lohmann *et al.*, 1999). Upon transfection of the human hepatocarcinoma cell line Huh-7 and G418 selection, only cells that support high-level HCV RNA replication mounted a sufficient G418 resistance, whereas non-transfected cells and cells in which the replicon did not amplify to sufficient levels were eliminated under this condition. By using this approach, a low number of G418-resistant cell clones could be established that carried high amounts of self-replicating HCV RNAs. In fact, replication was so efficient that positive- and negative-strand HCV RNAs could be detected by Northern-hybridization and viral RNA could be radiolabeled metabolically with [³H] uridine (Lohmann *et al.*, 1999). HCV proteins were detected by immunofluorescence, Western-blot, or immunoprecipitation after metabolic radiolabeling allowing for the first time detailed studies of HCV RNA replication.

Subsequent studies identified two parameters governing the efficiency of the replicon system. First, the permissiveness of the host cell and second, cell culture adaptive mutations (reviewed in Bartenschlager *et al.*, 2004). It turned out that only a minor fraction of transfected Huh-7 cells is capable of supporting high-level HCV RNA replication and these cells were selected out of a pool during passaging in the presence of G418. Consequently, when the replicons in these cells were removed by treatment with a selective drug, the resulting 'cured' cells supported HCV RNA replication more efficiently as compared to naive cells. Another important determinant of replication efficiency are cell culture adaptive mutations that were identified at several distinct positions in the nonstructural proteins. The mechanism(s) by which these mutations enhance RNA replication are not known.

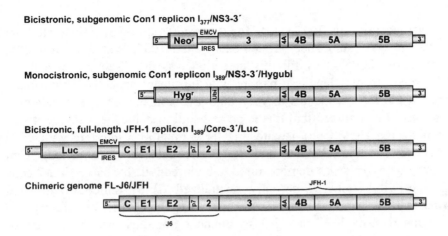

Figure 3. HCV replicons and recombinant full-length genomes. The bicistronic, subgenomic Con-1 replicon I_{377}/NS3-3' is composed of the 5' NTR plus nucleotides 342 to 377 of the core coding region, the *neo* gene (encoding the neomycin phosphotransferase), the IRES of the encephalomyocarditis virus (EMCV), the coding region of the nonstructural proteins NS3 to NS5B, and the 3' NTR (Lohmann *et al.*, 1999). The monocistronic, subgenomic Con-1 replicon I_{389}/Hygubi/NS3-3' is composed of the 5' NTR plus nucleotides 342 to 389 of the core coding region, the *hyg* gene (encoding the hygromycin phosphotransferase), the ubiquitin coding sequence (ubi), the coding region of the nonstructural proteins NS3 to NS5B, and the 3' NTR (Frese *et al.*, 2002). The design of the bicistronic, full-length JFH-1 replicon I_{389}/Core-3'/Luc is analogous to that of the Con1 replicon I_{377}/NS3-3', but the full-length construct contains the firefly luciferase gene (Luc) instead of the *neo* gene and the entire HCV coding sequence (Wakita *et al.*, 2005). The chimeric genome FL-J6/JFH-1 contains sequences of two different genotype 2a viruses (J6 and JFH-1). Both the NTRs and the coding sequence for the nonstructural proteins NS3 to 5B have been derived from the JFH-1 isolate whereas the coding sequence of the structural proteins up to NS2 are those of the J6 genome (Lindenbach *et al.*, 2005; T. Pietschmann and R. Bartenschlager, unpublished).

With the advent of highly permissive cell lines and cell culture adaptive mutations, a large set of replicon variants could be established (reviewed in Bartenschlager, 2002). One example is a replicon that carries a reporter instead of *neo*, allowing measurement of transient HCV replication in transfected cells. Another example is a monocistronic replicon in which a selectable marker is fused to NS3 *via* a protease cleavage site such as ubiquitin or the picornaviral 2A proteinase. In this replicon, both RNA translation and replication are controlled by the HCV 5' NTR which is an important prerequisite when considering NA-based antiviral strategies targeting this highly-conserved regulatory RNA element of the HCV genome.

2.3   Infectious HCV production systems

Very recently the first system for the production of infectious HCV in cell culture was established (Lindenbach et al., 2005; Wakita et al., 2005; Zhong et al., 2005). Key for this invention was the discovery of a novel HCV isolate from a Japanese patient with fulminant hepatitis. This isolate, designated JFH-1, replicates to very high levels in the absence of adaptive mutations. It is assumed that this is an essential prerequisite, because it was shown for the HCV Con1 isolate that these mutations interfere with virus production (T. Pietschmann and R. Bartenschlager, unpublished). Cell culture-grown virus was demonstrated to be infectious for naïve Huh-7 cells and infectivity could be partially neutralized (i) by antibodies directed against CD81, a presumed (co)receptor of HCV, (ii) by E2-specific monoclonal antibodies, and (iii) by immunoglobulins present in sera of chronically infected patients (Lindenbach et al., 2005; Wakita et al., 2005; Zhong et al., 2005). HCV generated in cell culture was also shown to be infectious in vivo demonstrating the authenticity of virus particles produced in the laboratory (Wakita et al., 2005). Finally, bicistronic HCV genomes were generated that carry the firefly luciferase reporter gene under the control of the HCV IRES, whereas translation of the HCV polyprotein is mediated by the EMCV IRES (Fig. 3). Fortunately, these genomes still replicate to very high levels and produce infectious HCV particles (Wakita et al., 2005). Since expression of the luciferase reporter gene is directly linked to RNA copy number, HCV replication can be easily measured in transfected and infected cells by simple luciferase assays. These new developments greatly broaden the scope for basic and applied HCV research and allow the evaluation of novel antiviral strategies that so far were not amenable.

2.4   *In vivo* propagation models for HCV

Until recently, reliable propagation of HCV *in vivo* has only been achieved in the chimpanzee. However, the chimpanzee is an endangered species and owing to ethical rules and costs, this is not a convenient animal model. A novel alternative that was first described by Mercer and coworkers are transgenic mice with human/murine chimeric livers (Mercer et al., 2001). Upon infection with serum obtained from HCV-infected patients, viremia was observed in approximately 75% of mice with efficient human hepatocyte engraftment, and titres were well in the range of those observed in infected humans (Mercer et al., 2001). HCV replication was confined to human liver cells and viremia was detectable for up to 35 weeks after inoculation. Although these findings have been confirmed recently by an independent group (Meuleman *et al.*, 2005), problems of this mouse model are its technical challenge, the difficult reproducibility and the high lethality associated with the *uPA* transgene in the case of homozygous mice. Nevertheless, this mouse model opens new avenues for the study of HCV *in vivo* and it provides an attractive alternative to chimpanzees for evaluation of HCV-specific therapies, including NA-based compounds.

3. NA-BASED ANTIVIRAL STRATEGIES

3.1 Antisense oligonucleotide-based approaches

AS-ONs operate in two different ways. The first one is cleavage of the target RNA by RNaseH, a ubiquitous cellular enzyme that is involved in DNA replication. RNaseH binds to DNA-RNA hybrids and specifically degrades the RNA moiety. Nucleolytic cleavage requires ONs with a natural phosphodiester or a synthetic phosphorothioate (PO) backbone (Fig. 4). Other chemical bonds are not a substrate of RnaseH, but artificial oligo-nucleotides with a central core that can be used by the enzyme and flanked by nucleotides with non-accepted bonds have been designed. Oligo-nucleotides with such mixed backbones have been designated gapmers, their advantage being the increase in flexibility to chemically modify the ON without losing the benefit of RNaseH-induced cleavage of the target sequence (Malchere *et al.*, 2000). Such chemical modifications confer higher bioavailability, improved uptake of the ON into cells, stability towards nucleolytic degradation, higher specificity for the target sequence and improved binding to it (reviewed in Kurreck, 2003). However, RNaseH is expressed in the nucleus, and therefore might not have a chance to interact with HCV RNAs in the cytoplasm.

The second mode of action that is probably of more relevance for HCV is a direct inhibition of RNA function. In this respect, AS-ONs targeting the 5' NTR can inhibit translation by preventing the binding and assembly of the translation machinery or by steric blockage of the ribosome. Likewise, AS-ONs directed against the 3' NTR or 5BSL3.2 could interfere with the assembly of the viral replication machinery required for synthesis of negative-strand RNA.

The first study on inhibition of the HCV IRES by AS-ONs was reported by Wakita and Wands in 1994. They generated a large set of phosphodiester ONs that were tested in rabbit reticulocyte lysates. Most efficient reduction of RNA translation was achieved with AS-ONs targeted to the region surrounding the initiator AUG codon of the large open reading frame encoding the polyprotein (Wakita *et al.,* 1994). Since only a minor fraction of RNAs was cleaved at the target site, which may be due to the low levels of RNaseH in rabbit reticulocyte lysates, repression of RNA translation was primarily caused by interference with 5' NTR function. Similar observations were made by Alt and coworkers with gapmer ONs (Alt *et al.,* 1999). Although inhibition of RNA translation was enhanced by RNaseH cleavage of the target sequence, it was not obligatory for the inhibition. Further support for an RNaseH-independent inhibition of the 5' NTR stems from studies with PO-ONs and 2'-methoxyethoxy-modified ONs (Fig. 4). By using a stable cell line expressing the 5'-terminal quarter of the HCV genome, a panel of 50 PO-ONs was screened (Hanecak *et al.,* 1996). Most efficient inhibition of IRES-dependent RNA translation (reduced to 20 to 30% of untreated controls) was found with ONs targeting the base of stem III, the IIId stem-loop and a region spanning the initiator AUG codon. Levels of HCV RNA were drastically decreased, indicating cleavage by RNaseH. However, RNaseH-incompetent 2'-methoxyethoxy ONs targeting the same site around the AUG codon reduced RNA translation with comparable potency, but did not affect HCV RNA levels. Thus, RNaseH-mediated cleavage is not absolutely essential for AS-ON activity (Hanecak *et al.,* 1996).

It is interesting to note that results from several independent studies in which sites that are most accessible for AS-ONs have been mapped identified similar regions in the 5' NTR (Alt *et al.,* 1995; Lima *et al.,* 1997; Mizutani *et al.,* 1995; Vidalin *et al.,* 1996; Wakita *et al.,* 1994). In essence, the IIId loop, the region around the initiator AUG codon and the intersection between loops III and IV represent the most promising target sequences (Fig. 1). In addition, target sequences in the NS3 coding region have been identified, but they appear to be less attractive as compared to the 5' NTR (Heintges *et al.,* 2001).

As described above, RNaseH-mediated cleavage of the target sequence is not a prerequisite for efficacy of AS-ONs. Therefore, RNaseH-acceptable ONs are not obligatory for interference with gene expression, providing the opportunity for chemical modifications that are more favourable for *in vivo* applications (reviewed in Herdewijn, 2000). One of the most frequently-used modifications is the introduction of a PO linkage (Fig. 4), resulting in an increase of the half-life from about 1 h in the case of unmodified ONs to 9 to 10 h in the case of PO-ONs. Nevertheless, PO-ONs retain their capacity to form regular Watson-Crick base pairs and therefore can activate RNaseH. However, PO-ONs also have several disadvantages, most importantly their affinity to proteins. Since PO-ONs retain a high negative charge, they bind primarily to proteins that interact with polyanions such as heparin, which could explain the toxicity associated with PO-ONs resulting from alterations of the complement and the clotting cascade. Moreover, PO-ONs have a reduced affinity to their cognate target sequence as compared to unmodified ONs, which is only in part compensated by an increased specificity of hybridization.

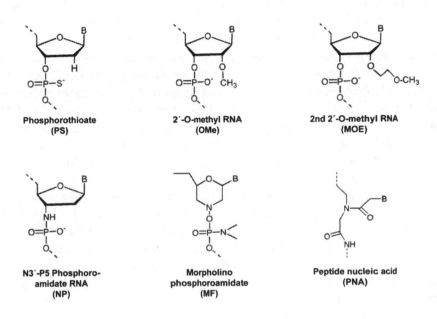

Figure 4. Chemical structure of nucleic acid analogs used for the construction of AS-ONs targeting the HCV genome.

To circumvent some of the disadvantages of PO-ONs, a second generation ONs were designed that carry alkyl modifications at the 2' position of the ribose, most importantly 2'-*O*-methyl and 2'-*O*-methoxy-ethyl groups (Fig. 4). These ONs no longer activate RNaseH and were shown to efficiently block HCV IRES function *in vitro* and in a cell-based assay when targeting the region surrounding the initiator AUG codon (Hanecak *et al.*, 1996; Vidalin *et al.*, 1996). Although the concept of steric block of IRES-mediated translation was confirmed in the *in vitro* study, a drawback of this approach is that not every target site identified with RNaseH activating ONs is equally useful for non-activating ONs. For instance, 2'-methoxyethoxy-ONs targeting stem-loop IIId or the pseudoknot were ineffective, whereas PO-ONs directed against the same sites were, indicating that the target sequences were accessible (Hanecak *et al.*, 1996). In agreement with this observation, a 2'-O-methyl-ON also targeting stem-loop IIId reduced IRES-mediated translation (Brown-Driver *et al.,* 1999), arguing that the lack of inhibition found with 2'-methoxyethoxy ONs can not be ascribed to the lack of RNaseH activation, but rather to the kind of modification introduced into the ON.

Morpholino ONs with demonstrated activity against the HCV 5' NTR have been described recently (Jubin *et al.,* 2000). Most efficient inhibition was found with ONs directed against stem-loop IIId of the IRES, corroborating the findings obtained in the studies described above. Since stem-loop IIId is involved in recruiting the ribosome to the IRES, AS-ONs binding to IIId abrogate the formation of ribosomal preinitiation complexes (Martinand-Mari *et al.,* 2003; Tallet-Lopez *et al.,* 2003). 2'-O-methyl ONs and cationic phosphoroamidate ONs as short as 12 nucleotides and 10-mer peptide nucleic acids (Fig. 4) targeting stem-loop IIId are sufficient to interfere with IRES function (Martinand-Mari *et al.,* 2003; Michel *et al.,* 2003; Tallet-Lopez *et al.,* 2003).

Based on mapping studies of the most promising target sequence in the 5' NTR described above, a clinical trial was initiated with a 20-mer PO-ON directed against a sequence surrounding the initiator AUG codon (Witherell, 2001; Zhang *et al.,* 1999). The compound called ISIS 14803 (HepaSense™) was found to be effective *in vitro*, in a cell-based assay and in a vaccinia virus-mouse model (Zhang *et al.,* 1999). A phase I dose escalation study with 24 patients suffering from a chronic genotype 1 HCV infection was conducted (Soler *et al.,* 2004). Two out of the 24 patients developed a more than 10-fold reduction of viremia and 9 patients showed a reduction of less than 1 log that was difficult to discriminate from normal fluctuations often found in persistent infections. No significant changes were found with the other patients and no resistance mutations in the target site or surrounding sequences were detected. Given the limited efficacy of this AS-ON, further clinical trials with this ON will most likely not be pursued (http://hcvadvocate.org/hepatitis/Drugs/approval.htm#ISIS14803).

3.2   Ribozyme-based approaches

Ribozymes are catalytically active RNA molecules acting as enzymes in the absence of proteins. They were first identified in the self-splicing group I intron of *Tetrahymena thermophila* and the RNA moiety of RNaseP (Guerrier-Takada *et al.,* 1983; Kruger *et al.,* 1982). A major advantage of ribozymes that makes them useful for therapy is the possibility to make these RNA *trans*-acting and to confer specificity to virtually any target sequence. This is achieved by fusing the ribozyme core element at the 5' and 3' ends with sequences that are complementary to the target sequence. Depending on the structure of the ribozyme core different classes can be distinguished: Hairpin ribozymes, hammerhead ribozymes, the ribozymes of the hepatitis delta virus (HDV) and the catalytic RNA present in the cellular enzyme RnaseP (reviewed in Scherer and Rossi, 2003). Hammerhead ribozymes, originally identified in single-stranded plant viroid and virusoid RNAs, are composed of about 30 nucleotides and have very simple requirements for the cleavage site in that virtually any motif with the dinucleotide sequence UU, UC, or UA can be targeted (Haseloff and Gerlach, 1992). These properties made hammerhead ribozymes very popular for the design of trans-acting ribozymes. In contrast, hairpin (also called 'paperclip') ribozymes have more complex structures and requirements for target sequences, with a preference for GUC with cleavage occuring directly upstream of the G residue. RNaseP is an enzyme that is found in organisms throughout nature. Although it is composed of RNA and protein, the RNA component can act as a site-specific cleavage enzyme under certain experimental conditions. Substrates cleaved by RNaseP have a structure resembling a segment of a transfer RNA molecule. This structure can be mimicked by using specifically-designed antisense RNAs hybridizing with the target sequence such that a tRNA-like structure is restored that can act as a substrate. This approach allows the design of ribozymes cleaving a given substrate in *trans* that have been used e.g. for inhibition of herpes viruses in cell culture (Dunn *et al.*, 2001; Kilani *et al.*, 2000; Trang *et al.*, 2000). Most complex is the HDV ribozyme that can also be engineered to cleave target RNAs in *trans*.

An advantage of ribozymes over AS-ONs is their catalytic mode of action, which should in principle require much lower concentrations of ribozymes as compared to non-catalytic AS-ONs. On the other hand, target sites are limited due to sequence requirements at the cleavage site and to structural constrains that interfere with ribozyme function to a higher extent as compared to AS-ONs. Therefore, the selection of appropriate target sites is of utmost importance which can not be predicted but must rather be determined empirically and which depends on the particular ribozyme used. For instance, Lieber and colleagues generated a hammerhead ribozyme library and used it for the screening of the most efficient cleavage sites in the HCV genome by using total liver RNA isolated from a chronically infected patient (Lieber *et al.*, 1996). Cleavage products were determined by sequence analysis of fragments amplified by RT-PCR, resulting in the identification of 6 ribozymes with high activities. Five hybridized to sequences of the HCV positive- or negative-strand RNA in close proximity to the initiator AUG codon, and one hybridized to a sequence in the core region. A dose-dependent reduction of viral RNA was observed upon expression with recombinant adenoviruses directing ribozyme expression in cell lines stably expressing a full length HCV genome in positive- or negative sense orientation. Inhibition of HCV was also observed in two HCV-infected hepatozyte cultures carrying genotype 1a or 1b viruses. Also in this culture system, HCV RNA levels were reduced in a dose-dependent manner with any of the recombinant adenoviruses. These results provide a proof-of-concept that inhibition of HCV replication is possible by HCV-specific ribozymes expressed intracellular.

In a random screening approach, Ryu and Lee constructed a ribozyme library containing the group I ribozyme of *T. thermophila* fused to a randomized internal guide sequence. With the help of this library, they mapped loop IIIb in the HCV IRES as the most susceptible target site (Ryu and Lee, 2004). A two-step selection approach with a hammerhead-derived library was also used by Romero-Lopez and coworkers (Romero-Lopez *et al.*, 2005). Their strategy was built on a hammerhead ribozyme described by Lieber and coworkers (Lieber *et al.*, 1996). This ribozyme was fused at the 3' terminus with a randomized 25 nucleotide-long RNA sequence which was used to select RNA aptamer motifs with high affinity for the HCV IRES region. After 6 selection rounds, several aptamer ribozyme RNAs were identified that efficiently blocked IRES-dependent translation in an *in vitro* translation assay. It will be interesting to see how these aptamer-ribozymes interfere with HCV replication in cell culture.

The feasibility of hammerhead ribozymes was evaluated in several other studies as well by using *in vitro* translation systems and by expression of reporter constructs in cell lines (Macejak *et al.*, 2000; Sakamoto *et al.*, 1996). Moreover, several ribozymes targeting various sites in the 5' NTR were shown to inhibit replication of the HCV-poliovirus chimera (see paragraph 2.1) by more than 50% (Macejak *et al.*, 2000). One of these ribozymes cleaving at nucleotide 195 lying in the IIIb loop region was characterized in more detail and used to optimize the ribozyme design (Macejak *et al.*, 2000). Stability of the ribozyme against nucleolytic degradation was increased by the introduction of several chemical modifications such as PO linkages in the 5' terminal binding arm and used to characterize pharmacokinetics and tissue distribution in mice (Lee *et al.*, 2000). After single injections, ribozyme 195 was cleared from the plasma with an average elimination half-life of about 25 min and accumulated in hepatocytes, endothelial cells lining the sinusoids, and within the subendothelial space of Disse. Moreover, it was found that ribozyme 195 enhanced the antiviral effect of IFN-α in cell culture studies performed with the HCV-poliovirus chimera (Macejak *et al.*, 2001). Based on these results, ribozyme 195 (Heptazyme) was taken to early clinical studies. In spite of initially encouraging results further development of Heptazyme will not be pursued due to limited efficacy and toxicity problems (http://www.hcvadvocate.org/ hepatitis/hepC/HCVDrugs.html).

Apart from hammerhead and group I intron ribozymes, more recently two alternative strategies have been developed: trans-cleaving ribozymes derived from HDV and cleavage of HCV RNA by RNase P. As described above, RNase P is a ribonucleoprotein complex catalyzing a hydrolysis reaction to remove the leader sequence of precursor tRNA (Robertson *et al.*, 1972). Substrate recognition by RNaseP does not rely on sequence requirements but rather on structural features of the substrate RNA. Surprisingly, purified RNaseP was found to cleave HCV RNA at two distinct sites, one of them residing in the 3' vicinity of the initiator AUG codon and one within the coding region for NS2 (Nadal *et al.*, 2002). Cleavage also occured with variant target sequences, supporting the notion that RNaseP recognizes RNA structures rather than primary sequences. Moreover, the results suggest that HCV RNA in the areas of cleavage form complex tRNA-like structures (Nadal *et al.*, 2003; Piron *et al.*, 2005). However, thus far it is unclear whether RNAseP can be engineered for efficient *trans*-cleavage of HCV RNA.

HDV-derived ribozymes capable of *trans*-cleavage of HCV RNAs *in vitro* have also been described, but their potential use for intracellular cleavage of viral RNA has not yet been evaluated (Yu *et al.*, 2002).

3.3   RNAi-based approaches

RNA-mediated, sequence-specific gene silencing (RNA silencing) is a general term used to describe post-transcriptional gene silencing (PTGS) in plants, quelling in fungi, and RNA interference (RNAi) in animals (Hannon, 2002). It has been speculated that RNA silencing is an ancient eukaryotic surveillance system for foreign nucleic acids and degenerated genetic elements. In plants, RNA silencing is a powerful antiviral defense mechanism which has provoked the evolution of counteracting strategies by many viruses (reviewed by Li and Ding, 2001). Furthermore, accumulating data indicate that RNA silencing plays an important role in gene regulation. It has been shown that plants as well as animals tune their protein expression by using the RNA silencing machinery to target their own mRNAs for cleavage and/or translational repression (reviewed by Bartel, 2004).

Specificity in RNA silencing is mediated by small RNAs that are called short interfering RNAs (siRNAs) or micro-RNAs (miRNAs). Both types of RNAs are generated by members of the Dicer family. This group of evolutionarily-conserved class III endoribonucleases cleaves double-stranded RNA into fragments with a length of 21 to 25 nucleotides,

3'overhangs of 2 or 3 nucleotides, and phosphorylated 5'ends. These small double-stranded molecules are then unwound, and the single-stranded RNA

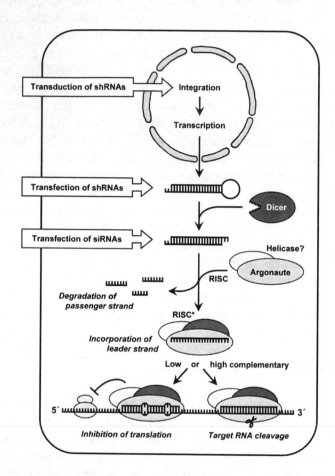

Figure 5. Induction of RNAi. HCV-specific RNAi has successfully been triggered in human cells by (i) the transduction of shRNAs, *i.e.* the infection with retroviral vectors that encode shRNAs under the control of a polymerase III promoter, (ii), the transfection of *in vitro*-transcribed shRNAs, and (iii) the transfection of synthetic siRNAs, *in vitro*-transcribed siRNAs, and endonuclease-prepared siRNAs. Nuclear shRNA transcripts are exported into the cytoplasm (probably by the same cellular machinery that translocates pre-miRNAs). Dicer is thought to bind to the stem of shRNAs and cleaves off the loop nucleotides, thereby generating functional siRNA duplexes. The strand with the less-tightly paired 5' end (leader strand) is loaded into RISC whereas the strand with the more-tightly paired 5' end (passenger strand) is degraded. Activated RISC (RISC*) may interfere with HCV RNA replication in two different ways. Imperfect base pairing between the leader siRNA strand and its target may result in the inhibition of viral protein translation. A more perfect base pairing might allow the formation of an A-form double-stranded RNA helix which has been shown to drive target RNA cleavage.

whose 5'end is less tightly paired to its complement is preferentially loaded into a protein complex called the RNA-induced silencing complex (RISC) (Grishok *et al.*, 2001; Hammond *et al.*, 2000). The number of Watson-Crick base pairings between the incorporated "leader" or "guide" strand and its target molecule determines the mode of silencing. Imperfect complementarity may result in the inhibition of protein translation (Doench *et al.*, 2003) whereas a more perfect basepairing might allow the formation of an A-form double-stranded RNA helix which would lead to the cleavage of the target RNA (Caudy *et al.*, 2003; Parker *et al.*, 2005).

Dicer is constitutively expressed by most mammalian cells but, for unknown reasons, the enzyme does not recognize viral RNAs as a substrate for the production of siRNAs (discussed in Caplen, 2003). It is tempting to speculate that the elaborate IFN-induced innate immunity has replaced more ancient antiviral defence mechanisms such as RNAi. Nevertheless, it has been demonstrated by many groups that the RNAi machinery of mammalian cells can be artificially triggered and directed towards almost any given target (Fig. 5). In most of the earlier studies, cells were transfected with chemically-synthesized siRNAs that have the potential to directly enter RISCs and to guide them to target RNA molecules. One disadvantage with transfected siRNAs is that the silencing effect usually wears off a few days after transfection. This problem is due to the lack of RNA-dependent RNA polymerases that have been found to amplify siRNAs in plants or invertebrate species such as *Caenorhabditis elegans*. This problem has been addressed by several approaches, e.g. by the construction of eukaryotic expression plasmids that allow a stable expression of "guide" and "passenger" strands from separate promoters (tandem type). Alternatively, plasmid vectors have been constructed that drive a stable expression of single-stranded RNAs with stem-loop structures resembling those of cellular pre-miRNAs (stem-loop type). The loop region of these so-called short-hairpin RNAs (shRNAs) is cleaved off by Dicer which leads to the production of small double-stranded RNAs that are indistinguishable from naturally occuring siRNAs/miRNAs. Although both strategies have been successfully used to knock down the expression of numerous mammalian genes (summerized in Mittal, 2004), it has been suggested that the stem-loop approach is more potent in inducing RNAi than is the tandem approach (Miyagishi and Taira, 2003; Siolas *et al.*, 2005). Problems associated with poor transfection efficiencies in interesting target cells such as primary cells or stem cells have led to the development of retroviral and adenoviral vector systems for the delivery of shRNAs (Arts *et al.*, 2003; Brummelkamp *et al.*, 2002; Rubinson *et al.*, 2003).

RNAi has been harnessed by virologists to block the replication of HIV-1, HCV, and other important human viral pathogens (reviewed by Caplen,

2003). Although rules for the design of siRNAs/shRNAs have been developed (*e.g.* Reynolds *et al.*, 2004), RNAi efficiency is still difficult to predict and an experimental assessment is required. Table 1 lists HCV sequences have have been targeted by siRNAs or shRNAs. In most cases, replicons were used to model true HCV replication. The majority of these experiments were performed in Huh-7 cells containing replicons lacking the coding sequence of the structural proteins. Thus, it is important to note that the replication of genomic replicons encoding the complete HCV polyprotein are also sensitive to RNAi (Krönke *et al.*, 2004; Randall *et al.*, 2003). This result suggests that HCV does not encode functions interfering with RNAi-mediated silencing.

Apart from replicons, in some studies, surrogate expression systems were used. For instance, Sen and coworkers utilized a cell line that stably expresses NS5A in the human hepatoma cell line HepG2 and measured RNAi-induced knock-down of NS5A expression as well as inhibition of NS5A-mediated interleukin-8 promoter activation (Sen *et al.*, 2003). Transient expression of a dual reporter construct in which translation of the downstream reporter was controlled by the HCV IRES was used by Wang and coworkers (Wang *et al.*, 2005). Upon coexpression of a shRNA targeting a sequence around the AUG start codon in the HCV IRES with a reporter plasmid, a dose-dependent reduction of HCV IRES activity was found. Moreover, the analogous coexpression of this shRNA with the reporter construct in mice after hydrodynamic injection led to a strong inhibition of HCV IRES activity suggesting that the shRNAs were also active *in vivo*. This result is somewhat surprising given the high nuclease levels in tissues and the fact that the shRNAs were not chemically modified. This may be due to the stability of the short hairpin structure because siRNAs with the same target sequence were found to be much less stable and repressed HCV IRES activity only transiently whereas repression was prolonged in case of shRNAs (Wang *et al.*, 2005).

As alluded to in section 1.3, high conservation of the target sequence is an important prerequisite that must be considered in order to achieve efficient knock-down of most, if not all HCV genotypes. For this reason, the 5' NTR and the first codons of the core coding sequence have long been in the focus of NA-based antiviral therapies and this is also the case with RNAi. However, the 5' NTR interacts with a number of different host cell factors, and therefore accessibility of a target sequence can not be predicted but must be determined empirically. For this reason, collections of siRNAs or shRNAs targeting various sequences in the 5' NTR were tested for their impact on inhibition of HCV replicons in Huh-7 cells. Comparable to what was described for AS-ONs and ribozymes, best inhibition was achieved in case of sequences in close proximity to the AUG start codon of the poly-

Table 1. HCV sequences that have been analyzed for their susceptability to RNAi

Target sequence and susceptibility[a]	RNAi approach[d]	Reference[i]
8933[+, b]	Hydrodynamic transfection of synthetic siRNAs[e]	McCaffrey *et al.*, 2002
360[+]	Electroporation of synthetic siRNAs	Randall *et al.*, 2003
286[+]	Lipofection of synthetic siRNAs	Seo *et al.*, 2003
3566[+], 7750[+]	Lipofection of *in vitro*-transcribed siRNAs	Kapadia *et al.*, 2003
3980[±], 4499[±], 7805[+], 7983[+], 8409[−]	Electroporation of synthetic siRNAs	Wilson *et al.*, 2003
7983[+]	Plasmid-delivered siRNAs	
12[−], 82[+], 189[+], 286[±], 331[+]	Lipofection of synthetic siRNAs	Yokota *et al.*, 2003
331[+]	Plasmid-delivered siRNAs and shRNAs	
6431[+], 7389[+, b]	Lipofection of *in vitro*-transcribed siRNAs[f]	Seo *et al.*, 2003
1-341[+], 5816-6901[+], 6930-8021[+], 8022-9106[+, c]	Lipofection of esiRNAs	Krönke *et al.*, 2004
38[−], 56[−], 71[−], 138[+], 156[−], 174[−], 279[±], 301[−], 321[+], 334[+], 360[+], 5879[±]	Retrovirus-delivered shRNAs	
286[±], 371[±], 2052[+], 2104[±], 7326[+]	Plasmid-delivered shRNAs	Takigawa *et al.*, 2004
286[±], 371[±], 2052[+], 2104[±], 7326[+]	Retrovirus-delivered shRNAs	
7805[+], 7983[+]	Electroporation of synthetic siRNAs	Wilson and Richardson, 2005
346[+]	Hydrodynamic transfection and lipofection of synthetic siRNAs[g]	Wang *et al.*, 2005b
346[+]	Hydrodynamic transfection and lipofection of *in vitro*-transcribed shRNAs[g]	
74[±], 156[±], 207[±], 9548[±], 9578[−]	Retrovirus-delivered shRNAs	Korf *et al.*, 2005
2426[+], 4498[+], 8666[+, b]	Plasmid-delivered shRNAs[h]	Prabhu *et al.*, 2005

[a] If not indicated otherwise, numbers refer to positions of the first nucleotide within the HCV Con1 genome (EMBL database accession number AJ238799) that is targeted by the specified RNA. Positions of important elements within the Con1 genome: 5' NTR, 1-341; core, 342-914; E1, 915-1490; E2, 1491-2579; p7, 2580-2768; NS2, 2769-3419; NS3, 3420-5312; NS4A; 5313-5474; NS4B, 5475-6257; NS5A, 6258-7598; NS5B, 7599-9371; and 3' NTR, 9372-9605. Increasing susceptibility of target sequences to RNAi is described by the following symbols: −, ±, and +.

[b] Numbers refer to nucleotide positions of the HCV H77 genome (EMBL database accession number AF009606).

[c] Numbers indicate the positions of *in vitro*-transcribed, double-stranded RNAs that were used to generate endonuclease-prepared siRNAs (esiRNAs).

[d] If not otherwise stated, the antiviral activitiy of siRNAs and shRNAs was analyzed by using Huh-7 cells containing HCV replicons.

[e] Sen *et al.* used stably transfected Hep5A cells that constitutively express NS5A and HepG2 cells that transiently express the entire H77 ORF for the analysis of HCV-specific siRNAs.

[f] McCaffrey *et al.* used a reporter plasmid encoding a luciferase-NS5B fusion protein to generate HCV-like target RNAs in mice.

[g] Wang *et al.* used an HCV IRES-driven luciferase reporter plasmid to generate HCV-like targed RNAs in 293 cells and in mice.

[h] Prabhu *et al.* used transfected Huh-7 cells that constitutively express the entire H77 genome under the control of the T7 promoter.

[i] Publications are listed in order by the date of appearence.

protein (domain IV) (Krönke *et al.*, 2004; Randall *et al.*, 2003; Seo *et al.*, 2003; Takigawa *et al.*, 2004; Wang *et al.*, 2005b; Yokota *et al.*, 2003).

For instance, Krönke and coworkers generated a panel of retroviral vectors that allow the transduction of shRNAs into Huh-7 cells carrying stably replicating HCV replicons (Krönke *et al.*, 2004). Short hairpin RNAs were designed on the basis of an extensive sequence comparison of the 5' terminal HCV genome region and highly conserved target sequences were selected. A total of 11 different shRNAs was constructed and tested for their inhibitory capacity. Four shRNAs turned out to be most efficient, 2 of them targeting sequences in close proximity of the AUG start codon, one complementary to a region in the 5' terminal core coding sequence and one targeting a stem structure in domain III (Fig. 1). The efficiency of the latter is in keeping with the notion that the RNA structure determined in the absence of interacting proteins (as shown in Fig. 1) is different from the one formed intracellular. Otherwise it would be difficult to envisage why a sequence present in a stable RNA double strand is such an efficient target for RNAi. Finally, these shRNAs not only inhibited HCV replicons that were already present in the cells but also conferred resistance to naïve cells against a subgenomic replicon introduced by RNA transfection (Krönke *et al.*, 2004).

Although efficient silencing of HCV replicons could be achieved with siRNAs targeting coding regions downstream of core, in most cases these sequences are not sufficiently conserved between various HCV genotypes. For instance, Takigawa and coworkers generated a shRNA targeting a sequence in NS5B and inhibiting replication of a subgenomic HCV replicon with high efficiency (Takigawa *et al.,* 2004). However, this target sequence is not well conserved among different HCV genotypes limiting the utility of this target sequence for therapeutic intervention on a broad scale. This dependency on high sequence conservation and the emergence of escape mutations is one of the major problems of RNAi- (and other NA-) based antiviral approaches. Indeed, it was found that 3 mismatches within a 21 nucleotides-long target sequence completely block RNAi (Randall *et al.,* 2003). Moreover, it was shown that HCV replicons escape RNAi by the accumulation of point mutations. Wilson and Richardson repeatedly transfected HCV replicon cells with siRNAs that target a region in the NS5B coding sequence and analyzed the target sequence of surviving replicons. Most cDNA clones contained two nucleotide substitutions, indicating that a single point mutation was not sufficient to confer complete resistance (Wilson *et al.,* 2005).

Different strategies have been explored to minimize the chances of escape mutants. In one approach, HCV-specific siRNAs were prepared by digestion of *in vitro*-transcribed, double-stranded RNAs with the RNAse III of *Escherichia coli.* These so-called esiRNAs simultaneously target multiple sites of the viral genome for degradation, a strategy that should block the evolution of escape mutants (Krönke *et al.,* 2004). Another approach is to target those parts of the viral genome that most likely have less freedom to mutate. As described above, this is best accomplished by using the 5' NTR and the first codons of the core sequence. Finally, combination of siRNAs targeting different regions in the HCV genome should reduce the probability that escape mutants develop. The combination of such siRNAs may at the same time increase the antiviral activity in an additive or eventually synergistic manner.

Another possibility to circumvent the selection for escape mutants is the knock-down of the expression of host cell factors required for HCV RNA translation and replication or virus particle formation. Thus far, several cellular proteins were found to be essential for replication of HCV replicons in Huh-7 cells. These are, amongst others, the human vesicle-associated membrane protein-associated protein A (hVAP-A), the geranylgeranylated cellular protein FBL2, and cyclophilin B (Evans *et al.,* 2004; Wang *et al.,* 2005a; Watashi *et al.,* 2005). RNAi-mediated knock-down of each of these proteins reduced replication of HCV replicons in Huh-7 cells to various extents supporting the notion that these proteins are important for the viral life cycle (Wang *et al.,* 2005a; Watashi *et al.,* 2005; Zhang *et al.,* 2004). From this point of view, silencing the expression of these, and other genes essential for HCV replication, is an attractive approach that would circumvent the problem of therapy resistance. However, it remains to be determined whether the sustained repression of these host cell factors is tolerated by the cell or causes adverse effects which would preclude these cellular genes as targets for HCV-specific therapy.

With the recent development of a cell culture system that supports the production of infectious HCV particles it became possible to determine whether siRNAs inhibiting replication of HCV replicons also interfere with infection and virus production. By using Huh-7 host cells that stably express a shRNA targeting a highly conserved region in the 5' NTR we found that both RNA replication and virus production were dramatically reduced (S. Sparacio and R. Bartenschlager, unpublished). Moreover, naïve Huh-7 cells expressing this shRNA were resistant to infection with HCV similar to what we found with subgenomic replicons. This result is in keeping with the notion that HCV does not express an antagonist of RNAi. The high susceptibility of HCV to siRNA therefore makes this approach very attractive for the development of a novel antiviral strategy.

4.   CONCLUSIONS

NA-based antiviral approaches are an attractive alternative to conventional small organic molecules interfering primarily with the NS3 proteinase and the NS5B RdRp of HCV. NAs, in particular siRNAs potently inhibit HCV replication, at least in cell culture, have a high selectivity and are easy and rather cheap to produce. For these reasons, several biotechnology companies are currently developing NAs for treatment of chronic hepatitis C. However, several problems such as stability, delivery and side-effects must be overcome before NA-based anti-HCV therapy can be considered for clinical application (reviewed in Leung and Whittaker, 2005). As described above, several chemical modifications have been developed in the past few years that greatly increase stability of NAs in tissue without affecting their biologic activity. Limitations in bioavailability can in part be overcome by coupling of therapeutic NAs to ligands such as cholesterol which stabilizes the molecules by increasing their binding to human serum albumin and at the same time increases their uptake by the liver (Soutschek *et al.,* 2004). Nevertheless, the therapeutic effect of NAs lasts only transiently and therefore, multiple doses at regular intervals are most likely required to achieve a sufficient level of the active compound. Such a prerequisite can be overcome by stable expression of shRNAs that are processed within the cells into siRNAs. For this purpose, viral vectors allowing the efficient transduction of shRNA-expression constructs with high efficiency into host cells can be used. However, efficient protection of the liver will require stable expression of the shRNA in the majority of hepatocytes and the use of the most popular integrating viral vectors bears the risk of insertional mutagenesis (Hacein-Bey-Abina *et al.*, 2003). Finally, undesired off-target effects must be carefully evaluated. For instance, it has been observed that siRNAs can tolerate one or even several mismatches and still retain silencing capacity (Leung *et al.*, 2005). Therefore, side-effects can be induced by cross-hybridization of the NA with cellular mRNAs. However, the algorithms to predict therapeutic NAs with least off-target effects are steadily improved which should be helpful for the design of NAs targeting HCV with high specificity.

In summary, NA-based therapy for the treatment of chronic hepatitis C holds some promise as a complement or even alternative to existing IFN/ribavirin combination therapy or drugs targeting the key viral enzymes. Several problems need to be solved before therapeutic NAs can be introduced into the clinic but the progress made in the last few years raises justified hopes that such NAs will become reality.

ACKNOWLEDGEMENTS

We would like to thank all the colleagues who have helped in many ways in the production of this chapter, in particular Nicole Appel, Kerry Mills, and Sandra Sparacio for their helpful suggestions and careful reading of the manuscript, and Fredy Huschmand for help in preparing the figures.

Work in the authors' laboratory was supported by grants from the European Union (VIRGIL European Network of Excellence on Antiviral Drug Resistance LSHM-CT-2004-503359), the Deutsche Forschungsgemeinschaft (Sonderforschungsbereich 638, Teilprojekt A5), the Bundesministerium für Bildung und Forschung (Kompetenznetz Hepatitis, Teil-projekt 13.4), and the Bristol-Myers Squibb Foundation.

REFERENCES

Ali, N. and Siddiqui, A., 1995, Interaction of polypyrimidine tract-binding protein with the 5' noncoding region of the hepatitis C virus RNA genome and its functional requirement in internal initiation of translation. *J. Virol.* **69**: 6367-6375

Ali, N. and Siddiqui, A., 1997, The La antigen binds 5' noncoding region of the hepatitis C virus RNA in the context of the initiator AUG codon and stimulates internal ribosome entry site-mediated translation. *Proc. Natl. Acad. Sci. U. S. A* **94**: 2249-2254

Alt, M., Eisenhardt, S., Serwe, M., Renz, R., Engels, J.W. and Caselmann, W.H., 1999, Comparative inhibitory potential of differently modified antisense oligodeoxynucleotides on hepatitis C virus translation. *Eur. J. Clin. Invest* **29**: 868-876

Alt, M., Renz, R., Hofschneider, P.H., Paumgartner, G. and Caselmann, W.H., 1995, Specific inhibition of hepatitis C viral gene expression by antisense phosphorothioate oligo-deoxynucleotides. *Hepatology* **22**: 707-717

Appel, N., Herian, U. and Bartenschlager, R., 2005, Efficient rescue of hepatitis C virus RNA replication by trans-complementation with nonstructural protein 5A. *J. Virol.* **79**: 896-909

Arts, G.J., Langemeijer, E., Tissingh, R., Ma, L., Pavliska, H., Dokic, K., Dooijes, R., Mesic, E., Clasen, R., Michiels, F., van der, S.J., Lambrecht, M., Herman, S., Brys, R., Thys, K., Hoffmann, M., Tomme, P. and van Es, H., 2003, Adenoviral vectors expressing siRNAs for discovery and validation of gene function. *Genome Res.* **13**: 2325-2332

Banerjee, R. and Dasgupta, A., 2001, Specific interaction of hepatitis C virus protease/ helicase NS3 with the 3'-terminal sequences of viral positive- and negative-strand RNA. *J. Virol.* **75**: 1708-1721

Bartel, D.P., 2004, MicroRNAs: genomics, biogenesis, mechanism, and function. *Cell* **116**: 281-297

Bartenschlager, R., 2002, Hepatitis C virus replicons: potential role for drug development. *Nat. Rev. Drug Discov.* **1**: 911-916

Bartenschlager, R., Ahlborn-Laake, L., Mous, J. and Jacobsen, H., 1994, Kinetic and structural analyses of hepatitis C virus polyprotein processing. *J. Virol.* **68**: 5045-5055

Bartenschlager, R., Frese, M. and Pietschmann, T., 2004, Novel insights into hepatitis C virus replication and persistence. *Adv. Virus Res.* **63**: 71-180

Bartosch, B., Dubuisson, J. and Cosset, F.L., 2003, Infectious hepatitis C virus pseudo-particles containing functional E1-E2 envelope protein complexes. *J. Exp. Med.* **197**: 633-642

Behrens, S.E., Tomei, L. and De Francesco, R., 1996, Identification and properties of the RNA-dependent RNA polymerase of hepatitis C virus. *EMBO J* **15**: 12-22

Blight, K.J., Kolykhalov, A.A. and Rice, C.M., 2000, Efficient initiation of HCV RNA replication in cell culture. *Science* **290**: 1972-1974

Blight, K.J. and Rice, C.M., 1997, Secondary structure determination of the conserved 98-base sequence at the 3' terminus of hepatitis C virus genome RNA. *J. Virol.* **71**: 7345-7352

Brown-Driver, V., Eto, T., Lesnik, E., Anderson, K.P. and Hanecak, R.C., 1999, Inhibition of translation of hepatitis C virus RNA by 2-modified antisense oligonucleotides. *Antisense Nucleic Acid Drug Dev.* **9**: 145-154

Brummelkamp, T.R., Bernards, R. and Agami, R., 2002, A system for stable expression of short interfering RNAs in mammalian cells. *Science* **296**: 550-553

Caplen, N.J., 2003, RNAi as a gene therapy approach. *Expert. Opin. Biol. Ther.* **3**: 575-586

Caudy, A.A., Ketting, R.F., Hammond, S.M., Denli, A.M., Bathoorn, A.M., Tops, B.B., Silva, J.M., Myers, M.M., Hannon, G.J. and Plasterk, R.H., 2003, A micrococcal nuclease homologue in RNAi effector complexes. *Nature* **425**: 411-414

Cheney, I.W., Lai, V.C., Zhong, W., Brodhag, T., Dempsey, S., Lim, C., Hong, Z., Lau, J.Y. and Tam, R.C., 2002, Comparative analysis of anti-hepatitis C virus activity and gene expression mediated by alpha, beta, and gamma interferons. *J. Virol.* **76**: 11148-11154

Contreras, A.M., Hiasa, Y., He, W., Terella, A., Schmidt, E.V. and Chung, R.T., 2002, Viral RNA mutations are region specific and increased by ribavirin in a full-length hepatitis C virus replication system. *J. Virol.* **76**: 8505-8517

Deleersnyder, V., Pillez, A., Wychowski, C., Blight, K., Xu, J., Hahn, Y.S., Rice, C.M. and Dubuisson, J., 1997, Formation of native hepatitis C virus glycoprotein complexes. *J. Virol.* **71**: 697-704

Doench, J.G., Petersen, C.P. and Sharp, P.A., 2003, siRNAs can function as miRNAs. *Genes Dev.* **17**: 438-442

Dunn, W., Trang, P., Khan, U., Zhu, J. and Liu, F., 2001, RNase P-mediated inhibition of cytomegalovirus protease expression and viral DNA encapsidation by oligonucleotide external guide sequences. *Proc. Natl. Acad. Sci. U. S. A* **98**: 14831-14836

Egger, D., Wolk, B., Gosert, R., Bianchi, L., Blum, H.E., Moradpour, D. and Bienz, K., 2002, Expression of hepatitis C virus proteins induces distinct membrane alterations including a candidate viral replication complex. *J. Virol.* **76**: 5974-5984

Evans, M.J., Rice, C.M. and Goff, S.P., 2004, Phosphorylation of hepatitis C virus nonstructural protein 5A modulates its protein interactions and viral RNA replication. *Proc. Natl. Acad. Sci. U. S. A* **101**: 13038-13043

Failla, C., Tomei, L. and De Francesco, R., 1994, Both NS3 and NS4A are required for proteolytic processing of hepatitis C virus nonstructural proteins. *J. Virol.* **68**: 3753-3760

Frese, M., Pietschmann, T., Moradpour, D., Haller, O. and Bartenschlager, R., 2001, Interferon-α inhibits hepatitis C virus subgenomic RNA replication by an MxA-independent pathway. *Journal of General Virology* **82**: 723-733

Frese, M., Schwärzle, V., Barth, K., Krieger, N., Lohmann, V., Mihm, S., Haller, O. and Bartenschlager, R., 2002. Interferon-γ inhibits replication of subgenomic and genomic hepatitis C virus RNAs. *Hepatology* **35**: 694-703

Friebe, P., Boudet, J., Simorre, J.P. and Bartenschlager, R., 2005, Kissing-loop interaction in the 3' end of the hepatitis C virus genome essential for RNA replication. *J. Virol.* **79**: 380-392

Friebe, P., Lohmann, V., Krieger, N. and Bartenschlager, R., 2001, Sequences in the 5' nontranslated region of hepatitis C virus required for RNA replication. *J. Virol.* **75**: 12047-12057

Fukushi, S., Okada, M., Kageyama, T., Hoshino, F.B., Nagai, K. and Katayama, K., 2001, Interaction of poly(rC)-binding protein 2 with the 5'-terminal stem-loop of the hepatitis C-virus genome. *Virus Res.* **73**: 67-79

Gosert, R., Egger, D., Lohmann, V., Bartenschlager, R., Blum, H.E., Bienz, K. and Moradpour, D., 2003, Identification of the hepatitis C virus RNA replication complex in Huh-7 cells harboring subgenomic replicons. *J. Virol.* **77**: 5487-5492

Grishok, A., Pasquinelli, A.E., Conte, D., Li, N., Parrish, S., Ha, I., Baillie, D.L., Fire, A., Ruvkun, G. and Mello, C.C., 2001, Genes and mechanisms related to RNA interference regulate expression of the small temporal RNAs that control C. elegans developmental timing. *Cell* **106**: 23-34

Guerrier-Takada, C., Gardiner, K., Marsh, T., Pace, N. and Altman, S., 1983, The RNA moiety of ribonuclease P is the catalytic subunit of the enzyme. *Cell* **35**: 849-857

Hacein-Bey-Abina, S., Von Kalle, C., Schmidt, M., McCormack, M.P., Wulffraat, N., Leboulch, P., Lim, A., Osborne, C.S., Pawliuk, R., Morillon, E., Sorensen, R., Forster, A., Fraser, P., Cohen, J.I., de Saint, B.G., Alexander, I., Wintergerst, U., Frebourg, T., Aurias, A., Stoppa-Lyonnet, D., Romana, S., Radford-Weiss, I., Gross, F., Valensi, F., Delabesse, E., Macintyre, E., Sigaux, F., Soulier, J., Leiva, L.E., Wissler, M., Prinz, C., Rabbitts, T.H., Le Deist, F., Fischer, A. and Cavazzana-Calvo, M., 2003, LMO2-associated clonal T cell proliferation in two patients after gene therapy for SCID-X1. *Science* **302**: 415-419

Hahm, B., Kim, Y.K., Kim, J.H., Kim, T.Y. and Jang, S.K., 1998, Heterogeneous nuclear ribonucleoprotein L interacts with the 3' border of the internal ribosomal entry site of hepatitis C virus. *J. Virol.* **72**: 8782-8788

Hammond, S.M., Bernstein, E., Beach, D. and Hannon, G.J., 2000, An RNA-directed nuclease mediates post-transcriptional gene silencing in Drosophila cells. *Nature* **404**: 293-296

Hanecak, R., Brown-Driver, V., Fox, M.C., Azad, R.F., Furusako, S., Nozaki, C., Ford, C., Sasmor, H. and Anderson, K.P., 1996,Antisense oligonucleotide inhibition of hepatitis C virus gene expression in transformed hepatocytes. *J. Virol.* **70**: 5203-5212

Hannon, G.J., 2002. RNA interference. *Nature* **418**: 244-251

Haseloff, J. and Gerlach, W.L., 1992, Simple RNA enzymes with new and highly specific endoribonuclease activities. 1988, *Biotechnology* **24**: 264-269

Heintges, T., Encke, J., zu Putlitz. J. and Wands, J.R., 2001, Inhibition of hepatitis C virus NS3 function by antisense oligodeoxynucleotides and protease inhibitor. *J. Med. Virol.* **65**: 671-680

Herdewijn, P., 2000, Heterocyclic modifications of oligonucleotides and antisense technology. *Antisense Nucleic Acid Drug Dev.* **10**: 297-310

Hinrichsen, H., Benhamou, Y., Wedemeyer, H., Reiser, M., Sentjens, R.E., Calleja, J.L., Forns, X., Erhardt, A., Cronlein, J., Chaves, R.L., Yong, C.L., Nehmiz, G. and Steinmann, G.G., 2004, Short-term antiviral efficacy of BILN 2061, a hepatitis C virus serine protease inhibitor, in hepatitis C genotype 1 patients. *Gastroenterology* **127**: 1347-1355

Honda, M., Beard, M.R., Ping, L.H. and Lemon, S.M., 1999, A phylogenetically conserved stem-loop structure at the 5' border of the internal ribosome entry site of hepatitis C virus is required for cap-independent viral translation. *J. Virol.* **73**: 1165-1174

Honda, M., Brown, E.A. and Lemon, S.M., 1996a, Stability of a stem-loop involving the initiator AUG controls the efficiency of internal initiation of translation on hepatitis C virus RNA. *RNA.* **2**: 955-968

Honda, M., Ping, L.H., Rijnbrand, R.C., Amphlett, E., Clarke, B., Rowlands, D. and Lemon, S.M., 1996b, Structural requirements for initiation of translation by internal ribosome entry within genome-length hepatitis C virus RNA. *Virology* **222**: 31-42

Hoofnagle, J.H., 1997, Hepatitis C: the clinical spectrum of disease. *Hepatology* **26(Suppl 1)**: 15S-20S

Horscroft, N., Lai, V.C., Cheney, W., Yao, N., Wu, J.Z., Hong, Z. and Zhong, W., 2005, Replicon cell culture system as a valuable tool in antiviral drug discovery against hepatitis C virus. *Antivir. Chem. Chemother.* **16**: 1-12

Hultgren, C., Milich, D.R., Weiland, O. and Sallberg, M., 1998, The antiviral compound ribavirin modulates the T helper (Th) 1/Th2 subset balance in hepatitis B and C virus-specific immune responses. *J. Gen. Virol.* **79**: 2381-2391

Jubin, R., Vantuno, N.E., Kieft, J.S., Murray, M.G., Doudna, J.A., Lau, J.Y. and Baroudy, B.M., 2000, Hepatitis C virus internal ribosome entry site (IRES) *stem-loop* IIId contains a phylogenetically conserved GGG triplet essential for translation and IRES folding. *J. Virol.* **74**: 10430-10437

Kapadia, S.B., Brideau-Andersen, A. and Chisari, F.V., 2003, Interference of hepatitis C virus RNA replication by short interfering RNAs. *Proc. Natl. Acad. Sci. U. S. A* **100:** 2014-2018

Kashiwagi, T., Hara, K., Kohara, M., Iwahashi, J., Hamada, N., Honda-Yoshino, H. and Toyoda, T., 2002, Promoter/origin structure of the complementary strand of hepatitis C virus genome. *J. Biol. Chem.* **277:** 28700-28705

Kato, N. and Shimotohno, K., 2000, Systems to culture hepatitis C virus. *Curr. Top. Microbiol. Immunol.* **242:** 261-278

Kieft, J.S., Zhou, K., Jubin, R. and Doudna, J.A., 2001, Mechanism of ribosome recruitment by hepatitis C IRES RNA. *RNA.* **7:** 194-206

Kieft, J.S., Zhou, K., Jubin, R., Murray, M.G., Lau, J.Y. and Doudna, J.A., 1999, The hepatitis C virus internal ribosome entry site adopts an ion-dependent tertiary fold. *J. Mol. Biol.* **292:** 513-529

Kilani, A.F., Trang, P., Jo, S., Hsu, A., Kim, J., Nepomuceno, E., Liou, K. and Liu, F., 2000, RNase P ribozymes selected *in vitro* to cleave a viral mRNA effectively inhibit its expression in cell culture. *J. Biol. Chem.* **275:** 10611-10622

Kim, Y.K., Kim, C.S., Lee, S.H. and Jang, S.K., 2002, Domains I and II in the 5' nontranslated region of the HCV genome are required for RNA replication. *Biochem. Biophys. Res. Commun.* **290:** 105-112

Klinck, R., Westhof, E., Walker, S., Afshar, M., Collier, A. and Aboul-Ela, F., 2000, A potential RNA drug target in the hepatitis C virus internal ribosomal entry site. *RNA.* **6:** 1423-1431

Kolupaeva, V.G., Pestova, T.V. and Hellen, C.U., 2000, An enzymatic footprinting analysis of the interaction of 40S ribosomal subunits with the internal ribosome entry site of hepatitis C virus. *J. Virol.* **74:** 6242-6250

Korf, M., Jarczak, D., Beger, C., Manns, M.P. and Kruger, M., 2005, Inhibition of hepatitis C virus translation and subgenomic replication by siRNAs directed against highly conserved HCV sequence and cellular HCV cofactors. *J. Hepatol.* **43:** 225-234

Krönke, J., Kittler, R., Buchholz, F., Windisch, M.P., Pietschmann, T., Bartenschlager, R. and Frese, M., 2004, Alternative approaches for efficient inhibition of hepatitis C virus RNA replication by small interfering RNAs. *J. Virol.* **78:** 3436-3446

Kruger, K., Grabowski, P.J., Zaug, A.J., Sands, J., Gottschling, D.E. and Cech, T.R., 1982, Self-splicing RNA: autoexcision and autocyclization of the ribosomal RNA intervening sequence of Tetrahymena. *Cell* **31:** 147-157

Kurreck, J., 2003, Antisense technologies. Improvement through novel chemical modifications. *Eur. J. Biochem.* **270:** 1628-1644

Lai, V.C., Dempsey, S., Lau, J.Y., Hong, Z. and Zhong, W., 2003, *In vitro* RNA replication directed by replicase complexes isolated from the subgenomic replicon cells of hepatitis C virus. *J. Virol.* **77:** 2295-2300

Lamarre, D., Anderson, P.C., Bailey, M., Beaulieu, P., Bolger, G., Bonneau, P., Bos, M., Cameron, D.R., Cartier, M., Cordingley, M.G., Faucher, A.M., Goudreau, N., Kawai, S.H., Kukolj, G., Lagace, L., LaPlante, S.R., Narjes, H., Poupart, M.A., Rancourt, J., Sentjens, R.E., St George, R., Simoneau, B., Steinmann, G., Thibeault, D., Tsantrizos, Y.S., Weldon, S.M., Yong, C.L. and Llinas-Brunet, M., 2003, An NS3 protease inhibitor with antiviral effects in humans infected with hepatitis C virus. *Nature* **426:** 186-189

Lanford, R.E., Guerra, B., Lee, H., Averett, D.R., Pfeiffer, B., Chavez, D., Notvall, L. and Bigger, C., 2003, Antiviral effect and virus-host interactions in response to alpha interferon, gamma interferon, poly(i)-poly(c), tumor necrosis factor alpha, and ribavirin in hepatitis C virus subgenomic replicons. *J. Virol.* **77:** 1092-1104

Lavillette, D., Tarr, A.W., Voisset, C., Donot, P., Bartosch, B., Bain, C., Patel, A.H., Dubuisson, J., Ball, J.K. and Cosset, F.L., 2005, Characterization of host-range and cell entry properties of the major genotypes and subtypes of hepatitis C virus. *Hepatology* **41:** 265-274

Lee, H., Shin, H., Wimmer, E. and Paul, A.V., 2004, cis-acting RNA signals in the NS5B C-terminal coding sequence of the hepatitis C virus genome. *J. Virol.* **78:** 10865-10877

Lee, P.A., Blatt, L.M., Blanchard, K.S., Bouhana, K.S., Pavco, P.A., Bellon, L. and Sandberg, J.A., 2000, Pharmacokinetics and tissue distribution of a ribozyme directed against hepatitis C virus RNA following subcutaneous or intravenous administration in mice. *Hepatology* **32:** 640-646

Leung, R.K. and Whittaker, P.A., 2005, RNA interference: from gene silencing to gene-specific therapeutics. *Pharmacol. Ther.* **107:** 222-239

Li, W.X. and Ding, S.W., 2001, Viral suppressors of RNA silencing. *Curr. Opin. Biotechnol.* **12:** 150-154

Lieber, A., He, C.Y., Polyak, S.J., Gretch, D.R., Barr, D. and Kay, M.A., 1996, Elimination of hepatitis C virus RNA in infected human hepatocytes by adenovirus-mediated expression of ribozymes. *J. Virol.* **70:** 8782-8791

Lima, W.F., Brown-Driver, V., Fox, M., Hanecak, R. and Bruice, T.W., 1997, Combinatorial screening and rational optimization for hybridization to folded hepatitis C virus RNA of oligonucleotides with biological antisense activity. *J Biol. Chem.* **272:** 626-638

Lindenbach, B.D., Evans, M.J., Syder, A.J., Wolk, B., Tellinghuisen, T.L., Liu, C.C., Maruyama, T., Hynes, R.O., Burton, D.R., McKeating, J.A. and Rice, C.M., 2005, Complete Replication of Hepatitis C Virus in Cell Culture. *Science* **309:** 623-626

Lohmann, V., Koch, J.O., Herian, U., Theilmann, L. and Bartenschlager, R., 1999, Replication of subgenomic hepatitis C virus RNAs in a hepatoma cell line. *Science* **285:** 110-113

Lohmann, V., Korner, F., Herian, U. and Bartenschlager, R., 1997, Biochemical properties of hepatitis C virus NS5B RNA-dependent RNA polymerase and identification of amino acid sequence motifs essential for enzymatic activity. *J. Virol.* **71:** 8416-8428

Lu, H.H. and Wimmer, E., 1996, Poliovirus chimeras replicating under the translational control of genetic elements of hepatitis C virus reveal unusual properties of the internal ribosomal entry site of hepatitis C virus. *Proc. Natl. Acad. Sci. U. S. A* **93:** 1412-1417

Lukavsky, P.J., Otto, G.A., Lancaster, A.M., Sarnow, P. and Puglisi, J.D., 2000, Structures of two RNA domains essential for hepatitis C virus internal ribosome entry site function. *Nat. Struct. Biol.* **7:** 1105-1110

Luo, G., Xin, S. and Cai, Z., 2003, Role of the 5'-proximal stem-loop structure of the 5' untranslated region in replication and translation of hepatitis C virus RNA. *J. Virol.* **77:** 3312-3318

Ma, H., Leveque, V., De Witte, A., Li, W., Hendricks, T., Clausen, S.M., Cammack, N. and Klumpp, K., 2005, Inhibition of native hepatitis C virus replicase by nucleotide and non-nucleoside inhibitors. *Virology* **332:** 8-15

Macejak, D.G., Jensen, K.L., Jamison, S.F., Domenico, K., Roberts, E.C., Chaudhary, N., von, C., I, Bellon, L., Tong, M.J., Conrad, A., Pavco, P.A. and Blatt, L.M., 2000, Inhibition of hepatitis C virus (HCV)-RNA-dependent translation and replication of a chimeric HCV poliovirus using synthetic stabilized ribozymes. *Hepatology* **31:** 769-776

Macejak, D.G., Jensen, K.L., Pavco, P.A., Phipps, K.M., Heinz, B.A., Colacino, J.M. and Blatt, L.M., 2001, Enhanced antiviral effect in cell culture of type 1 interferon and ribozymes targeting HCV RNA. *J Viral Hepat.* **8:** 400-405

Malchere, C., Verheijen, J., van, der, L.S., Bastide, L., van Boom, J., Lebleu, B. and Robbins, I., 2000, A short phosphodiester window is sufficient to direct RNase H-dependent RNA cleavage by antisense peptide nucleic acid. *Antisense Nucleic Acid Drug Dev.* **10:** 463-468

Martinand-Mari, C., Lebleu, B. and Robbins, I., 2003, Oligonucleotide-based strategies to inhibit human hepatitis C virus. *Oligonucleotides.* **13:** 539-548

McCaffrey, A.P., Meuse, L., Pham, T.T., Conklin, D.S., Hannon, G.J. and Kay, M.A., 2002, RNA interference in adult mice. *Nature* **418:** 38-39

McHutchison, J.G. and Fried, M.W., 2003, Current therapy for hepatitis C: pegylated interferon and ribavirin. *Clin. Liver Dis.* **7:** 149-161

Mercer, D.F., Schiller, D.E., Elliott, J.F., Douglas, D.N., Hao, C., Rinfret, A., Addison, W.R., Fischer, K.P., Churchill, T.A., Lakey, J.R., Tyrrell, D.L. and Kneteman, N.M., 2001, Hepatitis C virus replication in mice with chimeric human livers. *Nat. Med.* **7:** 927-933

Meuleman, P., Libbrecht, L., De Vos, R., de Hemptinne, B., Gevaert, K., Vandekerckhove, J., Roskams, T. and Leroux-Roels, G., 2005, Morphological and biochemical characterization of a human liver in a uPA-SCID mouse chimera. *Hepatology* **41:** 847-856

Michel, T., Martinand-Mari, C., Debart, F., Lebleu, B., Robbins, I. and Vasseur, J.J., 2003, Cationic phosphoramidate alpha-oligonucleotides efficiently target single-stranded DNA and RNA and inhibit hepatitis C virus IRES-mediated translation. *Nucleic Acids Res.* **31:** 5282-5290

Mittal, V., 2004, Improving the efficiency of RNA interference in mammals. *Nat. Rev. Genet.* **5:** 355-365

Miyagishi, M. and Taira, K., 2003, Strategies for generation of an siRNA expression library directed against the human genome. *Oligonucleotides.* **13:** 325-333

Miyanari, Y., Hijikata, M., Yamaji, M., Hosaka, M., Takahashi, H. and Shimotohno, K., 2003, Hepatitis C virus non-structural proteins in the probable membranous compartment function in viral genome replication. *J Biol. Chem.* **278:** 50301-50308

Mizutani, T., Kato, N., Hirota, M., Sugiyama, K., Murakami, A. and Shimotohno, K., 1995, Inhibition of hepatitis C virus replication by antisense oligonucleotide in culture cells. *Biochem. Biophys. Res. Commun.* **212:** 906-911

Mizutani, T., Kato, N., Saito, S., Ikeda, M., Sugiyama, K. and Shimotohno, K., 1996, Characterization of hepatitis C virus replication in cloned cells obtained from a human T-cell leukemia virus type 1-infected cell line, MT-2. *J. Virol.* **70:** 7219-7223

Moradpour, D., Gosert, R., Egger, D., Penin, F., Blum, H.E. and Bienz, K., 2003, Membrane association of hepatitis C virus nonstructural proteins and identification of the membrane alteration that harbors the viral replication complex. *Antiviral Res.* **60:** 103-109

Mottola, G., Cardinali, G., Ceccacci, A., Trozzi, C., Bartholomew, L., Torrisi, M.R., Pedrazzini, E., Bonatti, S. and Migliaccio, G., 2002, Hepatitis C virus nonstructural proteins are localized in a modified endoplasmic reticulum of cells expressing viral subgenomic replicons. *Virology* **293:** 31-43

Nadal, A., Martell, M., Lytle, J.R., Lyons, A.J., Robertson, H.D., Cabot, B., Esteban, J.I., Esteban, R., Guardia, J. and Gomez, J., 2002, Specific cleavage of hepatitis C virus RNA genome by human RNase P. *J. Biol. Chem.* **277:** 30606-30613

Nadal, A., Robertson, H.D., Guardia, J. and Gomez, J., 2003, Characterization of the structure and variability of an internal region of hepatitis C virus RNA for M1 RNA guide sequence ribozyme targeting. *J. Gen. Virol.* **84:** 1545-1548

Ni, Z.J. and Wagman, A.S., 2004, Progress and development of small molecule HCV antivirals. *Curr. Opin. Drug Discov. Devel.* **7:** 446-459

Okuse, C., Rinaudo, J.A., Farrar, K., Wells, F. and Korba, B.E., 2005, Enhancement of antiviral activity against hepatitis C virus *in vitro* by interferon combination therapy. *Antiviral Res.* **65:** 23-34

Otto, G.A., Lukavsky, P.J., Lancaster, A.M., Sarnow, P. and Puglisi, J.D., 2002, Ribosomal proteins mediate the hepatitis C virus IRES-HeLa 40S interaction. *RNA.* **8:** 913-923

Paolini, C., De Francesco, R. and Gallinari, P., 2000, Enzymatic properties of hepatitis C virus NS3-associated helicase. *J. Gen. Virol.* **81:** 1335-1345

Parker, J.S., Roe, S.M. and Barford, D., 2005, Structural insights into mRNA recognition from a PIWI domain-siRNA guide complex. *Nature* **434:** 663-666

Pause, A., Kukolj, G., Bailey, M., Brault, M., Do, F., Halmos, T., Lagace, L., Maurice, R., Marquis, M., McKercher, G., Pellerin, C., Pilote, L., Thibeault, D. and Lamarre, D., 2003, An NS3 serine protease inhibitor abrogates replication of subgenomic hepatitis C virus RNA. *J. Biol. Chem.* **278:** 20374-20380

Pavlovic, D., Neville, D.C., Argaud, O., Blumberg, B., Dwek, R.A., Fischer, W.B. and Zitzmann, N., 2003, The hepatitis C virus p7 protein forms an ion channel that is inhibited by long-alkyl-chain iminosugar derivatives. *Proc. Natl. Acad. Sci. U. S. A* **100:** 6104-6108

Pestova, T.V., Shatsky, I.N., Fletcher, S.P., Jackson, R.J. and Hellen, C.U., 1998, A prokaryotic-like mode of cytoplasmic eukaryotic ribosome binding to the initiation codon during internal translation initiation of hepatitis C and classical swine fever virus RNAs. *Genes Dev.* **12:** 67-83

Pileri, P., Uematsu, Y., Campagnoli, S., Galli, G., Falugi, F., Petracca, R., Weiner, A.J., Houghton, M., Rosa, D., Grandi, G. and Abrignani, S., 1998, Binding of hepatitis C virus to CD81. *Science* **282:** 938-941

Piron, M., Beguiristain, N., Nadal, A., Martinez-Salas, E. and Gomez, J., 2005, Characterizing the function and structural organization of the 5' tRNA-like motif within the hepatitis C virus quasispecies. *Nucleic Acids Res.* **33:** 1487-1502

Prabhu, R., Vittal, P., Yin, Q., Flemington, E., Garry, R., Robichaux, W.H. and Dash, S., 2005, Small interfering RNA effectively inhibits protein expression and negative strand RNA synthesis from a full-length hepatitis C virus clone. *J. Med. Virol.* **76:** 511-519

Quinkert, D., Bartenschlager, R. and Lohmann, V., 2005, Quantitative analysis of the hepatitis C virus replication complex. *J. Virol.,* in press

Randall, G., Grakoui, A. and Rice, C.M., 2003, Clearance of replicating hepatitis C virus replicon RNAs in cell culture by small interfering RNAs. *Proc. Natl. Acad. Sci. U. S. A* **100:** 235-240

Reynolds, A., Leake, D., Boese, Q., Scaringe, S., Marshall, W.S. and Khvorova, A., 2004, Rational siRNA design for RNA interference. *Nat. Biotechnol.* **22:** 326-330

Reynolds, J.E., Kaminski, A., Kettinen, H.J., Grace, K., Clarke, B.E., Carroll, A.R., Rowlands, D.J. and Jackson, R.J., 1995, Unique features of internal initiation of hepatitis C virus RNA translation. *EMBO J* **14:** 6010-6020

Rijnbrand, R., Bredenbeek, P., van der, S.T., Whetter, L., Inchauspe, G., Lemon, S. and Spaan, W., 1995, Almost the entire 5' non-translated region of hepatitis C virus is required for cap-independent translation. *FEBS Lett.* **365:** 115-119

Rijnbrand, R., Bredenbeek, P.J., Haasnoot, P.C., Kieft, J.S., Spaan, W.J. and Lemon, S.M., 2001, The influence of downstream protein-coding sequence on internal ribosome entry on hepatitis C virus and other flavivirus RNAs. *RNA.* **7:** 585-597

Robek, M.D., Boyd, B.S. and Chisari, F.V., 2005, Lambda interferon inhibits hepatitis B and C virus replication. *J. Virol.* **79:** 3851-3854

Robertson, H.D., Altman, S. and Smith, J.D., 1972, Purification and properties of a specific Escherichia coli ribonuclease which cleaves a tyrosine transfer ribonucleic acid presursor. *J. Biol. Chem.* **247:** 5243-5251

Romero-Lopez, C., Barroso-delJesus, A., Puerta-Fernandez, E. and Berzal-Herranz, A., 2005, Interfering with hepatitis C virus IRES activity using RNA molecules identified by a novel *in vitro* selection method. *Biol. Chem.* **386:** 183-190

Rubinson, D.A., Dillon, C.P., Kwiatkowski, A.V., Sievers, C., Yang, L., Kopinja, J., Rooney, D.L., Ihrig, M.M., McManus, M.T., Gertler, F.B., Scott, M.L. and Van Parijs, L., 2003, A lentivirus-based system to functionally silence genes in primary mammalian cells, stem cells and transgenic mice by RNA interference. *Nat. Genet.* **33**: 401-406

Ryu, K.J. and Lee, S.W., 2004, Comparative analysis of intracellular trans-splicing ribozyme activity against hepatitis C virus internal ribosome entry site. *J. Microbiol.* **42**: 361-364

Sakai, A., Claire, M.S., Faulk, K., Govindarajan, S., Emerson, S.U., Purcell, R.H. and Bukh, J., 2003, The p7 polypeptide of hepatitis C virus is critical for infectivity and contains functionally important genotype-specific sequences. *Proc. Natl. Acad. Sci. U. S. A* **100**: 11646-11651

Sakamoto, N., Wu, C.H. and Wu, G.Y., 1996, Intracellular cleavage of hepatitis C virus RNA and inhibition of viral protein translation by hammerhead ribozymes. *J. Clin. Invest* **98**: 2720-2728

Scherer, L.J. and Rossi, J.J., 2003, Approaches for the sequence-specific knockdown of mRNA. *Nat. Biotechnol.* **21**: 1457-1465

Schuster, C., Isel, C., Imbert, I., Ehresmann, C., Marquet, R. and Kieny, M.P., 2002, Secondary structure of the 3' terminus of hepatitis C virus minus-strand RNA. *J. Virol.* **76**: 8058-8068

Seo, M.Y., Abrignani, S., Houghton, M. and Han, J.H., 2003, Small interfering RNA-mediated inhibition of hepatitis C virus replication in the human hepatoma cell line Huh-7. *J. Virol.* **77**: 810-812

Shimizu, Y.K. and Yoshikura, H., 1994, Multicycle infection of hepatitis C virus in cell culture and inhibition by alpha and beta interferons. *J. Virol.* **68**: 8406-8408

Siolas, D., Lerner, C., Burchard, J., Ge, W., Linsley, P.S., Paddison, P.J., Hannon, G.J. and Cleary, M.A., 2005, Synthetic shRNAs as potent RNAi triggers. *Nat. Biotechnol.* **23**: 227-231

Smith, R.M., Walton, C.M., Wu, C.H. and Wu, G.Y., 2002, Secondary structure and hybridization accessibility of hepatitis C virus 3'-terminal sequences. *J. Virol.* **76**: 9563-9574

Snell, N.J., 2001, Ribavirin--current status of a broad spectrum antiviral agent. *Expert. Opin. Pharmacother.* **2**: 1317-1324

Soler, M., McHutchison, J.G., Kwoh, T.J., Dorr, F.A. and Pawlotsky, J.M., 2004, Virological effects of ISIS 14803, an antisense oligonucleotide inhibitor of hepatitis C virus (HCV) internal ribosome entry site (IRES), on HCV IRES in chronic hepatitis C patients and examination of the potential role of primary and secondary HCV resistance in the outcome of treatment. *Antivir. Ther.* **9**: 953-968

Soutschek, J., Akinc, A., Bramlage, B., Charisse, K., Constien, R., Donoghue, M., Elbashir, S., Geick, A., Hadwiger, P., Harborth, J., John, M., Kesavan, V., Lavine, G., Pandey, R.K., Racie, T., Rajeev, K.G., Rohl, I., Toudjarska, I., Wang, G., Wuschko, S., Bumcrot, D., Koteliansky, V., Limmer, S., Manoharan, M. and Vornlocher, H.P., 2004, Therapeutic silencing of an endogenous gene by systemic administration of modified siRNAs. *Nature* **432**: 173-178

Spahn, C.M., Kieft, J.S., Grassucci, R.A., Penczek, P.A., Zhou, K., Doudna, J.A. and Frank, J., 2001, Hepatitis C virus IRES RNA-induced changes in the conformation of the 40s ribosomal subunit. *Science* **291**: 1959-1962

Steinkuhler, C., Urbani, A., Tomei, L., Biasiol, G., Sardana, M., Bianchi, E., Pessi, A. and De Francesco, R., 1996, Activity of purified hepatitis C virus protease NS3 on peptide substrates. *J. Virol.* **70**: 6694-6700

Takigawa, Y., Nagano-Fujii, M., Deng, L., Hidajat, R., Tanaka, M., Mizuta, H. and Hotta, H., 2004, Suppression of hepatitis C virus replicon by RNA interference directed against the NS3 and NS5B regions of the viral genome. *Microbiol. Immunol.* **48:** 591-598

Tallet-Lopez, B., Aldaz-Carroll, L., Chabas, S., Dausse, E., Staedel, C. and Toulme, J.J., 2003, Antisense oligonucleotides targeted to the domain IIId of the hepatitis C virus IRES compete with 40S ribosomal subunit binding and prevent *in vitro* translation. *Nucleic Acids Res.* **31:** 734-742

Tam, R.C., Pai, B., Bard, J., Lim, C., Averett, D.R., Phan, U.T. and Milovanovic, T., 1999, Ribavirin polarizes human T cell responses towards a Type 1 cytokine profile. *J. Hepatol.* **30:** 376-382

Tellinghuisen, T.L., Marcotrigiano, J., Gorbalenya, A.E., and Rice, C.M. 2004, The NS5A protein of hepatitis C virus is a zinc metalloprotein. *J. Biol. Chem.* **279:** 48576-48587.

Tellinghuisen, T.L., Marcotrigiano, J. and Rice, C.M., 2005. Structure of the zinc-binding domain of an essential component of the hepatitis C virus replicase. *Nature* **435:** 374-379.

Theodore, D. and Fried, M.W., 2000, Natural history and disease manifestations of hepatitis C infection. *Current Topics in Microbiology and Immunology* **242:** 44-54

Trang, P., Lee, M., Nepomuceno, E., Kim, J., Zhu, H. and Liu, F., 2000, Effective inhibition of human cytomegalovirus gene expression and replication by a ribozyme derived from the catalytic RNA subunit of RNase P from Escherichia coli. *Proc. Natl. Acad. Sci. U. S. A* **97:** 5812-5817

Van Regenmortel, M. H. V., Fauquet, C. M., Bishop, D. H. L., Carstens, E. B., Estes, M. K., Lemon, S. M., Maniloff, J., Mayo, M. A., McGeoch, D. J., Pringle, C. R., and Wickner, R. B., 2000, *Virus Taxonomy: The VIIth Report of the International Committee on Taxonomy of Viruses.* Academic Press, San Diego

Varaklioti, A., Vassilaki, N., Georgopoulou, U. and Mavromara, P., 2002, Alternate translation occurs within the core coding region of the hepatitis C viral genome. *J. Biol. Chem.* **277:** 17713-17721

Vidalin, O., Major, M.E., Rayner, B., Imbach, J.L., Trepo, C. and Inchauspe, G., 1996, *In vitro* inhibition of hepatitis C virus gene expression by chemically modified antisense oligodeoxynucleotides. *Antimicrob. Agents Chemother.* **40:** 2337-2344

Wakita, T., Pietschmann, T., Kato, T., Date, T., Miyamoto, M., Zhao, Z., Murthy, K., Habermann, A., Krausslich, H.G., Mizokami, M., Bartenschlager, R. and Liang, T.J., 2005, Production of infectious hepatitis C virus in tissue culture from a cloned viral genome. *Nat. Med.* **11:** 791-796

Wakita, T. and Wands, J.R., 1994, Specific inhibition of hepatitis C virus expression by antisense oligodeoxynucleotides. *In vitro* model for selection of target sequence. *J Biol. Chem.* **269:** 14205-14210

Wang, C., Gale, M., Jr., Keller, B.C., Huang, H., Brown, M.S., Goldstein, J.L. and Ye, J., 2005a, Identification of FBL2 as a geranylgeranylated cellular protein required for hepatitis C virus RNA replication. *Mol. Cell* **18:** 425-434

Wang, C., Le, S.Y., Ali, N. and Siddiqui, A., 1995, An RNA pseudoknot is an essential structural element of the internal ribosome entry site located within the hepatitis C virus 5' noncoding region. *RNA.* **1:** 526-537

Wang, C., Sarnow, P. and Siddiqui, A., 1994, A conserved helical element is essential for internal initiation of translation of hepatitis C virus RNA. *J. Virol.* **68:** 7301-7307

Wang, Q., Contag, C.H., Ilves, H., Johnston, B.H. and Kaspar, R.L., 2005b, Small Hairpin RNAs Efficiently Inhibit Hepatitis C IRES-Mediated Gene Expression in Human Tissue Culture Cells and a Mouse Model. *Mol. Ther.* Epub ahead of print

Watashi, K., Ishii, N., Hijikata, M., Inoue, D., Murata, T., Miyanari, Y. and Shimotohno, K., 2005, Cyclophilin B is a functional regulator of hepatitis C virus RNA polymerase. *Mol. Cell* **19**: 111-122

Westaway, E.G., Mackenzie, J.M. and Khromykh, A.A., 2002. Replication and gene function in Kunjin virus. *Curr. Top. Microbiol. Immunol.* **267**: 323-351

Wilson, J.A., Jayasena, S., Khvorova, A., Sabatinos, S., Rodrigue-Gervais, I.G., Arya, S., Sarangi, F., Harris-Brandts, M., Beaulieu, S. and Richardson, C.D., 2003, RNA interference blocks gene expression and RNA synthesis from hepatitis C replicons propagated in human liver cells. *Proc. Natl. Acad. Sci. U. S. A* **100**: 2783-2788

Wilson, J.A. and Richardson, C.D., 2005, Hepatitis C virus replicons escape RNA interference induced by a short interfering RNA directed against the NS5b coding region. *J. Virol.* **79**: 7050-7058

Witherell, G.W., 2001, ISIS-14803 (Isis Pharmaceuticals). *Curr. Opin. Investig. Drugs* **2**: 1523-1529

World Health Organization, 2000, Hepatitis C: global prevalence (update). *Weekly Epidemiological Record* **75**: 18-19

Xu, Z., Choi, J., Yen, T.S., Lu, W., Strohecker, A., Govindarajan, S., Chien, D., Selby, M.J. and Ou, J., 2001, Synthesis of a novel hepatitis C virus protein by ribosomal frameshift. *EMBO J* **20**: 3840-3848

Yagnik, A.T., Lahm, A., Meola, A., Roccasecca, R.M., Ercole, B.B., Nicosia, A. and Tramontano, A., 2000, A model for the hepatitis C virus envelope glycoprotein E2. *Proteins* **40**: 355-366

Yanagiya, A., Ohka, S., Hashida, N., Okamura, M., Taya, C., Kamoshita, N., Iwasaki, K., Sasaki, Y., Yonekawa, H. and Nomoto, A., 2003, Tissue-specific replicating capacity of a chimeric poliovirus that carries the internal ribosome entry site of hepatitis C virus in a new mouse model transgenic for the human poliovirus receptor. *J. Virol.* **77**: 10479-10487

Yen, J.H., Chang, S.C., Hu, C.R., Chu, S.C., Lin, S.S., Hsieh, Y.S. and Chang, M.F., 1995, Cellular proteins specifically bind to the 5'-noncoding region of hepatitis C virus RNA. *Virology* **208**: 723-732

Yokota, T., Sakamoto, N., Enomoto, N., Tanabe, Y., Miyagishi, M., Maekawa, S., Yi, L., Kurosaki, M., Taira, K., Watanabe, M. and Mizusawa, H., 2003, Inhibition of intracellular hepatitis C virus replication by synthetic and vector-derived small interfering RNAs. *EMBO Rep.* **4**: 602-608

Yoo, B.J., Spaete, R.R., Geballe, A.P., Selby, M., Houghton, M. and Han, J.H., 1992, 5' end-dependent translation initiation of hepatitis C viral RNA and the presence of putative positive and negative translational control elements within the 5' untranslated region. *Virology* **191**: 889-899

You, S., Stump, D.D., Branch, A.D. and Rice, C.M., 2004, A cis-acting replication element in the sequence encoding the NS5B RNA-dependent RNA polymerase is required for hepatitis C virus RNA replication. *J. Virol.* **78**: 1352-1366

Yu, Y.C., Mao, Q., Gu, C.H., Li, Q.F. and Wang, Y.M., 2002, Activity of HDV ribozymes to trans-cleave HCV RNA. *World J. Gastroenterol.* **8**: 694-698

Zhang, H., Hanecak, R., Brown-Driver, V., Azad, R., Conklin, B., Fox, M.C. and Anderson, K.P., 1999, Antisense oligonucleotide inhibition of hepatitis C virus (HCV) gene expression in livers of mice infected with an HCV-vaccinia virus recombinant. *Antimicrob. Agents Chemother.* **43**: 347-353

Zhang, J., Yamada, O., Sakamoto, T., Yoshida, H., Iwai, T., Matsushita, Y., Shimamura, H., Araki, H. and Shimotohno, K., 2004, Down-regulation of viral replication by adenoviral-mediated expression of siRNA against cellular cofactors for hepatitis C virus. *Virology* **320**: 135-143

Zhang, J., Yamada, O., Yoshida, H., Iwai, T. and Araki, H., 2002, Autogenous translational inhibition of core protein: implication for switch from translation to RNA replication in hepatitis C virus. *Virology* **293:** 141-150

Zhao, W.D., Lahser, F.C. and Wimmer, E., 2000, Genetic analysis of a poliovirus/hepatitis C virus (HCV) chimera: interaction between the poliovirus cloverleaf and a sequence in the HCV 5' nontranslated region results in a replication phenotype. *J. Virol.* **74:** 6223-6226

Zhao, W.D., Wimmer, E. and Lahser, F.C., 1999, Poliovirus/Hepatitis C virus (internal ribosomal entry site-core) chimeric viruses: improved growth properties through modification of a proteolytic cleavage site and requirement for core RNA sequences but not for core-related polypeptides. *J. Virol.* **73:** 1546-1554

Zhong, J., Gastaminza, P., Cheng, G., Kapadia, S., Kato, T., Burton, D.R., Wieland, S.F., Uprichard, S.L., Wakita, T. and Chisari, F.V., 2005, Robust hepatitis C virus infection *in vitro*. *Proc. Natl. Acad. Sci. U. S. A* **102:** 9294-9

Chapter 1.4

INHIBITORS OF RESPIRATORY VIRUSES

E. KUECHLER[1] and J. SEIPELT[1]

[1] *Max F. Perutz Laboratories, University Departments at the Vienna Biocenter, Department of Medical Biochemistry, Medical University of Vienna, Dr. Bohr Gasse 9/3, A-1030 Vienna, Austria*

Abstract: Viral respiratory infections are caused by a variety of different virus species such as human rhinovirus, influenza virus, parainfluenza and respiratory syncytial virus. As the life cycles of these viruses are quite different, the development of a general treatment for respiratory infections seems impossible. For specific viruses however significant progress has been achieved, although only for influenza virus antiviral medication has reached the market. We present different approaches to inhibit respiratory viruses with a focus on human rhinoviruses, the most frequent cause of common cold.

1.   INTRODUCTION

1.1   Main Approaches

There are many viruses which cause diseases in the respiratory tract of humans and mammals. A variety of different virus families such as rhinovirus, influenza virus, parainfluenza virus, respiratory syncytial virus (RSV), coronaviruses and adenovirus are responsible for more than 95% of all human respiratory infections. The remaining 5% are due to bacterial infections. Typical symptoms of respiratory infections develop one to two days after viral inoculation and include nasal discharge, sneezing, sore throat, cough, headache and general weakness. In severity, the symptoms may vary from a mild "common cold" to a severe influenza virus infection, but they still make some 60 million people seek medical advice every year in the United States alone (Bertino 2002). The National Centre for Health

E. Bognar and A. Holzenburg (eds.), New Concepts of Antiviral Theapy, 87–114.

Statistics estimated that in 1996 about 62 million cases of viral respiratory tract infections (VRTI) required medical attention or resulted in restricted activity in the United States. Roughly one in six respiratory infections leads to a doctor's office visit and up to 50% of these visits result in an antibiotic prescription, which mainly treats secondary infections but fails to tackle the primary viral infection (Bertino, 2002). The market value relating to respiratory tract infections is illustrated by the fact that in 1998 more antibiotic prescriptions were written for presumed VRTIs than for bacterial infections, at a cost of approximately $ 726 million (Gonzales *et al.*, 2001).

The need for an effective treatment of VRTIs is obvious. In 1998, people in the United States spent around $ 5 billion on over-the-counter products that relieve symptoms of the common cold. This corresponds to a rise by more than 50% as compared to 1995, when the amount spent was $ 3.3 billion.

As viral respiratory infections are caused by a wide range of different viruses that require an equally great variety of treatments, establishing successful causative treatment is a challenging task. We will provide an overview on treatment approaches and discuss problems and potential solutions for successful therapy.

2.   RESPIRATORY VIRUSES

2.1   Epidemiology

Viral respiratory tract infections are caused by a heterogeneous group of viruses of different genera. The relative proportions of these viruses vary and depend on several factors such as season, age but also viral sampling and detection technique. About 30-50% of all infections – generally termed as "common cold" – are caused by rhinoviruses of the picornavirus group. This makes common cold the most widespread acute disease in individuals (Monto *et al.*, 1993, 2002a; Makela *et al.*, 1998). Approximately 5-15% of all viral respiratory infections are caused by influenza virus, which makes them more severe and sometimes lethal. Most people recover from the illness; however, the Centers for Disease Control and Prevention (CDC) estimates that in the United States an average of 36,000 persons die from influenza and its complications every year, most of them elderly people. Respiratory syncytial virus (RSV) is the main cause of infant bronchiolitis and pneumonia. The virus is ubiquitous, highly infectious and reaches epidemic proportions every year during the winter months. In the total population, approximately 5% of all VRTI cases are caused by RSV. Parainfluenza virus can cause upper and lower respiratory tract infections in both adults and children (3%). Coronaviruses are found in 7-18% of all adults suffering from VRTIs (Larson *et al.*, 1980; Nicholson *et al.*, 1997). Single study estimated that human coronaviruses account for up to 35% of cases of upper respiratory illness. While it is clear that human coronavirus can play an important role in respiratory outbreaks a much lower frequency is found in most studies (Vabret *et al.*, 2003; Louie *et al.*, 2005). The two remaining virus families, i.e. adenovirus and enterovirus, account for minor shares of respiratory infections. The fact that up to 30% of VRTIs are not assigned to a certain virus species is probably due to suboptimal sampling and detection methods in some cases (Heikkinen *et al.*, 2003).

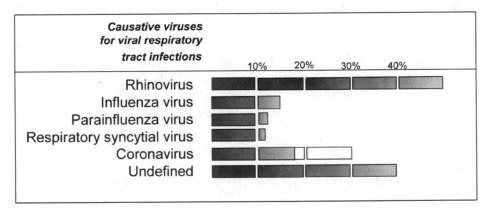

Figure 1. Causative viruses for viral respiratory tract infections (compiled from Monto 2002a and Gwaltney *et al.*, 1966).

An analysis of available epidemiology data has shown a distinct and consistent seasonal pattern in the occurrence of respiratory viruses (Monto 2002a). In temperate regions of the northern hemisphere, rhinoviruses account for up to 80% of all viruses circulating in early autumn (Arruda *et al.*, 1997; Makela *et al.*, 1998). In some years and some geographic areas, spring is an even more dangerous time for rhinovirus transmission. Although overall rates of respiratory diseases are lower in summer, rhinoviruses are the type of virus that is most frequently isolated at that time of the year. In winter, other viral agents, among them influenza viruses, parainfluenza virus and respiratory syncytial virus, predominate in the northern hemisphere (Monto 2002b).

2.2   Diagnostic Problem

From a clinical point of view, ascertaining upper respiratory tract virus infection is difficult, because many respiratory viruses cause similar symptoms. Temperatures above 37.8°C in the initial stage of the disease indicate influenza rather than other virus infections. Early identification of respiratory viruses is essential for effective diagnosis and patient management, as only for influenza virus causative treatment is available. Methods for identifying viruses include viral culture, antigen detection and highly sensitive molecular biology techniques based on polymerase chain reaction (PCR). Isolating viruses in cell culture is considered the gold standard but of limited relevance in clinical practice because it is too slow. Several antigen detection tests are available for diagnosing influenza A and B, parainfluenza, RSV and adenoviruses (Kim *et al.* 1983; Waris *et al.*, 1988; Nikkari *et al.*, 1989) though they are not widely used (Steininger *et al.*, 2001, 2002). Due to the high number of human rhinovirus (HRV) serotypes, detecting these viruses by immunological methods is not easy although PCR methods are available. A main disadvantage of the latter is the fact that they do not discriminate between infective viruses but rather detect the viral genome.

3.   INHIBITION OF RESPIRATORY VIRUSES

Currently, we lack a general treatment that addresses the underlying causes of viral respiratory tract infections, i.e. the virus infection.

In principle, antiviral drugs can be designed to either target a viral or a cellular protein. Targeting specific proteins of viruses basically has the advantage of being less toxic for cells with a narrow antiviral spectrum, while targeting cellular molecules might yield compounds with a broader antiviral activity. In the first case, the likelihood of generating drug-insensitive mutants is high, while this risk is unlikely in the latter case. The concept was proved for both strategies, e.g. in infections with human rhinoviruses. One example is the successful use of inhibitors targeting the viral proteinase 3C in HRV. A cellular approach was taken by the use of interferon to stimulate cellular defense against rhinovirus. For a thorough review of the development of antiviral strategies see a recent review by DeClerq (2004).

3.1 Rhinoviruses

Rhinovirus infections may account for up to one third of all "common colds". The virus family consists of more than 100 serotypes, is ubiquitous and can cause repeated episodes of infection throughout an individual's life time. Infection in otherwise healthy persons is unpleasant though usually self-limited. However, certain populations may be predisposed to severe manifestations, including bronchiolitis and pneumonia in infants and exacerbations of pre-existing airway disease in persons with chronic obstructive lung disease, asthma and cystic fibrosis (Hayden 2004).

3.1.1 Pathogenesis

The pathogenesis of a rhinovirus infection was elucidated by many studies of volunteers infected with rhinoviruses (Gwaltney *et al.*, 1966, 1975, 1977, 1978). Infection begins with the deposition of viral particles in the anterior nasal mucosa or in the eye, from where the virus can be transported to the nose via the lacrimal duct. At the mucosal surface, virions attach themselves to cellular receptors. Based on receptor specificity, rhinoviruses can be divided into two groups: the viruses belonging to the "major group" attach themselves to the cellular adhesion molecule-1 (ICAM-1), whereas "minor group" viruses bind to the LDL receptor or LDL-related proteins (Greve *et al.*, 1989; Hofer *et al.*, 1994). The serotype HRV 89 is reported to bind to heparan sulfate proteoglycan (Vlasak *et al.*, 2005). Following internalization and uncoating, the viral plus-strand RNA is translated into one large polyprotein, which is proteolytically processed by two viral proteases – 2A and 3C – into the individual viral proteins. Replication is mediated via a minus-strand RNA intermediate and catalyzed by the viral 3D polymerase. Resulting RNA molecules can then be used for translating viral proteins or are packaged into viral particles in a late phase of the infection. Viral proteinases specifically cleaving factors such as the eukaryotic initiation factor 4G required for cellular cap-dependent translation turn the host cell into a virus producing machine (host cell shutoff) (Etchison *et al.*, 1987). The viral mRNA is translated via a cap-independent mechanism involving direct binding of ribosomal subunits to internal ribosomal entry sites, which facilitates virus production. In addition, this host cell shutoff limits the defense response to the viral infection e.g., the interferon response, as cellular translation is inhibited (Weber *et al.*, 2004). Viral release is mediated via destruction of the host cell. Surprisingly and differently to patients infected with influenza virus, individuals infected with rhinoviruses show no major tissue destruction in nasal biopsies. This observation suggests that the clinical symptoms of rhinovirus infection might not be due to cell destruction mediated by virus replication (cytopathic effect) but are primarily caused by the immune response of the host. Several inflammatory mediators can be found in the nasal secretion of patients with common cold, e.g. interleukin 1, 6 and 8, histamines, leukotriens and kinins. Interleukin 6 and 8 levels correlate with the severity of symptoms. In contrast to the pro-inflammatory reaction, rhinovirus can even down-modulate an appropriate immune responses by inducing immunosuppressive cytokine interleukin 10 as was shown in monocytes (Stockl *et al.*, 1999). However, the immunological host response to rhinovirus infection is far from being fully explored. On average, colds take one week, although 25% of all cases last longer. Viral shedding can be observed up to three weeks, though the risk of transmission diminishes after three days (D'Alessio *et al.*, 1976).

3.1.2   Secondary infections

   With the help of modern PCR techniques, rhinoviruses have been detected more frequently at sites distant from the primary infection. This led to a greater appreciation of the role this pathogen plays in upper and lower respiratory tract disease. Specifically, rhinoviruses are supposed to be associated with otitis media (Pitkaranta *et al.*, 1998, 1999), sinusitis, asthma exacerbations, chronic obstructive pulmonary diseases and lower respiratory infections in elderly, neonates and in immunocompromised individuals (Greenberg 2003). In children, the most common bacterial complication is otitis media, which can occur in up to 20% of all children with viral upper respiratory tract infections. Using PCR techniques, various viruses were detected in middle ear fluid, suggesting a causative involvement of these viruses (Pitkaranta *et al.*, 1997, 1998, 1999, 2002). HRVs were also detected in the lower respiratory tract (Papadopoulos *et al.*, 1999; Mosser *et al.*, 2002, 2005).

3.1.3   Antiviral Agents

   Even though the common cold and its complications are medically important, efforts to find causative treatment have been rather futile. At present, there are no approved antiviral agents for treating HRV infection. Several treatment strategies were evaluated, albeit with limited success so far.

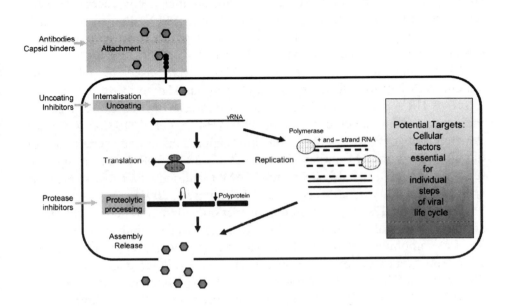

Figure 2. Rhinovirus life cycle. Current appr oaches for viral inhibition ar e indicated.

In the seventies of the last century Isaacs and Lindenmann reported a factor that could induce a virus resistant state that was termed "interferon" (Isaacs *et al.*, 1957a, 1957b; Lindenmann *et al.*, 1957). Interferon can trigger an antiviral response in the host cell based on activation of expression of more than 300 interferon induced genes that have antiviral, antiproliferative or immunomodulatory functions. The best studied interferon induced antiviral proteins are the 2'5' oligo adenylate synthetase, protein kinase R and the MX GTPases. From experiments in knock-out mice it became evident that also additional pathways exist (Zhou *et al.*, 1999). Although the importance of the interferon response to viral infections is clear, their mode of action is still incompletely understood (Weber *et al.*, 2004). Based on the broad induction of a cellular antiviral response inter-feron were used as "general" antiviral medication. In several studies, it proved to be an effective prophylactic but was associated with significant clinical and histo-pathological signs of nasal irritation (Hayden *et al.* , 1986). However, treatment with interferon neither cured experimentally induced nor naturally occurring colds (Farr *et al.*, 1984; Hayden *et al.*, 1984, 1988b; Turner *et al.*, 1986; Sperber *et al.*, 1989). Interferons are an example of molecules targeting a cellular rather than a viral function.

As the majority of HRVs bind to the intercellular adhesion molecule-1 (ICAM-1) as the receptor, strategies using soluble decoy receptors were developed to block attachment (Marlin *et al.*, 1990). Studies with such receptors such as "tremacamra" showed that it reduced the severity of the disease though the effect was modest (Turner *et al.*, 1999).

Monoclonal antibodies against ICAM-1 were used to inhibit the attachment of viruses to their cognate receptors. The antibody delayed the onset of HRV-induced colds in human volunteers but did not reduce the incidence of common cold in a clinical trial (Hayden *et al.*, 1988a). A further development of this receptor blocking approach is the generation of humanized anti-ICAM antibodies (Luo *et al.*, 2003) with higher affinity or the construction of multivalent anti-ICAM-1 antibody Fab fusion proteins with higher avidity (Charles *et al.*, 2003; Fang *et al.*, 2004).

Detailed analysis of virus-cell interaction led to the identification of another group of inhibitory substances, i.e. capsid-binding substances. Several compounds were identified (Rosenwirth *et al.*, 1995; Oren *et al.*, 1996) or developed over the last years, e.g. WIN-compounds such as disoxaril (McKinlay 1985; Andries *et al.*, 1992; Mallamo *et al.*, 1992) or pirodavir. However, in clinical trials, pirodavir was efficient from a pro-phylactic but not from a therapeutic point of view (Hayden *et al.*, 1995). This might be attributed to poor pharmacokinetics. Currently new molecules related to pirodavir are being evaluated (Barnard *et al.*, 2004). Pleconaril is the capsid-binding substance that has been tested most extensively. It is effective against a variety of rhino- and enteroviruses, orally bioavailable and was studied in more than 5,000 patients in clinical trials. If pleconaril treatment is started early, it reduces the duration and severity of the disease (Hayden *et al.*, 2002, 2003a Viropharma 2002). However, the US Food and Drug Administration did not approve pleconaril (Picovir™) in May 2002, mainly because of the observed induction of cytochrome P450 3A enzymes, which metabolize a variety of drugs, e.g. oral contraceptives. A new intranasal formulation of pleconaril is currently developed by Schering-Plough. Intranasal application should enable a more efficient delivery of the drug to the site of infection than the oral formulation, while limiting its systemic exposure and thereby minimizing the risk of drug interactions. Such other capsid-binding substances as BTA 798 are in the stage of early preclinical development.

Enviroxime is a compound inhibiting HRV multiplication. It targets a step in the RNA replication complex that depends on the availability of the viral proteins 3A and 3AB (Heinz *et al.*, 1995; Brown-Augsburger *et al.*,

1999). Similarly to 3C proteinase inhibitors, this class of compounds can be applied in tissue culture several hours after inoculation without loss of activity, as it targets a later step in the viral life cycle. Intolerance to oral dosing, poor pharmacokinetics, undesirable toxic side effects and modest benefits after intranasal administration hampered further development (Phillpotts *et al.*, 1981, 1983; Hayden *et al.*, 1982; Levandowski *et al.*, 1982; Miller *et al.*, 1985). However, derivatives are currently under preclinical evaluation.

Identification and molecular characterization of essential viral proteinases in HRV paved the way to the development of protease inhibitors to combat rhinoviral infection (Libby *et al.*, 1988; Sommergruber *et al.*, 1989).

The two viral proteinases 2A and 3C are both essential for virus multiplication. These enzymes are involved in processing the viral polyprotein into single viral proteins. Mutations in the active sites of these proteinases lead to total inhibition of virus multiplication. Structurally, both proteases are related to trypsin-like serine proteases. Functionally, however, the active site amino acid is replaced by a cystein residue. Sequence comparisons among several HRV serotypes have demonstrated a high degree of homology among amino acid residues involved in the active site of the molecule (Seipelt *et al.*, 1999). As these proteases are not closely related to such other cellular cysteine proteases as caspases, they are an attractive target for drug discovery (Matthews *et al.*, 1994).

Determination of the three-dimensional structure of the 3C protease and subsequent modeling led to the identification and development of specific peptide aldehyde inhibitors (Matthews *et al.*, 1994; Patick *et al.*, 1999; Kaiser *et al.*, 2000). AG-7088 is an irreversible peptidomimetic inhibitor with broad activity against several HRV serotypes, HRV clinical isolates and related picornaviruses *in vitro* (Patick *et al.*, 1999; Kaiser *et al.*, 2000; Binford *et al.*, 2005). Importantly, AG-7088 is also effective when added in cell culture post infection, which might be an advantage compared to capsid-binding substances. Data from *in vitro* and phase I and II studies suggest that AG-7088 is an effective and safe inhibitor of HRV replication (Hsyu *et al.*, 2002; Hayden *et al.*, 2003b). In phase II trials, a nasal spray of AG-7088 (rupintrivir) significantly reduced total cold symptoms and was well tolerated. Prophylaxis lowered the share of subjects with positive viral cultures and viral titers. Surprisingly, it did not reduce the frequency of colds, nor did early treatment decrease the frequency of the disease though it lowered the severity of daily symptoms. As rupintrivir was not able to significantly effect virus reduction and moderate disease severity in subsequent natural infection studies in patients, Agouron Pharmaceuticals terminated the development. Similarly, Eli Lilly designed a compound targeting the human rhinovirus 3C protease (LY 338387) but no recent development was reported. In addition, an orally bioavailable inhibitor of HRV 3C protease was identified. A Phase I trial showed bioavailability and lack of toxicity (Patick _et al._, 2005). Unfortunately, the development of this inhibitor for clinical use is not continued. It should be noted that 3C proteinase inhibitors are effective against a variety of HRV serotypes, as most of the amino acids critical for binding the protease inhibitor are conserved (Binford _et al._, 2005).

In principle, the 2A protease of HRV is also an attractive target for therapeutic intervention. The protease activity is essential for virus multiplication; it is highly specific to its cognate cleavage site. In the life cycle, the protease is directly involved in cleaving the translation factor eIF4G, which shuts off host cell translation. So far, no specific inhibitors for the HRV 2 protease are available. Recently, we could show that the methylated form of a commonly used caspase inhibitor, zVAD.fmk, inhibits HRV2 2A protease _in vitro_ and in cell culture, leading to an inhibition of viral multiplication (Deszcz _et al._, 2004). However, this is rather an undesired side effect discovered in experiments aimed at specific inhibition of caspases. Fluoromethyl ketone derivatized peptides used as inhibitors of caspases are usually employed as methyl-esters to facilitate cell permeation. Inside the cell endogenous esterases cause the demethylation of the inhibitors. We could show that the methylated form specifically inhibits 2A activity, whereas the de-methylated form does not. This is in good agreement with substrate requirements for the 2A protease (Skern _et al._, 1991; Sommergruber _et al._, 1992). However, these experiments clearly show that care must be taken when a "specific" inhibitor is used and exemplify, that inhibition of 2A protease does indeed lead to the block of HRV multiplication.

3.1.4   Resistance Problem

A major problem associated with common antiviral drugs targeting specific viral functions is the emergence of resistant escape mutants due to selective pressure and the high error rate in viral replication. Human rhinoviruses, as other RNA viruses, evolve as complex distributions of mutants termed viral quasispecies (Domingo _et al._, 2005). Inhibition of viral enzymes favours the selective advantage of some viral subpopulations over others (Vignuzzi _et al._, 2005).

Although patients treated with pleconaril show a rapid decrease of viral RNA, they occasionally continued to have positive cultures on study day 6 or later, though on a low level. In a study, viruses with at least tenfold reduced susceptibility to pleconaril were found in 10% of all patients who received this drug (Hayden *et al.*, 2003a). The genotype was evaluated for picornaviruses with reduced susceptibility to pleconaril following exposure to the drug. In all cases examined, the molecular basis of the reduced susceptibility involved amino acid changes in the drug-binding pocket of capsid protein VP1. Similarly, resistant mutants can be found when treatment is done with rupintrivir or other anti-rhinoviral substances (Heinz *et al.*, 1996; Nikolova *et al.*, 2003).

To combat resistance, two main strategies are available: (i) using combinations of antiviral substances with a different mode of action to increase the selective pressure on the virus or (ii) targeting cellular rather than viral proteins.

So far, several combinations of antiviral and anti-inflammatory agents were clinically tested, albeit with limited success (Gwaltney 1992; Sperber *et al.*, 1992; Stone *et al.*, 1992). Targeting cellular functions is difficult based on our limited understanding of the complex host virus interactions. Identification of essential cellular proteins for viral multiplication and a more complete picture regarding the antiviral capabilities of host cells will allow targeting these processes for future therapeutic strategies.

3.1.5   Unspecific agents

Self-medication includes decongestants (alpha-adrenergic agonists) such as pseudoephedrine (Sperber *et al.*, 2000) and phenylpropanolamine or anticholinergic agents such as ipratropium bromide (Winther *et al.*, 2001). These compounds are moderately active in relieving nasal obstructions and rhinorrhea. Nonsteroidal anti-inflammatory drugs such as ibuprofen significantly reduce fever, sneezing and headache associated with colds (Winther *et al.*, 2001). Antitusives, expectorants, mucolytics, antihistamine-decongestant combinations and other over-the-counter medicines are similar to placebo in relieving acute cough associated with upper respiratory tract infections (Smith *et al.*, 1993).

For several years, zinc has been considered a possible treatment of many illnesses (Hill *et al.*, 1987; Prasad *et al.*, 1989, 2002; Doerr *et al.*, 1997), amongst them respiratory tract infections. Zn salts are believed to have immunomodulatory effects and inhibit rhinovirus replication *in vitro*, possibly by inhibition of the viral 3C proteinase (Cordingley *et al.*, 1989) (Korant *et al.*, 1974; Geist *et al.*, 1987). Several trials have evaluated various preparations for treating respiratory illnesses (Farr *et al.*, 1987). Based on these results, one can conclude that benefit is lacking (Jackson *et al.*, 1997, 2000; Belongia *et al.*, 2001). The role of vitamin C in the prevention and treatment of common cold has been a subject of controversy for many years. However, a meta-analysis of studies reveals no major benefit for the public, except for minor subgroups (Douglas *et al.*, 2005). Studies with echinacea pose the problem of its varied composition in different preparations, as it stems from a natural product. Therefore, study results are not comparable and inconclusive. However, when compared to placebo, these supplements did not show a strong benefit for the patients (Grimm *et al.*, 1999; Turner *et al.*, 2000; Schroeder *et al.*, 2004). Perhaps the easiest though systematically not well-examined procedure with a soothing effect is inhalation of heated humified air. This was shown to reduce symptoms but had no influence on viral shedding (Singh 2004).

To a limited extent also prevention of rhinovirus infection was investigated experimentally. As rhinoviruses are sensitive to acid ph values, this property can be used to reduce person-to-person transmission. Disinfecting tissues with various compounds such as citric acid, malic acid and sodium lauryl sulfate have been used under experimental conditions but not in natural settings (Hayden *et al.*, 1985a; Hayden *et al.*, 1985b; Dick *et al.*, 1986).

3.2   Influenza

Amantidine is a ion channel blocker and has been used to treat and prevent influenza A for many years (1970). It blocks the viral M2 ion channel and thus prevents acidification and uncoating of the virus. Influenza B types do not have an M protein but have an NB protein which is resistant to amantadine. In the United States, rimantidine is frequently used because of its lesser side effects. In clinical trials, amantidine reduced the duration of the disease, though side effects and the frequent emergence of resistant viruses could be observed (Dolin *et al.*, 1982; Hayden *et al.*, 1989, 1991; Sweet *et al.*, 1991).

Based on the crystal structure of the influenza virus neuraminidase, zanamivir was developed as a specific inhibitor of this enzyme. Zanamivir is applied as inhalation and has been shown to be efficient and safe for treating influenza virus. Moreover, it is licensed for clinical use. Oseltamivir is an orally available prodrug and is also licensed. When treatment is initiated within 48 h after infection, the duration of the disease is reduced by 1 to 2 days. Since these drugs target influenza virus only, they do not protect patients against infections caused by other viruses involved in viral respiratory infections. This means, within this limited time frame, influenza has to be correctly diagnosed versus rhinovirus and other viral respiratory infections. Further developments include the cyclopentane peramivir (BCX-1812, RWJ-270201), a highly selective inhibitor of influenza A and B virus neuraminidases and a potent inhibitor of influenza A and B virus replication in cell culture (Smee *et al.*, 2001). In clinical trials with patients experimentally infected with influenza A or B viruses, oral treatment with peramivir significantly reduced nasal wash virus titers with no adverse effects. Phase III clinical trials are underway (Sidwell *et al.*, 2002).

Originally, resistant mutants against zanamivir and oseltamivir were rare. However, new results in children show the emergence of resistance in up to 18% (Kiso *et al.*, 2004).

For a more detailed discussion of neuraminidase inhibitors see chapter 1.5.

3.3   Coronaviruses

Human coronavirus infections such as 229 E were not considered serious enough to be controlled by vaccination or antiviral treatment. This view changed rapidly with the emergence of SARS, which has been associated with a newly discovered coronavirus (SARS-CoV) (Drosten *et al.*, 2003; Ksiazek *et al.*, 2003; Peiris *et al.*, 2003). Therapeutic strategies against SARS-CoV are described in chapter 3.1.

3.4   RSV

Research for an efficient treatment and vaccine against RSV was of limited success (Maggon *et al.*, 2004). Palivizumab is a humanized IgG1 monoclonal antibody that binds to the F-protein of RSV. It consists of human (95%) and murine (5%) antibody sequences and thus has a low rate of inducing immunogenic reactions. The drug exhibits neutralizing and fusion-inhibitory activity against RSV. The protection in children is transient and has to be repeated during the RSV season. Each immunization protects a baby for about 30 days, so a new vaccination is needed each month during the respiratory syncytial virus (RSV) season.

4. FUTURE APPROACHES

4.1 Antiviral molecules with unknown function

Our group has found that pyrrolidine dithiocarbamate (PDTC), a commonly used inhibitor of the transcription factor NF-κB, exerts a strong antiviral effect on the multiplication of human rhinovirus and poliovirus (Gaudernak *et al.*, 2002). Other groups recently confirmed these results with such other picornaviruses as the coxsackievirus (Si *et al.*, 2005). Interestingly, the antiviral property is not confined to the picornavirus family, as we and others have shown that PDTC is also active against influenza viruses, having a very different life cycle (Grassauer *et al.*, unpublished, Uchide *et al.*, 2002, 2005). However, other viruses such as tick-borne encephalitis virus are not inhibited by PDTC.

Currently, the striking property of PDTC and functionally related compounds is under investigation. We have shown so far that PDTC does not affect early steps in HRV infection such as virus attachment, internalization and uncoating. However, polyprotein processing and replication are severely impaired when PDTC is present (Krenn *et al.*, 2005).

What is the mechanistic basis of this inhibition?

PDTC is widely used as a specific inhibitor of the transcription factor NF-κB in eukaryotic cells. The molecular mechanism by which PDTC inhibits NF-κB is not yet elucidated and results obtained by different groups are contradictory. Depending on which cellular system is used, both antioxidative properties and prooxidative effects of PDTC have been described. Furthermore, inhibition of the ubiquitin-ligase system was discussed as being involved in the inhibition of NF-κB. An important question is whether the NF-κB inhibitory property of PDTC is related to its antiviral effect. Regarding this aspect, we believe that the inhibition of NF-κB is not important during rhinoviral infection. The transcription factor NF-κB is activated during picornaviral infection. However, as the viral proteinase 2A leads to a fast shut off of host cell transcription during infection, it seems unlikely that NF-κB - mediated genes have a major effect on the replication of HRVs, because the corresponding mRNAs cannot be translated. Certain NF-κB-induced genes may translate via a cap-independent mechanism, though experimental evidence for this theory is lacking. We have also shown that replication of HRV in cells, in which NF-κB is constitutively inactivated by overexpression of the cellular inhibitor IκB, cannot be distinguished from controls with functional NF-κB pathways. Other inhibitors such as aspirin also fail to exert an antiviral effect against HRV. From these data we conclude that the widely known NF-κB inhibitory property of PDTC is not important for the antiviral effect in HRV. However, in influenza virus the situation might be different.

Mechanistically, we have shown that metal ions such as copper and zinc ions are involved in the antiviral effect of PDTC (Krenn *et al.*, 2005). This is an agreement with recent reports obtained for the inhibition of coxsackievirus (van Kuppeveld F.J, pers. communication). It is tempting to speculate what the targets for metal ions in infected cells might be. In the case of rhinovirus, the proteases have been shown to be inhibited by zinc ions (Cordingley *et al.*, 1989). Similar results were obtained for the 3C polymerase *in vitro* (Hung *et al.*, 2002). However, as the metal ion balance in a eukaryotic cell is finely tuned, it is of major importance to analyze effects of metal ions on proteases in a cellular context.

A highly interesting property of these compounds is the fact that they can inhibit both + strand RNA viruses such as HRV and - strand RNA viruses as influenza. Although the mechanistic basis for inhibition might be different in both viruses, it is an interesting question whether these substances can trigger selected pathways of an unspecific antiviral response comparable to the induction of the "antiviral state" by interferon. Elucidation of these pathways would greatly deepen our understanding of virus-host interactions and facilitate therapeutic interventions by chemical molecules.

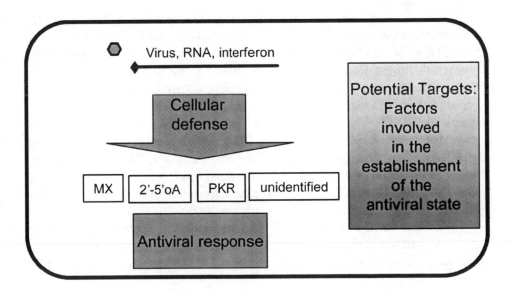

Figure 3. Induction of components of the antiviral defence system as an antiviral strategy

4.2   siRNA

As conventional antiviral strategies face several problems such as ineffectivity, toxicity and resistance, new approaches were taken also in the field of respiratory viruses.

RNA interference (RNAi) or RNA silencing are common designations of specific posttranscriptional gene silencing (PTGS). Small interfering RNAs (siRNAs) down-regulate gene expression by binding to complementary messenger RNAs and either triggering mRNA elimination or arresting mRNA translation into protein (miRNA) (Plasterk 2002; Carrington *et al.*, 2003; Denli *et al.*, 2003; Matzke *et al.*, 2003). This powerful technology has been widely employed to manipulate gene expression in diverse hosts and to identify gene function. RNAi is heavily used in basic research, and RNAi-based drugs are also being developed against human diseases, tumor and metabolic disorders. Consequently, this technique is now employed to inactivate viral genes and thus block viral replication (Carmichael 2002).

siRNA molecules mediating posttranscriptional gene silencing were originally discovered in plants and caenorhabditis elegans. In plants, there is evidence that RNAi has a role in the defense against viruses, as arabidopsis strains defective in posttranscriptional gene silencing are more susceptible to virus infections (Mourrain *et al.*, 2000). In addition, several plant viruses encode proteins that counteract RNAi-mediated silencing (Brigneti *et al.*, 1998; Voinnet 2001).

PTGS is mediated by siRNAs that are produced by type III endoribonuclease dicer. This enzyme digests large double-stranded RNAs into 21-23 nt double-strand RNA duplexes with 2 nt 3' overhangs. Importantly, these RNAs produced by dicer can be mimicked by synthetic RNAs (Elbashir *et al.*, 2001). In a second step, siRNAs are incorporated into a multicomponent nuclease complex termed the "RNA induced silencing complex" (RISC). The antisense-strand of the siRNA duplexes serves as a guide that directs RISC to the cognate RNA, which is subsequently degraded by RNAs. This process is highly efficient, as few RNAi molecules can trigger inactivation of continuously transcribed target genes over a prolonged period of time. Thus, siRNAs are usually more efficient than short antisense RNAs.

Guidelines for the choice of siRNAs are available. Preference should be given to siRNA target sequences not located in regions with heavy secondary structure such as the picornavirus IRES elements. siRNAs can be produced (i) synthetically, (ii) by transcription from DNA templates, (iii) enzymatically by digestion of duplex RNA by cloned dicer or E. coli RNase III (Yang *et al.*, 2002). However, siRNAs have to be transported into the target cells. This is achieved by the use of such synthetic carriers as cationic lipids and polymers or with the help of viral vectors. The transient nature of the siRNA-mediated silencing is not believed to be a significant limitation when targeting diseases caused by rapidly replicating viruses. This is in contrast to a more problematic situation in chronic diseases and infections by such latent or integrating viruses as HIV.

4.2.1   siRNA targeting antiviral genes

Attractive targets of siRNAs are plus-strand RNA viruses, as their genome is used both as an mRNA and as a template for replication. First experiments were carried out in poliovirus. For historic reasons, the latter is frequently used as a paradigm of picornaviruses. Based on their close relationship, results with poliovirus can be translated to human rhinovirus, albeit with caution. By targeting the capsid region and the 3D polymerase, virus titer was reduced by two orders of magnitude in a one step growth when cells were transfected before infection (Gitlin *et al.*, 2002). Virus inhibition occurred without need of interferon or such classical dsRNA-activated effectors as PKR and RNase L. This is in agreement with the view that RNA duplexes shorter than 30 bases do not activate the dsRNA-dependent protein kinase (Semizarov *et al.*, 2003).

However, poliovirus rapidly escapes highly effective siRNAs through unique point mutations within the targeted regions (Gitlin *et al.*, 2005). As picornaviruses exist as a quasispecies, it seems that pre-existing mutants can be selected rapidly. Combinations of siRNAs were more successful (Gitlin *et al.*, 2005). Another important animal virus of the picornavirus family, i.e. the foot and mouth disease virus (FMDV), could be inhibited by siRNAs against VP1 in cell culture and suckling mice (Chen *et al.*, 2004; Kahana *et al.*, 2004; Grubman *et al.*, 2005).

siRNAs have also been used to inhibit severe acute respiratory syndrome (SARS)-associated coronavirus replication. Several siRNAs were evaluated in cell culture against different viral genes (Zhang, R. *et al.*, 2003, 2004b; Qin *et al.*, 2004; Zhao *et al.*, 2005). siRNAs directed against spike sequences and the 3'-UTR can inhibit replication (Bitko *et al.*, 2005). siRNA targeting the leader sequence inhibited the replication of SARS-CoV more strongly than targeting the spike gene (Li *et al.*, 2005). Plasmid-mediated expression of siRNAs targeting the RNA polymerase reduced virus titer, RNA and protein levels (Wang *et al.*, 2004).

The non-segmented genomic and antigenomic RNAs of RSV are a difficult target for siRNAs as they are tightly wrapped in the nucleocapsid protein N, which makes them inaccessible (Barik 2004). However, siRNAs targeting the NS1 gene (siNS1) were successfully used in mice treated intranasally with siNS1 nanoparticles. Decreased virus titers in the lung and decreased inflammation and airway reactivity could be observed. Thus, siNS1 nanoparticles may effectively inhibit RSV infection in humans (Zhang *et al.*, 2005). Also, intranasal delivery of siRNA against RSV P was successful in preventing infection in mice and in reducing the severity of the disease when added after infection (Bitko *et al.*, 2005).

One of the advantages is that siRNA approaches do not rely on the immune system. Essentially, all genes of influenza were targeted by siRNAs, except for hemagglutinin and neuraminidase (Ge *et al.*, 2003). Inhibition of virus multiplication could be obtained *in vitro*, in chicken embryos and in mice. For example, cationic carrier siRNA complexes were administered i.v. or intranasally or via DNA vectors from which siRNAs could be transcribed (Ge *et al.*, 2004a; 2004b). These siRNAs can prevent and treat influenza virus infection in mice. Other groups used siNP nucleocapsid or siPA components to protect mice from lethal challenge with a variety of influenza A viruses including potential pandemic H5 and H7 subtypes (Tompkins *et al.*, 2004). These data clearly show the potential of siRNAs as prophylactic and therapy for influenza virus infections.

4.2.2   siRNA targeting cellular genes essential for virus multiplication

An attractive idea is not to pursue the classical approaches of targeting the viral RNA but to target cellular genes which are identified as being essential for virus multiplication. So far, these strategies were not yet applied to classic respiratory viruses; however there are successful reports in related virus groups. Examples for such methods are targeting the translation factor La, the polypyrimidine-binding protein or eukaryotic initiation factor 2B gamma. As these factors are involved in the cap-independent translation of hepatitis C virus, siRNA targeting of these molecules blocks viral replication in cell culture (Zhang, J. *et al.*, 2004a). Similar results were obtained with picornaviruses (polio- and encephalomyocaeditisvirus) when cells were depleted of the polypyrimidine-tract-binding protein (Florez *et al.*, 2005). These studies demonstrate that viral infections can be combated by targeting cellular co-factors required for replication. These approaches might lead to a more sustainable inhibition of viral replication as the emergence of resistant viruses is very unlikely.

The main limit of the siRNA technology is the sequence diversity of respiratory viruses. Resistant mutants were obtained in several viruses such as poliovirus. However, siRNA is a powerful technology that can be used to target viral genes or cellular co-factors and thus efficiently inhibit virus-induced pathogenesis.

5. CONCLUSIONS

Inhibiting respiratory viruses is a challenging task and many attempts have failed to show expected results. The diverse nature of the viruses involved is one reason for the problematic situation. Viruses have developed a number of strategies to evade common inhibition approaches. However, we believe that aiming at cellular rather than viral targets might improve the accessibility of viral diseases. It remains to be shown in further clinical trials whether new molecules are suitable for viral inhibition. New technologies such as siRNA techniques have yielded surprising results in the first years of development and offer a significant potential for treating viruses that have bothered people for many centuries.

ACKNOWLEDGEMENTS

This chapter is dedicated to the memory of our head and mentor Ernst Kuechler who was substantially involved in the research presented in this chapter. Sadly, Ernst Kuechler passed away in March 2005. We would like to thank Sylvia Trnka for editorial help and acknowledge support by grant P16642-B11 from the Austrian Science Foundation.

REFERENCES

Amantidine and influenza in general practice, 1970, *Lancet* **2** (7676): 761-2.
Andries, K., B. Dewindt, J. Snoeks, R. Willebrords, R., van Eemeren, K., Stokbroekx, R. and Janssen, P.A., 1992, *In vitro* activity of pirodavir (R 77975), a substituted phenoxy-pyridazinamine with broad-spectrum antipicornaviral activity. *Antimicrob. Agents Chemother.* **36**: 100-7.

Arruda, E., Pitkaranta, A., Witek, Jr., T.J., Doyle, C.A. and Hayden, F.G., 1997, Frequency and natural history of rhinovirus infections in adults during autumn. *J. Clin. Microbiol.* **35**: 2864-8.

Barik, S., 2004, Control of nonsegmented negative-strand RNA virus replication by siRNA. *Virus Res.* **102**: 27-35.

Barnard, D.L., Hubbard, V.D., Smee, D.F., Sidwell, R.W., Watson, K.G., Tucker, S.P. and Reece, P.A., 2004, *In vitro* activity of expanded-spectrum pyridazinyl oxime ethers related to pirodavir: novel capsid-binding inhibitors with potent antipicornavirus activity. *Antimicrob. Agents Chemother.* **48**: 1766-72.

Belongia, E.A., Berg; R. and Liu, K., 2001, A randomized trial of zinc nasal spray for the treatment of upper respiratory illness in adults. *Am. J. Med.* **111**: 103-8.

Bertino, J. S., 2002, Cost burden of viral respiratory infections: issues for formulary decision makers. *Am. J. Med.* **112 (Suppl 6A)**: 42S-49S.

Binford, S. L:, F. Maldonado, M. A. Brothers, P. T. Weady, L. S. Zalman, J. W. Meador, 3rd, D. A. Matthews, D.A. and Patick, A.K., 2005, Conservation of amino acids in human rhinovirus 3C protease correlates with broad-spectrum antiviral activity of rupintrivir, a novel human rhinovirus 3C protease inhibitor. *Antimicrob. Agents Chemother.* **49**: 619-26.

Bitko, V., A. Musiyenko, O. Shulyayeva and Barik, S., 2005, Inhibition of respiratory viruses by nasally administered siRNA. *Nat. Med.* **11**: 50-5.

Brigneti, G., O. Voinnet, W. X. Li, L. H. Ji, S. W. Ding and Baulcombe, D. C., 1998, Viral pathogenicity determinants are suppressors of transgene silencing in Nicotiana benthamiana. *EMBO J.* **17**: 6739-46.

Brown-Augsburger, P., L. M. Vance, S. K. Malcolm, H. Hsiung, D. P. Smith and Heinz, B. A., 1999, Evidence that enviroxime targets multiple components of the rhinovirus 14 replication complex. *Arch.Virol.* **144**: 1569-85.

Carmichael, G. G., 2002, Medicine: silencing viruses with RNA. *Nature* **418**: 379-80.

Carrington, J. C. and Ambros, V., 2003, Role of microRNAs in plant and animal development. *Science* **301**: 336-8.

Charles, C. H., G. X. Luo, L. A. Kohlstaedt, I. G. Morantte, E. Gorfain, L. Cao, J. H. Williams and Fang, F., 2003, Prevention of human rhinovirus infection by multivalent fab molecules directed against ICAM-1. *Antimicrob. Agents Chemother.* **47**: 1503-8.

Chen, W., W. Yan, Q. Du, L. Fei, M. Liu, Z. Ni, Z. Sheng and Zheng, Z., 2004, RNA interference targeting VP1 inhibits foot-and-mouth disease virus replication in BHK-21 cells and suckling mice. *J. Virol.* **78**: 6900-7.

Cordingley, M. G., R. B. Register, P. L. Callahan, V. M. Garsky and Colonno, R. J., 1989, Cleavage of small peptides *in vitro* by human rhinovirus 14 3C protease expressed in Escherichia coli. *J. Virol.* **63**: 5037-45.

D'Alessio, D. J., J. A. Peterson, C. R. Dick and Dick, E. C., 1976, Transmission of experimental rhinovirus colds in volunteer married couples. *J. Infect. Dis.* **133**: 28-36.

De Clercq, E., 2004, Antivirals and antiviral strategies. *Nat. Rev. Microbiol.* **2**: 704-20.

Denli, A. M. and Hannon, G. J., 2003, RNAi: an ever-growing puzzle. *Trends Biochem. Sci.* **28**: 196-201.

Deszcz, L., J. Seipelt, E. Vassilieva, A. Roetzer and Kuechler, E., 2004, Antiviral activity of caspase inhibitors: effect on picornaviral 2A proteinase. *FEBS Lett.* **560**: 51-5.

Dick, E. C., S. U. Hossain, K. A. Mink, C. K. Meschievitz, S. B. Schultz, W. J. Raynor and Inhorn, S. L., 1986, Interruption of transmission of rhinovirus colds among human volunteers using virucidal paper handkerchiefs. *J. Infect. Dis.* **153**: 352-6.

Doerr, T. D., A. S. Prasad, S. C. Marks, F. W. Beck, F. H. Shamsa, H. S. Penny and Mathog, R. H., 1997, Zinc deficiency in head and neck cancer patients. *J. Am. Coll. Nutr.* **16**: 418-22.

Dolin, R., R. C. Reichman, H. P. Madore, R. Maynard, P. N. Linton and Webber-Jones, J., 1982, A controlled trial of amantadine and rimantadine in the prophylaxis of influenza A infection. *N. Engl. J. Med.* **307**: 580-4.

Domingo, E., N. Pariente, A. Airaksinen, C. Gonzalez-Lopez, S. Sierra, M. Herrera, A. Grande-Perez, P. R. Lowenstein, S. C. Manrubia, E. Lazaro and Escarmis, C., 2005, Foot-and-mouth disease virus evolution: exploring pathways towards virus extinction. *Curr. Top. Microbiol. Immunol.* **288**: 149-73.

Douglas, R. M. and Hemila, H. 2005, Vitamin C for preventing and treating the common cold. *PLoS Med* **2**: e168.

Drosten, C., S. Gunther, W. Preiser, S. van der Werf, H. R. Brodt, S. Becker, H. Rabenau, M. Panning, L. Kolesnikova, R. A. Fouchier, A. Berger, A. M. Burguiere, J. Cinatl, M. Eickmann, N. Escriou, K. Grywna, S. Kramme, J. C. Manuguerra, S. Muller, V. Rickerts, M. Sturmer, S. Vieth, H. D. Klenk, A. D. Osterhaus, H. Schmitz and Doerr, H. W., 2003, Identification of a novel coronavirus in patients with severe acute respiratory syndrome. *N. Engl. J. Med.* **348**: 1967-76.

Elbashir, S. M., J. Harborth, W. Lendeckel, A. Yalcin, K. Weber and Tuschl, T., 2001, Duplexes of 21-nucleotide RNAs mediate RNA interference in cultured mammalian cells. *Nature* **411**: 494-8.

Etchison, D. and Etchison, J. R., 1987, Monoclonal antibody-aided characterization of cellular p220 in uninfected and poliovirus-infected HeLa cells: subcellular distribution and identification of conformers. *J. Virol.* **61**: 2702-10.

Fang, F. and Yu, M., 2004, Viral receptor blockage by multivalent recombinant antibody fusion proteins: inhibiting human rhinovirus (HRV) infection with CFY196. *J. Antimicrob. Chemother.* **53**: 23-5.

Farr, B. M., E. M. Conner, R. F. Betts, J. Oleske, A. Minnefor and Gwaltney Jr., J. M., 1987, Two randomized controlled trials of zinc gluconate lozenge therapy of experimentally induced rhinovirus colds. *Antimicrob. Agents Chemother.* **31**: 1183-7.

Farr, B. M., J. M. Gwaltney Jr., K. F. Adams and Hayden, F. G., 1984, Intranasal interferon-alpha 2 for prevention of natural rhinovirus colds. *Antimicrob. Agents Chemother.* **26**: 31-4.

Florez, P. M., O. M. Sessions, E. J. Wagner, M. Gromeier and Garcia-Blanco, M. A., 2005, The polypyrimidine tract binding protein is required for efficient picornavirus gene expression and propagation. *J. Virol.* **79**: 6172-9.

Gaudernak, E., J. Seipelt, A. Triendl, A. Grassauer and Kuechler, E., 2002, Antiviral effects of pyrrolidine dithiocarbamate on human rhinoviruses. *J. Virol.* **76**: 6004-6015.

Ge, Q., H. N. Eisen and Chen, J., 2004a, Use of siRNAs to prevent and treat influenza virus infection. *Virus Res.* **102**: 37-42.

Ge, Q., L. Filip, A. Bai, T. Nguyen, H. N. Eisen and Chen, J., 2004b, Inhibition of influenza virus production in virus-infected mice by RNA interference. *Proc. Natl. Acad. Sci. U.S.A.* **101**: 8676-81.

Ge, Q., M. T. McManus, T. Nguyen, C. H. Shen, P. A. Sharp, H. N. Eisen and Chen, J., 2003, RNA interference of influenza virus production by directly targeting mRNA for degradation and indirectly inhibiting all viral RNA transcription. *Proc. Natl. Acad. Sci. U.S.A.* **100**: 2718-23.

Geist, F. C., J. A. Bateman and Hayden, F. G., 1987, *In vitro* activity of zinc salts against human rhinoviruses. *Antimicrob. Agents Chemother.* **31**: 622-4.

Gitlin, L., S. Karelsky and Andino, R., 2002, Short interfering RNA confers intracellular antiviral immunity in human cells. *Nature* **418**: 430-4.

Gitlin, L., J. K. Stone and Andino, R., 2005, Poliovirus escape from RNA interference: short interfering RNA-target recognition and implications for therapeutic approaches. *J. Virol.* **79**: 1027-35.

Gonzales, R., D. C. Malone, J. H. Maselli and Sande, M. A., 2001, Excessive antibiotic use for acute respiratory infections in the United States. *Clin. Infect. Dis.* **33**: 757-62.

Greenberg, S. B., 2003, Respiratory consequences of rhinovirus infection. *Arch. Intern. Med.* **163**: 278-84.

Greve, J. M., G. Davis, A. M. Meyer, C. P. Forte, S. C. Yost, C. W. Marlor, M. E. Kamarck and McClelland, A., 1989, The major human rhinovirus receptor is ICAM-1. *Cell* **56**: 839-847.

Grimm, W. and Muller, H. H., 1999, A randomized controlled trial of the effect of fluid extract of Echinacea purpurea on the incidence and severity of colds and respiratory infections. *Am. J. Med.* **106**: 138-43.

Grubman, M. J. and de los Santos, T., 2005, Rapid control of foot-and-mouth disease outbreaks: is RNAi a possible solution? *Trends Immunol.* **26**: 65-8.

Gwaltney, J. M., 1975, Rhinoviruses. *Yale J. Biol. Med.* **48**: 17-45.

Gwaltney, J. M. and Hendley, J. O., 1977, Rhinovirus transmission: one if by air, two if by hand. *Trans. Am. Clin. Climatol. Assoc.* **89**: 194-200.

Gwaltney Jr, J. M., 1992, Combined antiviral and antimediator treatment of rhinovirus colds. *J. Infect. Dis.* **166**: 776-82.

Gwaltney Jr., J. M., J. O. Hendley, G. Simon and Jordan Jr., W. S., 1966, Rhinovirus infections in an industrial population. I. The occurrence of illness. *N. Engl. J. Med.* **275**: 1261-8.

Gwaltney Jr, J. M., P. B. Moskalski and Hendley, J. O., 1978, Hand-to-hand transmission of rhinovirus colds. *Ann. Intern. Med.* **88**: 463-7.

Hayden, F. G., 2004, Rhinovirus and the lower respiratory tract. *Rev. Med. Virol.* **14**: 17-31.

Hayden, F. G., J. K. Albrecht, D. L. Kaiser and Gwaltney Jr., J. M., 1986, Prevention of natural colds by contact prophylaxis with intranasal alpha 2-interferon. *N. Engl. J. Med.* **314**: 71-5.

Hayden, F. G., R. B. Belshe, R. D. Clover, A. J. Hay, M. G. Oakes and Soo, W., 1989, Emergence and apparent transmission of rimantadine-resistant influenza A virus in families. *N. Engl. J. Med.* **321**: 1696-702.

Hayden, F. G., T. Coats, K. Kim, H. A. Hassman, M. M. Blatter, B. Zhang and Liu, S., 2002, Oral pleconaril treatment of picornavirus-associated viral respiratory illness in adults: efficacy and tolerability in phase II clinical trials. *Antivir. Ther.* **7**: 53-65.

Hayden, F. G. and Gwaltney Jr., J. M., 1982, Prophylactic activity of intranasal enviroxime against experimentally induced rhinovirus type 39 infection. *Antimicrob. Agents Chemother.* **21**: 892-7.

Hayden, F. G. and Gwaltney Jr., J. M., 1984, Intranasal interferon-alpha 2 treatment of experimental rhinoviral colds. *J. Infect. Dis.* **150**: 174-80.

Hayden, F. G., J. M. Gwaltney, Jr. and Colonno, R. J., 1988a, Modification of experimental rhinovirus colds by receptor blockade. *Antiviral Res.* **9**: 233-47.

Hayden, F. G., D. T. Herrington, T. L. Coats, K. Kim, E. C. Cooper, S. A. Villano, S. Liu, S. Hudson, D. C. Pevear, M. Collett and McKinlay, M., 2003a, Efficacy and safety of oral pleconaril for treatment of colds due to picornaviruses in adults: results of 2 double-blind, randomized, placebo-controlled trials. *Clin. Infect. Dis.* **36**: 1523-32.

Hayden, F. G., G. J. Hipskind, D. H. Woerner, G. F. Eisen, M. Janssens, P. A. Janssen and Andries, K. 1995, Intranasal pirodavir (R77,975) treatment of rhinovirus colds. *Antimicrob. Agents Chemother.* **39**: 290-4.

Hayden, F. G., D. J. Innes, Jr., S. E. Mills and Levine, P. A., 1988b, Intranasal tolerance and histopathologic effects of a novel synthetic interferon, rIFN-alpha Con1. *Antiviral Res.* **10**: 225-34.

Hayden, F. G., S. J. Sperber, R. B. Belshe, R. D. Clover, A. J. Hay and Pyke, S., 1991, Recovery of drug-resistant influenza A virus during therapeutic use of rimantadine. *Antimicrob. Agents Chemother.* **35**: 1741-7.

Hayden, F. G., R. B. Turner, J. M. Gwaltney, K. Chi-Burris, M. Gersten, P. Hsyu, A. K. Patick, G. J. Smith, 3rd and Zalman, L. S., 2003b, Phase II, randomized, double-blind, placebo-controlled studies of ruprintrivir nasal spray 2-percent suspension for prevention and treatment of experimentally induced rhinovirus colds in healthy volunteers. *Antimicrob. Agents Chemother.* **47**: 3907-16.

Hayden, G. F., J. M. Gwaltney Jr., D. F. Thacker and Hendley, J. O., 1985a, Rhinovirus inactivation by nasal tissues treated with virucide. *Antiviral Res.* **5**: 103-9.

Hayden, G. F., J. O. Hendley and Gwaltney Jr., J. M., 1985b, The effect of placebo and virucidal paper handkerchiefs on viral contamination of the hand and transmission of experimental rhinoviral infection. *J. Infect. Dis.* **152**: 403-7.

Heikkinen, T. and Jarvinen, A., 2003, The common cold. *Lancet* **361**: 51-9.

Heinz, B. A. and Vance, L. M., 1995, The antiviral compound enviroxime targets the 3A coding region of rhinovirus and poliovirus. *J. Virol.* **69**: 4189-4197.

Heinz, B. A. and Vance, L. M., 1996, Sequence determinants of 3A-mediated resistance to enviroxime in rhinoviruses and enteroviruses. *J. Virol.* **70**: 4854-7.

Hill, G. M., G. J. Brewer, A. S. Prasad, C. R. Hydrick and Hartmann, D. E. 1987, Treatment of Wilson's disease with zinc. I. Oral zinc therapy regimens. *Hepatology* **7**: 522-8.

Hofer, F., M. Gruenberger, H. Kowalski, H. Machat, M. Huettinger, E. Kuechler and Blaas, D., 1994, Members of the low density lipoprotein receptor family mediate cell entry of a minor-Group common cold virus. *Proc. Nat. Acad. Sci. U.S.A.* **91**: 1839-1842.

Hsyu, P. H., Y. K. Pithavala, M. Gersten, C. A. Penning and Kerr, B. M. 2002, Pharmacokinetics and safety of an antirhinoviral agent, ruprintrivir, in healthy volunteers. *Antimicrob. Agents Chemother.* **46**: 392-7.

Hung, M., C. S. Gibbs and Tsiang, M. 2002, Biochemical characterization of rhinovirus RNA-dependent RNA polymerase. *Antiviral Res.* **56**: 99-114.

Isaacs, A. and Lindenmann, J., 1957a, Virus interference. I. The interferon. *Proc. R. Soc. Lond. B. Biol. Sci.* **147**: 258-67.

Isaacs, A., J. Lindenmann and Valentine, R. C., 1957b, Virus interference. II. Some properties of interferon. *Proc. R. Soc. Lond. B. Biol. Sci.* **147**(927): 268-73.

Jackson, J. L., E. Lesho and Peterson, C., 2000, Zinc and the common cold: a meta-analysis revisited. *J. Nutr.* **130** (5S Suppl): 1512S-5S.

Jackson, J. L., C. Peterson and Lesho, E., 1997, A meta-analysis of zinc salts lozenges and the common cold. *Arch. Intern. Med.* **157**: 2373-6.

Kahana, R., L. Kuznetzova, A. Rogel, M. Shemesh, D. Hai, H. Yadin and Stram, Y., 2004, Inhibition of foot-and-mouth disease virus replication by small interfering RNA. *J. Gen. Virol.* **85**: 3213-7.

Kaiser, L., C. E. Crump and Hayden, F. G., 2000, *In vitro* activity of pleconaril and AG7088 against selected serotypes and clinical isolates of human rhinoviruses. *Antiviral Res.* **47**: 215-20.

Kim, H. W., R. G. Wyatt, B. F. Fernie, C. D. Brandt, J. O. Arrobio, B. C. Jeffries and Parrott, R. H., 1983, Respiratory syncytial virus detection by immunofluorescence in nasal secretions with monoclonal antibodies against selected surface and internal proteins. *J. Clin. Microbiol.* **18**: 1399-404.

Kiso, M., K. Mitamura, Y. Sakai-Tagawa, K. Shiraishi, C. Kawakami, K. Kimura, F. G. Hayden, N. Sugaya and Kawaoka, Y., 2004, Resistant influenza A viruses in children treated with oseltamivir: descriptive study. *Lancet* **364**: 759-65.

Korant, B. D., J. C. Kauer and Butterworth, B. E., 1974, Zinc ions inhibit replication of rhinoviruses. *Nature* **248**: 588-90.

Krenn, B., B. Holzer, A. Triendl, E. Gaudernak and Seipelt, J., 2005, The inhibition of polyprotein processing and RNA replication of Human Rhinovirus by PDTC depends on metal ions. *J. Virol*: in press.

Ksiazek, T. G., D. Erdman, C. S. Goldsmith, S. R. Zaki, T. Peret, S. Emery, S. Tong, C. Urbani, J. A. Comer, W. Lim, P. E. Rollin, S. F. Dowell, A. E. Ling, C. D. Humphrey, W. J. Shieh, J. Guarner, C. D. Paddock, P. Rota, B. Fields, J. DeRisi, J. Y. Yang, N. Cox, J. M. Hughes, J. W. LeDuc, W. J. Bellini and Anderson, L. J., 2003, A novel coronavirus associated with severe acute respiratory syndrome. *N. Engl. J. Med.* **348**: 1953-66.

Larson, H. E., S. E. Reed and Tyrrell, D. A., 1980, Isolation of rhinoviruses and coronaviruses from 38 colds in adults. *J. Med. Virol.* **5**: 221-9.

Levandowski, R. A., C. T. Pachucki, M. Rubenis and Jackson, G. G., 1982, Topical enviroxime against rhinovirus infection. *Antimicrob. Agents Chemother.* **22**: 1004-7.

Li, T., Y. Zhang, L. Fu, C. Yu, X. Li, Y. Li, X. Zhang, Z. Rong, Y. Wang, H. Ning, R. Liang, W. Chen, L. A. Babiuk and Chang, Z., 2005, siRNA targeting the leader sequence of SARS-CoV inhibits virus replication. *Gene Ther.* **12**: 751-61.

Libby, R. T., D. Cosman, M. K. Cooney, J. E. Merriam, C. J. March and Hopp, T. P., 1988, Human Rhinovirus 3C Protease: Cloning and Expression of an Active Form in Escherichia coli. *Biochemistry* **27**: 6262-6268.

Lindenmann, J., D. C. Burke and Isaacs, A., 1957, Studies on the production, mode of action and properties of interferon. *Br. J. Exp. Pathol.* **38**: 551-62.

Louie, J. K., J. K. Hacker, R. Gonzales, J. Mark, J. H. Maselli, S. Yagi and Drew, W. L., 2005, Characterization of viral agents causing acute respiratory infection in a San Francisco University Medical Center Clinic during the influenza season. *Clin. Infect. Dis.* **41**: 822-8.

Luo, G. X., L. A. Kohlstaedt, C. H. Charles, E. Gorfain, I. Morantte, J. H. Williams and Fang, F., 2003, Humanization of an anti-ICAM-1 antibody with over 50-fold affinity and functional improvement. *J. Immunol. Methods* **275**: 31-40.

Maggon, K. and Barik, S., 2004, New drugs and treatment for respiratory syncytial virus. *Rev. Med. Virol.* **14**: 149-68.

Makela, M. J., T. Puhakka, O. Ruuskanen, M. Leinonen, P. Saikku, M. Kimpimaki, S. Blomqvist, T. Hyypia and Arstila, P. 1998, Viruses and bacteria in the etiology of the common cold. *J. Clin. Microbiol.* **36**: 539-42.

Mallamo, J. P., G. D. Diana, D. C. Pevear, F. J. Dutko, M. S. Chapman, K. H. Kim, I. Minor, M. Oliveira and Rossmann, M. G., 1992, Conformationally restricted analogues of disoxaril: a comparison of the activity against human rhinovirus types 14 and 1A. *J. Med. Chem.* **35**: 4690-5.

Marlin, S. D., D. E. Staunton, T. A. Springer, C. Stratowa, W. Sommergruber and Merluzzi, V. J., 1990, A soluble form of intercellular adhesion molecule-1 inhibits rhinovirus infection. *Nature* **344**: 70-2.

Matthews, D. A., W. W. Smith, R. A. Ferre, B. Condon, G. Budahazi, W. Sisson, J. E. Villafranca, C. A. Janson, H. E. Mcelroy, C. L. Gribskov and Worland, S., 1994, Structure of human rhinovirus 3C protease reveals a trypsin-like polypeptide fold, RNA-Binding site, and means for cleaving precursor polyprotein. *Cell* **77**: 761-771.

Matzke, M. and Matzke, A. J., 2003, RNAi extends its reach. *Science* **301**: 1060-1.

McKinlay, M. A., 1985, WIN 51711, a new systematically active broad-spectrum antipicornavirus agent. *J. Antimicrob. Chemother.* **16**: 284-6.

Miller, F. D., A. S. Monto, D. C. DeLong, A. Exelby, E. R. Bryan and Srivastava, S., 1985, Controlled trial of enviroxime against natural rhinovirus infections in a community. *Antimicrob. Agents Chemother.* **27**: 102-6.

Monto, A. S., 2002a, Epidemiology of viral respiratory infections. *Am. J. Med.* **112 Suppl 6A**: 4S-12S.

Monto, A. S., 2002b, The seasonality of rhinovirus infections and its implications for clinical recognition. *Clin. Ther.* **24**: 1987-97.

Monto, A. S. and Sullivan, K. M., 1993, Acute respiratory illness in the community. Frequency of illness and the agents involved. *Epidemiol. Infect.* **110**: 145-60.

Mosser, A. G., R. Brockman-Schneider, S. Amineva, L. Burchell, J. B. Sedgwick, W. W. Busse and Gern, J. E., 2002, Similar frequency of rhinovirus-infectible cells in upper and lower airway epithelium. *J. Infect. Dis.* **185**: 734-43.

Mosser, A. G., R. Vrtis, L. Burchell, W. M. Lee, C. R. Dick, E. Weisshaar, D. Bock, C. A. Swenson, R. D. Cornwell, K. C. Meyer, N. N. Jarjour, W. W. Busse and Gern, J. E., 2005, Quantitative and qualitative analysis of rhinovirus infection in bronchial tissues. *Am. J. Respir. Crit. Care Med.* **171**: 645-51.

Mourrain, P., C. Beclin, T. Elmayan, F. Feuerbach, C. Godon, J. B. Morel, D. Jouette, A. M. Lacombe, S. Nikic, N. Picault, K. Remoue, M. Sanial, T. A. Vo and Vaucheret, H., 2000, Arabidopsis SGS2 and SGS3 genes are required for posttranscriptional gene silencing and natural virus resistance. *Cell* **101**: 533-42.

Nicholson, K. G., J. Kent, V. Hammersley and Cancio, E., 1997, Acute viral infections of upper respiratory tract in elderly people living in the community: comparative, prospective, population based study of disease burden. *BMJ* **315**: 1060-4.

Nikkari, S., P. Halonen, I. Kharitonenkov, M. Kivivirta, M. Khristova, M. Waris and Kendal, A., 1989, One-incubation time-resolved fluoroimmunoassay based on monoclonal antibodies in detection of influenza A and B viruses directly in clinical specimens. *J. Virol. Methods* **23**: 29-40.

Nikolova, I. and Galabov, A. S., 2003, Development of resistance to disoxaril in Coxsackie B1 virus-infected newborn mice. *Antiviral Res.* **60**: 35-40.

Oren, D. A., A. Zhang, H. Nesvadba, B. Rosenwirth and Arnold, E., 1996, Synthesis and activity of piperazine-containing antirhinoviral agents and crystal structure of SDZ 880-061 bound to human rhinovirus 14. *J. Mol. Biol.* **259**: 120-34.

Papadopoulos, N. G., G. Sanderson, J. Hunter and Johnston, S. L., 1999, Rhinoviruses replicate effectively at lower airway temperatures. *J. Med. Virol.* **58**: 100-4.

Patick, A. K., S. L. Binford, M. A. Brothers, R. L. Jackson, C. E. Ford, M. D. Diem, F. Maldonado, P. S. Dragovich, R. Zhou, T. J. Prins, S. A. Fuhrman, J. W. Meador, L. S. Zalman, D. A. Matthews and Worland, S. T., 1999, In vitro antiviral activity of AG7088, a potent inhibitor of human rhinovirus 3C protease. *Antimicrob. Agents Chemother.* **43**: 2444-50.

Patick, A. K., M. A. Brothers, F. Maldonado, S. Binford, O. Maldonado, S. Fuhrman, A. Petersen, G. J. Smith, 3rd, L. S. Zalman, L. A. Burns-Naas and Tran, J. Q., 2005, In vitro antiviral activity and single-dose pharmacokinetics in humans of a novel, orally bioavailable inhibitor of human rhinovirus 3C protease. *Antimicrob. Agents Chemother* **49**: 2267-75.

Peiris, J. S., S. T. Lai, L. L. Poon, Y. Guan, L. Y. Yam, W. Lim, J. Nicholls, W. K. Yee, W. W. Yan, M. T. Cheung, V. C. Cheng, K. H. Chan, D. N. Tsang, R. W. Yung, T. K. Ng and Yuen, K. Y., 2003, Coronavirus as a possible cause of severe acute respiratory syndrome. *Lancet* **361**: 1319-25.

Phillpotts, R. J., R. W. Jones, D. C. Delong, S. E. Reed, J. Wallace and Tyrrell, D. A., 1981, The activity of enviroxime against rhinovirus infection in man. *Lancet* **1**: 1342-4.

Phillpotts, R. J., J. Wallace, D. A. Tyrrell and Tagart, V. B., 1983, Therapeutic activity of enviroxime against rhinovirus infection in volunteers. *Antimicrob. Agents Chemother.* **23**: 671-5.

Pitkaranta, A., E. Arruda, H. Malmberg and Hayden, F. G., 1997, Detection of rhinovirus in sinus brushings of patients with acute community-acquired sinusitis by reverse transcription-PCR. *J. Clin. Microbiol.* **35**: 1791-3.

Pitkaranta, A. and Hayden, F. G., 1999, Respiratory viruses and acute otitis media. *N. Engl. J. Med.* **340**: 2001-2.

Pitkaranta, A., H. Rihkanen, O. Carpen and Vaheri, A., 2002, Rhinovirus RNA in children with longstanding otitis media with effusion. *Int. J. Pediatr. Otorhinolaryngol.* **66**: 247-50.

Pitkaranta, A., A. Virolainen, J. Jero, E. Arruda and Hayden, F. G., 1998, Detection of rhinovirus, respiratory syncytial virus, and coronavirus infections in acute otitis media by reverse transcriptase polymerase chain reaction. *Pediatrics* **102**: 291-5.

Plasterk, R. H., 2002, RNA silencing: the genome's immune system. *Science* **296**: 1263-5.

Prasad, A. S., J. Kaplan, G. J. Brewer and Dardenne, M., 1989, Immunological effects of zinc deficiency in sickle cell anemia. *Prog. Clin. Biol. Res.* **319**: 629-47.

Prasad, A. S. and Kucuk, O., 2002, Zinc in cancer prevention. *Cancer Metastasis Rev.* **21**: 291-5.

Qin, Z. L., P. Zhao, X. L. Zhang, J. G. Yu, M. M. Cao, L. J. Zhao, J. Luan and Qi Z. T., 2004, Silencing of SARS-CoV spike gene by small interfering RNA in HEK 293T cells. *Biochem. Biophys. Res. Commun.* **324**: 1186-93.

Rosenwirth, B., D. A. Oren, E. Arnold, Z. L. Kis and Eggers, H. J., 1995, SDZ 35-682, a new picornavirus capsid-binding agent with potent antiviral activity. *Antiviral Res.* **26**: 65-82.

Schroeder, K. and Fahey, T., 2004, Over-the-counter medications for acute cough in children and adults in ambulatory settings. *Cochrane Database Syst. Rev.* (4): CD001831.

Seipelt, J., A. Guarne, E. Bergmann, M. James, W. Sommergruber, I. Fita and Skern, T., 1999, The structures of picornaviral proteinases. *Virus Res.* **62**: 159-168.

Semizarov, D., L. Frost, A. Sarthy, P. Kroeger, D. N. Halbert and Fesik, S. W., 2003, Specificity of short interfering RNA determined through gene expression signatures. *Proc. Natl. Acad. Sci. U.S.A.* **100**: 6347-52.

Si, X., B. M. McManus, J. Zhang, J. Yuan, C. Cheung, M. Esfandiarei, A. Suarez, A. Morgan and Luo, H., 2005, Pyrrolidine dithiocarbamate reduces coxsackievirus B3 replication through inhibition of the ubiquitin-proteasome pathway. *J. Virol.* **79**: 8014-23.

Sidwell, R. W. and Smee, D. F., 2002, Peramivir, BCX-1812 (RWJ-270201): potential new therapy for influenza. *Expert Opin. Investig. Drugs* **11**: 859-69.

Singh, M., 2004, Heated, humidified air for the common cold. *Cochrane Database Syst. Rev.*(2): CD001728.

Skern, T., W. Sommergruber, H. Auer, P. Volkmann, M. Zorn, H. D. Liebig, F. Fessl, D. Blaas and Kuechler, E., 1991, Substrate requirements of a human rhinoviral 2A proteinase. *Virology* **181**: 46-54.

Smee, D. F., J. H. Huffman, A. C. Morrison, D. L. Barnard and Sidwell, R. W., 2001, Cyclopentane neuraminidase inhibitors with potent *in vitro* anti-influenza virus activities. *Antimicrob. Agents Chemother.* **45**: 743-8.

Smith, M. B. H. and Feldman, W., 1993, Over-the-Counter cold medications - a critical review of clinical trials between 1950 and 1991. *JAMA* **269**: 2258-2263.

Sommergruber, W., H. Ahorn, A. Zöphel, I. Maurer-Fogy, F. Fessl, G. Schnorrenberg, H. D. Liebig, D. Blaas, E. Kuechler and Skern, T., 1992, Cleavage specifity on synthetic peptide substrates of human rhinovirus 2 proteinase 2A. *J. Biol. Chem.* **267**: 22639-22644.

Sommergruber, W., M. Zorn, D. Blaas, F. Fessl, P. Volkmann, F. I. Maurer, P. Pallai, V. Merluzzi, M. Matteo, T. Skern and Kuechler, E., 1989, Polypeptide 2A of human rhinovirus type 2: identification as a protease and characterization by mutational analysis. *Virology* 169: 68-77.

Sperber, S. J., J. O. Hendley, F. G. Hayden, D. K. Riker, J. V. Sorrentino and Gwaltney Jr., J. M., 1992, Effects of naproxen on experimental rhinovirus colds. A randomized, double-blind, controlled trial. *Ann. Intern. Med.* 117: 37-41.

Sperber, S. J., P. A. Levine, J. V. Sorrentino, D. K. Riker and Hayden, F. G., 1989, Ineffectiveness of recombinant interferon-beta serine nasal drops for prophylaxis of natural colds. *J. Infect. Dis.* 160: 700-5.

Sperber, S. J., R. B. Turner, J. V. Sorrentino, R. R. O'Connor, J. Rogers and Gwaltney Jr., J. M., 2000, Effectiveness of pseudoephedrine plus acetaminophen for treatment of symptoms attributed to the paranasal sinuses associated with the common cold. *Arch. Fam. Med.* 9: 979-85.

Steininger, C., S. W. Aberle and Popow-Kraupp, T., 2001, Early detection of acute rhinovirus infections by a rapid reverse transcription-PCR assay. *J. Clin. Microbiol.* 39: 129-33.

Steininger, C., M. Kundi, S. W. Aberle, J. H. Aberle and Popow-Kraupp, T., 2002, Effectiveness of reverse transcription-PCR, virus isolation, and enzyme-linked immuno-sorbent assay for diagnosis of influenza A virus infection in different age groups. *J. Clin. Microbiol.* 40: 2051-6.

Stockl, J., H. Vetr, O. Majdic, G. Zlabinger, E. Kuechler and Knapp, W., 1999, Human major group rhinoviruses downmodulate the accessory function of monocytes by inducing IL-10. *J. Clin. Invest.* 104: 957-65.

Stone, A. A., D. H. Bovbjerg, J. M. Neale, A. Napoli, H. Valdimarsdottir, D. Cox, F. G. Hayden and Gwaltney Jr., J. M., 1992, Development of common cold symptoms following experimental rhinovirus infection is related to prior stressful life events. *Behav. Med.* 18: 115-20.

Sweet, C., F. G. Hayden, K. J. Jakeman, S. Grambas and Hay, A. J., 1991, Virulence of rimantadine-resistant human influenza A (H3N2) viruses in ferrets. *J. Infect. Dis.* 164: 969-72.

Tompkins, S. M., C. Y. Lo, T. M. Tumpey and Epstein, S. L., 2004, Protection against lethal influenza virus challenge by RNA interference *in vivo. Proc. Natl. Acad. Sci. U.S.A.* 101: 8682-6.

Turner, R. B., A. Felton, K. Kosak, D. K. Kelsey and Meschievitz, C. K., 1986, Prevention of experimental coronavirus colds with intranasal alpha-2b interferon. *J. Infect. Dis.* 154: 443-7.

Turner, R. B., D. K. Riker and Gangemi, J. D., 2000, Ineffectiveness of echinacea for prevention of experimental rhinovirus colds. *Antimicrob. Agents Chemother.* 44: 1708-9.

Turner, R. B., M. T. Wecker, G. Pohl, T. J. Witek, E. McNally, R. St George, B. Winther and Hayden, F. G., 1999, Efficacy of tremacamra, a soluble intercellular adhesion molecule 1, for experimental rhinovirus infection: a randomized clinical trial. *JAMA* 281: 1797-804.

Uchide, N., K. Ohyama, T. Bessho and Toyoda, H., 2005, Inhibition of influenza-virus-induced apoptosis in chorion cells of human fetal membranes by nordihydroguaiaretic Acid. *Intervirology* 48: 336-40.

Uchide, N., K. Ohyama, T. Bessho, B. Yuan and Yamakawa, T., 2002, Effect of antioxidants on apoptosis induced by influenza virus infection: inhibition of viral gene replication and transcription with pyrrolidine dithiocarbamate. *Antiviral Res.* 56: 207-17.

Vabret, A., T. Mourez, S. Gouarin, J. Petitjean and Freymuth, F., 2003, An outbreak of coronavirus OC43 respiratory infection in Normandy, France. *Clin. Infect. Dis.* 36: 985-9.

Vignuzzi, M., J. K. Stone and Andino, R., 2005, Ribavirin and lethal mutagenesis of poliovirus: molecular mechanisms, resistance and biological implications. *Virus Res.* **107**: 173-81.

Viropharma, 2002. Antiviral Drugs Advisory Committee Briefing Document (PICOVIR, PLECONARIL) NDA 21-245.

Vlasak, M., I. Goesler and Blaas, D., 2005, Human rhinovirus type 89 variants use heparan sulfate proteoglycan for cell attachment. *J. Virol.* **79**: 5963-70.

Voinnet, O., 2001, RNA silencing as a plant immune system against viruses. *Trends Genet.* **17**: 449-59.

Wang, Z., L. Ren, X. Zhao, T. Hung, A. Meng, J. Wang and Chen, Y. G., 2004, Inhibition of severe acute respiratory syndrome virus replication by small interfering RNAs in mammalian cells. *J. Virol.* **78**: 7523-7.

Waris, M., P. Halonen, T. Ziegler, S. Nikkari and Obert, G., 1988, Time-resolved fluoroimmunoassay compared with virus isolation for rapid detection of respiratory syncytial virus in nasopharyngeal aspirates. *J. Clin. Microbiol.* **26**: 2581-5.

Weber, F., G. Kochs and Haller, O., 2004, Inverse interference: how viruses fight the interferon system. *Viral Immunol.* **17**: 498-515.

Winther, B. and Mygind, N., 2001, The therapeutic effectiveness of ibuprofen on the symptoms of naturally acquired common colds. *Am. J. Rhinol.* **15**: 239-42.

Yang, D., F. Buchholz, Z. Huang, A. Goga, C. Y. Chen, F. M. Brodsky and Bishop, J. M., 2002, Short RNA duplexes produced by hydrolysis with Escherichia coli RNase III mediate effective RNA interference in mammalian cells. *Proc. Natl. Acad. Sci. U.S.A.* **99**: 9942-7.

Zhang, J., O. Yamada, T. Sakamoto, H. Yoshida, T. Iwai, Y. Matsushita, H. Shimamura, H. Araki and Shimotohno, K., 2004a, Down-regulation of viral replication by adenoviral-mediated expression of siRNA against cellular cofactors for hepatitis C virus. *Virology* **320**: 135-43.

Zhang, R., Z. Guo, J. Lu, J. Meng, C. Zhou, X. Zhan, B. Huang, X. Yu, M. Huang, X. Pan, W. Ling, X. Chen, Z. Wan, H. Zheng, X. Yan, Y. Wang, Y. Ran, X. Liu, J. Ma, C. Wang and Zhang, B., 2003, Inhibiting severe acute respiratory syndrome-associated coronavirus by small interfering RNA. *Chin. Med. J.* **116**: 1262-4.

Zhang, W., H. Yang, X. Kong, S. Mohapatra, H. San Juan-Vergara, G. Hellermann, S. Behera, R. Singam, R. F. Lockey and Mohapatra, S. S., 2005, Inhibition of respiratory syncytial virus infection with intranasal siRNA nanoparticles targeting the viral NS1 gene. *Nat. Med.* **11**: 56-62.

Zhang, Y., T. Li, L. Fu, C. Yu, Y. Li, X. Xu, Y. Wang, H. Ning, S. Zhang, W. Chen, L. A. Babiuk and Chang, Z., 2004b, Silencing SARS-CoV Spike protein expression in cultured cells by RNA interference. *FEBS Lett.* **560**: 141-6.

Zhao, P., Z. L. Qin, J. S. Ke, Y. Lu, M. Liu, W. Pan, L. J. Zhao, J. Cao and Qi, Z. T., 2005, Small interfering RNA inhibits SARS-CoV nucleocapsid gene expression in cultured cells and mouse muscles. *FEBS Lett.* **579**: 2404-10.

Zhou, A., J. M. Paranjape, S. D. Der, B. R. Williams and Silverman, R. H., 1999, Interferon action in triply deficient mice reveals the existence of alternative antiviral pathways. *Virology* **258**: 435-40.

Chapter 1.5

ANTI-VIRAL APPROACHES AGAINST INFLUENZA VIRUSES

S. PLESCHKA[1], S. LUDWIG[2], T. WOLFF[3] and O. PLANZ[4]

[1]Institute of Medical Virology, Justus-Liebig-University, Frankfurter Str. 107, D-35392 Gießen, Germany; [2]Institute of Molecular Virology (IMV), ZMBE, Westfälische-Wilhelms-Universität, Von-Esmarch Str. 56, D-48149 Münster, Germany; [3]Robert-Koch-Institute (RKI), Nordufer 20, D-13353 Berlin, Germany; [4]Friedrich-Loeffler-Institute (FLI), Paul-Ehrlich Str. 28, D-72076 Tübingen, Germany

Abstract: Influenza viruses are a continuous and severe global threat to mankind and many animal species. The re-emerging disease gives rise to thousands of deaths and enormous economic losses each year. The devastating results of the recent outbreaks of avian influenza in Europe and south East Asia demonstrate this immanent danger. The major problem in fighting the flu is the high genetic variability of the virus. This results in the rapid formation of variants that escape the acquired immunity against previous virus strains or confer resistance to anti-viral agents. Despite successful vaccination against circulating strains causing annual epidemics the number of admitted measures to fight acute infection in risk patients is limited and quarantine is of limited help as the virus transmitted before onset of symptoms. This poses an even greater challenge when a completely new virus should hit the human population without preexisting immunity and start a pandemic. Therefore the development of effective drugs against viral functions or essential cellular activities supporting viral replication is of outmost importance today.

1.   INTRODUCTION

Influenza is a highly contagious, acute respiratory disease with global significance that affects all age groups and can occur repeatedly in any individual. The etiological agent of the disease, influenza virus is responsible for an average between three and five Mill. cases of severe influenza leading

115

E. Bogner and A. Holzenburg (eds.), New Concepts of Antiviral Therapy, 115–167.

to about 250,000-500,000 mortalities annually in the industrialized world according to WHO estimations. Compared to otherwise healthy persons, death rates in patients of risk groups (s. 2.4) are 50-100 fold higher in patients with cardiovascular or pulmonary disease as compared to healthy individuals. Annual health cost, costs, e.g. due to work absenteeism (also related to parental care of infected children) or costs related to death, increased disabilities etc. can be higher than 40 Mil. € in European countries. Furthermore. For a pandemic outbreak the Centers for Disease Control (CDC) estimates that in the USA that 85% of all death will be caused by 15% of the population which are at high-risk. This will result in a financial burden of up to 166.5 billion US$ not including the commercial impact. The death rate would be up to 207,000 accompanied by up to 734,000 hospitalizations, 18-42 million outpatient visits and 20-47 million additional illnesses (Cox *et al.*, 2004; Wilschut and McElhaney, 2005). This clearly would overrun the capacity of current supply and management of vaccines available.

Since waterfowl represents the natural reservoir for the virus (Lamb and Krug, 2001; Webster, 1999; Wilschut and McElhaney, 2005; Wright and Webster, 2001) and many other animal species can be infected, the eradication of the virus is impossible and a constant reemergence of the disease will continue to occur. Epidemics appear almost annually and are due to an antigenic change of the viral surface glycoproteins (Fig. 1). Furthermore, highly pathogenic strains of influenza-A-virus have emerged unpredictably but repeatedly in recent history as pandemics like the "Spanish-Flu" that caused the death of 20-40 millions people worldwide (Taubenberger *et al.*, 2000; Webster, 1999). Since these pandemic virus strains usually possess different antigenic characteristics, current vaccines will be ineffective once such a virus emerges. Regarding the vast possibilities for such a strain to "travel" around the world (Hufnagel *et al.*, 2004) it becomes evident that effective countermeasures are required for the fight against these foes. In recent outbreaks of avian viruses that infected humans (1997, 1999, 2003/4/5) (Chen *et al.*, 2004; Hatta and Kawaoka, 2002; Li *et al.*, 2004) from a total of 108 confirmed cases 54 people died (07/2005) (World Health Organization, 2005). Fortunately, until now these particular viruses have not acquire the ability to spread in the human population. However, any novel virus strain emerging in the future may have such a capability (Webby and Webster, 2003).

Here, we give an overview of current and new anti-influenza strategies, such as immunization methods and drugs against the virus. Since every virus depends on its host cell, cellular functions essential for viral replication may also be suitable targets for anti-viral therapy. In this respect intra-cellular signaling cascades activated by the virus, in particular MAPK pathways, have recently come into focus (Ludwig *et al.*, 2003; Ludwig *et al.*, 1999).

2. THE VIRUS AND ITS REPLICATION

2.1 Viral components

Influenza viruses belong to the order of the *Orthomyxoviridae*. They possess a segmented, single stranded RNA-genome with negative orientation. They are divided into three types, A, B and C based on genetic and antigenic differences. Among the three types influenza-A-viruses are clinically the most important pathogens since they have been responsible for severe epidemics in humans and domestic animals in the past. Thus the focus of this chapter will be on type-A influenza viruses. A detailed description of the viral proteins and the replication cycle of influenza-A-viruses can be found elsewhere (Lamb and Krug, 2001; Wright and Webster, 2001). Therefore we will only give a brief overview on these topics without referring to individual references.

The influenza-A-virus particle is composed of a lipid envelope derived from the host cell and of 9-10 structural virus proteins (Figure 1 and Table 1). The components of the RNA-dependent RNA-polymerase complex (RDRP), PB2, PB1 and PA are associated with the ribonucleoprotein complex (RNP) and are encoded by the vRNA segments 1-3. The PB1 segment of many, but not all, influenza-A-virus strains also contains a +1-reading frame encoding the recently discovered PB1-F2 protein (Chen *et al.*, 2001). The viral surface glycoproteins hemagglutinin (HA) and neuraminidase (NA) are expressed from vRNA segments 4 and 6, respectively. The nucleoprotein (NP) is encoded by segment 5 and associates with the vRNA segments. It is the major component of the RNPs. The two smallest vRNA segments each code for two proteins. The matrix protein (M1) is colinear translated from the mRNA of segment 7 and forms an inner layer within the virion. A spliced version of the mRNA gives rise to a third viral transmembrane component, the M2 protein, which functions as a pH-dependent ion channel. Employing a similar coding strategy segment 8 harbors the sequence information for the nonstructural NS1 protein and the nuclear export protein NS2/NEP. NS2/NEP is a minor component of the virion and is found associated with the M1 protein.

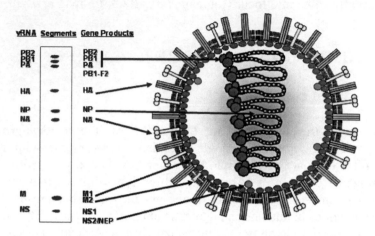

Figure 1. THE INFLUENZA-A-VIRUS PARTICLE Schematic representation of the spherical influenza-A-virus particle that has a diameter of about 100nm. The eight viral RNA segments were separated by urea-polyacrylamide gel electrophoresis and visualized by silver staining (left). The corresponding gene products and their presumed location in the virus particle are indicated (right). NS1 is not a structural part of the mature virion. For details see text.

Table 1 summarizes details of the genome segments, the encoded viral proteins and their according function.

Table 1. Influenza-A-virus Genome (strain A/PR/8/34)

Segment	vRNA	Protein	AA	Function(s)
1	2341	PB2	759	Subunit of viral RNA polymerase; cap-binding
2	2341	PB1	757	Catalytic subunit of viral RNA polymerase
		PB1-F2	87-91	Pro-apoptotic activity
3	2233	PA	716	Subunit of viral RNA polymerase
4	1778	HA	566	Surface glycoprotein; receptor binding, membrane fusion
5	1565	NP	498	Nucleoprotein; encapsidation of viral genomic and anti-genomic RNA
6	1413	NA	454	Neuraminidase
7	1027	M1	252	Matrixprotein
		M2	97	Ion channel activity, protecting HA conformation
8	890	NS1	230	Regulation of viral RDRP activity Interferon antagonist; Enhancer of viral mRNA translation; inhibition of (i) pre-mRNA splicing, (ii) cellular mRNA-polyadenylation, (iii) PKR activity,
		NEP	121	Nuclear export factor

2.2   The influenza replication cycle

The viral replication cycle is initiated by binding of the HA to sialic-acid (neuraminic acid) containing cellular receptors and subsequent endocytosis of the virus (Figure 2) (For references: (Lamb and Krug, 2001; Wright and Webster, 2001)). The active HA molecule consists of two subunits (HA_1 / HA_2) derived from the uncleaved precursor HA_0, which becomes proteolytically processed after release of the virion by extra-cellular proteases. This cleavage is absolutely essential for HA-function and cell infection. Virus disassembly occurs in the acidic environment of late endosomal vesicles and involves two crucial events. First, the conformation of the HA is changed to a low-pH form, which results in exposure of a fusion active protein sequence within the HA_2. This fusion peptide is thought to contact the endosomal membrane and to initiate fusion with the viral envelope. Second, the low pH in the endosomes activates the viral M2 ion channel protein resulting in a flow of protons into the interior of the virion. Acidification facilitates dissociation of the RNPs from the M1 protein. The RNPs are subsequently released into the cytoplasm and rapidly imported into the nucleus through the nuclear pore complexes. The viral genomic segments are replicated and transcribed by the viral RDRP associated with the RNPs in the nucleus of the infected cell. The vRNA is directly transcribed to mRNA and, in addition, serves as a template for a complementary copy (cRNA), which itself is the template for new vRNA. In the late phase of infection newly synthesized viral RNPs are exported to the cytoplasm. NS1 protein functions as a regulatory factor in the virus infected cell. The NA, the M2 and the precursor HA (HA_0) proteins follow the exocytotic transport pathway from the rER via the Golgi complex and the trans Golgi network. The mature HA and NA glycoproteins and the nonglycosylated M2 are finally integrated into the plasma membrane as trimers (HA) or tetramers (NA, M2), respectively.

M1 assembles in patches at the cell membrane. It is thought to associate with the glycoproteins (HA and NA) and to recruit the RNPs to the plasma membrane in the late phase of the replication cycle. Finally the viral RNPs become enveloped by a cellular bilipid layer carrying the HA, NA and M2 proteins resulting in budding of new virus particles from the apical cell surface.

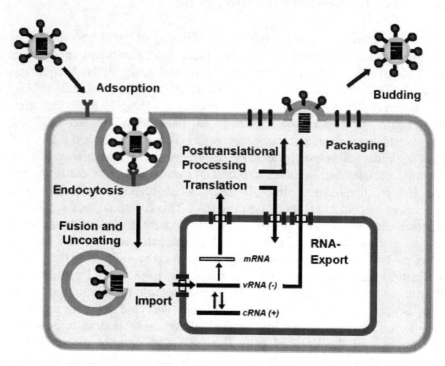

Figure 2. THE REPLICATION CYCLE OF INFLUENZA VIRUSES The virion attaches to the cellular receptor determinant. The receptor-bound particle enters the cell via endocytosis. After fusion of the viral and the endosomal membrane the viral genome is released into the cytoplasm. The RNPs are transported into the nucleus where replication and transcription of the viral RNA segments occurs. The mRNAs are exported into the cytoplasm and are translated into viral proteins. The viral glycoproteins enter the exocytotic transport pathway to the cell surface. Replicative viral proteins enter the nucleus to amplify the viral genome. In the late stage of the infection cycle newly synthesized RNPs are exported from the nucleus and are assembled into progeny virions that bud from the cell surface.

2.3   Antigenic drift and antigenic shift

The polymerase complex of influenza viruses does not possess a proof reading activity, thus numerous mutations accumulate in the viral genome during ongoing replication (Lamb and Krug, 2001) leading to changes in all proteins. This includes conformational alteration of HA- and NA-epitopes against which neutralizing antibodies are generated. Influenza-A-viruses are categorized by antigenic differences of the HA- and NA-proteins. The high mutation rate combined with the high replication rate results in a multitude of new variants produced in each replication cycle, thus allowing the virus to rapidly adapt to changes in the environment. This results in an escape of the existing immunity and in resistance to drugs acting directly against viral functions. Gradual changes of the antigenic properties that make existing vaccines less or non effective are described as *antigenic drift* and demand for new compositions of the yearly vaccines.

Due to the nature of their segmented genome influenza virus can independently recombine segments upon the infection of a cell with two different viruses. This is described as genetic reassortment. Today 16 HA-subtypes (H1-H16) and 9 NA-subtypes (N1-N9) are known, which can mix and lead to new antigenic properties. (Lamb and Krug, 2001; Webster *et al.*, 1992; Wright and Webster, 2001). Not all combination will ultimately be advantageous, but can lead to the generation of a virus that combines the ability to replicate in humans with novel antigenic properties (*antigenic shift*). This has happened at least three times in the last century resulting in the pandemics of 1918 ("Spanish Flu"), 1957 ("Asian Flu") and 1968 ("Hong Kong Flu") that caused up to 40 million death. Therefore, the question is not "if" but "when" will such a pandemic occur again (Horimoto and Kawaoka, 2001; Webby and Webster, 2003; Webster, 1997b). A vaccine against such "new" viruses can not be generated in advance and as vaccine production would need significantly more time than it takes for a pandemic virus to spread around the world (Hufnagel *et al.*, 2004), alternative weapons in the fight against these enemies are urgently needed. Besides pandemic variants that can occur when human and avian influenza virus reassort in porcine hosts (regarded as "mixing vessels") (Webster, 1997a; Webster *et al.*, 1995; Webster *et al.*, 1997), avian influenza virus strains have directly infected humans, as happened in Hong Kong in 1997 (Claas *et al.*, 1998; de Jong *et al.*, 1997; Subbarao *et al.*, 1998) and recently (2004/2005) (Fouchier *et al.*, 2004; Koopmans *et al.*, 2004) during vast outbreaks of avian influenza. These viruses show an extremely high virulence in humans with case fatality rates up to 70%.

2.4   The disease

The virus that normally causes a respiratory disease (for references: (Wilschut and McElhaney, 2005)) is transmitted by aerosol droplets and contaminated hands and can already be shed before onset of symptoms (Cox *et al.*, 2004). Therefore, high population density and dry air leading to reduced protection of respiratory epithelium by the mucus are conditions that promote transmission of the virus.

The infection with influenza viruses is normally limited to the respiratory tract. Here proteases released by Clara cells in the epithelium are present that activate the HA to allow further infections (s. 2.2) (for review (Ludwig *et al.*,

1999)). Innate immunity as well as the adaptive immune system will normally restrict virus propagation. Therefore population groups, that have a less protective immune system, such as young children up to two years and older persons over 65 as well as immunocompromised or chronically diseased persons are especially of risk. The replication of the virus leads to the lysis of the epithelial cells and enhanced mucus production causing running nose and cough. Furthermore, inflammation and oedema at the replication site are due to cytokines released. This can lead to fever and related symptoms. Bacterial super-infections of the harmed tissue can further complicate the situation. Normally onset of systemic (fever, myaglia, headaches, severe malaise) and respiratory (coughing, sore throat, rhinitis) symptoms occur after about two days incubation period and can last for about seven to ten days. Coughing and overall weakness can persist for up to two weeks. If the virus spreads from the bronchiolar tract to the aveolars, viral pneumonia and interstitial pneumonitis with mononuclear and haemorrhage infiltration and finally lysis of the inter-aveolar space is possible (Wilschut and McElhaney, 2005).

This scenario is a likely picture in case of infection with a pandemic influenza strain, where the individual has not had a prior exposure to the virus and the innate immunity reaction can lead to a strong immun-pathogenesis. High virus replication will induce secretion of large quantities of cytokines by the infected epithelia and will stimulate inflammatory processes. Together with the destruction of the epithelia this results in an influx of fluid into the aveolars leading to hypoxia and acute respiratory distress syndrome, that may cause the death within a short period of time (1-2 days after onset). This scenario might also be caused by additional viral factors enhancing pathogenicity. Such factors that are yet not well defined probably have contributed to the devastating outcome of the "Spanish Flu" (Wilschut and McElhaney, 2005).

Accurate and rapid diagnosis of the disease is essential for an effective treatment, especially with anti-viral substances, as virus replication and therefore illness progresses rapidly. Samples can be tested serologically, by cell culture or RT-PCR for strain typing and should be done within four days after onset of symptoms (Wilschut and McElhaney, 2005).

3.   TARGETING THE VIRUS

There are two main methods of influenza prophylaxis: the use of anti-viral drugs and vaccines. Several drugs are available for influenza prophylaxis functioning either as M2-ion channel inhibitors (amantadine and rimantadine) or as inhibitors of the NA (zanamivir and oseltamivir). Despite these anti-viral drugs, which are a useful adjunct to influenza vaccines, vaccination itself remains the cornerstone of prophylaxis. Vaccination induces a good degree of protection and is in general well tolerated by the recipient. Nevertheless, while resistant virus variants can emerge after anti-viral drug treatment the disadvantage of vaccination is that immunization needs to be refreshed almost every year, since the vaccine must reformulated to take account of the changing virus.

3.1   Host immunity: old and new vaccine approaches

In the immune response to influenza infection both the humoral and cell mediated immunity are involved. From the side of the humoral immune system, both the mucosal and the systemic immunity contribute to resistance to influenza infection. The cellular immune response is involved in recovery from influenza virus infection by eliminating virus-infected cells and by providing help for antibody production (Cox *et al.*, 2004; Woodland *et al.*, 2001). Consequently, the humoral immune response is the primary target of vaccination. After influenza virus infection antibodies directed against all major viral proteins can be detected in humans and the level of serum antibodies correlate with resistance to disease (Couch, 2003; Couch and Kasel, 1983; Coulter *et al.*, 2003; Nichol *et al.*, 1998; Potter and Oxford, 1979). Only antibodies specific for the surface glycoproteins HA and NA are associated with resistance to infection. In contrast, antibodies to the conserved internal antigens M and NP are not protective (de Jong *et al.*, 2003; Tamura and Kurata, 2004). The mucosal tissues of the respiratory system are the main portal entry of influenza virus and consequently the mucosal immune system functions as the first line of defense against infection apart from innate immunity (see paragraph 4). Antibodies secreted locally in the upper respiratory tract are a major factor in resistance to natural infection. Secretory immunoglobulin A (SIgA) and to some extent IgM are the major neutralizing antibodies directed against the entering virus. Furthermore, these antibodies can function intra-cellular to inhibit influenza replication. IgA and IgM are involved in protection of the upper respiratory tract while serum IgG acts in protection of the lower respiratory tract (Cox *et al.*, 2004). An anti-HA antibody response (haemagglutination-inhibition (HI) titre ≥ 40) can be detected in approximately 80% of subjects after natural influenza virus infection and correlates with protection against the flu. Plasma cells producing all three major Ig classes are present in the peripheral blood in normal subjects (Cox and Subbarao, 1999; LaForce *et al.*, 1994).

The immune response induced by infection protects against reinfection with the same virus or an antigenically similar viral strain.

Cell mediated immunity plays a role in recovery from influenza virus infection and may also prevent flu-associated complications, but it does not seem to contribute significantly in preventing infection. Influenza specific cellular T cells have been detected in the blood and the lower respiratory tract secretions of infected subjects (Cox *et al.*, 2004). Influenza virus-specific cytotoxic T-lymphocytes (CTL) regognize both external and internal proteins of virus on infected cells. In humans a major component of this response is directed toward the NP- and M1-protein. Even though influenza virus specific CTL's are not able to protect against the infection, these cells are important for the clearance of the virus. Futhermore, cytolysis of influenza virus-infected cells can be mediated by influenza virus-specific antibodies and complement (Cox *et al.*, 2004; McMichael *et al.*, 1983; McMichael *et al.*, 1986; Townsend *et al.*, 1989). CD4+ T cells function as helper cells for antibody production. Moreover, it is suggested, that CD4 cells might act as direct effectors in protection against influenza virus-infection (Brown *et al.*, 2004).

3.1.1 Inactivated influenza vaccines

Inactivated vaccines (IV) are availeble for about 60 years. Because of the antigenic drift observed in influenza HA- and NA-glycoproteins these vaccines need to be matched with the randomly mutating molecular structure of the new occurring "drift" strain. Besides these vaccines there are various new approaches for influenza vaccines in promising developmental stages. These new stratagies include vaccines with immunomodulators, virosomes and DNA-vaccines.

IVs vaccines are administered world wide each year with millions of doses. These vaccines have good safety and tolerance profiles, with very low number of adverse reactions reported. These reactions are tenderness and redness that arise locally at the injection site and are more frequent in healthy (<50%) than in elderly recipients (25%) (Cox *et al.*, 2004). IVs are produced by propagation of the virus in embryonated chicken eggs. The currently used bacterial endotoxin-free trivalent IVs (TIV) are formulated with 15µg HA each from a current influenza virus A/H1N1, A/H3N2, and a B-virus strain. The seed strain is prepared by co-infecting the allantoic sac of the chicken embryo with a laboratory-adapted high-growth phenotype of H1N1 (A/PR/8/34) and the epidemic strain. This results in viral replication and genetic re-assortment leading to high growth reassortants. Thereafter the new hybrid viruses are screened for the absence of genes encoding PR/8 or PR/8-like surface glycoproteins. The selected seed strain containing HA- and NA-components of the epidemic strain is mass propagated in chicken eggs to obtain sufficient quantities of vaccine virus. The allantoic fluid is harvested, and the virus is concentrated and highly purified by zonal centrifugation. As a next step the virus is inactivated. Depending on the nature of inactivation the vaccine is used as whole inactivated vaccine after treatment with formalydehyde or β-propiolactone or as split vaccine (chemically disrupted by ethyl either or SDS). Furthermore, the vaccine is used as subunit vaccines (purified surface glycoproteins). Even though influenza vaccines have excellent tolerant profiles, since propagation in chicken eggs may lead to contamination of the vaccine with trace amounts of residual egg proteins, they should not be administered to persons who have anaphylactic hyper-sensitivity to eggs. Whole inactivated influenza vaccine is more immunogenic than split vaccine or subunit vaccine, but is also associated with more frequent side reactions. Consequently, split or subunit are given to children younger than age 9 and two half doses are recommended given at least 1 month apart for naïve persons to develop protective immunity (Bridges *et al.*, 2003). Protection after vaccination against influenza virus infection is dependent on the antigenic match between the vaccine strains and circulating the influenza virus strain. Moreover, protection is also dependent the age and the previous exposition to influenza of the vaccine recipient. If IVs have a good antigenic match they are 60-90% effective in the prevention of morbidity and mortality among healthy adults (Beyer *et al.*, 2002). In elderly people the effect of protection is reduced to 50-70% because of decreased immune function. Since the immune system is naïve in young children, they also show a reduction in protection against influenza vaccination (Nichol *et al.*, 1998).

Vaccines with immunopotentiators

Immunosuppressed individuals, elderly people and subjects with underlying chronic diseases are at increased risk for influenza and related complications. For these people conventional influenza vaccines provide only limited protection. In order to enhance the immune reaction after influenza vaccination, several adjuvants (Latin verb: adjuvare - to help) that function as immunopotentiators have been evaluated.

The liposomal influenza vaccine (INFLUSOME-VAC) consists of liposomes containing the viral surface proteins HA- and NA-derived from various influenza strains and IL-2 or granulocyte-macrophage colony-stimulating factor (GM-CSF), as an adjuvant (Babai *et al.*, 2001). In clinical trails with either young adults or elderly vaccination of INFLUSOME-VAC appeared to be both safe and more immunogenic than the currently used vaccine (Ben-Yehuda *et al.*, 2003a; Ben-Yehuda *et al.*, 2003b). Furthermore adjuvant emulsions combined with subunit influenza antigens are in use, such as the "oil in water"-emulsion containing squalene, MF59, (FLUAD).

This commercially available product was tested in clinical trials in comparison with non-adjuvanted conventional vaccines. Again in elderly individuals the addition of the MF59-adjuvant to subunit influenza vaccines enhances significantly the immune response without causing clinically important changes in the safety profile of the influenza vaccine (Podda, 2001). Other adjuvants that increase immunoreactivity after influenza vaccination are immune stimulating complexes (iscoms). They are 30-40 nm cage-like structures, which consist of glycoside molecules of the adjuvant Quil A, cholesterol and phospholipids in which the antigen can be integrated. (Osterhaus and Rimmelzwaan, 1998). In animal models, even in the presence of pre-existing antibodies they function as a potent adjuvant system by inducing cellular and humoral immune responses. (Coulter et al., 2003; Rimmelzwaan et al., 2001; Windon et al., 2001).

As mentioned, GM-CSF has a potential role as a vaccine adjuvant. It may enhance the response to vaccination in immunosuppressed individuals. GM-CSF stimulates maturation of hematopoietic progenitor cells, induces class II major histocompatibility complex antigen expression on the surface of macrophages, and enhances dendritic cell migration and maturation (Jones et al., 1994). Nevertheless, in various clinical trails with immunosuppressed individuals and cancer patents it was shown, that it is unlikely that GM-CSF improves the immune response (Ramanathan et al., 2002).

Influenza vaccine production in mammalian cells

For production of influenza vaccines in large-scale cell culture systems several continuous cell lines have been tested for the production of influenza vaccines (Kistner et al., 1998; Pau et al., 2001; Seo et al., 2001; Youil et al., 2004). Production of influenza vaccine in mammalian cell lines has some advantages but also has disadvantages compared to production in chicken egg. (Tree et al., 2001; Youil et al., 2004). Process controllability, scalability and supply of substrates are much easier in cell culture systems. Furthermore, cell culture production reduces the risk of microbial contamination. In contrast, the greatest disadvantage of cell culture based influenza vaccine is the relative low viral yield. On the other hand and a major disadvantage of production in chicken eggs is their supply and possible bacterial contaminations. Additionally the lethality of H5N1 influenza virus to chicken embryos (s. 3.1.3). At the present (2005) two cell line derived vaccines have been licensed in Europe (Kemble and Greenberg, 2003). Estimated time for production of such vaccines is about 6 months. The power of this time gaining approach to generate a great variety of specific influenza-vaccines under controlled safety conditions is achieved by the direct use of field strains (Kistner et al., 1998) as well as seed strains specifically designed by reverse genetics systems and the large scale cell culture system. Nevertheless, the application of these techniques largely depends on meeting the needs of high viral yield, appropriate permissiveness, and ability to support replication of all influenza virus strains to high titers in short time (reviewed in: Bardiya and Bae, 2005).

Virosomes

Immunopotentiating reconstituted influenza virosomes (IRIVs) possess several characteristics defining them as vaccine adjuvants. They are a liposomal carrier system. These reconstituted virus-like particles (VLP; diameter 150nm) contain a lipid bilayer of phosphatidylcholine and phosphatidylethanolamine. HA and NA are intercalated into the lipid bilayer and give the IRIVs their fusogenic activity, but lack the viral genetic material. IRIVs are able to deliver proteins, RNA/DNA and peptides to immunocompetent cells. In addition, virosomes, as vaccine delivery systems, have been shown to be safe and not to engender any antibodies against the phospholipid components. Therefore, their use in vaccination of children and elderly people is recommended. The system is already registered for human use and allows a specific targeting of antigens by a cellular or a humoral immune response. A virosome vaccine, Inflexal-V, is used in Switzerland and Italy. (Gluck *et al.*, 2004; Langley and Faughnan, 2004; Zurbriggen, 2003).

DNA-vaccines

DNA-vaccines are non-infectious and non-replicative plasmid constructs that encode either only the proteins of interest or the protein of interest in combination with immunomodulatory proteins. This kind of vaccination by direct intra-muscular injection of DNA was first demonstrated in 1990 in a mouse model system *et al.* (Wolff *et al.*, 1990). Directed intra-muscular DNA-vaccination is not very common. The creation of recombinant influenza vaccines based on DNA-plasmids is more appropriate. With this technique rapid and flexible construction of DNA-plasmid vectors can be achieved, which can address the problems of antigenic drift induced by the circulating influenza virus strain (Ljungberg *et al.*, 2000).

These above described techniques have a potential for the development of live and inactivated vaccines. The efficacy of the plasmid based DNA-vaccines expressing the immunogenic influenza virus genes alone or in combination with DNA encoding various cytokines has also been demonstrated in several animal models (for detail see: Bardiya and Bae, 2005). During DNA-vaccination, the foreign genes are endogenously expressed in the host, the proteins subsequently processed, and recognized by the immune system of the host. DNA-vaccines elicit a broad-based humoral and cellular immunity against influenza virus proteins (Justewicz *et al.*, 1995). In addition, alterations in the vector, dose of the DNA, inclusion of CpG-ODN motifs, fusion with influenza virus-specific helper T cell or CTL-epitopes, and appropriate vaccine delivery mechanisms will further improve the efficacy of these vaccines (Bowersock and Martin, 1999; Joseph *et al.*, 2002).

3.1.2   Live vaccines: Cold adapted virus strains and NS1-variants

An alternative to IVs are attenuated "live" vaccines such as cold-adapted vaccines (CAV: CAIV-T, FluMist®) and NS1-defective strains used as intra-nasal influenza vaccine, that may lead to long-lasting, broader immune response (humoral and cellular) that resembles more closely the natural immunity derived from viral infection. For example, CTLs, which are important for the clearance of the virus are activated during an productive infection (Cox *et al.*, 2004). Additional cytokines produced by the infected cells during the innate immunity response enhance and support reaction of the humoral system. Compared to IVs, that are strain- and subtype-specific the CAVs (that also have to be adapted to circulating strains) can provide a broader immunity against circulating viruses (Belshe *et al.*, 2000; King *et al.*, 1998; Nichol, 2001; Stepanova *et al.*, 2002; Treanor *et al.*, 1999; Wareing and Tannock, 2001).

CAVs that already have been used successfully in Russia and are now licensed in the USA (Cox *et al.*, 2004; Kendal, 1997) can be administered intra-nasally for example as aerosols (Abramson, 1999). This results in a limited viral replication in the upper and lower respiratory tract and circumvents the need for syringes. It also supports protective mucosal immunity, which is an important property of nasally applied live influenza virus vaccines. For the generation of a CAV a donor and a wild type strain are reassorted in such a way, that the HA- and NA-segments are wild type (wt) derived and the remaining six segments originate from the donor strain. For this purpose two master strains are currently used as donors in the USA. One to generate A-type and one B-type influenza CAVs (Mendelman *et al.*, 2001; Murphy and Coelingh, 2002). These strains are cold adapted (25°C) (Kendal, 1997; Maassab and Bryant, 1999) and therefore temperature sensitive (ts) and attenuated, meaning that these viruses will not propagate efficiently at body temperatures. To prevent easy reversion of the genetic markers, that encode the ts-defect and allowing the virus to regain full virulence, all six donor-derived segments carry mutations.

For the production of such CAV strains embryonated eggs are infected with both viruses (wild type and donor strain) under the selection of antibodies directed against the HA and NA of the donor strain. The attenuated donor strain by itself is unable to cause significant illness in humans, but is able to donate the HA- and NA-proteins of the contemporary epidemic strain to produce live attenuated vaccine by the traditional egg-based process (Belshe, 2004; Clements and Murphy, 1986; Jin *et al.*, 2003). The live attenuated vaccines were shown to be safe and effective in the general population (Belshe *et al.*, 2004; Kendal, 1997; Langley and Faughnan, 2004). CAV are trivalent like the IVs and are composed according to the WHO recommendations (Mendelman *et al.*, 2001). New possibilities of reverse genetic techniques will certainly improve production of vaccine strains in time and quality (s. 3.1.3).

Even though one should consider the possibility of reassortment with another human strain in the vaccinated person, which could produce an aggressive virus, CAVs have been successfully used in Russia without reports of severe side effects and seem to be safe. They show a comparable effectiveness to trivalent IV's (TIVs) and both vaccines can also be used in combination (Belshe *et al.*, 1998; Belshe *et al.*, 2004; Boyce and Poland, 2000; Edwards *et al.*, 1994; Glezen, 2004; Jackson *et al.*, 1999; Mendelman *et al.*, 2001; Swierkosz *et al.*, 1994; Treanor and Betts, 1998; Treanor *et al.*, 1992).

In addition to the traditional live attenuated vaccines, production by reverse genetics (s. 3.1.3) of replication-incompetent influenza virus-like particles (VLPs) by deletion of either the entire NS gene (encoding both the NS1 and NS2 protein) or only the NS2 gene has also been reported. These VLPs were entirely produced from cDNAs (Watanabe *et al.*, 2002b). Although, these technologies are in the very early stages of development and so far only tested in animal models, the VLP incapable of replication and spread to other cells due to deletion of a major protion of the NS1 or M2, are expected to be good novel influenza vaccine candidates (Galarza *et al.*, 2005; Watanabe *et al.*, 2002a). A variation of the theme is presented by influenza virus strains (generated by reverse genetics (s. 3.1.3)) that express a modified NS1 (Palese and Garcia-Sastre, 2002; Palese *et al.*, 1999; Talon *et al.*, 2000). This non-structural protein is the major viral interferon (IFN)-antagonist (s. Tab. 1). Even though NS1 is a multifunctional viral protein that supports viral replication it seems to be an accessory protein as a virus without the NS1-gene can replicate in IFN-deficient systems (Garcia-Sastre *et al.*, 1998a; Garcia-Sastre *et al.*, 1998b; Ludwig *et al.*, 1999). IFNα/β are two important cytokines expressed in primary infected epithelia cells, that induce innate immunity. IFNα/β-induction severely reduces viral replication even in the presence of NS1. Therefore recombinant viruses expressing altered NS1 with reduced capacity to suppress cellular IFN-induction could raise protective immunity and might represent interesting attenuated live vaccine candidates (Talon et al., 2000). Such viruses have been generated by reverse genetic techniques (s. 3.1.3) and have been successfully tested in experimental settings (Ferko et al., 2004).

3.1.3   Plasmid-based reverse genetic techniques

After initial experiments that implied the in vivo reconstitution of RNPs from plasmid-expressed RDRP, NP and vRNA (Pleschka et al., 1996) it became possible to generate recombinant influenza virus de novo totally from plasmid DNA (Fodor et al., 1999; Neumann et al., 1999), allowing complete genetic manipulation. This manipulation can either concern the combination/mixture of the genomic RNA-segments and/or the gene-sequences themselves. The technique involves the transfection of four plasmids expressing the viral RDRP and the NP together with eight plasmids (for all eight genomic RNAs) that generate a vRNA-like transcript. This again results in the in vivo reconstitution of active RNP-complexes, which will replicated and transcribe the vRNAs. Thereby all viral RNAs and proteins are generated and the viral replication cycle is established resulting in the production of infectious influenza viruses (for review: Garcia-Sastre, 1998; Neumann and Kawaoka, 1999; Palese et al., 1996).

Reverse genetics technique can be used to produce influenza vaccines based on recombinant virus (for detail see: Bardiya and Bae, 2005). These methods do not require selection procedures and eliminate the need for multiple passages in eggs, thereby reducing the time required for vaccine production. It is known that interference among the vaccine viruses of type-A and -B can occur that affect the efficacy of the live attenuated vaccines by restricting their replication. To overcome that problem a chimeric virus (A/B) possessing chimeric (A/B) HA, and full-length B-type NA in the background of a type-A vaccine virus was created (Horimoto et al., 2004). This study provided a novel method for creating live attenuated vaccine from a single donor strain.

For different reasons the technique of reverse genetics has become highly relevant for anti-viral vaccine approaches. (I) For the production of regular IVs against wild type strains, that either grow poorly or are too pathogenic in eggs (s. later) one can generate strains carrying the HA and NA needed in the background of an egg adapted virus. This is normally done by reassortment of the wild type with the egg-adapted strain and can not be well controlled. This problem can be circumvented by plasmid based reverse genetics that allow the controlled design of the reassortant. (II) As mentioned the CAV are composed of HA- and NA-genes from the wild type strain and a mixture of the other six segments from wild type and donor virus. By choice of the according plasmids one can compose a CAV-strain that carries all six segments from the donor strain each with an adaptive mutation. This way it is less likely that a revertant virus will arise by mutation in one of the donor strain segments (s. 3.1.2) (Maassab and Bryant, 1999; Schickli *et al.*, 2001). (III) It is possible to specifically design viruses with altered NS1-genes that could be used as highly attenuated life vaccines (s. 3.1.2), additionally modification of other viral genes (Murphy *et al.*, 1997; Parkin *et al.*, 1997) or of the replications efficiency of the gene-segment (Muster *et al.*, 1991) can be applied to further attanuate the virus. (IV) Viruses could be produced that lack an essential gene (e.g. NEP) (Watanabe *et al.*, 2002b). The missing gene-product can be trans-complemented from an expression-plasmid in the transfected cell during virus generation. The recombinant viruses would be still infectious and lead to expression and presentation of viral proteins, but could not themselves establish a productive propagation as they are lacking the according gene.

Currently used IVs are prepared from egg-grown viruses (Wilschut and McElhaney, 2005). This method is not without limitations but has proven to be efficient. As mentioned earlyer (3.1.1), one particular problem that could arise would be production of a vaccine strain against an highly pathogenic avain influenza virus like the types that have recently infected humans. Besides bio-safety questions they pose further problems. The HA of these viruses is activated within the infected cell by ubiquitous proteases allowing the virus to spread through out the organism. Due to the special HA-characteristics these viruses themselves are highly pathogenic birds and eggs as well as a vaccine strain that would carry the according HA. Therefore efficient virus production in embryonated eggs will be problematic (Lipatov *et al.*, 2004). By plasmid based reverse genetic techniques recombinant viruses can be produced that have lost the pathogenic character of the HA and can replicate well in eggs (Chen *et al.*, 2003; Li *et al.*, 1999; Liu *et al.*, 2003; Subbarao *et al.*, 2003). This could additionally be combined with virus production in cell culture systems (s. p. 12) (Ozaki *et al.*, 2004; Romanova *et al.*, 2004) thereby overcoming the limitation posed by the number of embryonated eggs available at a given time (Stephenson *et al.*, 2004).

It should also be mentioned that not only type-A influenza viruses but also type-B influenza viruses can be generated and manipulated by reverse genetic systems and can therefore also be engineered to fit the circulating wild type strains (Dauber *et al.*, 2004; Hatta *et al.*, 2004; Jackson *et al.*, 2004; Maassab and Bryant, 1999).

3.2   Inhibitors of viral functions (Treatment and anti-viral chemoprophylaxis of influenza)

3.2.1   M2-Inhibitors

Anti-viral treatment is generally considered a supporting measure to prevent and control outbreaks of epidemic influenza in addition to immuno-prophylaxis. However, chemotherapy is the only option to combat the disease when there is no type-specific vaccine available as for instance upon the emergence of a pandemic shift variant. Two classes of substances are currently licensed in many countries for the treatment and/or prophylaxis of influenza, which includes the adamantane compounds amantadine and rimantadine, and the NA-inhibitors oseltamivir and zanamivir. Other small inhibitory compounds that target the viral polymerase complex are also introduced in this section, although none of them has been converted into a pharmaceutical product so far.

3.2.2   M2-Inhibitors (Amantadine and Rimantadine)

Amantadine (1-amino adamantane hydrochloride) and its derivative rimantadine (α-methyl-1-adamantane methylamine hydrochloride) have potent anti-viral activity against most influenza-A-viruses, because they block the viral M2 ion channel protein during the early stage of viral uncoating (Pinto et al., 1995). Specifically, the adamantane compounds inhibit the acidification of the virion inside the endosome, which prevents the intra-cellular release of the viral RNPs. The 50% inhibitory concentration (IC_{50}) of most natural influenza-A-virus strains against adamantane compounds is in the range of 0.2 to 0.4 µg/ml as determined by plaque reduction assay (Appleyard et al., 1977; Hayden et al., 1980; Scholtissek and Faulkner, 1979). Amantadine and rimantadine have proven effectiveness in the treatment of uncomplicated influenza-A-virus infection. They can reduce the duration of fever and system symptoms by approximately one day when given within two days after onset of disease signs (Demicheli et al., 2000; Tominack and Hayden, 1987). Furthermore, both substances also have prophylactic effectiveness in reducing influenza-associated morbidity and clinical symptoms. A survey of studies undertaken with healthy adults demonstrated average effectiveness of 61% for amantadine and 72% for rimantadine in preventing laboratory confirmed influenza (Demicheli et al., 2000). During long-term prophylaxis amantadine was found to cause mild reversible adverse effects in a small proportion of the recipients, which involved central nervous system (CNS) and minor gastrointestinal complaints. No increase in side effects was observed during treatment with rimantadine compared to placebo (N.N., 1985).

An early recognized limitation for widespread clinical use of M2-blockers is the rapid emergence of drug-resistant viruses in tissue culture, in animal models and in patients (Appleyard *et al.*, 1977; Hayden *et al.*, 1989; Oxford *et al.*, 1970). One study found that a total of 27% of children with laboratory-confirmed influenza shed resistant viruses after seven days of treatment with rimantadine (Hall *et al.*, 1987). Unfortunately, such selected drug-resistant viruses are virulent, as they can transmit to family members and cause disease even when the contact persons were treated prophylactically with rimantadine (Hayden *et al.*, 1989). Viruses that become insensitive to amantadine show complete cross-resistance to rimantadine and *vice versa*. Thus, the clinical usage of adamantane amine compounds has been limited by the reported adverse effects, the induction of viral drug resistance and the inactivity towards influenza-B-viruses. Nevertheless, these drugs are still recommended as a cost-effective choice particularly in influenza chemoprophylaxis (Harper *et al.*, 2004). It is noteworthy, that amantadine resistance has also been detected in the highly-pathogenic H5N1 viruses currently circulating in South East Asia (Puthavathana *et al.*, 2005). Thus, adamantane compounds are not an option to treat such infections.

Clinical use for treatment and prophylaxis

Amantadine and rimantadine are approved for treatment of adults and children older than 12 years at two daily 100 mg doses. The substances should carefully be used in individuals above 64 years in age and patients with impaired renal functions and halving of the daily doses is recommended. Only amantadine is licensed for treatment of children between 1 and 9 years and should be dosed with 5 mg/kg per day. In order to avoid emergence and transmission of drug-resistent viruses, treatment should be kept to a minimal time of 3 to 5 days until disease symptoms disappear. Chemoprophylaxis can be considered for protection among high-risk groups including children and adults with chronic pulmonary or cardiac disease, immunocompromised persons with a reduced response to vaccines or in the case of a poor match between an epidemic virus strain and the current vaccine. Since the adamantane compounds do not interfere with the development of neutralizing antibodies (Tominack and Hayden, 1987), they can also be used for the protection of persons at high risk to bridge the time gap between vaccination and the establishment of an efficient immune status. For adults and children older than 9 years two 100 mg doses of amantadine or rimantadine per day are recommended. Children between

1 and 9 years should receive a maximum of 150 mg per day in two divided doses.

3.2.3   Neuraminidase (NA)-inhibitors

Two anti-viral drugs that inhibit both influenza-A and B-viruses, zanamivir (Relenza™, GlaxoSmithCline) and oseltamivir (Tamiflu™, Roche Pharmaceuticals) have recently been approved for general use in the USA, Australia, Europe and Japan. The current knowledge suggests that NA-inhibitors (NI) will have a better clinical utility than the M2-blockers, because these substances are broadly effective against type-A and -B influenza viruses including highly virulent avian virus strains. Further, they appear to have a very low frequency of adverse effects and are less prone to induce drug resistance. Zanamivir and oseltamivir function as slow binding, substrate competitive inhibitors that strongly reduce viral NA-activity by interacting with five sub-sites close to the enzymatic pocket of the NA. The IC_{50} values of these inhibitors were found to be in the range of 0.8 – 8.8 nM depending on the virus types and subtypes (McKimm-Breschkin et al., 2003).

Targeting of the viral NA does not require the delivery of an inhibitor into the cell interior, because the enzyme is a surface glycoprotein. Influenza viruses attach to the host cell through binding of the viral HA to sialic acid moieties that are conjugated to cellular glycoproteins. By the time of progeny virus budding these receptor determinants need to be removed to allow efficient release from the host cell. This is accomplished by the viral NA (acylneuraminyl hydrolase, EC 3.2.1.18) that hydrolyzes glycosidic linkages adjacent to N-acetyl-neuraminic acid (Neu5Ac, sialic acid). Thus, blockade of NA-activity by antibodies, temperature-sensitive mutation or inhibitory substances results in the aggregation of budding virions at the cell membrane and, hence, reduction of virus release (Compans et al., 1969; Palese and Compans, 1976; Palese et al., 1974). In infected animals or humans, NA probably also enhances penetration of the virion through the viscous mucus on respiratory epithelia, which contains sialic acids (Matrosovich et al., 2004). Thus, inhibition of viral NA-activity was the rationale behind several efforts to identify substances that would reduce influenza virus spread and replication.

The development of the current NIs was based on early characterizations of the sialic acid transition state analogue 2-deoxy-2,3 dehydro N-acetylneuraminc acid (Neu 5Ac2en) (Meindl et al., 1974) and the determination of the three-dimensional structure of the NA by X-ray crystallography (Colman et al., 1983; Varghese et al., 1983; Varghese et al., 1992). Neu 5Ac2en had been shown to inhibit viral NA-acitvity but was not protective in a mouse model of influenza (Palese and Schulman, 1977). Based upon computer-assisted drug design, von Itzstein *et al.* demonstrated that the introduction of positively charged amino- or guanidino moieties at position 4 of Neu 5Ac2en increased NA inhibition by two to four orders of magnitude (von Itzstein *et al.*, 1993). Importantly, the inhibition of NA-activity by 4 guanidino-Neu5Ac2en that is now also termed zanamivir translated into efficient reduction of viral replication of type-A and B-influenza viruses in the nanomolar range *in vitro* and dose-dependent decrease of viral titers in infected animals (von Itzstein *et al.*, 1993; Woods *et al.*, 1993). Zanamivir has low oral bioavailability, but shows high anti-viral activity in humans or animals when administered topically by inhalation of dry-powder aerosol (Cass *et al.*, 1999). The second currently approved NA inhibitor compound oseltamivir (3R,4R,5S-4acetamido-5-amino-3-(1-ethylpropoxyl)-1-cyclohexene-1carboxylic acid, also termed GS4071/Ro64-0802) has similarly potent activities against type A and B influenza viruses (Kim *et al.*, 1998). Oseltamivir emerged from an independent NA structure-based study and is based on a cyclohexen ring structure in which the polar glycerol side chain of the sialic acid analogues is replaced by a lipophilic 3-pentyloxy moiety (Kim *et al.*, 1997). Importantly, oseltamivir has high oral anti-viral activity when administered as its methylester pro-drug, GS4071/oseltamivir phosphate, that is converted to the active drug by hepatic enzymes (Hayden *et al.*, 1999b; Li *et al.*, 1998; Mendel *et al.*, 1998).

Effectiveness

Zanamivir and oseltamivir have potent anti-viral effectiveness against community-acquired influenza and are in general safe to use in healthy adults (Abramson, 1999; Boivin *et al.*, 2000; Hayden *et al.*, 1997; Makela *et al.*, 2000; Monto *et al.*, 1999; N.N., 1985). In clinical trials the NIs significantly shortened disease duration and reduced symptoms and viral loads when treatment was initiated within 26 hours post infection (Hayden *et al.*, 1996; Hayden *et al.*, 1999b). Even, when inhalation of zanamivir was begun within 30 hours after onset of symptoms the time to alleviation of major disease signs (cough, myalgias, fever, headache) was shortened by one to two days and patients were able to resume normal activities earlier (Hayden *et al.*, 1997; Monto *et al.*, 1999). Initiation of therapy later than 30 hours after disease onset still reduced viral loads but was less beneficial for symptom recovery. Two 75 mg daily doses of oseltamivir for five days were shown to reduce shedding of virus and the severity and duration of influenza symptoms by one to two days when therapy was begun within 36 hours after onset of disease signs (Nicholson *et al.*, 2000; Treanor *et al.*, 2000). Some side effects that included diarrhea, nausea and nasal symptoms were observed during clinical testings of zanamivir but were similar in placebo groups (GlaxoWellcome, 2001). The NI substances are also highly effective to prevent spread of the disease. A post-exposure protection study with zanamivir demonstrated 79% efficacy in preventing transmission of influenza to family members, when the index case was treated with zanamivir (Hayden et al., 2000). Oseltamivir had a comparably high efficacy in preventing laboratory-confirmed influenza by 74% and influenza with fever by 82% (Hayden et al., 1999a). Within households, one 75 mg dose oseltamivir per day was 89% protective against clinical influenza even when the index cases were not treated (Welliver et al., 2001). Thus, to prevent the spread of the flu within household contacts the NIs appear to be preferable compared to the M2-blockers that can induce the emergence and transmission of virulent drug-resistant viruses.

Resistance to NA-inhibitors

During the development of NIs for clinical use it was recognized that viruses with a reduced drug sensitivity could be selected in tissue culture (summarized in (McKimm-Breschkin, 2000; Tisdale, 2000)). Resistance can be characterized by various methods including IC_{50}-determination of the viral NA, by plaque reduction assays (number and size) and yield reduction assay in tissue culture (Matrosovich et al., 2003; Tisdale, 2000). Under laboratory conditions several passages are usually required to select such variants, which is different to amantadine-resistant viruses that can emerge in a single cycle experiment. Drug-resistant viruses were also isolated from diseased persons treated with NIs (Gubareva et al., 2001; Gubareva et al., 1998; Kiso et al., 2004; Zambon and Hayden, 2001). However, the available data on the pathogenicity of these mutant viruses in animal models suggest that they have reduced replication capability in vivo and may therefore be clinically less relevant in humans.

Resistance to NIs was found to be complex, because it can be associated with mutations in the NA, the HA or synergistically in both genes. NA-mutations that confer reduced drug sensitivity were identified at amino acid residues 119, 152, 274, 292 and 294 (based on N2-NA numbering) (Gubareva, 2004; Zambon and Hayden, 2001). These amino acids are part of or cluster around the conserved catalytic pocket and their mutation can decrease the enzymatic activity to below 5% and some also destabilize the enzyme (Varghese et al., 1998). The various NI-molecules slightly differ in their interactions with the enzyme. Thus, a given NA-mutant enzyme may show a range of sensitivity against different inhibitors (Gubareva et al., 2001). Interestingly, some viruses with a reduced sensitivity to NIs were found to carry mutations in the HA, which affected the receptor binding site in the globular head region, the stalk region and the HA_2-subunit (McKimm-

Breschkin, 2000). Apparently, the HA-mutations reduce drug sensitivity by decreasing the affinity for cellular sialic acid receptor molecules and thereby easing the release of budding viruses from the plasma membrane. These findings corroborate the concept that efficient viral replication requires a carefully balanced interplay between the strength of HA/receptor binding and the activity of the NA that removes these receptor determinants (Wagner *et al.*, 2000).

Clinical use for treatment and chemoprophylaxis

The use of zanamivir (Relenza™) and oseltamivir (Tamiflu™) is recommended for the treatment of uncomplicated influenza caused by type-A and B-viruses (Harper *et al.*, 2004). Therapy with either drug should be initiated within 48 hours after the onset of disease signs and should be continued for five days (GlaxoWellcome, 2001; Roche, 2001). It is important to consider that bacterial superinfections may occur that would not be affected by these anti-virals. Neither substance has been shown to prevent serious complications of influenza like pneumonia. Zanamivir is approved for treatment of influenza in persons aged 7 years and older. The recommended dosage is two inhalations of 5 mg doses twice a day using the inhalation device provided by the manufacturer. Zanamivir is not recommended for persons with underlying respiratory conditions like asthma or chronic obstructive pulmonary disease, because of the risk of precipitating bronchospasm in such patients (GlaxoWellcome, 2001). Oseltamivir can be used for treatment of patients of 1 year or older. Depending on the age, the recommended doses for children above 12 years and adults are two 75 mg capsules a day. Two daily doses of 15 – 30 mg is recommended for children under 15 kg, 2 x 45 mg for children between 15-23 kg and 2 x 60 mg for persons weighing >23-40 kg. Currently, Tamiflu™ but not Relenza™ is licensed for chemoprophylaxis in children older than 12 years and in adults. For persons with creatinine clearance of 10-30 ml/min, halving of the usual dosage for therapy or prophylaxis is recommended. Two approaches are possible, a seasonal prophylaxis that provides a 92% reduction of confirmed influenza infection in a vaccinated population of frail elderly persons (McClellan and Perry, 2001), and a short-term prophylaxis for controlling institutional outbreaks by breaking the virus circulation.

Several further compounds that inhibit the influenza virus NA were identified in independent efforts and have been evaluated as anti-influenza agents. Thus, the cyclopentane derivatives BCX-1812 (RWJ-270201), BCX-1827, BCX-1898 and BCX-1923 (from BioCryst Pharmaceuticals) as well as the pyrrolidine-based A315675 (from Abbott Laboratories) showed strong potent anti-viral activies at least *in vitro* (Kati *et al.*, 2002; Smee *et al.*,

2001). Thus, although development of BCX-1812 has been halted after showing a lack of activity in a phase III clinical trial (Chand *et al.*, 2005), additional NIs may emerge as anti-influenza drugs in the future.

3.2.4   RDRP- and endonuclease-inhibitors

Two unique properties of the trimeric RNA-dependent RNA-polymerase of influenza viruses, which are not shared by cellular enzymes, provide attractive opportunities for anti-viral interference with possibly little disturbances of the host cell. First, the polymerase exhibits an endonuclease activity that cleaves the first 10 – 13 nucleotides including the 5'-cap structure from nascent host RNA-polymerase II cap transcripts and use them to prime viral mRNAs (Lamb and Krug, 2001). Second, the viral polymerase replicates the negative-sense viral RNA-segments via unprimed synthesis of a complementary positive-strand RNA-intermediate. For both of these activities, inhibitory small molecule compounds have been identified, some of which were also shown to reduce viral propagation in tissue culture and/or in infected mice. However, further clinical development has not been reported for any of those substances so far.

The viral endonuclease activity is associated with the PB1-subunit and depends on binding of the polymerase to the terminal ends of the vRNA-template and the cap structures of nascent mRNA-transcripts (Li *et al.*, 2001). The endonuclease most likely utilizes a two metal ion mechanism for cleavage of the cellular nucleic acid (Klumpp, 2004a). It has been shown that derivatives of the fungal metabolite flutimide as well as a class of 4-substituted 2,4-dioxobutanoic acids specifically inhibited the cap-dependent endonuclease, presumably by interaction with the active catalytic site of the enzyme (Hastings *et al.*, 1996; Parkes *et al.*, 2003; Tomassini *et al.*, 1994; Tomassini, 1996). The most potent compounds of these two classes had IC_{50} values in the range of 0.2 – 6 μM when tested in virus yield assays in tissue culture experiments. Further, intranasal instillation of the L-735,882 compound was reported to inhibit viral titers in nasal washes of mice infected with influenza virus A/PR/8/34 virus, but the effects on disease progression were not studied (Hastings *et al.*, 1996).

Another screening effort has identified T-705 (6-fluoro-3-hydroxy-2-pyrazinecarboxamide) to have potent and selective anti-influenza activity. T-705 showed IC_{50} values of less than 0.5 μg/ml in virus yield assays in MDCK cells against all three influenza virus types (A, B, C) with no signs of cytotoxicity (Furuta *et al.*, 2002). Importantly, T-705 was also orally active in a mouse model and shown to significantly reduce viral lung titers and enhance survival rates from 20% to 100% after infection with influenza virus A/PR/8/34 virus at a dose of 200 mg/kg per day (Furuta *et al.*, 2002;

Takahashi *et al.*, 2003). Although the basis for its anti-viral activity was unclear at that time, T-705 was found to inhibit replication of an oseltamivir-resistant mutant virus *in vitro* suggesting that this inhibitor targets a different viral function (Takahashi *et al.*, 2003). Indeed, recent analyses showed that the compound is metabolized inside the cell into T-705-ribofuranosyl-5'-triphosphate (T-705-RTP), which is a potent and selective inhibitor of ApG-primed viral RNA-polymerase activity (Furuta *et al.*, 2005). These findings show that T-705 may have the potential to become a novel oral anti-influenza drug that targets a viral function not blocked by one of the currently licensed NIs or M2-blockers.

4. TARGETING HOST CELL FUNCTIONS AND FACTORS

4.1 Inhibitors of cell signaling and apoptosis

Influenza viruses only have a limited coding capacity. Thus, these viruses employ functions of their host-cell for efficient replication. These dependencies create opportunities to design novel anti-viral strategies by targeting specific host cell functions.

Cell fate decisions in response to extra-cellular agents, including pathogenic invaders are commonly mediated by intra-cellular signaling cascades that transduce signals into stimulus specific actions, e.g. changes in gene expression patterns, alterations in the metabolic state of the cell or induction of programmed cell death (apoptosis). Thus, these signaling molecules are at the bottleneck of the control of cellular responses. In this section we will review the recent advances in the analysis of influenza virus induced signaling pathways and first attempts to use signaling mediators as targets for anti-viral approaches.

4.1.1 Intra-cellular signaling cascades – MAP-kinases and the IKK/NFκB-module

Mitogen activated protein kinase (MAPK)-cascades have gained much attention as being critical transducers to convert a variety of extra-cellular signals into a multitude of responses (English *et al.*, 1999; Hazzalin and Mahadevan, 2002; Widmann *et al.*, 1999) Thereby, these pathways regulate numerous cellular decision processes, such as proliferation and differentiation, but also cell activation and immune responses (Dong *et al.*, 2002). Four different members of the MAPK-family that are organized in separate cascades have been identified so far: ERK (extra-cellular signal regulated kinase), JNK (Jun-N-terminal kinase), p38 and ERK5/BMK-1 (Big MAP kinase) (Garrington and Johnson, 1999; Widmann *et al.*, 1999). These MAPKs are activated by a dual phosphorylation event on threonine and tyrosine mediated by MAPK-kinases (MAPKK also termed MEKs or MKKs). The MAPK "ERK" is activated by the dual-specific MAPKK MEK1 and -2 that are controlled by the upstream serine threonine MAPKK-kinase Raf. Raf, MEK and ERK form the prototype module of a MAPK-pathway and are also known as the classical mitogenic cascade. The MAPK p38 and JNK are activated by MKK3/6 and MKK4/7, respectively, and are predominantly activated by pro-inflammatory cytokines and certain environmental stress conditions. The MEK5/ERK5 module is both activated by mitogens and certain stress inducers. There is evidence that all these different MAPK-cascades are activated upon infection with RNA-viruses, including influenza viruses. Thus, these signaling cascades may serve different functions in viral replication and host cell response.

Another important signaling pathway, which is commonly activated upon virus infection is the IκB-kinase (IKK)/NFκB-signaling module (Hiscott *et al.*, 2001). The NFκB/IκB family of transcription factors promote the expression of well over 150 different genes, such as cytokine or chemokine genes, or genes encoding for adhesion molecules or anti- and pro-apoptotic protein (Pahl, 1999). The canonical mechanism of NFκB activation includes activation of IκB-kinase (IKK) that phosphorylates the inhibitor of NFκB (IκB) and targets the protein for subsequent degradation (Delhase and Karin, 1999; Karin, 1999b). This leads to the release and migration of the transcriptionally active NFκB factors to the nucleus (Ghosh, 1999; Karin and Ben-Neriah, 2000). The IKK-complex consists of at least three isozymes of IKK: (I) IKK1/IKKα, (II) IKK2/IKKβ and (III) NEMO/IKKγ. The most important isozyme for NFκB-activation via the degradation of IκB is IKK2 (Karin, 1999a). NEMO acts as a scaffolding protein for the large IKK complex (Courtois *et al.*, 2001) that contains still other kinases such as MEKK1 (MAPKK-kinase 1) (Lee *et al.*, 1998), NIK (NFκB inducing kinase) (Nemoto *et al.*, 1998; Woronicz *et al.*, 1997) and the dsRNA-activated protein-kinase (PKR) (Gil *et al.*, 2000; Zamanian-Daryoush *et al.*, 2000).

Both NFκB and the JNK MAPK-pathway regulate one of the most important anti-viral gene expression events, the transcriptional induction of interferon beta (IFNβ) (Maniatis *et al.*, 1998). IFNβ is one of the first anti-viral cytokines to be expressed upon virus infection, initiating an auto-amplification loop to cause an efficient and strong type-I IFN response. The IFNβ enhanceosome, which mediates the inducible expression of IFNβ,

carries binding sites for transcription factors of three families, namely the
AP-1 family members and JNK targets c-Jun and ATF-2, the NFκB factors
p50 and p65, and the interferon-regulatory factors (IRFs) (Hiscott *et al.*,
1999; Thanos and Maniatis, 1995). In the initial phase of a virus infection
this promoter element specifically binds the constitutively expressed and
specifically activated IRF3-dimer (Taniguchi and Takaoka, 2002). AP-1-
and NFκB-transcription factors are activated by a variety of stimuli.
However, a strong IRF3-activation is selectively induced upon infection with
several RNA-viruses, in particular by the dsRNA, which accumulates during
replication (Lin *et al.*, 1998; Yoneyama *et al.*, 1998). Thus, IRF3 is the
major determinant of a strong virus- and dsRNA-induced IFNβ-response.

4.1.2   MAP Kinase-cascades and influenza virus-infection: The ERK-pathway

Interestingly all four so far defined MAPK-family members are activated
upon an influenza virus infection (Kujime *et al.*, 2000; Ludwig *et al.*, 2001;
Pleschka *et al.*, 2001; Virginia Korte and S.L., unpublished) and recent work
has helped to get a clearer picture of the importance of the ERK-signaling
pathway for influenza virus replication.

The activation of the MAP-kinase ERK upon productive influenza virus
infection (Kujime *et al.*, 2000) appears to serve a mechanism that is
beneficial for the virus (Pleschka *et al.*, 2001). Strikingly, blockade of the
pathway by specific inhibitors of the upstream kinase MEK and dominant-
negative mutants of ERK or the MEK-activator Raf resulted in a strongly
impaired growth of both, influenza A- and B-type viruses (Pleschka *et al.*,
2001). Conversely, virus titers are enhanced in cells expressing active
mutants of Raf or MEK (Ludwig *et al.*, 2004; Olschlager *et al.*, 2004). This
has not only been demonstrated in cell culture but also *in vivo* in infected
mice expressing a constitutively active form of the Raf-kinase in the alveolar
epithelial cells of the lung (Olschlager *et al.*, 2004). While in the wt-situation
influenza viruses primarily infect bronchiolar epithelial cells, there is
efficient replication in the alveolar layer most exclusively in the cells
carrying the transgene. As a consequence this results in an earlier death
of the transgenic animals (Olschlager *et al.*, 2004). This indicates that
activation of the Raf/MEK/ERK pathway is required for efficient virus
growth. Noticeably, inhibition of the pathway did not significantly affect
viral RNA- or protein-synthesis (Pleschka *et al.*, 2001). The pathway rather
appears to control the active nuclear export of the viral RNP-complexes that
are readily retained in the nucleus upon blockade of the signaling pathway.
Most likely this is due to an impaired activity of the viral nuclear export
protein NEP (Pleschka *et al.*, 2001). This indicates that active RNP-export is
an induced rather than a constitutive event, a hypothesis supported by a late

activation of ERK in the viral life cycle. So far the detailed mechanism of how ERK regulates export of the RNPs is unsolved. There are two likely scenarios: Either it does occur directly via phosphorylation of a viral protein involved in RNP-transport or by control of a cellular export factor. Although in the initial studies no alteration of the overall phosphorlyation status of the NP, M and NEP proteins was observed (Pleschka et al., 2001) there are now first indications that certain phosphorylation sites of the NP indeed are affected by MEK-inhibition (S.P., unpublished data). It remains to be shown, whether this is of functional relevance for the RNP-export process. It is striking that MEK-inhibitors are not toxic for the cell, while more general blockers of the active transport machinery, such as leptomycin-B exert a high toxicity even in quite low concentrations. This may indicate that MEK-inhibitors are no general export blockers but only block a distinct nuclear export pathway. Indeed there are first evidences that the classical mitogenic cascade specifically regulates nuclear export of certain cellular RNA-protein complexes. In LPS-treated mouse macrophages MEK-inhibition results in a specific retention of the TNF-mRNA in the nucleus (Dumitru et al., 2000). This is also observed in cells deficient for Tpl-2, an activator of MEK and ERK. In these cells the failure to activate MEK and ERK by LPS again correlated with TNF mRNA retention while other cytokines are normally expressed (Dumitru et al., 2000). Thus the ERK-pathway may regulate a specific cellular export process but leaves other export mechanisms unaffected. It is likely that such a specific export pathway is employed by influenza-A and B-viruses.

The finding of an anti-viral action of MEK-inhibitors prompted further research showing that replication of other viruses, such as Borna disease virus (Planz et al., 2001), Visna virus (Barber et al., 2002) or Coxsackie B3 virus (Luo et al., 2002) is also impaired upon MEK-inhibition.

Requirement of Raf/MEK/ERK-activation for efficient influenza virus replication may suggest that this pathway may be a cellular target for anti-viral approaches. Besides the anti-viral action against both, A- and B-type viruses (Ludwig et al., 2004), MEK-inhibitors meet two further criteria which are a prerequisite for a potential clinical use. Although targeting an important signaling pathway in the cell the inhibitors showed a surprisingly little toxicity (a) in cell culture (Ludwig et al., 2004; Planz et al., 2001; Pleschka et al., 2001) (b) in an in vivo mouse model (Sebolt-Leopold et al., 1999) and (c) in clinical trials for the use as anti-cancer agent (Cohen, 2002). In the light of these findings it was hypothesized that the mitogenic pathway may only be of major importance during early development of an organism and may be dispensable in adult tissues (Cohen, 2002). Another very important feature of MEK-inhibitors is that they showed no tendency to induce formation of resistant virus variants (Ludwig et al., 2004). Although targeting of a cellular factor may still raise the concern about side effects of a drug, it appears likely that local administration of an agent such as a MEK-inhibitor to the primary site of influenza virus infection, the lung, is well tolerated. Here the drug primarily affects differentiated lung epithelial cells for which a proliferative signaling cascade like the Raf/MEK/ERK-cascade may be dispensable. Following this approach it was recently demonstrated that the MEK inhibitor U0126 is effective in reducing virus titers in the lung of infected mice after local administration (O.P., S.P. and S.L., unpublished).

4.1.3   Protein kinase C: A viral entry regulator

Activation of the classical mitogenic Raf/MEK/ERK-cascade is initiated by yet other phosphorylation events. The kinase Raf is known to be regulated by phosphorylation of different upstream kinases including members of the protein kinase C (PKC)-family (Cai *et al.*, 1997; Kolch *et al.*, 1993).

The PKC-superfamily consists of at least 12 different PKC-isoforms that carry out diverse regulatory roles in cellular processes by linking into several downstream signaling pathways (Toker, 1998). Beside a regulation of the Raf/MEK/ERK-cascade and other downstream pathways, PKCs may have additional functions during viral replication. A role of PKCs in the process of entry of several enveloped viruses has been proposed based on the action of protein kinase inhibitors H7 and staurosporine (Constantinescu *et al.*, 1991) as well as by the calcium-channel blocker verapamil (Nugent and Shanley, 1984). Influenza virus infection or treatment of cells with purified viral HA results in rapid activation of PKCs upon binding to host-cell surface receptors (Arora and Gasse, 1998; Kunzelmann *et al.*, 2000; Rott *et al.*, 1995). In a recent study it was shown that the pan PKC-inhibitor bisindolylmalcimide-I prevented influenza virus entry and subsequent infection in a dose dependent and reversible manner (Root *et al.*, 2000). Using a dominant-negative mutant approach this function was assigned to the PKCßII-isoform. Overexpression of a phosphorylation-deficient mutant of PKCßII revealed that the kinase is a regulator of late endosomal sorting. Accordingly, expression of the PKCßII-mutant resulted in a block of virus entry at the level of late endosomes (Sieczkarski *et al.*, 2003; Sieczkarski and Whittaker, 2002). Thus, a specific inhibition of PKCßII may be a suitable approach to blunt virus replication at a very early time point in the replication cycle.

4.1.4   Influenza virus and the IKK/NFκB-pathway

Activation of the transcription factor NFκB is a hallmark of most infections by viral pathogens (Hiscott *et al.*, 2001) including influenza viruses (reviewed in: Julkunen *et al.*, 2000; Ludwig *et al.*, 2003; Ludwig *et al.*, 1999). Influenza viral NFκB-induction involves activation of IκB-kinase (IKK) (Wurzer *et al.*, 2004) and is also achieved with isolated influenza virus components. This includes dsRNA (Chu *et al.*, 1999) or over-expression of the viral HA, NP or M1 proteins (Flory *et al.*, 2000). Since gene expression of many pro-inflammatory or anti-viral cytokines, such as IFNβ or TNFα, is controlled by NFκB the concept emerged that IKK and NFκB are essential components in the innate immune response to virus infections (Chu *et al.*, 1999). Accordingly, influenza virus-induced IFNβ-promoter activity is impaired in cells expressing transdominant negative mutants of IKK2 or IκBα (Wang *et al.*, 2000; Wurzer *et al.*, 2004).

Nevertheless, IKK and NFκB might not only have anti-viral functions as two recent studies demonstrate that influenza viruses replicate much better in cells where NFκB is pre-activated (Nimmerjahn *et al.*, 2004; Wurzer *et al.*, 2004). Conversely, influenza virus titers from different host cells in which NFκB-signaling was impaired by means of specific inhibitors or dominant-negative mutants, a dramatic reduction could be observed (Nimmerjahn *et al.*, 2004; Wurzer *et al.*, 2004). Thus, in the context of an influenza virus infection a function of NFκB to support virus replication appears to be dominant over the function as a transcription factor in the anti-viral response. On a molecular basis this was shown to be due to the NFκB-dependent expression of pro-apoptotic factors, such as TNF-related apoptosis inducing ligand (TRAIL) or FasL (Wurzer *et al.*, 2004). Inhibition of virus induced expression of these factors results in strongly impaired viral growth. This links the pro-viral action of NFκB to the induction of apoptosis, a process that will be discussed in the next section. Finally, viral need for NFκB-activity suggests that this pathway may be suitable as a target for anti-viral intervention. To this end we have shown recently that several pharmacological inhibitors of NFκB act anti-viral *in vivo*, without toxic side effects or the tendency to induce resistant virus variants (I. Mazur, W. Wurzer, C. Erhardt, T. Silberzahn, T.W., O.P., S.P. and S.L, unpublished).

4.1.5   Influenza virus-induced programmed cell death (Apoptosis)

Another cellular signaling response commonly observed upon virus infections, including influenza virus is the induction of the apoptotic cascade. Apoptosis is a morphological and biochemical defined form of cell death (Kerr *et al.*, 1972) and has been demonstrated to play a role in a variety of diseases, including virus infections (Razvi and Welsh, 1995). Apoptosis is mainly regarded to be a host cell defense against virus infections since many viruses express anti-apoptotic proteins to prevent this cellular response. The central component of the apoptotic machinery is a proteolytic system consisting of a family of cysteinyl proteases, termed caspases (for review see: Cohen, 1997; Thornberry and Lazebnik, 1998). Two groups of caspases can be distinguished: upstream initiator caspases such as caspase-8 or caspase-9, which cleave and activate other caspases and downstream effector caspases, including caspase-3, -6 and -7, that cleave a variety of cellular substrates, thereby disassembling cellular structures or inactivating enzymes (Thornberry and Lazebnik, 1998). Caspase-3 is the most intensively studied effector caspase. Work on caspase-3 deficient MCF-7 breast carcinoma cells has revealed a caspase-3 driven feedback loop, that is crucial to mediate the apoptotic process (Janicke *et al.*, 1998; Slee *et al.*, 1999). Thus, caspase-3 is a central player in apoptosis regulation and the level of pro-caspase-3 in the cell determines the impact of a given apoptotic stimulus.

It is long known that influenza virus infection with A- and B-type viruses results in the induction of apoptosis both in permissive and un-permissive cultured cells as well as *in vivo* (Fesq *et al.*, 1994; Hinshaw *et al.*, 1994; Ito *et al.*, 2002; Mori *et al.*, 1995; Takizawa *et al.*, 1993). Interestingly, viral activation of MAPKs or upstream kinases has been linked to the onset of apoptosis. In a mouse model for a neurovirulent influenza infection, JNK-activity correlated with apoptosis induction in the infected brain (Mori *et al.*, 2003). In embryonic fibroblasts deficient for the MAPKK-kinase ASK-1 the virus-induced JNK-activation was blunted concomitant with an inhibition of caspase-3 activation and virus-induced apoptosis (Maruoka *et al.*, 2003). As an extrinsic mechanism of viral apoptosis induction it has been noted quite early on that the Fas receptor/FasL-apoptosis inducing system (Fujimoto *et al.*, 1998; Takizawa *et al.*, 1995; Takizawa *et al.*, 1993; Wada *et al.*, 1995) is expressed in a PKR-dependent manner in infected cells (Takizawa *et al.*, 1996). This most likely contributes to virus-induced cell death via a receptor mediated FADD/caspase-8-dependent pathway (Balachandran *et al.*, 2000). Another mode of viral apoptosis induction might occur via activation of TGF-β that is converted from its latent form by the viral NA (Schultz-Cherry and Hinshaw, 1996). Within the influenza virus infected cell the apoptotic program is mediated by activation of caspases (Lin *et al.*, 2002; Takizawa *et al.*, 1999; Zhirnov *et al.*, 1999) with a most crucial role of caspase-3 (Wurzer *et al.*, 2003).

Although it is now well established that influenza virus infection induces caspses and subsequent apoptosis, the consequence for virus replication or host cell defense is still under a heavy debate (reviewed in (Lowy, 2003; Ludwig *et al.*, 1999; Schultz-Cherry *et al.*, 1998).

With the identification of PB1-F2, a new influenza virus protein expressed from a +1 reading frame of the PB1 polymerase gene segment, a pro-apoptotic influenza virus protein has been discovered (Chen *et al.*,

2001). PB1-F2 induces apoptosis via the mitochondrial pathway if added to cells and infection with recombinant viruses lacking the protein results in reduced apoptotic rates of lymphocytes (Chen *et al.*, 2001). However, most of the avian virus strains are lacking the PB1-F2 reading frame and PB1-F2-deficient viruses do not affect apoptosis in a variety of other host cells (Chen *et al.*, 2001). These results have let to the assumption that apoptosis induction by PB1-F2 may be required for the specific depletion of lymphocytes during an influenza virus infection, a process which is observed in infected animals (Tumpey *et al.*, 2000; Van Campen *et al.*, 1989a; Van Campen *et al.*, 1989b).

A recent study adds a new aspect to the open discussion by the surprising observation that influenza virus propagation was strongly impaired in the presence of caspase inhibitors (Wurzer *et al.*, 2003). This dependency on caspase activity was most obvious in cells where caspase-3 was partially knocked-down by siRNA (Wurzer *et al.*, 2003). Consistent with these findings, poor replication efficiencies of influenza-A-viruses in cells deficient for caspase-3 could be boosted 30-fold by ectopic expression of the protein. Mechanistically, the block in virus propagation appeared to be due to the retention of viral RNP-complexes in the nucleus preventing formation of progeny virus particles (Wurzer *et al.*, 2003). Interestingly the findings are consistent with a much earlier report showing that upon infection of cells over expressing the anti-apoptotic protein Bcl-2 the viral RNP-complexes were retained in the nucleus (Hinshaw *et al.*, 1994) resulting in repressed virus titers (Olsen *et al.*, 1996). Furthermore the recently identified pro-apoptotic PB1-F2 (Chen *et al.*, 2001) is only expressed in later phases of replication consistent with a later step in the virus life cycle that requires caspase activity (Wurzer *et al.*, 2003). The observation of a caspase requirement for RNP-nuclear export was quite puzzling since this export process was shown before to be mediated by the active cellular export machinery involving the viral nuclear export protein (NS2/NEP) (Neumann *et al.*, 2000; O'Neill *et al.*, 1998) and the anti-apoptotic Raf/MEK/ERK-cascade (Pleschka *et al.*, 2001). Caspase activation does not support, but rather inhibit the active nuclear export machinery by cleavage of transport proteins. This suggests an alternate strategy by which caspases may regulate RNP-export, e.g. by directly or indirectly increase the diffusion limit of nuclear pores (Faleiro and Lazebnik, 2000) to allow passive diffusion of larger proteins. Such a scenario is supported by the finding that isolated NPs or RNP-complexes, which are nuclear if ectopically expressed, can partially translocate to the cytoplasm upon stimulation with an apoptosis inducer in a caspase-3-dependent manner (Wurzer *et al.*, 2003). These findings can be merged into a model in which the RNPs are transported via an active export mechanism in intermediate steps of the virus life cycle. Once caspase activity increases in the cells, proteins of the transport machinery get destroyed, however, widening of nuclear pores may allow the viral RNPs to use a second mode of exit from the nucleus (Faleiro and Lazebnik, 2000). That would be a likely mechanism to further enhance RNP-migration to the cytoplasm in late phase of the viral life cycle and thereby support virus replication. Such a complementary use of both "active" (Raf/MEK/ERK-dependent) and "passive" (caspase-dependent) transport mechanisms is supported by the observation, that at concentrations of MEK- and caspase-inhibitors, which only poorly block influenza virus replication alone, efficiently impaired virus propagation if used in combination (Wurzer *et al.*, 2003). Thus, while both pathways do not interfere with each other (Wurzer *et al.*, 2003) they appear to synergize to mediate RNP-export via different routes.

Therefore one may conclude that influenza virus has acquired the capability to take advantage of supposedly anti-viral host cell responses to support viral propagation. This includes early induction of caspase activity but not necessarily execution of the full apoptotic process that most likely is an anti-viral response. This dual role of "early" versus "late" apoptotic events during virus replication may exclude the use of caspase-inhibitors as anti-flu agents, although in cell culture these inhibitors may have a beneficial outcome for the host cell.

4.1.6   Other cellular targets: Glyocosidases and proteases

Besides mediators of signaling and apoptosis a variety of other cellular enzymes are required for efficient virus growth. There are also some initial attempts to use these components as target for an anti-viral intervention.

The viral glycoproteins are glycosylated in the endoplasmic reticulum (ER) and the ER-α-glycosidase-I is responsible for the removal of terminal α-1,2 glucose residues from precursor oligosaccharides in the ER. A variety of viruses such as HIV, HSV and Dengue-virus have been shown to be highly sensitive to inhibitors of these enzymes (Mehta *et al.*, 2001; Mehta *et al.*, 2002). One of these inhibitors, castanopermine, has been demonstrated to inhibit replication of influenza virus A/Hongkong/11/88 in MDCK cells with an IC_{50} value of <6 μM (Klumpp, 2004b). The inhibitor also acted anti-viral *in vivo* in a mouse model and reduced lung titers of A/PR8/34 infected mice by tenfold when administered intranasal. The compound has reached Phase II clinical trials for the treatment of HIV and has been licensed for a potential treatment of Hepatitis-C-virus infections (reviewed in (Klumpp, 2004b)).

Another important requirement for a cellular enzyme is the proteolytic cleavage of the HA by proteases. The infectivity and pathogenicity of influenza virus is based on the proteolytic cleavage of the precursor HA_0 into HA_1 and HA_2 chains by an arginine-specific, trypsin-like host protease.

Several exogenous protease inhibitors were investigated with respect to their anti-influenza activity: Camostat, a serine protease inhibitor; was shown to exhibit strong anti-influenza effects *in vitro* and *in vivo* in mice and in chicken embryos. The compound also showed strong anti-influenza effects in amantadine-resistant type-A and -B virus infection *in vitro* (Lee *et al.*, 1996).

Other protease inhibitors Nafamostat mesilate, camostat mesilate, gabexate mesilate and aprotinin also inhibited virus replication *in vitro* (Ovcharenko and Zhirnov, 1994) (reviewed in: Luscher-Mattli, 2000). The protease inhibitors Gordox, Contrycal and epsilonaminocapronic acid were tested in both animal and clinical experiments. Inhalation of aminocapronic acid-containing aerosols exerted the most effective therapeutic effect, reducing the duration of viral antigen in the nasopharyngeal epithelium 1 1/2 to 2 fold (Zhirnov *et al.*, 1984). Recombinant human mucus protease inhibitor (MPI) was investigated for its anti-viral activity in rat lungs *in vitro*. The C-, but not the N-terminal domain of MPI was shown to inhibit the proteolytic activity of tryptase Clara and of virus activation at nM concentrations (Beppu *et al.*, 1997; Kido *et al.*, 1999).

However, the current understanding is that protease inhibitors – mainly used in HIV therapy – may produce serious toxic side effects. Recent investigations showed that protease inhibitors can cause diabetes, hepatic and renal failures and mutagenic (potentially carcinogenic) effects. A further disadvantage of protease inhibitors is the rapid development of viral resistance, and a variable strain sensitivity to these anti-viral agents (discussed in: Luscher-Mattli, 2000). Nevertheless protease inhibitors are still under evaluation as potential anti-influenza therapeutics (Kido *et al.*, 2004; Savarino, 2005).

5.   CONCLUSIONS

Regarding the continues threat caused by seasonal flu-epidemics and the immanent danger of re-occurring pandemic outbreaks that both impose a great burden on human and animal health, and considering the fact, that influenza viruses can not be eradicated, the possibilities to fight this disease have been greatly improved by novel molecular biological techniques in recent years. Vaccination is still by far the best prophylactic measure, but new drugs, which attack the virus directly, will further support to combat these foes. Nevertheless the viral tactic to escape direct intervention by resistance is a major drawback of current therapeutic interventions. This problem might be overcome by innovative methods that target cellular functions essential for efficient virus replication

ACKNOWLEDGEMENTS

We would like to thank all the colleagues who have helped in many invaluable ways in the production of this chapter, in particular, Dr. C. Erhardt, Dr. W. Wurzer, H. Marjuki, B. Daubner, V. Oehlschlaeger, J. Lampe and K. Oesterle.

NOTES

This work is dedicated to Prof. Dr. C. Scholtissek's 75th Birthday.

REFERENCES

Abramson, J. S., 1999, Intranasal, cold-adapted, live, attenuated influenza vaccine. *Pediatr Infect Dis J* **18**, 1103-4.

Appleyard, G., Monto, A. S., Gunn, R. A., Bandyk, M. G., and King, C. L., 1977, Amantadine-resistance as a genetic marker for influenza viruses. *J Gen Virol* **36**, 249-255.

Arora, D. J., and Gasse, N., 1998, Influenza virus hemagglutinin stimulates the protein kinase C activity of human polymorphonuclear leucocytes. *Arch Virol* **143**, 2029-37.

Babai, I., Barenholz, Y., Zakay-Rones, Z., Greenbaum, E., Samira, S., Hayon, I., Rochman, M., and Kedar, E., 2001, A novel liposomal influenza vaccine (INFLUSOME-VAC) containing hemagglutinin-neuraminidase and IL-2 or GM-CSF induces protective anti-neuraminidase antibodies cross-reacting with a wide spectrum of influenza A viral strains. *Vaccine* **20**, 505-15.

Balachandran, S., Roberts, P. C., Kipperman, T., Bhalla, K. N., Compans, R. W., Archer, D. R., and Barber, G. N., 2000, Alpha/beta interferons potentiate virus-induced apoptosis through activation of the FADD/Caspase-8 death signaling pathway. *J Virol* **74**, 1513-23.

Barber, S. A., Bruett, L., Douglass, B. R., Herbst, D. S., Zink, M. C., and Clements, J. E., 2002, Visna virus-induced activation of MAPK is required for virus replication and correlates with virus-induced neuropathology. *J Virol* **76**, 817-28.

Bardiya, N., and Bae, J. H., 2005, Influenza vaccines: recent advances in production technologies. *Appl Microbiol Biotechnol* **67**, 299-305.

Belshe, R. B., 2004, Current status of live attenuated influenza virus vaccine in the US. *Virus Research* **103**, 177.

Belshe, R. B., Gruber, W. C., Mendelman, P. M., Mehta, H. B., Mahmood, K., Reisinger, K., Treanor, J., Zangwill, K., Hayden, F. G., Bernstein, D. I., Kotloff, K., King, J.,

Piedra, P. A., Block, S. L., Yan, L., and Wolff, M., 2000, Correlates of immune protection induced by live, attenuated, cold-adapted, trivalent, intranasal influenza virus vaccine. *J Infect Dis* **181**, 1133-7.

Belshe, R. B., Mendelman, P. M., Treanor, J., King, J., Gruber, W. C., Piedra, P., Bernstein, D. I., Hayden, F. G., Kotloff, K., Zangwill, K., Iacuzio, D., and Wolff, M., 1998, The efficacy of live attenuated, cold-adapted, trivalent, intranasal influenzavirus vaccine in children. *N Engl J Med* **338**, 1405-12.

Belshe, R. B., Nichol, K. L., Black, S. B., Shinefield, H., Cordova, J., Walker, R., Hessel, C., Cho, I., and Mendelman, P. M., 2004, Safety, efficacy, and effectiveness of live, attenuated, cold-adapted influenza vaccine in an indicated population aged 5-49 years. *Clin Infect Dis* **39**, 920-7.

Ben-Yehuda, A., Joseph, A., Barenholz, Y., Zeira, E., Even-Chen, S., Louria-Hayon, I., Babai, I., Zakay-Rones, Z., Greenbaum, E., Galprin, I., Gluck, R., Zurbriggen, R., and Kedar, E., 2003a, Immunogenicity and safety of a novel IL-2-supplemented liposomal influenza vaccine (INFLUSOME-VAC) in nursing-home residents. *Vaccine* **21**, 3169-78.

Ben-Yehuda, A., Joseph, A., Zeira, E., Even-Chen, S., Louria-Hayon, I., Babai, I., Zakay-Rones, Z., Greenbaum, E., Barenholz, Y., and Kedar, E., 2003b, Immunogenicity and safety of a novel liposomal influenza subunit vaccine (INFLUSOME-VAC) in young adults. *J Med Virol* **69**, 560-7.

Beppu, Y., Imamura, Y., Tashiro, M., Towatari, T., Ariga, H., and Kido, H., 1997, Human mucus protease inhibitor in airway fluids is a potential defensive compound against infection with influenza A and Sendai viruses. *J Biochem (Tokyo)* **121**, 309-16.

Beyer, W. E. P., Palache, A. M., de Jong, J. C., and Osterhaus, A. D. M. E., 2002, Cold-adapted live influenza vaccine versus inactivated vaccine: systemic vaccine reactions, local and systemic antibody response, and vaccine efficacy A meta-analysis. *Vaccine* **20**, 1340.

Boivin, G., Goyette, N., Hardy, I., Aoki, F., Wagner, A., and Trottier, S., 2000, Rapid antiviral effect of inhaled zanamivir in the treatment of naturally occurring influenza in otherwise healthy adults. *J Infect Dis* **181**, 1471-1474.

Bowersock, T. L., and Martin, S., 1999, Vaccine delivery to animals. *Adv Drug Deliv Rev* **38**, 167-194.

Boyce, T. G., and Poland, G. A., 2000, Promises and challenges of live-attenuated intranasal influenza vaccines across the age spectrum: a review. *Biomed Pharmacother* **54**, 210-8.

Bridges, C. B., Harper, S. A., Fukuda, K., Uyeki, T. M., Cox, N. J., Singelton, J. A., and Practices, A. C. o. I., 2003, Prevention and control of influenza. Recommendations of the Advisory Committee on Immunization Practices (ACIP). *MMWR Recomm Rep* **52(PR-8)**, 1-34.

Brown, D. M., Roman, E., and Swain, S. L., 2004, CD4 T cell responses to influenza infection. *Semin Immunol* **16**, 171-7.

Cai, H., Smola, U., Wixler, V., Eisenmann-Tappe, I., Diaz-Meco, M. T., Moscat, J., Rapp, U., and Cooper, G. M., 1997, Role of diacylglycerol-regulated protein kinase C isotypes in growth factor activation of the Raf-1 protein kinase. *Mol Cell Biol* **17**, 732-41.

Cass, L. M., Efthymiopoulos, C., and Bye, A., 1999, Pharmacokinetics of zanamivir after intravenous, oral, inhaled or intranasal administration to healthy volunteers. *Clin Pharmacokinet* **36 Suppl 1**, 1-11.

Chand, P., Bantia, S., Kotian, P. L., El-Kattan, Y., Lin, T. H., and Babu, Y. S., 2005, Comparison of the anti-influenza virus activity of cyclopentane derivatives with oseltamivir and zanamivir in vivo. *Bioorg Med Chem* **13**, 4071-7.

Chen, H., Deng, G., Li, Z., Tian, G., Li, Y., Jiao, P., Zhang, L., Liu, Z., Webster, R. G., and Yu, K., 2004, The evolution of H5N1 influenza viruses in ducks in southern China. *Proc Natl Acad Sci U S A* **101**, 10452-7.

Chen, H., Subbarao, K., Swayne, D., Chen, Q., Lu, X., Katz, J., Cox, N., and Matsuoka, Y., 2003, Generation and evaluation of a high-growth reassortant H9N2 influenza A virus as a pandemic vaccine candidate. *Vaccine* **21**, 1974-9.

Chen, W., Calvo, P. A., Malide, D., Gibbs, J., Schubert, U., Bacik, I., Basta, S., O'Neill, R., Schickli, J., Palese, P., Henklein, P., Bennink, J. R., and Yewdell, J. W., 2001, A novel influenza A virus mitochondrial protein that induces cell death. *Nat Med* **7**, 1306-12.

Chu, W. M., Ostertag, D., Li, Z. W., Chang, L., Chen, Y., Hu, Y., Williams, B., Perrault, J., and Karin, M., 1999, JNK2 and IKKbeta are required for activating the innate response to viral infection. *Immunity* **11**, 721-31.

Claas, E. C., Osterhaus, A. D., van Beek, R., De Jong, J. C., Rimmelzwaan, G. F., Senne, D. A., Krauss, S., Shortridge, K. F., and Webster, R. G., 1998, Human influenza A H5N1 virus related to a highly pathogenic avian influenza virus. *Lancet* **351**, 472-7.

Clements, M. L., and Murphy, B. R., 1986, Development and Persistence of Local and Systemic Antibody-Responses in Adults Given Live Attenuated Or Inactivated Influenza-A Virus-Vaccine. *Journal of Clinical Microbiology* **23**, 66.

Cohen, G. M., 1997, Caspases: the executioners of apoptosis. *Biochem J* **326**, 1-16.

Cohen, P. (2002): Protein kinases--the major drug targets of the twenty-first century? *Nat Rev Drug Discov* **1**, 309-15.

Colman, P. M., Varghese, J. N., and Laver, W. G., 1983, Structure of the catalytic and antigenic sites in influenza virus neuraminidase. *Nature* **303**, 41-44.

Compans, R. W., Dimmock, N. J., and Meier-Ewert, H., 1969, Effect of antibody to neuraminidase on the maturation and hemagglutinating activity of an influenza A2 virus. *J Virol* **4**, 528-534.

Constantinescu, S. N., Cernescu, C. D., and Popescu, L. M., 1991, Effects of protein kinase C inhibitors on viral entry and infectivity. *FEBS Lett* **292**, 31-3.

Couch, R. B., 2003, An overview of serum antibody responses to influenza virus antigens. *Dev Biol* **115**, 25-30.

Couch, R. B., and Kasel, J. A., 1983, Immunity to influenza in man. *Annu Rev Microbiol* **37**, 529-49.

Coulter, A., Harris, R., Davis, R., Drane, D., Cox, J., Ryan, D., Sutton, P., Rockman, S., and Pearse, M., 2003, Intranasal vaccination with ISCOMATRIX adjuvanted influenza vaccine. *Vaccine* **21**, 946-9.

Courtois, G., Smahi, A., and Israel, A., 2001, NEMO/IKK gamma: linking NF-kappa B to human disease. *Trends Mol Med* **7**, 427-30.

Cox, N. J., and Subbarao, K., 1999, Influenza. *Lancet* **354**, 1277.

Cox, R. J., Brokstad, K. A., and Ogra, P., 2004, Influenza virus: immunity and vaccination strategies. Comparison of the immune response to inactivated and live, attenuated influenza vaccines. *Scand J Immunol* **59**, 1-15.

Dauber, B., Heins, G., and Wolff, T., 2004, The influenza B virus nonstructural NS1 protein is essential for efficient viral growth and antagonizes beta interferon induction. *J Virol* **78**, 1865-72.

de Jong, J. C., Claas, E. C., Osterhaus, A. D., Webster, R. G., and Lim, W. L., 1997, A pandemic warning? *Nature* **389**, 554.

de Jong, J. C., Palache, A. M., Beyer, W. E., Rimmelzwaan, G. F., Boon, A. C., and Osterhaus, A. D., 2003, Haemagglutinin-inhibiting antibody to influenza virus. *Dev Biol* **115**, 63-73.

Delhase, M., and Karin, M., 1999, The I kappa B kinase: a master regulator of NF-kappa B, innate immunity, and epidermal differentiation. *Cold Spring Harb Symp Quant Biol* **64**, 491-503.

Demicheli, V., Jefferson, T., Rivetti, D., and Deeks, J., 2000, Prevention and early treatment of influenza in healthy adults. *Vaccine* **18**, 957-1030.

Dong, C., Davis, R. J., and Flavell, R. A., 2002, MAP kinases in the immune response. *Annu Rev Immunol* **20**, 55-72.

Dumitru, C. D., Ceci, J. D., Tsatsanis, C., Kontoyiannis, D., Stamatakis, K., Lin, J. H., Patriotis, C., Jenkins, N. A., Copeland, N. G., Kollias, G., and Tsichlis, P. N., 2000, TNF-alpha induction by LPS is regulated posttranscriptionally via a Tpl2/ERK-dependent pathway. *Cell* **103**, 1071-83.

Edwards, K. M., Dupont, W. D., Westrich, M. K., Plummer, W. D., Jr., Palmer, P. S., and Wright, P. F., 1994, A randomized controlled trial of cold-adapted and inactivated vaccines for the prevention of influenza A disease. *J Infect Dis* **169**, 68-76.

English, J., Pearson, G., Wilsbacher, J., Swantek, J., Karandikar, M., Xu, S., and Cobb, M. H., 1999, New insights into the control of MAP kinase pathways. *Exp Cell Res* **253**, 255-70.

Faleiro, L., and Lazebnik, Y., 2000, Caspases disrupt the nuclear-cytoplasmic barrier. *J Cell Biol* **151**, 951-9.

Ferko, B., Stasakova, J., Romanova, J., Kittel, C., Sereinig, S., Katinger, H., and Egorov, A., 2004, Immunogenicity and protection efficacy of replication-deficient influenza A viruses with altered NS1 genes. *J Virol* **78**, 13037-45.

Fesq, H., Bacher, M., Nain, M., and Gemsa, D., 1994, Programmed cell death (apoptosis) in human monocytes infected by influenza A virus. *Immunobiology* **190**, 175-82.

Flory, E., Kunz, M., Scheller, C., Jassoy, C., Stauber, R., Rapp, U. R., and Ludwig, S., 2000, Influenza virus-induced NF-kappaB-dependent gene expression is mediated by overexpression of viral proteins and involves oxidative radicals and activation of IkappaB kinase. *J Biol Chem* **275**, 8307-14.

Fodor, E., Devenish, L., Engelhardt, O. G., Palese, P., Brownlee, G. G., and Garcia-Sastre, A., 1999, Rescue of influenza A virus from recombinant DNA. *J Virol* **73**, 9679-82.

Fouchier, R. A., Schneeberger, P. M., Rozendaal, F. W., Broekman, J. M., Kemink, S. A., Munster, V., Kuiken, T., Rimmelzwaan, G. F., Schutten, M., Van Doornum, G. J., Koch, G., Bosman, A., Koopmans, M., and Osterhaus, A. D., 2004, Avian influenza A virus (H7N7) associated with human conjunctivitis and a fatal case of acute respiratory distress syndrome. *Proc Natl Acad Sci U S A* **101**, 1356-61.

Fujimoto, I., Takizawa, T., Ohba, Y., and Nakanishi, Y., 1998, Co-expression of Fas and Fas-ligand on the surface of influenza virus-infected cells. *Cell Death Differ* **5**, 426-31.

Furuta, Y., Takahashi, K., Fukuda, Y., Kuno, M., Kamiyama, T., Kozaki, K., Nomura, N., Egawa, H., Minami, S., Watanabe, Y., Narita, H., and Shiraki, K., 2002, In vitro and in vivo activities of anti-influenza virus compound T-705. *Antimicrob Agents Chemother* **46**, 977-81.

Furuta, Y., Takahashi, K., Kuno-Maekawa, M., Sangawa, H., Uehara, S., Kozaki, K., Nomura, N., Egawa, H., and Shiraki, K., 2005, Mechanism of action of T-705 against influenza virus. *Antimicrob Agents Chemother* **49**, 981-6.

Galarza, J. M., Latham, T., and Cupo, A., 2005, Virus-like particle vaccine conferred complete protection against a lethal influenza virus challenge. *Viral Immunol* **18**, 365-72.

Garcia-Sastre, A., 1998, Negative-strand RNA viruses: applications to biotechnology. *Trends Biotechnol* **16**, 230-5.

Garcia-Sastre, A., Durbin, R. K., Zheng, H., Palese, P., Gertner, R., Levy, D. E., and Durbin, J. E., 1998a, The role of interferon in influenza virus tissue tropism. *J Virol* **72**, 8550-8.

Garcia-Sastre, A., Egorov, A., Matassov, D., Brandt, S., Levy, D. E., Durbin, J. E., Palese, P., and Muster, T., 1998b, Influenza A virus lacking the NS1 gene replicates in interferon-deficient systems. *Virology* **252**, 324-30.

Garrington, T. P., and Johnson, G. L., 1999, Organization and regulation of mitogen-activated protein kinase signaling pathways. *Curr Opin Cell Biol* **11**, 211-8.

Ghosh, S. (1999): Regulation of inducible gene expression by the transcription factor NF-kappaB. *Immunol Res* **19**, 183-9.

Gil, J., Alcami, J., and Esteban, M., 2000, Activation of NF-kappa B by the dsRNA-dependent protein kinase, PKR involves the I kappa B kinase complex. *Oncogene* **19**, 1369-78.

GlaxoWellcome, 2001, Relenza™ product information.

Glezen, W. P., 2004, Control of influenza. *Tex Heart Inst J* **31**, 39-41.

Gluck, R., Moser, C., and Metcalfe, I. C., 2004, Influenza virosomes as an efficient system for adjuvanted vaccine delivery. *Expert Opin Biol Ther* **4**, 1139-45.

Gubareva, L. V., 2004, Molecular mechanisms of influenza virus resistance to neuraminidase inhibitors. *Virus Res* **103**, 199-203.

Gubareva, L. V., Kaiser, L., Matrosovich, M. N., Soo-Hoo, Y., and Hayden, F. G., 2001, Selection of influenza virus mutants in experimentally infected volunteers treated with oseltamivir. *J Infect Dis* **183**, 523-531.

Gubareva, L. V., Matrosovich, M. N., Brenner, M. K., Bethell, R. C., and Webster, R. G., 1998, Evidence for zanamivir resistance in an immunocompromised child infected with influenza B virus. *J Infect Dis* **178**, 1257-1262.

Hall, C. B., Dolin, R., Gala, C. L., Markovitz, D. M., Zhang, Y. Q., Madore, P. H., Disney, F. A., Talpey, W. B., Green, J. L., Francis, A. B., and *et al.*, 1987, Children with influenza A infection: treatment with rimantadine. *Pediatrics* **80**, 275-282.

Harper, S. A., Fukuda, K., Uyeki, T. M., Cox, N. J., and Bridges, C. B., 2004, Prevention and control of influenza: recommendations of the Advisory Committee on Immunization Practices (ACIP). *MMWR Recomm Rep* **53**, 1-40.

Hastings, J. C., Selnick, H., Wolanski, B., and Tomassini, J. E., 1996, Anti-influenza virus activities of 4-substituted 2,4-dioxobutanoic acid inhibitors. *Antimicrob Agents Chemother* **40**, 1304-7.

Hatta, M., Goto, H., and Kawaoka, Y., 2004, Influenza B virus requires BM2 protein for replication. *J Virol* **78**, 5576-83.

Hatta, M., and Kawaoka, Y., 2002, The continued pandemic threat posed by avian influenza viruses in Hong Kong. *Trends Microbiol* **10**, 340-4.

Hayden, F. G., Atmar, R. L., Schilling, M., Johnson, C., Poretz, D., Paar, D., Huson, L., Ward, P., and Mills, R. G., 1999a, Use of the selective oral neuraminidase inhibitor oseltamivir to prevent influenza. *N Engl J Med* **341**, 1336-1343.

Hayden, F. G., Belshe, R. B., Clover, R. D., Hay, A. J., Oakes, M. G., and Soo, W., 1989, Emergence and apparent transmission of rimantadine-resistant influenza A virus in families. *N Engl J Med* **321**, 1696-1702.

Hayden, F. G., Cote, K. M., and Douglas, R. G., Jr., 1980, Plaque inhibition assay for drug susceptibility testing of influenza viruses. *Antimicrob Agents Chemother* **17**, 865-870.

Hayden, F. G., Gubareva, L. V., Monto, A. S., Klein, T. C., Elliot, M. J., Hammond, J. M., Sharp, S. J., and Ossi, M. J., 2000, Inhaled zanamivir for the prevention of influenza in families. Zanamivir Family Study Group. *N Engl J Med* **343**, 1282-1289.

Hayden, F. G., Osterhaus, A. D., Treanor, J. J., Fleming, D. M., Aoki, F. Y., Nicholson, K. G., Bohnen, A. M., Hirst, H. M., Keene, O., and Wightman, K., 1997, Efficacy and safety of the neuraminidase inhibitor zanamivir in the treatment of influenzavirus infections. GG167 Influenza Study Group. *N Engl J Med* **337**, 874-880.

Hayden, F. G., Treanor, J. J., Betts, R. F., Lobo, M., Esinhart, J. D., and Hussey, E. K., 1996, Safety and efficacy of the neuraminidase inhibitor GG167 in experimental human influenza. *Jama* **275**, 295-299.

Hayden, F. G., Treanor, J. J., Fritz, R. S., Lobo, M., Betts, R. F., Miller, M., Kinnersley, N., Mills, R. G., Ward, P., and Straus, S. E., 1999b, Use of the oral neuraminidase inhibitor oseltamivir in experimental human influenza: randomized controlled trials for prevention and treatment. *Jama* **282**, 1240-1246.

Hazzalin, C. A., and Mahadevan, L. C., 2002, MAPK-regulated transcription: a continuously variable gene switch? *Nat Rev Mol Cell Biol* **3**, 30-40.

Hinshaw, V. S., Olsen, C. W., Dybdahl-Sissoko, N., and Evans, D., 1994, Apoptosis: a mechanism of cell killing by influenza A and B viruses. *J Virol* **68**, 3667-73.

Hiscott, J., Kwon, H., and Genin, P., 2001, Hostile takeovers: viral appropriation of the NF-kappaB pathway. *J Clin Invest* **107**, 143-51.

Hiscott, J., Pitha, P., Genin, P., Nguyen, H., Heylbroeck, C., Mamane, Y., Algarte, M., and Lin, R., 1999, Triggering the interferon response: the role of IRF-3 transcription factor. *J Interferon Cytokine Res* **19**, 1-13.

Horimoto, T., Iwatsuki-Horimoto, K., Hatta, M., and Kawaoka, Y. (2004): Influenza A viruses possessing type B hemagglutinin and neuraminidase: potential as vaccine components. *Microbes Infect* **6**, 579-83.

Horimoto, T., and Kawaoka, Y., 2001, Pandemic threat posed by avian influenza A viruses. *Clin Microbiol Rev* **14**, 129-49.

Hufnagel, L., Brockmann, D., and Geisel, T., 2004, Forecast and control of epidemics in a globalized world. *Proc Natl Acad Sci U S A* **101**, 15124-9.

Ito, T., Kobayashi, Y., Morita, T., Horimoto, T., and Kawaoka, Y., 2002, Virulent influenza A viruses induce apoptosis in chickens. *Virus Res* **84**, 27-35.

Jackson, D., Zurcher, T., and Barclay, W., 2004, Reduced incorporation of the influenza B virus BM2 protein in virus particles decreases infectivity. *Virology* **322**, 276-85.

Jackson, L. A., Holmes, S. J., Mendelman, P. M., Huggins, L., Cho, I., and Rhorer, J., 1999, Safety of a trivalent live attenuated intranasal influenza vaccine, FluMist, administered in addition to parenteral trivalent inactivated influenza vaccine to seniors with chronic medical conditions. *Vaccine* **17**, 1905-9.

Janicke, R. U., Sprengart, M. L., Wati, M. R., and Porter, A. G., 1998, Caspase-3 is required for DNA fragmentation and morphological changes associated with apoptosis. *J Biol Chem* **273**, 9357-60.

Jin, H., Lu, B., Zhou, H., Ma, C. H., Zhao, J., Yang, C. F., Kemble, G., and Greenberg, H,. 2003, Multiple amino acid residues confer temperature sensitivity to human influenza virus vaccine strains (FluMist) derived from cold-adapted A/Ann Arbor/6/60. *Virology* **306**, 18.

Jones, T., Stern, A., and Lin, R., 1994, Potential role of granulocyte-macrophage colony-stimulating factor as vaccine adjuvant. *Eur J Clin Microbiol Infect Dis* **13 Suppl 2**, S47-53.

Joseph, A., Louria-Hayon, I., Plis-Finarov, A., Zeira, E., Zakay-Rones, Z., Raz, E., Hayashi, T., Takabayashi, K., Barenholz, Y., and Kedar, E., 2002, Liposomal immunostimulatory DNA sequence (ISS-ODN): an efficient parenteral and mucosal adjuvant for influenza and hepatitis B vaccines. *Vaccine* **20**, 3342-54.

Julkunen, I., Melen, K., Nyqvist, M., Pirhonen, J., Sareneva, T., and Matikainen, S., 2000, Inflammatory responses in influenza A virus infection. *Vaccine* **19**, S32-7.

Justewicz, D. M., Morin, M. J., Robinson, H. L., and Webster, R. G., 1995, Antibody-forming cell response to virus challenge in mice immunized with DNA encoding the influenza virus hemagglutinin. *J Virol* **69**, 7712-7.

Karin, M., 1999a, The beginning of the end: IkappaB kinase (IKK) and NF-kappaB activation. *J Biol Chem* **274**, 27339-42.

Karin, M., 1999b, How NF-kappaB is activated: the role of the IkappaB kinase (IKK) complex. *Oncogene* **18**, 6867-74.

Karin, M., and Ben-Neriah, Y., 2000, Phosphorylation meets ubiquitination: the control of NF-[kappa]B activity. *Annu Rev Immunol* **18**, 621-63.

Kati, W. M., Montgomery, D., Carrick, R., Gubareva, L., Maring, C., McDaniel, K., Steffy, K., Molla, A., Hayden, F., Kempf, D., and Kohlbrenner, W., 2002, In vitro characterization of A-315675, a highly potent inhibitor of A and B strain influenza virus neuraminidases and influenza virus replication. *Antimicrob Agents Chemother* **46**, 1014-21.

Kemble, G., and Greenberg, H., 2003, Novel generations of influenza vaccines. *Vaccine* **21**, 1789.

Kendal, A. P., 1997, Cold-adapted live attenuated influenza vaccines developed in Russia: can they contribute to meeting the needs for influenza control in other countries? *Eur J Epidemiol* **13**, 591-609.

Kerr, J. F., Wyllie, A. H., and Currie, A. R., 1972, Apoptosis: a basic biological phenomenon with wide-ranging implications in tissue kinetics. *Br J Cancer* **26**, 239-57.

Kido, H., Beppu, Y., Imamura, Y., Chen, Y., Murakami, M., Oba, K., and Towatari, T., 1999, The human mucus protease inhibitor and its mutants are novel defensive compounds against infection with influenza A and Sendai viruses. *Biopolymers* **51**, 79-86.

Kido, H., Okumura, Y., Yamada, H., Mizuno, D., Higashi, Y., and Yano, M., 2004, Secretory leukoprotease inhibitor and pulmonary surfactant serve as principal defenses against influenza A virus infection in the airway and chemical agents up-regulating their levels may have therapeutic potential. *Biol Chem* **385**, 1029-34.

Kim, C. U., Lew, W., Williams, M. A., Liu, H., Zhang, L., Swaminathan, S., Bischofberger, N., Chen, M. S., Mendel, D. B., Tai, C. Y., Laver, W. G., and Stevens, R. C., 1997, Influenza neuraminidase inhibitors possessing a novel hydrophobic interaction in the enzyme active site: Design, synthesis, and structural analysis of carbocyclic sialic acid analogues with potent anti-influenza activity. *J. Am. Chem. Soc.* **119**, 681-690.

Kim, C. U., Lew, W., Williams, M. A., Wu, H., Zhang, L., Chen, X., Escarpe, P. A., Mendel, D. B., Laver, W. G., and Stevens, R. C., 1998, Structure-activity relationship studies of novel carbocyclic influenza neuraminidase inhibitors. *J Med Chem* **41**, 2451-2460.

King, J. C., Jr., Lagos, R., Bernstein, D. I., Piedra, P. A., Kotloff, K., Bryant, M., Cho, I., and Belshe, R. B., 1998, Safety and immunogenicity of low and high doses of trivalent live cold-adapted influenza vaccine administered intranasally as drops or spray to healthy children. *J Infect Dis* **177**, 1394-7.

Kiso, M., Mitamura, K., Sakai-Tagawa, Y., Shiraishi, K., Kawakami, C., Kimura, K., Hayden, F. G., Sugaya, N., and Kawaoka, Y., 2004, Resistant influenza A viruses in children treated with oseltamivir: descriptive study. *Lancet* **364**, 759-65.

Kistner, O., Barrett, P. N., Mundt, W., Reiter, M., Schober-Bendixen, S., and Dorner, F., 1998, Development of a mammalian cell (Vero) derived candidate influenza virus vaccine. *Vaccine* **16**, 960.

Klumpp, K., 2004a, Recent advances in the discovery and development of anti-influenza drugs. *Expert Opinion on Therapeutic Patents* **14**, 1153-1168.

Klumpp, K., 2004b, Recent advances in the discovery and development of anti-influenza drugs. *Expert Opin. Ther. Patents* **14**, 1153-1168.

Kolch, W., Heidecker, G., Kochs, G., Hummel, R., Vahidi, H., Mischak, H., Finkenzeller, G., Marme, D., and Rapp, U. R., 1993, Protein kinase C alpha activates RAF-1 by direct phosphorylation. *Nature* **364**, 249-52.

Koopmans, M., Wilbrink, B., Conyn, M., Natrop, G., van der Nat, H., Vennema, H., Meijer, A., van Steenbergen, J., Fouchier, R., Osterhaus, A., and Bosman, A., 2004, Transmission of H7N7 avian influenza A virus to human beings during a large outbreak in commercial poultry farms in the Netherlands. *Lancet* **363**, 587-93.

Kujime, K., Hashimoto, S., Gon, Y., Shimizu, K., and Horie, T., 2000, p38 mitogen-activated protein kinase and c-jun-NH2-terminal kinase regulate RANTES production by influenza virus-infected human bronchial epithelial cells. *J Immunol* **164**, 3222-8.

Kunzelmann, K., Beesley, A. H., King, N. J., Karupiah, G., Young, J. A., and Cook, D. I., 2000, Influenza virus inhibits amiloride-sensitive Na+ channels in respiratory epithelia. *Proc Natl Acad Sci U S A* **97**, 10282-7.

LaForce, F. M., Nichol, K. L., and Cox, N. J., 1994, Influenza: virology, epidemiology, disease, and prevention. *Am J Prev Med* **10 Suppl**, 31-44.

Lamb, R. A., and Krug, R. M., 2001, Orthomyxoviridae: The viruses and their replication, pp. 1487-1531. In D. M. Knipe, P. M. Howley, and D. E. Griffin (Eds): *Fields Virology*, Lippencott Williams & Williams.

Langley, J. M., and Faughnan, M. E., 2004, Prevention of influenza in the general population. *Canadian Medical Association Journal* **171**, 1213.

Lee, F. S., Peters, R. T., Dang, L. C., and Maniatis, T., 1998, MEKK1 activates both IkappaB kinase alpha and IkappaB kinase beta. *Proc Natl Acad Sci U S A* **95**, 9319-24.

Lee, M. G., Kim, K. H., Park, K. Y., and Kim, J. S., 1996, Evaluation of anti-influenza effects of camostat in mice infected with non-adapted human influenza viruses. *Arch Virol* **141**, 1979-89.

Li, K. S., Guan, Y., Wang, J., Smith, G. J., Xu, K. M., Duan, L., Rahardjo, A. P., Puthavathana, P., Buranathai, C., Nguyen, T. D., Estoepangestie, A. T., Chaisingh, A., Auewarakul, P., Long, H. T., Hanh, N. T., Webby, R. J., Poon, L. L., Chen, H., Shortridge, K. F., Yuen, K. Y., Webster, R. G., and Peiris, J. S., 2004, Genesis of a highly pathogenic and potentially pandemic H5N1 influenza virus in eastern Asia. *Nature* **430**, 209-13.

Li, M. L., Rao, P., and Krug, R. M., 2001, The active sites of the influenza cap-dependent endonuclease are on different polymerase subunits. *Embo J* **20**, 2078-86.

Li, S. Q., Liu, C. G., Klimov, A., Subbarao, K., Perdue, M. L., Mo, D., Ji, Y. Y., Woods, L., Hietala, S., and Bryant, M., 1999, Recombinant influenza A virus vaccines for the pathogenic human A Hong Kong 97 (H5N1) viruses. *Journal of Infectious Diseases* **179**, 1132.

Li, W., Escarpe, P. A., Eisenberg, E. J., Cundy, K. C., Sweet, C., Jakeman, K. J., Merson, J., Lew, W., Williams, M., Zhang, L., Kim, C. U., Bischofberger, N., Chen, M. S., and Mendel, D. B., 1998, Identification of GS 4104 as an orally bioavailable prodrug of the influenza virus neuraminidase inhibitor GS 4071. *Antimicrob Agents Chemother* **42**, 647-653.

Lin, C., Holland, R. E., Jr., Donofrio, J. C., McCoy, M. H., Tudor, L. R., and Chambers, T. M., 2002, Caspase activation in equine influenza virus induced apoptotic cell death. *Vet Microbiol* **84**, 357-65.

Lin, R., Heylbroeck, C., Pitha, P. M., and Hiscott, J., 1998, Virus-dependent phosphorylation of the IRF-3 transcription factor regulates nuclear translocation, transactivation potential, and proteasome-mediated degradation. *Mol Cell Biol* **18**, 2986-96.

Lipatov, A. S., Govorkova, E. A., Webby, R. J., Ozaki, H., Peiris, M., Guan, Y., Poon, L., and Webster, R. G. (2004): Influenza: emergence and control. *J Virol* **78**, 8951-9.

Liu, M., Wood, J. M., Ellis, T., Krauss, S., Seiler, P., Johnson, C., Hoffmann, E., Humberd, J., Hulse, D., Zhang, Y., Webster, R. G., and Perez, D. R., 2003, Preparation of a standardized, efficacious agricultural H5N3 vaccine by reverse genetics. *Virology* **314**, 580-90.

Ljungberg, K., Wahren, B., Almqvist, J., Hinkula, J., Linde, A., and Winberg, G., 2000, Effective construction of DNA vaccines against variable influenza genes by homologous recombination. *Virology* **268**, 244-50.

Lowy, R. J., 2003, Influenza virus induction of apoptosis by intrinsic and extrinsic mechanisms. *Int Rev Immunol* **22**, 425-49.

Ludwig, S., Ehrhardt, C., Neumeier, E. R., Kracht, M., Rapp, U. R., and Pleschka, S., 2001, Influenza virus-induced AP-1-dependent gene expression requires activation of the JNK signaling pathway. *J Biol Chem* **276**, 10990-8.

Ludwig, S., Planz, O., Pleschka, S., and Wolff, T., 2003, Influenza-virus-induced signaling cascades: targets for antiviral therapy? *Trends Mol Med* **9**, 46-52.

Ludwig, S., Pleschka, S., and Wolff, T., 1999, A fatal relationship–influenza virus interactions with the host cell. *Viral Immunol* **12**, 175-96.

Ludwig, S., Wolff, T., Ehrhardt, C., Wurzer, W. J., Reinhardt, J., Planz, O., and Pleschka, S., 2004, MEK inhibition impairs influenza B virus propagation without emergence of resistant variants. *FEBS Lett* **561**, 37-43.

Luo, H., Yanagawa, B., Zhang, J., Luo, Z., Zhang, M., Esfandiarei, M., Carthy, C., Wilson, J. E., Yang, D., and McManus, B. M., 2002, Coxsackievirus B3 replication is reduced by inhibition of the extracellular signal-regulated kinase (ERK) signaling pathway. *J Virol* **76**, 3365-73.

Luscher-Mattli, M., 2000, Influenza chemotherapy: a review of the present state of art and of new drugs in development. *Arch Virol* **145**, 2233-48.

Maassab, H. F., and Bryant, M. L., 1999, The development of live attenuated cold-adapted influenza virus vaccine for humans. *Rev Med Virol* **9**, 237-44.

Makela, M. J., Pauksens, K., Rostila, T., Fleming, D. M., Man, C. Y., Keene, O. N., and Webster, A., 2000, Clinical efficacy and safety of the orally inhaled neuraminidase inhibitor zanamivir in the treatment of influenza: a randomized, double-blind, placebo-controlled European study. *J Infect* **40**, 42-48.

Maniatis, T., Falvo, J. V., Kim, T. H., Kim, T. K., Lin, C. H., Parekh, B. S., and Wathelet, M. G., 1998, Structure and function of the interferon-beta enhanceosome. *Cold Spring Harb Symp Quant Biol* **63**, 609-20.

Maruoka, S., Hashimoto, S., Gon, Y., Nishitoh, H., Takeshita, I., Asai, Y., Mizumura, K., Shimizu, K., Ichijo, H., and Horie, T., 2003, ASK1 regulates influenza virus infection-induced apoptotic cell death. *Biochem Biophys Res Commun* **307**, 870-6.

Matrosovich, M., Matrosovich, T., Carr, J., Roberts, N. A., and Klenk, H. D., 2003, Over-expression of the alpha-2,6-sialyltransferase in MDCK cells increases influenza virus sensitivity to neuraminidase inhibitors. *J Virol* **77**, 8418-25.

Matrosovich, M. N., Matrosovich, T. Y., Gray, T., Roberts, N. A., and Klenk, H. D., 2004, Neuraminidase is important for the initiation of influenza virus infection in human airway epithelium. *J Virol* **78**, 12665-7.

McClellan, K., and Perry, C. M., 2001, Oseltamivir: a review of its use in influenza. *Drugs* **61**, 263-283.

McKimm-Breschkin, J., Trivedi, T., Hampson, A., Hay, A., Klimov, A., Tashiro, M., Hayden, F., and Zambon, M., 2003, Neuraminidase sequence analysis and susceptibilities of influenza virus clinical isolates to zanamivir and oseltamivir. *Antimicrob Agents Chemother* **47**, 2264-72.

McKimm-Breschkin, J. L., 2000, Resistance of influenza viruses to neuraminidase inhibitors– a review. *Antiviral Res* **47**, 1-17.

McMichael, A. J., Gotch, F. M., Noble, G. R., and Beare, P. A., 1983, Cytotoxic T-cell immunity to influenza. *N Engl J Med* **309**, 13-7.

McMichael, A. J., Michie, C. A., Gotch, F. M., Smith, G. L., and Moss, B., 1986, Recognition of influenza A virus nucleoprotein by human cytotoxic T lymphocytes. *J Gen Virol* **67 (Pt 4)**, 719-26.

Mehta, A., Carrouee, S., Conyers, B., Jordan, R., Butters, T., Dwek, R. A., and Block, T. M., 2001, Inhibition of hepatitis B virus DNA replication by imino sugars without the inhibition of the DNA polymerase: therapeutic implications. *Hepatology* **33**, 1488-95.

Mehta, A., Ouzounov, S., Jordan, R., Simsek, E., Lu, X., Moriarty, R. M., Jacob, G., Dwek, R. A., and Block, T. M., 2002, Imino sugars that are less toxic but more potent as antivirals, in vitro, compared with N-n-nonyl DNJ. *Antivir Chem Chemother* **13**, 299-304.

Meindl, P., Bodo, G., Palese, P., Schulman, J., and Tuppy, H., 1974, Inhibition of neuraminidase activity by derivatives of 2-deoxy-2,3-dehydro-N-acetylneuraminic acid. *Virology* **58**, 457-463.

Mendel, D. B., Tai, C. Y., Escarpe, P. A., Li, W., Sidwell, R. W., Huffman, J. H., Sweet, C., Jakeman, K. J., Merson, J., Lacy, S. A., Lew, W., Williams, M. A., Zhang, L., Chen, M. S., Bischofberger, N., and Kim, C. U., 1998, Oral administration of a prodrug of the influenza virus neuraminidase inhibitor GS 4071 protects mice and ferrets against influenza infection. *Antimicrob Agents Chemother* **42**, 640-646.

Mendelman, P. M., Cordova, J., and Cho, I., 2001, Safety, efficacy and effectiveness of the influenza virus vaccine, trivalent, types A and B, live, cold-adapted (CAIV-T) in healthy children and healthy adults. *Vaccine* **19**, 2221-6.

Monto, A. S., Fleming, D. M., Henry, D., de Groot, R., Makela, M., Klein, T., Elliott, M., Keene, O. N., and Man, C. Y., 1999, Efficacy and safety of the neuraminidase inhibitor zanamivir in the treatment of influenza A and B virus infections. *J Infect Dis* **180**, 254-261.

Mori, I., Goshima, F., Koshizuka, T., Koide, N., Sugiyama, T., Yoshida, T., Yokochi, T., Nishiyama, Y., and Kimura, Y., 2003, Differential activation of the c-Jun N-terminal kinase/stress-activated protein kinase and p38 mitogen-activated protein kinase signal transduction pathways in the mouse brain upon infection with neurovirulent influenza A virus. *J Gen Virol* **84**, 2401-2408.

Mori, I., Komatsu, T., Takeuchi, K., Nakakuki, K., Sudo, M., and Kimura, Y., 1995, In vivo induction of apoptosis by influenza virus. *J Gen Virol* **76**, 2869-73.

Murphy, B. R., and Coelingh, K., 2002, Principles underlying the development and use of live attenuated cold-adapted influenza A and B virus vaccines. *Viral Immunol* **15**, 295-323.

Murphy, B. R., Park, E. J., Gottlieb, P., and Subbarao, K., 1997, An influenza A live attenuated reassortant virus possessing three temperature-sensitive mutations in the PB2 polymerase gene rapidly loses temperature sensitivity following replication in hamsters. *Vaccine* **15**, 1372-8.

Muster, T., Subbarao, E. K., Enami, M., Murphy, B. R., and Palese, P., 1991, An influenza A virus containing influenza B virus 5' and 3' noncoding regions on the neuraminidase gene is attenuated in mice. *Proc Natl Acad Sci U S A* **88**, 5177-81.

N.N., 1985, Current status of amantadine and rimantadine as anti-influenza-A agents: memorandum from a WHO meeting. *Bull World Health Organ* **63**, 51-56.

Nemoto, S., DiDonato, J. A., and Lin, A., 1998, Coordinate regulation of IkappaB kinases by mitogen-activated protein kinase kinase kinase 1 and NF-kappaB-inducing kinase. *Mol Cell Biol* **18**, 7336-43.

Neumann, G., Hughes, M. T., and Kawaoka, Y., 2000, Influenza A virus NS2 protein mediates vRNP nuclear export through NES-independent interaction with hCRM1. *Embo J* **19**, 6751-8.

Neumann, G., and Kawaoka, Y., 1999, Genetic engineering of influenza and other negative-strand RNA viruses containing segmented genomes. *Adv Virus Res* **53**, 265-300.

Neumann, G., Watanabe, T., Ito, H., Watanabe, S., Goto, H., Gao, P., Hughes, M., Perez, D. R., Donis, R., Hoffmann, E., Hobom, G., and Kawaoka, Y., 1999, Generation of influenza A viruses entirely from cloned cDNAs. *Proc Natl Acad Sci U S A* **96**, 9345-50.

Nichol, K. L., 2001, Live attenuated influenza virus vaccines: new options for the prevention of influenza. *Vaccine* **19**, 4373-7.

Nichol, K. L., Wuorenma, J., and von Sternberg, T., 1998, Benefits of influenza vaccination for low-, intermediate-, and high-risk senior citizens. *Archives of Internal Medicine* **158**, 1769.

Nicholson, K. G., Aoki, F. Y., Osterhaus, A. D., Trottier, S., Carewicz, O., Mercier, C. H., Rode, A., Kinnersley, N., and Ward, P., 2000, Efficacy and safety of oseltamivir in treatment of acute influenza: a randomised controlled trial. Neuraminidase Inhibitor Flu Treatment Investigator Group. *Lancet* **355**, 1845-1850.

Nimmerjahn, F., Dudziak, D., Dirmeier, U., Hobom, G., Riedel, A., Schlee, M., Staudt, L. M., Rosenwald, A., Behrends, U., Bornkamm, G. W., and Mautner, J., 2004, Active NF-kappaB signalling is a prerequisite for influenza virus infection. *J Gen Virol* **85**, 2347-56.

Nugent, K. M., and Shanley, J. D., 1984, Verapamil inhibits influenza A virus replication. *Arch Virol* **81**, 163-70.

O'Neill, R. E., Talon, J., and Palese, P., 1998, The influenza virus NEP (NS2 protein) mediates the nuclear export of viral ribonucleoproteins. *Embo J* **17**, 288-96.

Olschlager, V., Pleschka, S., Fischer, T., Rziha, H. J., Wurzer, W., Stitz, L., Rapp, U. R., Ludwig, S., and Planz, O., 2004, Lung-specific expression of active Raf kinase results in increased mortality of influenza A virus-infected mice. *Oncogene* **23**, 6639-46.

Olsen, C. W., Kehren, J. C., Dybdahl-Sissoko, N. R., and Hinshaw, V. S., 1996, bcl-2 alters influenza virus yield, spread, and hemagglutinin glycosylation. *J Virol* **70**, 663-6.

Osterhaus, A. D., and Rimmelzwaan, G. F. (1998): Induction of virus-specific immunity by iscoms. *Dev Biol Stand* **92**, 49-58.

Ovcharenko, A. V., and Zhirnov, O. P., 1994 Aprotinin aerosol treatment of influenza and paramyxovirus bronchopneumonia of mice. *Antiviral Res* **23**, 107-18.

Oxford, J. S., Logan, I. S., and Potter, C. W., 1970, In vivo selection of an influenza A2 strain resistant to amantadine. *Nature* **226**, 82-83.

Ozaki, H., Govorkova, E. A., Li, C., Xiong, X., Webster, R. G., and Webby, R. J., 2004, Generation of high-yielding influenza A viruses in African green monkey kidney (Vero) cells by reverse genetics. *J Virol* **78**, 1851-7.

Pahl, H. L., 1999, Activators and target genes of Rel/NF-kappaB transcription factors. *Oncogene* **18**, 6853-66.

Palese, P., and Compans, R. W., 1976, Inhibition of influenza virus replication in tissue culture by 2-deoxy-2,3-dehydro-N-trifluoroacetylneuraminic acid (FANA): mechanism of action. *J Gen Virol* **33**, 159-163.

Palese, P., and Garcia-Sastre, A., 2002, Influenza vaccines: present and future. *J Clin Invest* **110**, 9-13.

Palese, P., Muster, T., Zheng, H., O'Neill, R., and Garcia-Sastre, A., 1999, Learning from our foes: a novel vaccine concept for influenza virus. *Arch Virol Suppl* **15**, 131-8.

Palese, P., and Schulman, J., 1977, Inhibitors of viral neuraminidase as potential antiviral drugs, pp. 189-205. In J. S. Oxford (Ed.): *Chemoprophylaxis and virus infections of the upper respiratory tract*, CRC Press, Boca Raton.

Palese, P., Tobita, K., Ueda, M., and Compans, R. W., 1974, Characterization of temperature sensitive influenza virus mutants defective in neuraminidase. *Virology* **61**, 397-410.

Palese, P., Zheng, H., Engelhardt, O. G., Pleschka, S., and Garcia-Sastre, A., 1996, Negative-strand RNA viruses: genetic engineering and applications. *Proc Natl Acad Sci U S A* **93**, 11354-8.

Parkes, K. E., Ermert, P., Fassler, J., Ives, J., Martin, J. A., Merrett, J. H., Obrecht, D., Williams, G., and Klumpp, K., 2003, Use of a pharmacophore model to discover a new class of influenza endonuclease inhibitors. *J Med Chem* **46**, 1153-64.

Parkin, N. T., Chiu, P., and Coelingh, K., 1997, Genetically engineered live attenuated influenza A virus vaccine candidates. *J Virol* **71**, 2772-8.

Pau, M. G., Ophorst, C., Koldijk, M. H., Schouten, G., Mehtali, M., and Uytdehaag, F., 2001 The human cell line PER.C6 provides a new manufacturing system for the production of influenza vaccines. *Vaccine* **19**, 2716.

Pinto, L. H., Lamb, R. A., and Holsinger, L. J., 1995, Understanding the mechanism of action of the anti-influenza virus drug amantadine. *Trends Microbiol* **3**, 271. The wild-type M2 channel was found to be regulated by pH. The wild-type M2 ion channel activity is proposed to have a pivotal role in the biology of influenza virus infection.

Planz, O., Pleschka, S., and Ludwig, S., 2001, MEK-specific inhibitor U0126 blocks spread of Borna disease virus in cultured cells. *J Virol* **75**, 4871-7.

Pleschka, S., Jaskunas, R., Engelhardt, O. G., Zurcher, T., Palese, P., and Garcia-Sastre, A., 1996, A plasmid-based reverse genetics system for influenza A virus. *J Virol* **70**, 4188-92.

Pleschka, S., Wolff, T., Ehrhardt, C., Hobom, G., Planz, O., Rapp, U. R., and Ludwig, S., 2001, Influenza virus propagation is impaired by inhibition of the Raf/MEK/ERK signalling cascade. *Nat Cell Biol* **3**, 301-5.

Podda, A., 2001, The adjuvanted influenza vaccines with novel adjuvants: experience with the MF59-adjuvanted vaccine. *Vaccine* **19**, 2673-80.

Potter, C. W., and Oxford, J. S., 1979, Determinants of Immunity to Influenza Infection in Man. *British Medical Bulletin* **35**, 69.

Puthavathana, P., Auewarakul, P., Charoenying, P. C., Sangsiriwut, K., Pooruk, P., Boonnak, K., Khanyok, R., Thawachsupa, P., Kijphati, R., and Sawanpanyalert, P., 2005, Molecular characterization of the complete genome of human influenza H5N1 virus isolates from Thailand. *J Gen Virol* **86**, 423-33.

Ramanathan, R. K., Potter, D. M., Belani, C. P., Jacobs, S. A., Gravenstein, S., Lim, F., Kim, H., Savona, S., Evans, T., Buchbarker, D., Simon, M. B., Depee, J. K., and Trump,

D. L., 2002, Randomized trial of influenza vaccine with granulocyte-macrophage colony-stimulating factor or placebo in cancer patients. *J Clin Oncol* **20**, 4313-8.

Razvi, E. S., and Welsh, R. M., 1995, Apoptosis in viral infections. *Adv Virus Res* **45**, 1-60.

Rimmelzwaan, G. F., Baars, M., van Amerongen, G., van Beek, R., and Osterhaus, A. D., 2001, A single dose of an ISCOM influenza vaccine induces long-lasting protective immunity against homologous challenge infection but fails to protect Cynomolgus macaques against distant drift variants of influenza A (H3N2) viruses. *Vaccine* **20**, 158-63.

Roche, 2001, Tamiflu™ patient information.

Romanova, J., Katinger, D., Ferko, B., Vcelar, B., Sereinig, S., Kuznetsov, O., Stukova, M., Erofeeva, M., Kiselev, O., Katinger, H., and Egorov, A., 2004, Live cold-adapted influenza A vaccine produced in Vero cell line. *Virus Res* **103**, 187-93.

Root, C. N., Wills, E. G., McNair, L. L., and Whittaker, G. R., 2000, Entry of influenza viruses into cells is inhibited by a highly specific protein kinase C inhibitor. *J Gen Virol* **81**, 2697-705.

Rott, O., Charreire, J., Semichon, M., Bismuth, G., and Cash, E., 1995, B cell super-stimulatory influenza virus (H2-subtype) induces B cell proliferation by a PKC-activating, Ca(2+)-independent mechanism. *J Immunol* **154**, 2092-103.

Savarino, A., 2005, Expanding the frontiers of existing antiviral drugs: Possible effects of HIV-1 protease inhibitors against SARS and avian influenza. *J Clin Virol.*

Schickli, J. H., Flandorfer, A., Nakaya, T., Martinez-Sobrido, L., Garcia-Sastre, A., and Palese, P., 2001, Plasmid-only rescue of influenza A virus vaccine candidates. *Philos Trans R Soc Lond B Biol Sci* **356**, 1965-73.

Scholtissek, C., and Faulkner, G. P., 1979, Amantadine-resistant and -sensitive influenza A strains and recombinants. *J Gen Virol* **44**, 807-815.

Schultz-Cherry, S., and Hinshaw, V. S., 1996, Influenza virus neuraminidase activates latent transforming growth factor beta. *J Virol* **70**, 8624-9.

Schultz-Cherry, S., Krug, R. M., and Hinshaw, V. S., 1998, Induction of apoptosis by influenza virus. *Semin Virol* **8**, 491-495.

Sebolt-Leopold, J. S., Dudley, D. T., Herrera, R., Van Becelaere, K., Wiland, A., Gowan, R. C., Tecle, H., Barrett, S. D., Bridges, A., Przybranowski, S., Leopold, W. R., and Saltiel, A. R., 1999, Blockade of the MAP kinase pathway suppresses growth of colon tumors in vivo. *Nat Med* **5**, 810-6.

Seo, S. H., Goloubeva, O., Webby, R., and Webster, R. G., 2001, Characterization of a porcine lung epithelial cell line suitable for influenza virus studies. *Journal of Virology* **75**, 9517.

Sieczkarski, S. B., Brown, H. A., and Whittaker, G. R., 2003, Role of protein kinase C betaII in influenza virus entry via late endosomes. *J Virol* **77**, 460-9.

Sieczkarski, S. B., and Whittaker, G. R., 2002, Dissecting virus entry via endocytosis. *J Gen Virol* **83**, 1535-45.

Slee, E. A., Adrain, C., and Martin, S. J., 1999, Serial killers: ordering caspase activation events in apoptosis. *Cell Death Differ* **6**, 1067-74.

Smee, D. F., Huffman, J. H., Morrison, A. C., Barnard, D. L., and Sidwell, R. W., 2001, Cyclopentane neuraminidase inhibitors with potent in vitro anti-influenza virus activities. *Antimicrob Agents Chemother* **45**, 743-8.

Stepanova, L., Naykhin, A., Kolmskog, C., Jonson, G., Barantceva, I., Bichurina, M., Kubar, O., and Linde, A., 2002, The humoral response to live and inactivated influenza vaccines administered alone and in combination to young adults and elderly. *J Clin Virol* **24**, 193-201.

Stephenson, I., Nicholson, K. G., Wood, J. M., Zambon, M. C., and Katz, J. M., 2004, Confronting the avian influenza threat: vaccine development for a potential pandemic. *Lancet Infect Dis* **4**, 499-509.

Subbarao, K., Chen, H., Swayne, D., Mingay, L., Fodor, E., Brownlee, G., Xu, X., Lu, X., Katz, J., Cox, N., and Matsuoka, Y., 2003, Evaluation of a genetically modified reassortant H5N1 influenza A virus vaccine candidate generated by plasmid-based reverse genetics. *Virology* **305**, 192-200.

Subbarao, K., Klimov, A., Katz, J., Regnery, H., Lim, W., Hall, H., Perdue, M., Swayne, D., Bender, C., Huang, J., Hemphill, M., Rowe, T., Shaw, M., Xu, X., Fukuda, K., and Cox, N., 1998, Characterization of an avian influenza A (H5N1) virus isolated from a child with a fatal respiratory illness. *Science* **279**, 393-6.

Swierkosz, E. M., Newman, F. K., Anderson, E. L., Nugent, S. L., Mills, G. B., and Belshe, R. B., 1994, Multidose, live attenuated, cold-recombinant, trivalent influenza vaccine in infants and young children. *J Infect Dis* **169**, 1121-4.

Takahashi, K., Furuta, Y., Fukuda, Y., Kuno, M., Kamiyama, T., Kozaki, K., Nomura, N., Egawa, H., Minami, S., and Shiraki, K., 2003, In vitro and in vivo activities of T-705 and oseltamivir against influenza virus. *Antivir Chem Chemother* **14**, 235-41.

Takizawa, T., Fukuda, R., Miyawaki, T., Ohashi, K., and Nakanishi, Y., 1995, Activation of the apoptotic Fas antigen-encoding gene upon influenza virus infection involving spontaneously produced beta-interferon. *Virology* **209**, 288-96.

Takizawa, T., Matsukawa, S., Higuchi, Y., Nakamura, S., Nakanishi, Y., and Fukuda, R., 1993, Induction of programmed cell death (apoptosis) by influenza virus infection in tissue culture cells. *J Gen Virol* **74**, 2347-55.

Takizawa, T., Ohashi, K., and Nakanishi, Y., 1996, Possible involvement of double-stranded RNA-activated protein kinase in cell death by influenza virus infection. *J Virol* **70**, 8128-32.

Takizawa, T., Tatematsu, C., Ohashi, K., and Nakanishi, Y., 1999, Recruitment of apoptotic cysteine proteases (caspases) in influenza virus-induced cell death. *Microbiol Immunol* **43**, 245-52.

Talon, J., Salvatore, M., O'Neill, R. E., Nakaya, Y., Zheng, H., Muster, T., Garcia-Sastre, A., and Palese, P., 2000, Influenza A and B viruses expressing altered NS1 proteins: A vaccine approach. *Proc Natl Acad Sci U S A* **97**, 4309-14.

Tamura, S., and Kurata, T., 2004, Defense Mechanisms against Influenza Virus Infection in the Respiratory Tract Mucosa. *Jpn. J. Infect. Dis.* **57**, 236-247.

Taniguchi, T., and Takaoka, A., 2002, The interferon-alpha/beta system in antiviral responses: a multimodal machinery of gene regulation by the IRF family of transcription factors. *Curr Opin Immunol* **14**, 111-6.

Taubenberger, J. K., Reid, A. H., and Fanning, T. G., 2000, The 1918 influenza virus: A killer comes into view. *Virology* **274**, 241-5.

Thanos, D., and Maniatis, T., 1995, Virus induction of human IFN beta gene expression requires the assembly of an enhanceosome. *Cell* **83**, 1091-100.

Thornberry, N. A., and Lazebnik, Y., 1998, Caspases: enemies within. *Science* **281**, 1312-6.

Tisdale, M., 2000: Monitoring of viral susceptibility: new challenges with the development of influenza NA inhibitors. *Rev Med Virol* **10**, 45-55.

Toker, A., 1998, Signaling through protein kinase C. *Front Biosci* **3**, D1134-47.

Tomassini, J., Selnick, H., Davies, M. E., Armstrong, M. E., Baldwin, J., Bourgeois, M., Hastings, J., Hazuda, D., Lewis, J., McClements, W., and et al., 1994, Inhibition of cap (m7GpppXm)-dependent endonuclease of influenza virus by 4-substituted 2,4-dioxobutanoic acid compounds. *Antimicrob Agents Chemother* **38**, 2827-37.

Tomassini, J. E., 1996, Expression, purification, and characterization of orthomyxovirus: influenza transcriptase. *Methods Enzymol* **275**, 90-9.

Tominack, R. L., and Hayden, F. G., 1987, Rimantadine hydrochloride and amantadine hydrochloride use in influenza A virus infections. *Infect Dis Clin North Am* **1**, 459-478.

Townsend, A., Bastin, J., Bodmer, H., Brownlee, G., Davey, J., Gotch, F., Gould, K., Jones, I., McMichael, A., Rothbard, J., and et al., 1989, Recognition of influenza virus proteins by cytotoxic T lymphocytes. *Philos Trans R Soc Lond B Biol Sci* **323**, 527-33.

Treanor, J. J., and Betts, R. F., 1998, Evaluation of live, cold-adapted influenza A and B virus vaccines in elderly and high-risk subjects. *Vaccine* **16**, 1756-60.

Treanor, J. J., Hayden, F. G., Vrooman, P. S., Barbarash, R., Bettis, R., Riff, D., Singh, S., Kinnersley, N., Ward, P., and Mills, R. G., 2000, Efficacy and safety of the oral neuraminidase inhibitor oseltamivir in treating acute influenza: a randomized controlled trial. US Oral Neuraminidase Study Group. *Jama* **283**, 1016-1024.

Treanor, J. J., Kotloff, K., Betts, R. F., Belshe, R., Newman, F., Iacuzio, D., Wittes, J., and Bryant, M., 1999, Evaluation of trivalent, live, cold-adapted (CAIV-T) and inactivated (TIV) influenza vaccines in prevention of virus infection and illness following challenge of adults with wild-type influenza A (H1N1), A (H3N2), and B viruses. *Vaccine* **18**, 899-906.

Treanor, J. J., Mattison, H. R., Dumyati, G., Yinnon, A., Erb, S., O'Brien, D., Dolin, R., and Betts, R. F., 1992, Protective efficacy of combined live intranasal and inactivated influenza A virus vaccines in the elderly. *Ann Intern Med* **117**, 625-33.

Tree, J. A., Richardson, C., Fooks, A. R., Clegg, J. C., and Looby, D., 2001, Comparison of large-scale mammalian cell culture systems with egg culture for the production of influenza virus A vaccine strains. *Vaccine* **19**, 34-44.

Tumpey, T. M., Lu, X., Morken, T., Zaki, S. R., and Katz, J. M., 2000, Depletion of lymphocytes and diminished cytokine production in mice infected with a highly virulent influenza A (H5N1) virus isolated from humans. *J Virol* **74**, 6105-16.

Van Campen, H., Easterday, B. C., and Hinshaw, V. S., 1989a, Destruction of lymphocytes by a virulent avian influenza A virus. *J Gen Virol* **70**, 467-72.

Van Campen, H., Easterday, B. C., and Hinshaw, V. S., 1989b, Virulent avian influenza A viruses: their effect on avian lymphocytes and macrophages in vivo and in vitro. *J Gen Virol* **70**, 2887-95.

Varghese, J. N., Laver, W. G., and Colman, P. M., 1983, Structure of the influenza virus glycoprotein antigen neuraminidase at 2.9 A resolution. *Nature* **303**, 35-40.

Varghese, J. N., McKimm-Breschkin, J. L., Caldwell, J. B., Kortt, A. A., and Colman, P. M., 1992, The structure of the complex between influenza virus neuraminidase and sialic acid, the viral receptor. *Proteins* **14**, 327-332.

Varghese, J. N., Smith, P. W., Sollis, S. L., Blick, T. J., Sahasrabudhe, A., McKimm-Breschkin, J. L., and Colman, P. M., 1998, Drug design against a shifting target: a structural basis for resistance to inhibitors in a variant of influenza virus neuraminidase. *Structure* **6**, 735-746.

von Itzstein, M., Wu, W. Y., Kok, G. B., Pegg, M. S., Dyason, J. C., Jin, B., Van Phan, T., Smythe, M. L., White, H. F., Oliver, S. W., and *et al,*. 1993, Rational design of potent sialidase-based inhibitors of influenza virus replication. *Nature* **363**, 418-423.

Wada, N., Matsumura, M., Ohba, Y., Kobayashi, N., Takizawa, T., and Nakanishi, Y., 1995, Transcription stimulation of the Fas-encoding gene by nuclear factor for interleukin-6 expression upon influenza virus infection. *J Biol Chem* **270**, 18007-12.

Wagner, R., Wolff, T., Herwig, A., Pleschka, S., and Klenk, H. D., 2000, Interdependence of hemagglutinin glycosylation and neuraminidase as regulators of influenza virus growth: a study by reverse genetics. *J Virol* **74**, 6316-6323.

Wang, X., Li, M., Zheng, H., Muster, T., Palese, P., Beg, A. A., and Garcia-Sastre, A., 2000, Influenza A virus NS1 protein prevents activation of NF-kappaB and induction of alpha/beta interferon. *J Virol* **74**, 11566-73.

Wareing, M. D., and Tannock, G. A., 2001, Live attenuated vaccines against influenza; an historical review. *Vaccine* **19**, 3320-30.

Watanabe, T., Watanabe, S., Kida, H., and Kawaoka, Y., 2002a, Influenza A virus with defective M2 ion channel activity as a live vaccine. *Virology* **299**, 266-70.

Watanabe, T., Watanabe, S., Neumann, G., Kida, H., and Kawaoka, Y., 2002b, Immunogenicity and protective efficacy of replication-incompetent influenza virus-like particles. *J Virol* **76**, 767-73.

Webby, R. J., and Webster, R. G., 2003, Are we ready for pandemic influenza? *Science* **302**, 1519-22.

Webster, R. G., 1997a Influenza virus: transmission between species and relevance to emergence of the next human pandemic. *Arch Virol Suppl* **13**, 105-13.

Webster, R. G., 1997b Predictions for future human influenza pandemics. *J Infect Dis* **176 Suppl 1**, S14-9.

Webster, R. G., 1999, 1918 Spanish influenza: the secrets remain elusive. *Proc Natl Acad Sci U S A* **96**, 1164-6.

Webster, R. G., Bean, W. J., Gorman, O. T., Chambers, T. M., and Kawaoka, Y., 1992, Evolution and ecology of influenza A viruses. *Microbiol Rev* **56**, 152-79.

Webster, R. G., Sharp, G. B., and Claas, E. C., 1995, Interspecies transmission of influenza viruses. *Am J Respir Crit Care Med* **152**, S25-30.

Webster, R. G., Shortridge, K. F., and Kawaoka, Y., 1997, Influenza: interspecies transmission and emergence of new pandemics. *FEMS Immunol Med Microbiol* **18**, 275-9.

Welliver, R., Monto, A. S., Carewicz, O., Schatteman, E., Hassman, M., Hedrick, J., Jackson, H. C., Huson, L., Ward, P., and Oxford, J. S., 2001, Effectiveness of oseltamivir in preventing influenza in household contacts: a randomized controlled trial. *Jama* **285**, 748-754.

Widmann, C., Gibson, S., Jarpe, M. B., and Johnson, G. L., 1999, Mitogen-activated protein kinase: conservation of a three-kinase module from yeast to human. *Physiol Rev* **79**, 143-80.

Wilschut, J., and McElhaney, J. E., 2005, *Influenza*. Mosby Elsevier Limited.

Windon, R. G., Chaplin, P. J., McWaters, P., Tavarnesi, M., Tzatzaris, M., Kimpton, W. G., Cahill, R. N., Beezum, L., Coulter, A., Drane, D., Sjolander, A., Pearse, M., Scheerlinck, J. P., and Tennent, J. M., 2001, Local immune responses to influenza antigen are synergistically enhanced by the adjuvant ISCOMATRIX. *Vaccine* **20**, 490-7.

Wolff, J. A., Malone, R. W., Williams, P., Chong, W., Acsadi, G., Jani, A., and Felgner, P. L., 1990, Direct gene transfer into mouse muscle in vivo. *Science* **247**, 1465-8.

Woodland, D. L., Hogan, R. J., and Zhong, W., 2001, Cellular immunity and memory to respiratory virus infections. *Immunol Res* **24**, 53-67.

Woods, J. M., Bethell, R. C., Coates, J. A., Healy, N., Hiscox, S. A., Pearson, B. A., Ryan, D. M., Ticehurst, J., Tilling, J., Walcott, S. M., and *et al.*, 1993, 4-Guanidino-2,4-dideoxy-2,3-dehydro-N-acetylneuraminic acid is a highly effective inhibitor both of the sialidase (neuraminidase) and of growth of a wide range of influenza A and B viruses in vitro. *Antimicrob Agents Chemother* **37**, 1473-1479.

World Health Organization, 2005, Communicable Disease Surveillance & Response (CSR). Avian Influenza.

Woronicz, J. D., Gao, X., Cao, Z., Rothe, M., and Goeddel, D. V., 1997, IkappaB kinase-beta: NF-kappaB activation and complex formation with IkappaB kinase-alpha and NIK. *Science* **278**, 866-9.

Wright, P. F., and Webster, R. G., 2001, Orthomyxoviruses, pp. 1533-1579. In D. M. Knipe, P. M. Howley, and D. E. Griffin (Eds): *Fields Virology*, Lippencott Williams & Williams.

Wurzer, W. J., Ehrhardt, C., Pleschka, S., Berberich-Siebelt, F., Wolff, T., Walczak, H., Planz, O., and Ludwig, S., 2004, NF-kappaB-dependent induction of tumor necrosis factor-related apoptosis-inducing ligand (TRAIL) and Fas/FasL is crucial for efficient influenza virus propagation. *J Biol Chem* **279**, 30931-7.

Wurzer, W. J., Planz, O., Ehrhardt, C., Giner, M., Silberzahn, T., Pleschka, S., and Ludwig, S., 2003, Caspase 3 activation is essential for efficient influenza virus propagation. *Embo J* **22**, 2717-28.

Yoneyama, M., Suhara, W., Fukuhara, Y., Fukuda, M., Nishida, E., and Fujita, T., 1998, Direct triggering of the type I interferon system by virus infection: activation of a transcription factor complex containing IRF-3 and CBP/p300. *Embo J* **17**, 1087-95.

Youil, R., Su, Q., Toner, T. J., Szymkowiak, C., Kwan, W. S., Rubin, B., Petrukhin, L., Kiseleva, I., Shaw, A. R., and DiStefano, D., 2004, Comparative study of influenza virus replication in Vero and MDCK cell lines. *Journal of Virological Methods* **120**, 23.

Zamanian-Daryoush, M., Mogensen, T. H., DiDonato, J. A., and Williams, B. R., 2000, NF-kappaB activation by double-stranded-RNA-activated protein kinase (PKR) is mediated through NF-kappaB-inducing kinase and IkappaB kinase. *Mol Cell Biol* **20**, 1278-90.

Zambon, M., and Hayden, F. G., 2001, Position statement: global neuraminidase inhibitor susceptibility network. *Antiviral Res* **49**, 147-156.

Zhirnov, O. P., Konakova, T. E., Garten, W., and Klenk, H., 1999, Caspase-dependent N-terminal cleavage of influenza virus nucleocapsid protein in infected cells. *J Virol* **73**, 10158-63.

Zhirnov, O. P., Ovcharenko, A. V., Bukrinskaia, A. G., Ursaki, L. P., and Ivanova, L. A., 1984, [Antiviral and therapeutic action of protease inhibitors in viral infections: experimental and clinical observations]. *Vopr Virusol* **29**, 191-7.

Zurbriggen, R., 2003, Immunostimulating reconstituted influenza virosomes. *Vaccine* **21**, 921-4.

Chapter 1.6

A NEW APPROACH TO AN INFLUENZA LIVE VACCINE: MODIFICATION OF CLEAVAGE SITE OF THE HAEMAGGLUTININ BY REVERSE GENETICS

J. STECH[1], H. GARN [2], and H.-D. KLENK[1]

[1]*Institute of Virology, Philipps-University Marburg, Germany;* [2]*Institute for Clinical Chemistry and Molecular Diagnostics, Philipps-University Marburg, Germany*

Abstract: A promising approach to reduce the impact of influenza is the use of an attenuated live virus as a vaccine. Using reverse genetics, we generated a mutant of strain A/WSN/33 with a modified cleavage site within its haemagglutinin which depends on proteolytic activation by elastase. Unlike the wild-type requiring trypsin, this mutant is strictly dependent on elastase. Both viruses grow equally well in cell culture. In contrast to the lethal wild-type, the mutant is entirely attenuated in mice. At a dose of 10^5 pfu it induced complete protection against lethal challenge. This approach allows the conversion of any epidemic strain into a genetically homologous attenuated virus.

1.   INTRODUCTION

Due to annually recurring epidemics and the enduring pandemic threat, influenza remains a serious public health problem, despite the availability of inactivated vaccines. A live vaccine FluMist™ manufactured by MedImmune Inc. is now commercially available in some countries. This vaccine consists of two influenza A and one influenza B reassortants of a cold-adapted master strain. These reassortants have temperature-sensitive mutations in the polymerase subunits PB2 and PB1 and the nucleoprotein genes (Jin *et al.*, 2003) and are decorated with the haemagglutinin and the neuraminidase of the circulating epidemic strain. Such a vaccine virus might

E. Bogner and A. Holzenburg (eds.), New Concepts of Antiviral Therapy, 169–187.

provide especially the internal genes without temperature-sensitive mutation for reassortment with an epidemic virus and, thus, give rise to a new strain with unpredictable traits (Scholtissek *et al.*, 1979, Yamnikova *et al.*, 1993).

An important step in the replication cycle of the influenza virus is cleavage of the haemagglutinin by a host protease in order to gain infectivity (Klenk *et al.*, 1975, Lazarowitz and Choppin, 1975) by activating the fusion potential (Maeda and Ohnishi, 1980, Huang *et al.*, 1980, White *et al.*, 1981). The cleavage site contains a conserved arginine or a stretch of basic amino acids like the highly-pathogenic avian strains (Bosch *et al.*, 1981, Garten *et al.*, 1981, Suarez *et al.*, 1998).

Using reverse genetics we generated a mutant of strain A/WSN/33 (H1N1) with a modified haemagglutinin cleavage site that requires elastase instead of trypsin for activation. This virus is attenuated *in vivo*, but grows *in vitro* as well as the wild-type, and induces a strong protection against lethal infection with the wild-type. Methods and experimental details are described in (Stech *et al.*, 2005). Such an approach allows the conversion of any epidemic strain as a whole into an attenuated live vaccine virus. It is genetically homologous to the wild-type and, thus, avoids the risk of generating pathogenic reassortants.

2.   GENERATION OF AN ELASTASE-DEPENDENT INFLUENZA A VIRUS

We sought to make a virus that is no longer susceptible to *in vivo* activation at the basic haemagglutinin cleavage site but can be activated *in vitro* by a protease not available under natural conditions. Wild-type haemagglutinin is cleaved by trypsin-like serine proteases, and cleavage generates two fragments, HA1 and HA2. Two requirements had to be considered for the choice of the cleavage motif of such a protease. First, the glycine had to be retained at position P1 because this amino acid is essential at the aminoterminus of the HA2 part for induction of fusion and cannot be replaced without compromising or even abolishing fusion (Garten *et al.*, 1981, Cross *et al.*, 2001, Qiao *et al.*, 1999, Steinhauer *et al.*, 1995). Second, for reducing the probability of reversion, an amino acid should be selected whose codon differs by two nucleotides from *any* arginine or lysine codon (Gunther *et al.*, 1993, Kawaoka *et al.*, 1990). We therefore exchanged AG at positions 1059 and 1060 of the A/WSN/33 haemagglutinin for GT, replacing arginine 343 by valine and generating a cleavage site for porcine pancreatic elastase (Gunther *et al.*, 1993, Kawaoka *et al.*, 1990) (Figure 1). Using the reverse genetics system established by (Hoffmann *et al.*, 2000), we generated recombinant viruses containing either authentic (WSNwt) or mutated (cleavable by elastase, WSN-E) haemagglutinin.

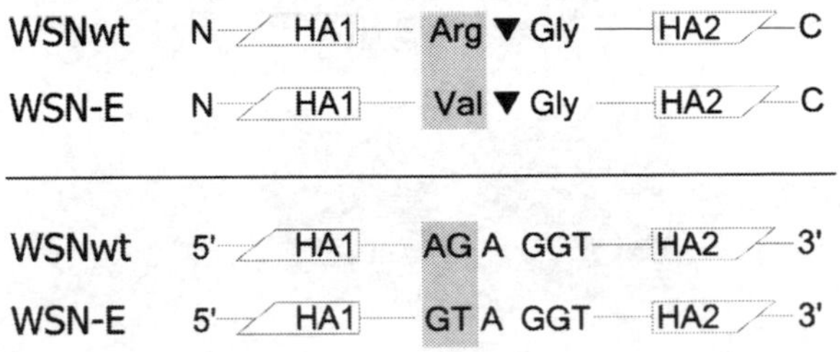

WSNwt   N — HA1 — Arg ▼ Gly — HA2 — C

WSN-E   N — HA1 — Val ▼ Gly — HA2 — C

_____

WSNwt   5' — HA1 — AG A GGT — HA2 — 3'

WSN-E   5' — HA1 — GT A GGT — HA2 — 3'

Figure 1. Cleavage sites of WSNwt and WSN-E. The monobasic cleavage site of the WSNwt haemagglutinin contains an arginine, which has been replaced by a valine in the case of WSN-E.

3.   *IN VITRO* PROPERTIES

Multicycle replication and, thus, plaque formation require proteolytic activation (Klenk *et al.*, 1975). Therefore, we performed plaque assays on MDCK cells in the presence of either trypsin or elastase in the plaque overlay or in the absence of an exogenous protease for control. With WSNwt, clear plaques were only visible in the presence of trypsin. In contrast, WSN-E was exclusively activated by elastase (Figure 2). These results demonstrate that neither plasmin shown to activate WSNwt (Lazarowitz *et al.*, 1973, Lazarowitz and Choppin, 1975) nor another protease provided by MDCK cells or fetal calf serum (from the cell culture medium) is able to cleave WSN-E haemagglutinin.

We then analyzed the dependance of the haemagglutinin cleavage in WSNwt on trypsin and in WSN-E on elastase by Western blotting. The haemagglutinin of WSN-E is cleaved by elastase, but is resistant to trypsin (Figure 3). After incubation of WSNwt with elastase, a weak HA1 band appeared which can be attributed to incomplete cleavage between amino acids glycine and leucine (Gunther *et al.*, 1993, Kawaoka *et al.*, 1990). This cleavage does not cause proteolytic activation as demonstrated by the plaque assay.

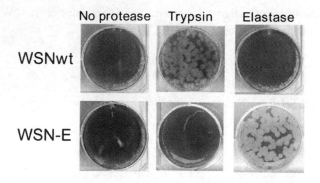

Figure 2. Plaque assay of WSNwt and WSN-E in the absence of an exogenous protease, in the presence of trypsin or elastase.

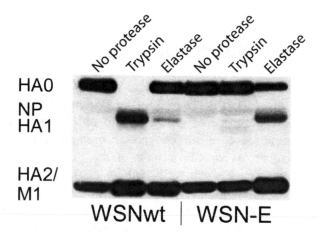

Figure 3. Western blot of WSNwt and WSN-E in the absence of an exogenous protease, in the presence of trypsin or elastase.

In presence of the appropriate protease, both viruses grew to similar titers although kinetics were somewhat slower with WSN-E (Figure 4).

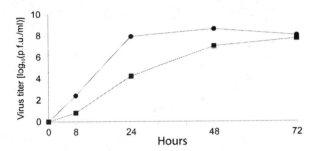

Figure 4. Growth curves of WSNwt (circles) in the presence of trypsin and WSN-E (squares) in the presence of elastase.

4. *IN VIVO* PROPERTIES

4.1 Attenuation of WSN-E

To analyze WSN-E *in vivo*, we infected mice with 10^6 pfu of either WSNwt or WSN-E and observed them for 14 d. When infected with wild-type virus, mice ($n=3$) showed signs of disease and died on day seven. However, in the case of WSN-E, all mice ($n=4$) survived without weight loss or any visible symptoms of disease (Figures 5 - 6).

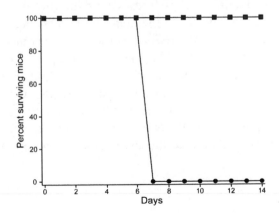

Figure 5. Survival rates of mice following infection with WSNwt (circles) or WSN-E (squares).

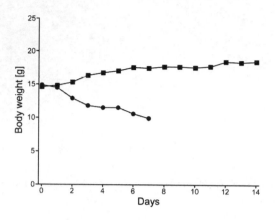

Figure 6. Average weight loss following infection with WSNwt (circles) or WSN-E (squares).

4.2   Restricted replication in lung

For comparison of the pathogenic traits of both viruses *in vivo*, we inoculated mice with either PBS, 10^6 pfu WSNwt, or 10^6 pfu WSN-E. After 12 h, 1 d, and 3 d, we removed lung, brain, and heart. Beginning with undiluted organ homogenate, we performed plaque titrations with WSNwt in the presence of trypsin and with WSN-E in the presence of elastase. We found WSNwt in the lungs up to the third day, showing a rise in titer. Wild-type virus was also detectable in brain and heart. However, mice inoculated with WSN-E revealed a different picture. We did not find WSN-E in brain or heart at any time point analyzed. In the lung, viral titers stagnated from 12 h to 1 d past inoculation. On day three, we could not detect any WSN-E virus (Figures 7- 9). The lung titer of WSN-E at 12 h was approximately 1.4×10^6 pfu per mouse lung which is close to the inoculum dose. For WSNwt, the result is strikingly different. The lung titer at 12 h was approximately 2.2×10^8 pfu per mouse lung, showing that WSNwt readily multiplies in the lung. In contrast, replication of WSN-E is abortive because it is restricted *in vivo* to one replication cycle due to the absence of the appropriate protease.

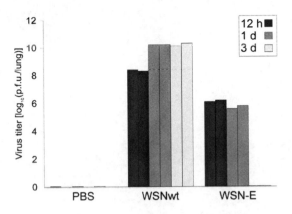

Figure 7. Plaque titers of WSNwt and WSN-E in mice from entire lungs removed at 12 h, 1 d or 3 d.

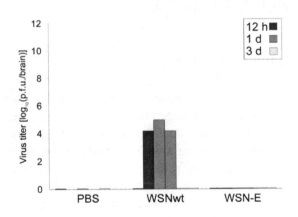

Figure 8. Plaque titers of WSNwt and WSN-E in mice from entire brains removed at 12 h, 1 d or 3 d.

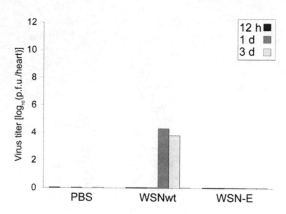

Figure 9. Plaque titers of WSNwt and WSN-E in mice from entire hearts removed at 12 h, 1 d or 3 d.

5.   GENETIC STABILITY

5.1   Mouse lung passages

To check for the emergence of revertants, we carried out sequential lung passages in mice. At the first passage, we inoculated the animals with 10^6 pfu WSN-E. For the subsequent infections, we used 50 µl lung homogenate from the previous passage. During five lung passages of either 1 d or 3 d duration, the mice remained unaffected. We determined the virus titers by plaque assay in the presence of elastase. Because WSNwt is activated by plasmin (Lazarowitz *et al.*, 1973, Lazarowitz and Choppin, 1975), this virus can be detected in absence of trypsin if only one wash is performed before inoculation. We detected no virus even in undiluted lung homogenate obtained from the second to the fifth of the 1 d passages and from the first to the fifth of the 3 d passages. This demonstrates the absence of WSN-E and of revertants.

5.2   Reversion of WSN-E *in vitro*

After the first passage in mouse lung, the entire amount of WSN-E was in the range of 10^5 to 10^6 pfu. Such virus populations are too small for generation of double-point revertants having an equilibrium frequency of approximately 10^{-5} to 10^{-8} (Ribeiro *et al.*, 1998). Therefore, we passaged the elastase-dependent WSN-E on MDCK cells in the presence of trypsin, beginning with different inocula of 10^8, 10^7, or 10^6 pfu each in 10 parallel cell cultures. From all 10^8 pfu inocula and from six out of ten 10^7 pfu inocula, we found trypsin-dependent virus with lysine at its cleavage site. We could not obtain any revertants from inocula of 10^6 pfu. Therefore, the reversion frequency within the WSN-E stock is approximately 10^{-7}. The low reversion rate and the small viral loads of WSN-E in the mouse lung explain the absence of revertants during mouse passages. Another reason for the genetic stability of WSN-E *in vivo* is the restriction to one replication cycle due to the absence of the appropriate protease.

5.3   Genetic stability and vaccine production

A frequent objection against the use of influenza virus live vaccines is the possibility of reversion to pathogenicity. Because of the double mutation (Arg→Val) within the cleavage site, two nucleotides at once have to be replaced for back-mutation; suppressor mutants outside of the cleavage region seem to be impossible. This explains the low reversion frequency in cell culture. In hen eggs, a factor X-like protease is present (Gotoh *et al.*, 1990) which should cause a considerably higher proportion of revertants. Therefore, eggs might not be suitable for vaccine production. However, in cell culture the substitution of trypsin by elastase for propagation of WSN-E leads both to positive selection for elastase-dependent virus and to negative selection against revertants.

6.   IMMUNIZATION

6.1   Protection against lethal challenge

To investigate the potential of WSN-E to serve as a live vaccine, we immunized five groups of mice with WSN-E: four groups received live virus at dosages of 10^3 ($n=6$), 10^4 ($n=6$), 10^5 ($n=6$) or 10^6 pfu ($n=5$); one received formalin-inactivated virus ($n=4$). An additional group of non-immunized mice ($n=6$) served as a positive control during challenge. The animals tolerated the immunization without any signs of illness as in the previously described survival experiment. Four weeks later, we challenged the mice with 10^6 pfu WSNwt and monitored survival and weight loss. The challenge was lethal for both groups vaccinated with formalin-inactivated virus or 10^3

pfu WSN-E, and for the non-immunized control animals. From the mice immunized with 10^4 pfu WSN-E, four out of six mice survived and partially recovered from disease. Although some animals from the group vaccinated with 10^5 pfu WSN-E developed temporary weight loss and milder disease symptoms, they eventually recovered. All animals vaccinated with the highest dose of 10^6 WSN-E survived the challenge and did not develop any weight loss or other visible symptoms of illness (Figures 10-11). 3 d past challenge, we removed the lungs of two mice per group for plaque assay. Remarkably, after vaccination with 10^6 pfu WSN-E, no plaques were seen even in undiluted lung homogenates. This contrasts strikingly with the plaque titers of the other groups (Figure 12).

When inoculated with 10^6 pfu inactivated WSN-E, mice did neither survive a challenge with WSNwt nor develop a detectable antibody response. It may be argued that the formalin-inactivation had been too harsh. The protein amount used was approximately 80 ng per immunization dose. Moreover, we prepared the formalin-inactivated WSN-E from the same non-concentrated virus stock used for life immunization and inoculated it intranasally just once (like life WSN-E). However, such inactivated vaccines are made from concentrated virus and usually administered 3 times to the mice (Takada *et al.*, 2003). The failure of immunization with formalin-inactivated virus demonstrated that WSN-E replication was required for protection.

The challenge experiment indicates that the degree of protection against disease increases with the immunization dose. Additionally, the failure of the formalin-inactivated virus to prevent death shows that WSN-E must replicate in order to induce protection. Taken together, these results demonstrate that WSN-E is an attenuated virus that is able to prevent lethal influenza virus infection.

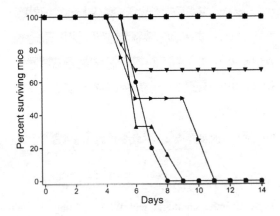

Figure 10. Protection against lethal challenge with WSNwt. Survival rates of challenged mice immunized with formalin-inactivated WSN-E (right-facing triangles), 10^3 pfu WSN-E (triangles), 10^4 pfu WSN-E (inverted triangles), 10^5 pfu WSN-E (diamonds), or 10^6 pfu WSN-E (squares), and non-immunized (circles).

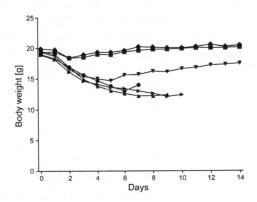

Figure 11. Protection against lethal challenge with WSNwt. Average weight loss of challenged mice immunized with formalin-inactivated WSN-E (right-facing triangles), 10^3 pfu WSN-E (triangles), 10^4 pfu WSN-E (inverted triangles), 10^5 pfu WSN-E (diamonds), or 10^6 pfu WSN-E (squares), and non-immunized (circles).

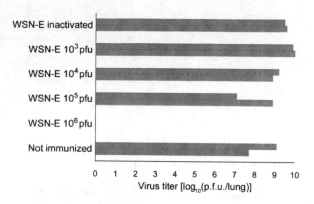

Figure 12. Protection against lethal challenge with WSNwt. Lung plaque titers from the third day.

6.2   Serum and mucosal antibody responses

We immunized mice with 10^3 (n=6, n=5), 10^4 (n=5, n=4), 10^5 (n=5, n=6) or 10^6 pfu (n=4, n=6) WSN-E, 10^6 pfu formalin-inactivated WSN-E (n=5, n=4), WSNwt at a nonlethal dosage of 10^3 pfu (n=5, n=7), or with PBS (n=6, n=6). Four weeks later, we sacrificed the animals. 3 d prior to analysis, we challenged a subgroup of each treatment cohort with 10^6 pfu WSNwt. To investigate the virus-specific antibody response, we determined the IgG titers of sera and IgA titers of bronchoalveolar (BAL) fluid and nasal wash samples by ELISA. Additionally, we performed serum haemagglutination inhibition (HI) tests. Animals which received 10^3 pfu WSN-E, formalin-inactivated WSN-E or PBS did not show any antibody response neither before nor after challenge. By contrast, the groups immunized with 10^4, 10^5, or 10^6 pfu WSN-E showed substantial levels of virus-specific IgG and HI titers in serum as well as IgA titers in BAL and nasal wash that increased with the immunization dose (Figure 13-16). We achieved the highest antibody titers with 10^3 pfu WSNwt. With 10^6 pfu WSN-E, the HI titer before challenge was 1:40 (Figure 14). In comparison with non-challenged animals, challenged mice of the WSN-E 10^6 pfu group showed elevated antibody titers (Figure 13-16), especially in the nasal wash (Figure 16).

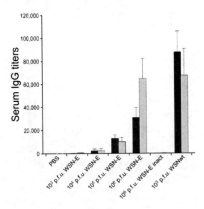

Figure 13. Serum IgG titers of non-challenged (black bars) and challenged (grey bars) mice immunized with PBS, 10^3 pfu WSN-E, 10^4 pfu WSN-E, 10^5 pfu WSN-E, 10^6 pfu WSN-E, 10^6 pfu formalin-inactivated WSN-E, and 10^3 pfu WSNwt.

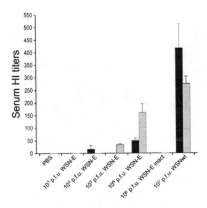

Figure 14. Serum haemagglutination inhibition titers of non-challenged (black bars) and challenged (grey bars) mice immunized with PBS, 10^3 pfu WSN-E, 10^4 pfu WSN-E, 10^5 pfu WSN-E, 10^6 pfu WSN-E, 10^6 pfu formalin-inactivated WSN-E, and 10^3 pfu WSNwt.

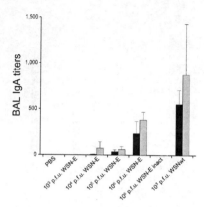

Figure 15. Bronchoalveolar lavage IgA titers of non-challenged (black bars) and challenged (grey bars) mice immunized with PBS, 10^3 pfu WSN-E, 10^4 pfu WSN-E, 10^5 pfu WSN-E, 10^6 pfu WSN-E, 10^6 pfu formalin-inactivated WSN-E, and 10^3 pfu WSNwt.

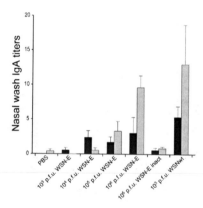

Figure 16. Nasal wash IgA titers of non-challenged (black bars) and challenged (grey bars) mice immunized with PBS, 10^3 pfu WSN-E, 10^4 pfu WSN-E, 10^5 pfu WSN-E, 10^6 pfu WSN-E, 10^6 pfu formalin-inactivated WSN-E, and 10^3 pfu WSNwt.

One intranasal inoculation of WSN-E induced a substantial, dose-dependent local and systemic immune response despite very limited presence in lung. A dosage of 10^5 or 10^6 pfu induced remarkable HI titers, serum IgG and mucosal IgA titers. They were lower than those induced by 10^3 pfu WSNwt because its longer replication enables antigenic stimulation. However, the challenged animals showed almost comparable systemic and mucosal Ig titers if immunized with 10^6 pfu WSN-E. This indicates that a notable immunological memory had already been induced in these animals.

7.   CLEAVAGE SITE MUTANTS AS LIVE VACCINE

Live influenza vaccines presently approved for human application are reassortants generated by coinfection of a cold-adapted temperature-sensitive master strain from which the six segments coding for the internal virion components are derived and the circulating strain which provides the haemagglutinin and neuraminidase genes (Maassab, 1967, Murphy *et al.*, 1979). The faster generation of such reassortants by reverse genetics entirely from plasmids (Jin *et al.*, 2003, Hoffmann *et al.*, 2000, Neumann *et al.*, 1999) is feasible as well. These live vaccines are well-tolerated and effective (Belshe *et al.*, 1992, Gruber, 1998, Beyer *et al.*, 2002). However, such an attenuated virus may give rise to a new viral strain with unpredictable traits by exchanging the internal genes especially those without temperature-sensitive mutations (the polymerase subunit PA, matrix (M) and non-structural (NS) proteins genes) (Jin *et al.*, 2003) with the circulating strain. Experimental evidence for generation of a pathogenic virus from reassortment of two apathogenic strains has indeed been obtained (Scholtissek *et al.*, 1979, Yamnikova *et al.*, 1993). Such a scenario is avoided when a cleavage site mutant is used containing all eight genes of the circulating strain. A cleavage site mutant would deliver all antigens identical to the circulating strain and, therefore, be the most authentic vaccine. The possible advantage of this feature is indicated by studies demonstrating that the internal influenza virion components prime a helper response cooperating in the antibody response against the haemagglutinin (Russell and Liew, 1979, Scherle and Gerhard, 1988). The immunogenic relevance of the internal components is also underlined by the observation that live virus and to some extent inactivated whole virus vaccine can induce heterotypic protection in contrast to subunit vaccines (Webster and Askonas, 1980). Furthermore, it has been reported (O'Neill *et al.*, 2000), that mice could be protected successfully from lethal infection with A/HongKong/156/97 (H5N1) by prior immunization with the A/Quail/HongKong/G1/97 (H9N2) isolate that harbors internal genes 98% homologous to the H5N1 isolate.

However, in vaccine production, some circulating strains may grow to inadequate titers. Therefore, the propagation of such a seed virus rescued from genes of the epidemic strain may be delayed unpredictably. A solution would be to adapt an epidemic strain to cell lines suitable for vaccine production and to use its internal genes as a backbone each year. The internal genes evolve considerably more slowly than the surface glycoproteins (Webster *et al.*, 1992). Thereby, both high growth properties and sufficient antigenic homology of the internal viral proteins can be provided. Moreover, this backbone can carry additional attenuating mutations.

8.   CONCLUSIONS

The goal of this study was to generate influenza A virus with an atypical haemagglutinin cleavage site that is resistant to activation during natural infection but can readily be activated under *in vitro* conditions. We have accomplished this by replacing the original trypsin-specific cleavage site Arg-Gly by the elastase-sensitive one Val-Gly. Elastase mutants have previously been obtained after conventional cell culture passages in the presence of this enzyme (Orlich *et al.*, 1995). This study demonstrates, however, that by reverse genetics generation of such mutants has become a fast and reproducible procedure suitable for routine production. WSNwt and the elastase-substituted WSN-E grew to similar titers in cell culture. In mouse lung, WSN-E was present only temporarily and did not cause any disease. But after infection with wild-type virus, we observed much higher lung titers, spread of virus to other organs, and 100% lethality. Thus, the cleavage site mutant proved to be equivalent to wild-type virus regarding growth rates *in vitro*, but was completely attenuated *in vivo* (Stech *et al.*, 2005). Because of these properties, WSN-E is a promising candidate for a live vaccine.

In order to demonstrate that our approach is generalizable to highly-pathogenic influenza strains, we recently generated an elastase-dependent mutant of the strain SC35M. This virus is an H7N7 isolate, carries a poly-basic cleavage site and is highly-pathogenic both for chickens and mice (Scheiblauer *et al.*, 1995). Like WSN-E, the SC35M mutant is strictly dependent on elastase and grows to similar titers in cell culture like the wild-type (unpublished).

The absence of appropriate proteases for WSN-E *in vivo* allows only one (or just very few) replication cycle(s) leading to self-limiting replication. This is the main difference to other attenuated viruses undergoing many replication cycles *in vivo*. For cold-adapted viruses, a duration of viral shedding up to 11 d in susceptible humans was reported (Wright *et al.*, 1975). An important advantage of the short self-limiting replication is the decreased probability of any reversion including the cleavage site motif itself and other attenuating mutations.

Because proteolytic activation is essential for the replication of each influenza virus, the conversion of any epidemic strain or of viruses with pandemic potential, such as highly pathogenic H5N1 strains, into a live vaccine by altering the cleavage site is possible. Major assets of cleavage site mutants are antigenic identity to the parent strain, nonexisting risk of generating new pathogenic reassortants, complete attenuation *in vivo*, and *in vitro* growth equivalent to wild-type. Such an attenuated virus is an ideal candidate for a live vaccine.

ACKNOWLEDGEMENTS

We are very grateful to E. Hoffmann and R. G. Webster for providing us the plasmids of the reverse genetics system. This life vaccine approach has been published first in a Nature Medicine technical report (Stech *et al.*, 2005).

REFERENCES

Belshe, R. B., Swierkosz, E. M., Anderson, E. L., Newman, F. K., Nugent, S. L. and Maassab, H. F., 1992, Immunization of infants and young children with live attenuated trivalent cold-recombinant influenza A H1N1, H3N2, and B vaccine *J Infect Dis.* **165**: 727-32.

Beyer, W. E., Palache, A. M., de Jong, J. C. and Osterhaus, A. D., 2002, Cold-adapted live influenza vaccine versus inactivated vaccine: systemic vaccine reactions, local and systemic antibody response, and vaccine efficacy. A meta-analysis. *Vaccine.* **20**: 1340-53.

Bosch, F. X., Garten, W., Klenk, H. D. and Rott, R., 1981, Proteolytic cleavage of influenza virus hemagglutinins: primary structure of the connecting peptide between HA1 and HA2 determines proteolytic cleavability and pathogenicity of Avian influenza viruses. *Virology* **113**: 725-35.

Cross, K. J., Wharton, S. A., Skehel, J. J., Wiley, D. C. and Steinhauer, D. A., 2001, Studies on influenza haemagglutinin fusion peptide mutants generated by reverse genetics. *Embo J.* **20**: 4432-42.

Garten, W., Bosch, F. X., Linder, D., Rott, R. and Klenk, H. D., 1981, Proteolytic activation of the influenza virus hemagglutinin: The structure of the cleavage site and the enzymes involved in cleavage. *Virology* **115**: 361-74.

Gotoh, B., Ogasawara, T., Toyoda, T., Inocencio, N. M., Hamaguchi, M. and Nagai, Y., 1990, An endoprotease homologous to the blood clotting factor X as a determinant of viral tropism in chick embryo. *Embo J.* **9**: 4189-95.

Gruber, W. C., 1998, Children as target for immunization In *Textbook of Influenza* (Eds, Nicholson, K. G., Webster, R. G. and Hay, A. J.) Blackwell Science, Oxford, UK, pp. 435-444.

Gunther, I., Glatthaar, B., Doller, G. and Garten, W., 1993, A H1 hemagglutinin of a human influenza A virus with a carbohydrate-modulated receptor binding site and an unusual cleavage site. *Virus Res.* **27**: 147-60.

Hoffmann, E., Neumann, G., Kawaoka, Y., Hobom, G. and Webster, R. G., 2000, A DNA transfection system for generation of influenza A virus from eight plasmids. *Proc Natl Acad Sci U S A* **97**: 6108-13.

Huang, R. T. C., Wahn, K., Klenk, H. D. and Rott, R., 1980, Fusion between cell membranes and liposomes containing the glycoprotein of influenza virus *Virology* **104**: 294-302.

Jin, H., Lu, B., Zhou, H., Ma, C., Zhao, J., Yang, C., Kemble, G. and Greenberg, H.,2003, Multiple amino acid residues confer temperature sensitivity to human influenza virus vaccine strains (flumist) derived from cold-adapted A/Ann Arbor/6/60. *Virology* **306**: 18-24.

Kawaoka, Y., Yamnikova, S., Chambers, T. M., Lvov, D. K. and Webster, R. G., 1990, Molecular characterization of a new hemagglutinin, subtype H14, of influenza A virus. *Virology* **179**: 759-67.

Klenk, H. D., Rott, R., Orlich, M. and Blodorn, J., 1975, Activation of influenza A viruses by trypsin treatment. *Virology* **68**: 426-39.

Lazarowitz, S. G. and Choppin, P. W., 1975, Enhancement of the infectivity of influenza A and B viruses by proteolytic cleavage of the hemagglutinin polypeptide. *Virology* **68**: 440-54.

Lazarowitz, S. G., Goldberg, A. R. and Choppin, P. W., 1973, Proteolytic cleavage by plasmin of the HA polypeptide of influenza virus: host cell activation of serum plasminogen. *Virology* **56**: 172-80.

Maassab, H. F., 1967, Adaptation and growth characteristics of influenza virus at 25 degrees C. *Nature* **213**: 612-4.

Maeda, T. and Ohnishi, S., 1980, Activation of influenza virus by acidic media causes hemolysis and fusion of erythrocytes. *FEBS Lett.* **122**: 283-7.

Murphy, B. R., Holley, H. P., Jr., Berquist, E. J., Levine, M. M., Spring, S. B., Maassab, H. F., Kendal, A. P. and Chanock, R. M., 1979, Cold-adapted variants of influenza A virus: evaluation in adult seronegative volunteers of A/Scotland/840/74 and A/Victoria/3/75 cold-adapted recombinants derived from the cold-adapted A/Ann Arbor/6/60 strain. *Infect Immun.* **23**: 253-9.

Neumann, G., Watanabe, T., Ito, H., Watanabe, S., Goto, H., Gao, P., Hughes, M., Perez, D. R., Donis, R., Hoffmann, E., Hobom, G. and Kawaoka, Y., 1999, Generation of influenza A viruses entirely from cloned cDNAs. *Proc Natl Acad Sci U S A* **96**: 9345-50.

O'Neill, E., Krauss, S. L., Riberdy, J. M., Webster, R. G. and Woodland, D. L., 2000, Heterologous protection against lethal A/HongKong/156/97 (H5N1) influenza virus infection in C57BL/6 mice. *J Gen Virol* **81**: 2689-96.

Orlich, M., Linder, D. and Rott, R., 1995, Trypsin-resistant protease activation mutants of an influenza virus. *J Gen Virol* **76**: 625-33.

Qiao, H., Armstrong, R. T., Melikyan, G. B., Cohen, F. S. and White, J. M., 1999, A specific point mutant at position 1 of the influenza hemagglutinin fusion peptide displays a hemifusion phenotype. *Mol Biol Cell* **10**: 2759-69.

Ribeiro, R. M., Bonhoeffer, S. and Nowak, M. A., 1998, The frequency of resistant mutant virus before antiviral therapy. *Aids* **12**: 461-5.

Russell, S. M. and Liew, F. Y., 1979, T cells primed by influenza virion internal components can cooperate in the antibody response to haemagglutinin. *Nature* **280**: 147-8.

Scheiblauer, H., Kendal, A. P. and Rott, R., 1995, Pathogenicity of influenza A/Seal/Mass/1/80 virus mutants for mammalian species. *Arch Virol.* **140**: 341-8.

Scherle, P. A. and Gerhard, W., 1988, Differential ability of B cells specific for external vs. internal influenza virus proteins to respond to help from influenza virus-specific T-cell clones in vivo. *Proc Natl Acad Sci U S A,* **85**: 4446-50.

Scholtissek, C., Vallbracht, A., Flehmig, B. and Rott, R., 1979, Correlation of pathogenicity and gene constellation of influenza A viruses. II. Highly neurovirulent recombinants derived from non-neurovirulent or weakly neurovirulent parent virus strains. *Virology* **95**: 492-500.

Stech, J., Garn, H., Wegmann, M., Wagner, R. and Klenk, H. D., 2005, A new approach to an influenza live vaccine: modification of the cleavage site of hemagglutinin *Nat Med.,* **epub**, doi:10.1038/nm1256.

Steinhauer, D. A., Wharton, S. A., Skehel, J. J. and Wiley, D. C., 1995, Studies of the membrane fusion activities of fusion peptide mutants of influenza virus hemagglutinin. *J Viro.,* **69**: 6643-51.

Suarez, D. L., Perdue, M. L., Cox, N., Rowe, T., Bender, C., Huang, J. and Swayne, D. E., 1998, Comparisons of highly virulent H5N1 influenza A viruses isolated from humans and chickens from Hong Kong. *J Viro.,* **72**: 6678-88.

Takada, A., Matsushita, S., Ninomiya, A., Kawaoka, Y. and Kida, H., 2003, Intranasal immunization with formalin-inactivated virus vaccine induces a broad spectrum of heterosubtypic immunity against influenza A virus infection in mice. *Vaccine* **21**: 3212-8.

Webster, R. G. and Askonas, B. A., 1980, Cross-protection and cross-reactive cytotoxic T cells induced by influenza virus vaccines in mice. *Eur J Immunol,* **10**: 396-401.

Webster, R. G., Bean, W. J., Gorman, O. T., Chambers, T. M. and Kawaoka, Y., 1992, Evolution and ecology of influenza A viruses. *Microbiol Rev.* **56**: 152-79.

White, J. M., Matlin, K. and Helenius, A., 1981, Cell fusion by Semliki Forest, influenza, and vesicular stomatitis viruses. *J Cell Biol.* **89**: 674-9.

Wright, P. F., Sell, S. H., Shinozaki, T., Thompson, J. and Karzon, D. T., 1975, Safety and antigenicity of influenza A/Hong Kong/68-ts-1 (E) (H3N2). *J Pediatr.* **87**, 1109-16.

Yamnikova, S. S., Mandler, J., Bekh-Ochir, Z. H., Dachtzeren, P., Ludwig, S., Lvov, D. K. and Scholtissek, C., 1993, A reassortant H1N1 influenza A virus caused fatal epizootics among camels in Mongolia. *Virology.* **197**: 558-63.

Chapter 1.7

NEW CONCEPTS IN ANTI-HIV THERAPIES

J. STEBBING[1], M. BOWER[1] and B. GAZZARD[1]
[1] *The Chelsea and Westminster Hospital, 369 Fulham Road, London, UK*

Abstract: Fewer than one million HIV infected individuals are currently receiving anti-
              retroviral therapy. The limitations of such treatment have underscored the need
              to develop more effective strategies to control the spread and pathogenesis of
              infection. In 1996, the era of highly active anti-retroviral therapy (HAART)
              commenced in established market economies, causing a dramatic reduction in
              morbidity and mortality in those infected individuals who received these
              medicines. These agents target aspects of the viral life cycle and there are now
              20 approved therapeutic agents for licensed for treatment of infection with the
              human immunodeficiency virus (HIV), a pathogen that in the 1980s was
              uniformly fatal. These advances have been associated with significant
              toxicities and drug resistance. Antiviral potency and durability causing
              suppression of viremia has been the cornerstone of the initial success of
              HAART regimens. Following this, the restoration of immune function, the
              prevention of the emergence of resistance, and ultimately the prevention of
              disease progression has been the focus of treatment. Future progress will allow
              greater choices for physicians and patients.

1.   INTRODUCTION

The treatment of HIV infection has been revolutionised in developed
countries as a result of the introduction of highly active anti-retroviral
therapy (HAART), which has reduced short-term mortality and markedly
increased quality of life by preventing opportunistic diseases. A major
challenge has been linking the potency of HAART with the other desirable

189

E. Bogner and A. Holzenburg (eds.), New Concepts of Antiviral Therapy, 189–212.

aspects of a therapeutic regimen: low pill burden, excellent tolerability, absence of major drug interactions, absence of long-term toxic effects, and absence of cross-resistance to other agents (Hammer, 2002).

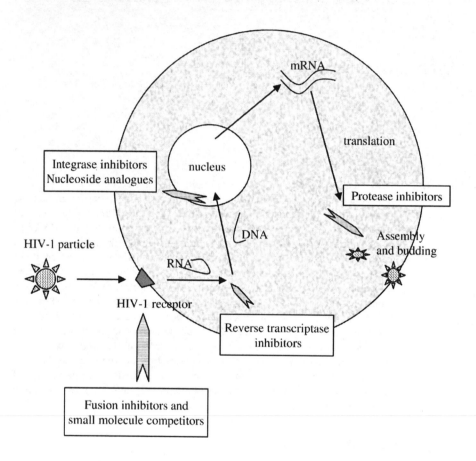

Figure 1. The HIV life cycle and drug targets.

While previous agents target the reverse transcriptase and viral protease, newer agents attack different stages of the HIV life cycle (figure 1).

2.   NUCLEOSIDE ANALOGUES

The first drugs to be used clinically were the nucleoside analogues which act as chain terminators of HIV viral reverse transcriptase and require tri-phosphorylation within cells to become active. For most drugs the rate-limiting step involves the initial phosphorylation of Zidovudine (ZDV) during the conversion of ZDV monophosphate to diphosphate by thymidine kinase.

Zidovudine was the first drug to be used clinically. In short term studies there was a dramatic improvement in mortality and reduction in AIDS defining events in patients with symptomatic HIV disease (Fischl *et al.*, 1989; Fischl *et al.*, 1990a; Fischl *et al.*, 1990b). Subsequently studies indicated that using ZDV as monotherapy in asymptomatic individuals offered no advantage over treating symptomatic disease only (Concorde, 1994). Subsequently two large randomised controlled trials with clinical end points demonstrated that using two nucleoside analogues together (Zidovudine and Didanosine or Zidovudine and Zalcitabine) produced a 33% survival advantage compared with use of ZDV monotherapy (Fischl *et al.*, 1995; Collier *et al.*, 1993). Subsequently a dual nucleoside analogue backbone has been the commonest component of the highly effective antiretroviral therapy regimes (HAART) currently used to treat HIV infection.

The drugs include

Zidovudine This drug produces a viral load drop of between 0.4 and 0.7 of a Log in HIV-1 RNA. Its major side effect is anaemia although a myopathy has also rarely been described. In the initial phases of therapy nausea, headache and sometimes diarrhoea occur. Initial studies used 1.2 g of Zidovudine spaced throughout the day but more recently, 250 mg and 300 mg of the drug twice a day are exclusively used following clinical studies suggesting that this dose was as effective as the larger dose but had fewer side effects.

Didanosine Monotherapy with this adenine nucleoside analogue used in the ACTG 175 trial was shown to be as effective as dual nucleoside combinations (Ragni *et al.*, 1995). The viral load drop produced in monotherapy studies is similar or greater than that seen with ZDV. A major side effect of this drug is pancreatitis which may be fatal and the drug is contra- indicated in those with pre-existing pancreatic damage. Peripheral neuropathy also probably occurs with didanosine, particularly in those who are pre-disposed although some large trials have failed to show an increased incidence of this complication which is sometimes a direct effect of HIV infection (De la Monte *et al.*, 1988). The dose normally prescribed is 400 mg per day as one dose taken on an empty stomach. There is evidence that higher doses are associated with increased toxicity including hepatic damage.

Stavudine   Like ZDV this is a thymidine analogue. *In vitro* studies indicate that stavudine may antagonize the effects of ZDV as it competes with the phosphorylation pathway mentioned above (Hoggard *et al.*, 1997; Fischl *et al.*, 1993a). This was confirmed in early clinical studies which showed a fall in CD4 count when AZT and stavudine were used together. A major side effect of stavudine is a peripheral neuropathy which may be irreversible as it produces axonal degeneration. The dose normally prescribed is 40 mg twice a day although an unlicensed extended release preparation was shown to have a similar effect on viral load given once a day. This preparation is likely to be licensed soon and there is a possibility that this may have fewer side effects as the peak level of drug in the plasma is lower.

Zalcitabine   This drug is rarely used now because one unpublished study showed it to be less effective than ZDV as monotherapy and peripheral neuropathy is a major side effect (Fischl *et al.*, 1993b). It is given three times a day which can be inconvenient for patients.

Abacavir   This guanidine analogue was introduced more recently than the other compounds and in monotherapy studies was shown to have a greater effect on plasma HIV viral load than some other nucleosides in short term studies with a viral load drop of more than 1.2 logs. A major side effect of this drug is an idiosyncratic hypersensitivity reaction which may be fatal if the patient is re-challenged following cessation of therapy. The hyper-sensitivity reaction is easy to recognize with prominent respiratory and gastrointestinal symptoms, a fever and usually a rash. Discontinuation of therapy results in rapid improvement in symptomatology but rechallenge is strongly contra-indicated. The CPK is often greatly raised during an attack. A widespread physician and patient educational programme has ensured that this condition is recognized. The hypersensitivity reaction is very strongly associated with the ancestral haplotype B5701 (Hetherington *et al.*, 2002). Whether this should be used as a screening test prior to introduction of therapy is unclear and the association is not sufficiently strong to use this as a test to diagnose a hypersensitivity reaction. This drug is prescribed twice a day but intracellular pharmacokinetics indicate that the triphosphate levels of the drug are maintained throughout a 24 hour period and as studies comparing twice a day with once a day regimens indicated a similar drop in plasma HIV viral load, it may be licensed for once a day use in the future.

Lamivudine (an analogue of cytosine) This drug was almost not developed because although significant plasma viral load drops were seen in patients treated in monotherapy studies, these were evanescent. However, sustained viral load drops were subsequently shown when Lamivudine was combined with other nucleoside analogues, particularly ZDV (Randomised trial, 1997). It is relatively free of side effects. Although the licensed dose is 150 mg twice a day, the long intracellular half-life of the triphosphate component indicates it can probably be used once a day. Again studies indicate that once a day regimes produced similar viral load drops to those when the drug is given twice a day and it is likely to be licensed for this indication shortly.

There are a number of new nucleoside analogues in development to treat HIV infection. These are summarized in Table 1. A number of nucleoside analogues which have been in development over the last two or three years have fallen by the wayside because of toxicities or unexpected pharmacokinetic difficulties. DAPG is continuing its Phase 3 development programme. Like Tenofovir DAPG is also active against Hepatitis B.

As the mechanisms for resistance to nucleoside analogues becomes clarified, it is likely that medicinal chemistry will start to develop drugs which can either overcome the reduced sensitivity of the virus or avoid it. Two mechanisms of resistance development are now clearly understood. With some drugs like 3TC the development of resistance by the virus involves mutations which increase the affinities to the natural nucleoside substrates at the expense of the drug. For other drugs resistance involves conformational changes which encourage the removal of the chain terminating nucleoside analogue once it is attached. The process of reverse transcription and elongation of the DNA chain is a reversible process. The nucleoside is substituted for by a phosphorus molecule either donated from pyrophosphate (pyrophosphorylisis) or from ATP. This process of reversal of chain termination is exaggerated in resistant mutant viruses. Tenofovir still acts as a chain terminator with some AZT resistant viruses because this reversal process is inefficient, presumably because of the particular structure of tenofovir.

Table 1. New nucleoside analogs for HIV

Drug	Stage of development	Comment
Emtricitabine [(-)-FTC]	Approved in July 2003	Similar in many ways to lamivudine (cross resistant to M184V) with once daily dosing
Alovudine (MIV-310, FLT)	Phase II	*In vitro*, it has potent activity against NRTI-resistant viral strains of HIV-1, including zidovudine-resistant viruses
Amdoxovir (DAPD)	Phase II	A guanosine analogue NRTI that is active *in vitro* against both HIV-1 and HBV
Racivir [(±)-FTC]	Phase II	See below
Reverset (D-d4FC, DPC-817)	Phase II	A cytidine nucleoside analog with potent activity against both wild-type and NRTI-resistant HIV-1, including lamivudine- and zidovudine-associated mutants
SPD 754 (dOTC)	Phase II	An investigational cytosine analogue NRTI
Elvucitabine (L-d4FC	Phase Ia/IIb	A L-cytidine analog with activity against HIV resistant to several other nucleoside analogs, including zidovudine and lamivudine

3.   PROTEASE INHIBITORS (PIS)

The development of the protease inhibitors was a major advance in the treatment of HIV infection. Following the formation of HIV DNA as a result of reverse transcription, this is incorporated into the host genome.

Subsequent host cell activation is associated with transcription of this HIV DNA and a polyprotein is produced which is cleaved into its active constituents by a virally encoded protease. As this is an aspartate protease, it is dissimilar to mammalian proteases and was an obvious drug target (Andreeva *et al.*, 1995). Clinical utility was rapidly confirmed by three clinical end point studies. One showed an improved outcome in late disease when ritonavir was added to otherwise failing therapy. A second showed a reduced frequency in clinical end points when a regimen of indinavir and two nucleoside analogues was compared with using the nucleoside analogues alone (Hammer *et al.*, 1997). Although protease inhibitors are potent drugs, they have a number of potential disadvantages. Most have short half-lives and have to be taken several times during the day. Absorption through the gut mucosa is often variable producing wide intra-patient and inter-patient variability in plasma levels. Most are metabolized by the microsomal enzyme system and some induce and others inhibit their own metabolism via cytochrome P450 which produces a wide variety of drug interactions.

For many protease inhibitors plasma levels can be enhanced by blockade of the cytochrome P450 system. This improves absorption, which is reduced in the gut as a result of the presence of cytochrome P450 and inhibits metabolism of these drugs. The most commonly used inhibitor of liver cytochrome P450 is ritonavir which, because of its side effect profile and its potent effects on cytochrome P450, is used in small doses which do not have an anti HIV effect per se. Few randomised trials are available to assess the effectiveness of boosted PIs although one study has demonstrated that lopinavir boosted by ritonavir produces superior surrogate marker results at 60 weeks when compared with nelfinavir therapy (Ruane *et al.*, 2001). Most guidelines now recommend the use of boosted PI regimens because of convenience of administration and high plasma levels. A number of different PIs are now licensed.

Saquinavir This was the first PI to be licensed and was in the form of a hard capsule where absorption was sub-optimal although this is significantly enhanced by Ritonavir. Although the soft-gel formulation of saquinavir has better absorption characteristics, it is mainly used clinically in conjunction with ritonavir in twice daily regimens.

Ritonavir This produces major gastrointestinal side effects when it is used as an anti HIV agent and so is little used other than as a pharmacokinetic (PK) booster.

Nelfinavir This is the only protease inhibitor which remains widely used without a PK enhancer. The drug can be administered twice a day, diarrhoea being the major side effect. Nelfinavir itself is an inhibitor if cytochrome P450 and can be used to boost the levels of saquinavir (Moyle *et al.*, 2000). This combination is rarely used in clinical practice. To maximise absorption nelfinavir is taken with a fatty meal. Nelfinavir has been shown to improve surrogate marker outcome at 48 weeks when compared with the use of dual nucleoside analogues alone (PENTA 5 trial, 2002) but to be inferior to lopinavir boosted with ritonavir over a similar period.

Indinavir As a single PI is taken three times a day on an empty stomach. The regime is inconvenient and is now little used to initiate therapy. Indinavir's major side effect is the development of renal damage, mainly because of the formation of Indinavir calculi in the collecting system. This side effect occurs in 10% to 15% of individuals and there have been reports of progressive renal damage. Indinavir's pharmacokinetics are improved by administration with ritonavir although in the doses first used, (100 mg Ritonavir and 800 mg of Indinavir twice a day) the instance of renal complications was unacceptably high at 20% at 24 weeks (Gatell *et al.*, 2000). Other unlicensed dosage regimes such as 400 mg twice a day + 100 mg of Ritonavir twice a day may be an effective anti HIV regimen and have fewer side effects.

Amprenavir This drug induces its own metabolism and thus if drug regimes are started with a standard dose, many patients suffer from gastrointestinal side effects during the initiation of therapy. Thus relatively few patients initiate treatment with amprenavir. Amprenavir boosted by ritonavir may have a role in treating patients who have failed previous protease inhibitor therapy.

Lopinavir Lopinavir is a potent drug *in vitro*. When used alone, because it is rapidly metabolised, plasma levels fall quickly. However, when boosted with ritonavir, high plasma levels can be maintained with twice daily treatment. For this reason many viruses with reduced sensitivity to other protease inhibitors remain susceptible to this drug combination. This was confirmed clinically in cohort studies of patients taking lopinavir and non nucleoside reverse transcriptase inhibitors (see below), although the relative importance of these two drugs in the favourable outcome is unclear. Recent data have supported the efficacy of the lopinavir-ritonavir combination (Walmsley *et al.*, 2002).

Atazanavir is an azapeptide which has potent inhibitory effects on HIV *in vitro* and because of its long plasma half life, can be given once daily. Phase

2 dose ranging studies in antiretroviral naïve patients indicates that at 48 weeks this agent is at least as potent as Nelfinavir (Cahn *et al.,* 2001). In individuals failing protease inhibitor containing regimes, using atazanavir is as effective in subsequent therapy as ritonavir/saquinavir combinations when combined with a new, nucleoside analogue regime (Haas *et al.,* 2001). Unfortunately this latter data set is relatively difficult to interpret as the majority of people failing the initial PI did not have resistant mutations to this class of drugs but presumably were failing because of poor adherence. Nevertheless atazanavir represents a potentially important new drug as 48 week studies have shown very little effect on triglyceride or cholesterol blood levels in treated individuals when compared with those treated with other PI containing regimens.

It may be that some individuals experiencing failure of protease inhibitor containing regimens who do have resistant mutations will remain sensitive to atazanavir particularly when plasma levels of this drug are boosted by additional ritonavir; doing this, however, is likely to reduce one of the major advantages of atazanavir which is its freedom from lipid abnormalities. This drug does now have a license for naïve patients in the US though it is used in conjunction with ritonavir boosting strategies in Europe.

TMC126 This drug which is in early development with Tibotec/Virco (as is TMC125) is active *in vitro* against a wide variety of viruses with reduced sensitivity to virtually all known PIs. Encouraging Phase 2 studies indicated a favourable pharmacokinetic profile when this drug is used with a small boosting dose of ritonavir (Erickson *et al.*, 2001).

Tipranavir Tipranavir is now widely available on compassionate release and has been developed as *in vitro*. This drug continues to be active against viruses with widespread mutations to the PIs. Recent studies have confirmed this *in vivo* although Fuzeon, the effects of Tipranavir are evanescent in that other drugs which are also active can be combined in the regime. Tipranavir is dosed twice a day with 200 mg of Ritonavir on each occasion because of its otherwise unfavourable pharmacokinetics.

TMC114 This drug is in phase 3 development. It has been developed again because of its ability *in vitro* to inhibit viruses which contain mutations to the presently available PIs. It is also dosed with ritonavir either once or twice a day and should be licensed in 2006 if all goes well.

Other new protease inhibitors for HIV include R944 (in phase 1; Roche) and TMC-114 (in phase 2, Johnson and Johnson).

4.   NON NUCLEOSIDE REVERSE TRANSCRIPTASE INHIBITORS (NNRTIS)

The clinical utility of this class of compounds has been assessed by surrogate marker trials. Like Lamivudine their early development was hampered by the rapid development of resistance and the prototype compound, TIBO, only produced transient rises in CD4 count because of the rapid selection of resistant mutations (Larder *et al.,* 1993). The first important positive controlled trial of non nucleoside reverse transcriptase inhibitors was with nevirapine which demonstrated that in a regimen also containing zidovudine and didanosine, there was a superior surrogate marker outcome compared to the use of the two nucleosides alone. This study was followed by a comparison of zidovudine and lamivudine with efavirenz and indinavir with the same nucleoside analogues which showed an equivalent or superior surrogate marker response at 48 weeks for the NNRTI containing regimen (Staszewski *et al.,* 1999). Nevertheless both efavirenz and nevi-rapine have established an important role in the initial treatment of HIV infection because of their freedom from irritating toxicities associated with drug administration and their relatively long half life (efavirenz is taken once a day and nevirapine which is licensed for twice a day is often used once a day). This long plasma half-life also gives considerable latitude around the time of dosing to maintain an antiviral effect. Two drugs in this class are currently licensed in Europe and three in the United States.

Nevirapine This drug is administered twice a day in the dose of 200 mg although pharmacokinetic data confirms that it can be given in a single 400 mg dose and this is widely used. As the drug induces its own metabolism, the initial dose is 200 mg a day for two weeks followed by the full dose which is said to minimise toxicity. Rash affects 20% to 30% of individuals taking it and is occasionally serious with a Stevens-Johnson reaction (also known as erythema multiforme major). Another serious side effect is hepato-toxicity. Fulminant hepatic failure requiring liver transplantation has occurred in HIV negative individuals given this drug in an unlicensed indication as post exposure prophylaxis (Sha *et al.,* 2000). This effect appears to be much less common in HIV seropositive patients and the frequency appears to be inversely related to the CD4 count. Hepatotoxicity is commoner in those with pre-existing abnormal liver function tests, those with other forms of liver disease, particularly Hepatitis C and B, and in older patients. Fulminant hepatic failure may not be possible to predict even when liver function tests are frequently monitored.

Efavirenz This drug is administered once a day in a dose of 600 mg. Drug absorption is enhanced by food although it is usually taken on an empty stomach last thing at night which is thought to minimise the vivid dreams which are an early side effect of this drug. It is also associated with skin rash although this is less likely to be severe than that seen with Nevirapine and continuing treatment is usually associated with resolution of the rash. A major side effect of Efavirenz is the central nervous system disturbance which usually wanes with continuing use although it is sometime persistent. Discontinuation because of this side effect is rare. Efavirenz acts both as an inducer and inhibitor of the cytochrome P450 complex and has the potential to cause a number of drug-drug interactions.

Delavirdine This is the only NNRTI which acts as a pure inhibitor of cytochrome P450 and allows dose reduction of protease inhibitors when used in combination. Rash which is common shortly after administration of this drug, is very rarely serious and most patients continue therapy. Hepato-toxicity is unusual. The drug has to be taken three times a day. The few surrogate marker trials performed with Delavirdine have been relatively more difficult to interpret than studies with other NNRTIs which is why it was not licensed in Europe.

5. SURROGATE MARKERS

The era of comparative studies with clinical end points as the major outcome came to an end with the initial studies on protease inhibitors. Clinicians and patient groups felt that treatment had improved so much that further studies using death or major deterioration as an end point were unethical (ACTG 320, 1997).

Both the fall in plasma viral load and rise in CD4 count following the introduction of antiretroviral therapy fulfil many of the criteria of clinical outcome (Lagakos, 1993). Thus both the levels of HIV, RNA and CD4 count prior to treatment predict outcome. Changes in both of these markers have a biological plausibility that they would affect outcome and changes in these markers can explain most but not all of the treatment effects (Carosi *et al.*, 2001). All recent studies of antiretroviral therapy have used these surrogate markers to assess likely clinical effectiveness. Preventing HIV viral replication as completely as possible has always been regarded as important and, therefore, particular attention has been paid to the ability of drugs in combinations to reduce plasma levels of HIV below detectable limits of sensitive assays, (currently less than 50 copies per ml). Many clinicians thus believe that the most important outcome of a clinical trial is the proportion of individuals who have a below 50 copy HIV-1 RNA assay using an intent to treat analysis at 48 weeks. This does not assess the durability of treatment and, therefore, time to treatment of virological failure is becoming an important yardstick for successful combinations. Monoclonal antibodies can be categorised according to type.

6.   WHAT TREATMENT TO START WITH?

There have been no definitive controlled trials to demonstrate the clinical superiority of any one HAART regimen containing three active drugs used as initial therapy i.e. protease inhibitor first, NNRTI first or three nucleoside reverse transcriptase inhibitors first. For patients with very high viral loads there is some suggestion that more than three active drugs may result in a more rapid decline in viral load (Hoetelmans *et al.*, 1998). Studies are in progress to determine whether this will lead to better long term outcomes compared with standard, three drug HAART.

With currently available antiretroviral agents, eradication of HIV infection is not likely to be possible (Chun *et al.*, 1997). The aim of treatment is thus to prolong life and improve quality of life by maintaining suppression of virus replication for as long as possible.

The three groups of treatment naïve patients for whom treatment guidelines are required are: patients with primary HIV infection, patients with asymptomatic HIV infection and patients with symptomatic HIV disease or AIDS.

6.1   Primary HIV infection

There is one placebo-controlled study of zidovudine (ZDV) monotherapy in primary HIV infection (PHI) (Kinloch-de Loes *et al.*, 1995) and it showed short-term benefit only. As yet there is no evidence of long-term clinical benefit from any study of treatment of PHI compared with deferring treatment until later, however. If it is recognized clinically, the diagnosis of PHI may represent a unique opportunity for therapeutic intervention. It is likely that, at the time of PHI: (1) there is a narrowing of the genetic diversity of the infecting virus compared with the virus in the index case

(Zhang *et al.*, 1999); (2) viral ability to infect different cell types may be limited; and (3) the capacity to mount an immune response is usually greater than it is later on. Therefore, the treatment of PHI may preserve HIV specific immune responses.

6.2 Symptomatic HIV infection

All patients with late disease and/or symptomatic HIV infection with a CD4 lymphocyte count consistently <200 cells/mm^3, or who have been diagnosed with AIDS or severe/recurrent HIV related illnesses or tumour at any CD4 count, should start therapy. This is because of the high risk of further opportunistic infections which, although treatable, may cause irreversible damage or be life threatening.

6.3 Asymptomatic HIV infection

There are no ongoing or planned controlled studies that sufficiently address the optimum time to start therapy. Current guidelines are therefore based upon previous studies of monotherapy and data from large clinical cohorts. Since the quality of evidence is relatively poor, opinion is divided on this question. In the UK, patients are diagnosed with HIV infection at a late stage. Over 30% present with a CD4 count of <200 cells/mm^3 (Gupta *et al.*, 2000) and, consequently, the early vs. late debate is irrelevant to many.

The decision on when to start treatment will be influenced principally by two considerations: the short term risk of developing AIDS prior to treatment and the potential efficacy of starting treatment at various CD4 counts. Data from several cohort studies with short term follow-up have suggested that patients who initiate therapy when the CD4 count is <200 cells/mm^3 have an increased mortality (Hogg and Wood, 2001; Sterling *et al.*, 2001; Kaplan and Karon, 2001) compared with those above this level, but were unable to show any difference in those starting at any CD4 level >200 cells/mm^3. However, data from other cohort studies (Phillips *et al.*, 2000) suggest that patients who delay therapy until the CD4 lymphocyte count is <200 cells/mm^3 may have a similar virological and immunological response to those starting earlier. This is in contrast to data from prospective clinical studies (Wood and Team, 2000; Opravil *et al.*, 2001), although the effect of baseline CD4 response on therapy may not be the same for all drugs (Nelson *et al.*, 2001). One study (Nelson *et al.*, 2001) has suggested that patients who commenced therapy with a CD4 count >350 cclls/mm^3 were less likely than those who commenced later to experience disease progression or death.

These data suggest that, ideally, patients should start therapy earlier, before the CD4 count has fallen to <200 cells/mm^3. A number of factors need to be considered when making decisions with each individual patient. Patients with a rapidly falling CD4 count (e.g. falling >80 cells/mm^3 per year on repeated testing) have an increased risk of CD4 cell count decline to <200 cells/mm^3 in the next 6 months. This group many thus be considered for initiation of therapy relatively earlier within the CD4 count range 200-350 cells/mm^3. Previous guidelines have suggested starting therapy relatively early in patients with a high plasma viral load (Carpenter *et al.*, 2000). There are three reasons why viral load measurement should help guide decisions about when to start antiretroviral therapy. First, a viral load >55000 copies/ml (Mellors *et al.* (2000) predicts a faster rate of decline in CD4 cells. Second, this level of viral load is an independent risk factor for subsequent disease progression and death. However, these data are from an era before the introduction of highly active antiretroviral therapy (HAART), and may not be relevant. Furthermore, recent cohort studies (Sterling *et al.*, 2001; Kaplan *et al.*, 2001) have suggested that baseline viral load does not predict subsequent mortality independently of the baseline CD4 count after starting therapy. Third, some data have suggested that the baseline viral load adversely affects the virological response to treatment in some prospective studies (Staszewski *et al.*, 1999; Moyle and Opravil, 2000).

In asymptomatic patients with established chronic HIV infection and CD4 cell counts consistently above 350 cells/mm^3, few data support starting therapy.

6.4   Management of treatment failure

The possibility of using cyclical treatment using drugs before they become resistant to limit toxicity is an important concept and there have been recent successes with use of trizivir and tenofovir in this setting. Only a limited amount of clinical controlled trial data helps the clinician to decide what therapy to switch to following initial treatment. Many of the controlled trials which do exist are relatively unhelpful because they have studied patients who received sub-optimal initial therapy. Nevertheless a number of important general points can be made (for more detail refer chapter 1.9).

Resistance to the NNRTI class of drugs is caused by mutations in a pocket of the reverse transcriptase enzyme adjacent to the catalytic site. Resistance to one of the NNRTIs produces cross resistance to other members of this class and so the currently available drugs are not used in sequence. Promising data with TMC-125, a diarylpyrimidine derivative, suggests that this NNRTI is able to overcome resistance to other NNRTIs.

It was initially thought that nucleoside analogues induced distinct mutational patterns in the viral reverse transcriptase associated with reduced sensitivity and that substitution of one nucleoside analogue for another would be successful. While this remains broadly true it is now appreciated that mutations reducing sensitivity to one nucleoside analogue also reduce the sensitivity to other members of this class. This is particularly true of the thymidine analogue mutations producing reduced sensitivity to zidovudine and also display reduced sensitivity to stavudine. Tenofovir is a nucleotide analogue closely related to the nucleosides with a similar mode of action. Most viruses with reduced sensitivity to the other nucleoside analogues remain sensitive to tenfovir *in vitro* and viral load drops in individuals who harbour such viruses have been demonstrated *in vivo*.

Proteinase Inhibitors Virological failure of initial therapies containing protease inhibitors is often associated with only a few mutations in the protease gene. With some mutations such as at codon 30 for nelfinavir, a good response to a subsequent proteinase inhibitor can be expected.

7. USE OF RESISTANCE TESTING

Development of rapid sequencing techniques which allow the demonstration of changes in the viral genome associated with reduced sensitivity to drugs has had major importance in our understanding of the way in which such drugs work and the causes of drug failure. However, resistance testing demonstrates which drugs are unlikely to be effective rather than those that will. When patients fail an initial regime, most clinicians would wish to change all components of that regime even if resistance testing indicated continuing sensitivity to some of them. The primary value of resistance testing at this stage is to ensure that all the switches have a reasonable chance of working and, at this stage, may prove of value when trying to construct successful 3rd and 4th line regimens.

A failing NNRTI regime Trials to test optimum policies in this situation are difficult to undertake because failure of such regimes is normally associated with poor adherence rather than virological failure despite continued drug use. Nevertheless as NNRTI resistant mutations are likely to remain protease inhibitor sensitive, most clinicians would switch to two different nucleoside analogues and a boosted PI containing regimen. Clinical experience indicates that this is usually successful in completely suppressing plasma viral load.

A Failing PI containing regime   The obvious drug class to use in this situation is an NNRTI. This will only be successful providing the total regimen is capable of stopping viral replication completely. Otherwise resistance to the NNRTI will develop rapidly. It is for this reason that most clinicians prescribe nucleosides and an NNRTI and add an alternative PI. This has been shown to be the most successful policy in individuals failing two nucleoside analogues (Deeks *et al.*, 2000). The second PI to be added will depend partly upon the resistance test and also on which PI has been used first. Probably the most successful PI to use in this situation is Lopinavir boosted with Ritonavir (Ruane *et al.*, 2001).

Subsequent therapies There is a growing number of patients who have become resistant to all the present medication. Even in these patients the death rate is low providing the CD4 count is maintained above 50 cells/mm^3. It is clear that continuing therapy is beneficial compared with discontinuing, presumably because viral mutations which are less sensitive to drugs are also less virulent and therefore have less effect on reducing the CD4 count. Thus in these individuals, there is a paradigm switch of treatment from trying to make the viral load undetectable (which is not possible) to one in which the CD4 count is kept as high as possible for as long as possible. In this situation some clinicians believe that the minimum number of drugs to continue to give those viruses which are resistant already in the circulation a continuing survival advantage, whereas others believe that large numbers of drugs, even though they may have little effect individually are beneficial overall. The disadvantage of this latter mega HAART approach is increased toxicity, a large pill burden and unexpected pharmacokinetic interactions. It is likely that this is the stage at which T20 would be prescribed but it only has an evanescent effect if it is the only active drug in the regime. Thus it is probably better to use T20 in the last regime at which undetectability is likely to be achieved. Alternatively T20 can be used so that the patient can be treated with new drugs as they come on stream.

8. NEW DRUGS

New drugs in development are of two sorts. There are a number of drugs which are developments of currently available classes which are developed because of improved pharmacokinetics, a reduced toxicity profile or are active against viruses which have resistance mutations to presently available drugs.

New drug classes acting against different parts of the life cycle of HIV are under development as well.

Attachment Inhibitors The viral attachment process has been very thoroughly elucidated and involves initially loose binding between the CD4 receptor of T cells and the B3 loop of the GP120 of the viral coat. Interestingly during this process constant regions of GP120 are exposed and although the exposure of these constant regions of GP120 are extremely evanescent the neutralising antibodies to this region might have a beneficial effect either as a vaccine or in HIV infected individuals. Bristol Myers Squibb have a drug in the early stages of development which inhibits this interaction although there is considerable intrinsic variability in the sensitivity of viruses to this class of compound. As they are active in molecule concentrations, this natural variability may not prevent further development.

The second part of the attachment process is tighter binding of the virus envelope to the cell surface by means of interaction with a chemokine receptor (see chapter 1.8). The commonest chemokine receptor utilised is CCR5 and a number of compounds capable of inhibiting this interaction are now in various stages of development, the most advanced of which is now in phase 3 study. Obviously only individuals who harbour viruses which is CCR5 trophic are likely to respond to this drug and the long term side effects of inhibiting one of the body's receptors is unknown although a large deletion within this receptor which renders it inactive is a common balanced polymorphism in European communities without major untoward effects. The virus also is capable of utilising the CXCR4 receptor present particularly on CD4 cells and such viruses are associated with a more rapid progression to AIDS. Although there have been worries that the use of CCR5 inhibitors may encourage the virus to mutate to become CXCR4 trophic, there is limited evidence in vivo as yet that this is the case. CXCR4 inhibitors are also being developed but these are at an earlier stage of evolution.

Table 2. New attachment inhibitors to treat HIV

Phase I	Phase I/II	Phase II
AMD-070 (CXCR4)	BMS-9043 (Anti-gp120)	Pro-542 (attachment inhibitor)
AMD-887 (CCR5)	SCH D (CCR5)	SP-01A
GSK-873140 (CCR5)	TNX335 (Anti-CD4)	UK-427/857 (CCR5)
CCR5mAb004 (CCR5)		

Fusion inhibitors represent an excellent example of where detailed knowledge of the processes involved in HIV replication have led to specific drug design. T20 represents the first of a new series of fusion inhibitors. This peptide was specifically designed to interact with helical portions of GP41 which contract during the process of fusion to draw the surface of GP120 into close proximity with the host cell, allowing interaction with the T cell receptor and the chemokine receptors. T20 is administered subcutaneously twice a day and is effective in salvage therapy. It was used as additional therapy to standard care in patients who failed all three classes of drugs where marked viral load drops (1 log) which persisted for up to 48 weeks were seen. Virological failure in patients taking T20 is associated with mutations in the relevants portions of GP41. A new form of this drug which can be administered once a day is also being developed. As often happens during drug development the optimum role for this drug which is likely to be licensed soon remains unclear. Most clinicians would like to combine T20 with other drugs which are also likely to be effective to reduce viral replication to undetectable levels eg tenofovir and lopinavir boosted with ritonavir in individuals following initial PI failure.

A variety of small molecules which inhibit various phases in the fusion process are also under development. Thus the interaction between the V3 loop of GP120 and the CCR5 receptor can be inhibited by a series of products made by Schering Plough. Product C which is in the most advanced stage of development was shown recently to produce viral load drops of nearly 1 log in short term human studies. Interestingly this occurred after a one to two day lag period following administration. As a result of experiments *in vitro*, worries have been expressed that the inhibition of the interaction between the virus and CCR5 would encourage mutations to a more virulent form capable of interacting with the alternative chemokine receptor, CXCR4. More recent *in vitro* studies and limited human experiments have not indicated that this is likely to happen. Unfortunately the main drug which is being developed to inhibit the interaction between HIV and CXCR4, AMD310, which has to be administered intravenously, showed very little effect in Phase 1 human studies (Dameta *et al.*, 1996). Its future development must be in doubt.

As well as chemokine V3 loop interactions, fusion is brought about as a result of interactions with the TCR. During the process of fusion a normally hidden area of envelope is exposed by confirmational change which interacts with the TCR. This highly conserved area is an obvious target for the development of neutralising antibodies although it is only evanescently exposed to the environment. A number of drug molecules are being developed (Bristol Myers Squibb) which are capable of inhibiting this interaction and some quasi species of HIV are sensitive to these molecules while others are not. As these compounds are extremely potent, the relative insensitivity of some HIV variants may be dealt with by increasing the dose of such drugs.

New drugs in the NNRTI class are likely to be developed either to improve upon the pharmacokinetics of the present drugs (which would be difficult) or to reduce toxicity. Other important reasons for developing new NNRTIs would be that they would be active against HIV viruses with mutations rendering them insensitive to present members of the class.

BMS083 This drug was in development by duPont Pharma and its future progress will now be decided by Bristol Myers Squibb. Initial studies suggest that 083 is equally potent as Efavirenz when used in antiretroviral naïve patients with a relatively similar toxicity profile (Jeffrey *et al.*, 2000). Results in individuals who are resistant to efavirenz and then treated with 083 remain difficult to interpret and further studies in these patients will make it easier to define whether this drug has a future.

TMC120 This drug was specifically designed by Tibotec/Virco to have activity against viruses which contain mutations rendering them insensitive to the present NNRTIs. Eight day Phase 1 studies performed in Russia indicated that this drug was highly active *in vitro* (De Bethune *et al.*, 2001). A Proof of concept of Phase 1 study has also looked at viral activity in a group of individuals failing efavirenz and nevirapine containing regimens and has shown a viral load drop over 8 days of nearly 1 log in such individuals who all had very high levels of phenotypic and genotypic resistance to both drugs. At present large numbers of pills must be given three times a day to produce these effects but the company is working to reduce the pill burden to an acceptable level.

Capavirene This drug is being developed for similar reasons to TMC120 (Hernandez *et al.*, 2000). It has activity against viruses with reduced sensitivity to the presently available NNRTI. It is likely to be given three times a day and will require ritonavir boosting to produce acceptable drug levels. Drug development was suspended until recently because of a vasculitis noticed at high dosages in dogs. However, in lower dosages comparable to those given to patients, no vasculitis was seen and so development has now been allowed to continue. The role of this drug which requires remains unclear but one obvious use would be in individuals failing initial NNRTIs in whom standard therapy at the moment would be a ritonavir boosted PI, and adding capavirene to this regime might increase potency.

Our understanding of the biochemical processes which take place in the cell during virus replication are increasing rapidly. Of particular importance is the role of many of the regulatory proteins produced by the virus which should, in the relatively near future, provide new targets to inhibit viral replication. Particularly important are the understandings of the role of Rev S and the clarification of the biochemical mechanism involved in the ability of tat to enable sufficient RNA transcriptation following activation of the LTR.

It is also likely that as the chemical process involved in viral assembly can more clearly understand a variety of inhibitors might be developed. However, in the short-term future, it is likely that two new targets to prevent HIV replication will receive most attention.

Two companies are developing Di Keto compounds which inhibit HIV replication *in vitro*. It is now clear that these compounds act as integrase inhibitors as viruses with reduced sensitivity to them have mutations in the integrase gene. Initial Phase 1 studies indicate that they have a favourable pharmacokinetic profile and Phase II dose ranging studies in HIV seropositive individuals are planned for the near future.

9. CONCLUSIONS

In recent years, the demand for new antiviral strategies has increased markedly. There are many contributing factors to this increased demand, including the ever-increasing prevalence of chronic viral infections such as HIV and hepatitis B as well as the emergence of new viruses such as the SARS coronavirus. The weaknesses of current drugs in the treatment of HIV are being tackled with new targeted therapies. Because of their early stage of development, the question of improved tolerance remains largely unanswered for most of these compounds and many such drugs will undoubtedly fall by the wayside, however some, will become new and valuable treatments.

REFERENCES

Andreeva N.S., Bochkarev A., and Pechik I., 1995, A new way of looking at aspartic proteinase structures: a comparison of pepsin structure to other aspartic proteinases in the near active site region. *Advances in Experimental Medicine & Biology* **362**:19-3

ACTG 320: last of the body count trials? PI Perspect. 1997:7

Cahn P., Percival L., and Phanuphak P., 2001, Phase II 24-week data from study AI 424-008: comparative results of BMS 232-632, stavudine, lamivudine as HAART for treatment naïve HIV infected patients. Abstract 5. Program and abstracts of The 1st IAS Conference on HIV Pathogenesis and Treatment. Buenos Aires, Argentina

Carosi G., Castelli F., and Suter F., 2001, Antiviral potency of HAART regimens and clinical success are not strictly coupled in real life conditions: evidence from the MASTER-1 study. *HIV Clin Trials* **2**:399-407

Carpenter C., Cooper D., and Fischl M., 2000, Antiretroviral therapy in adults. Updated recommendations of the International AIDS Society-USA panel. *JAMA* **283**:381-390

Chun T.W., Carruth L., Finzi D., Shen X., DiGiuseppe J.A., Taylor H., Hermankova M., Chadwick K., Margolick J., Quinn T.C., Kuo Y.H., Brookmeyer R., Zeiger M.A., Barditch-Crovo P., and Siliciano R.F., 1997, Quantification of latent tissue reservoirs and total body viral load in HIV-1 infection [see comments]. *Nature* **387**:183-188

Collier A.C, Coombs R.W., Fischl M.A., Skolnik P.R., Northfelt D., Boutin P., Hooper C.J., Kaplan L.D., Volberding P.A., Davis L.G., et al., 1993, Combination therapy with zidovudine and didanosine compared with zidovudine alone in HIV-1 infection. *Annals of Internal Medicine* **119**:786-793

Comparison of dual nucleoside-analogue reverse-transcriptase inhibitor regimens with and without nelfinavir in children with HIV-1 who have not previously been treated: the PENTA 5 randomised trial. *The Lancet* 2002 **359**:733-740

Concorde: MRC/ANRS randomised double-blind controlled trial of immediate and deferred zidovudine in symptom-free HIV infection. Concorde Coordinating Committee. *The Lancet* 1994 **343**:871-881

Dameta R., Rabin L., and Hincenbergs M., 1996, Antiviral efficacy in vivo of the anti-human immunodeficiency virus bicyclam SDZ SID 791 (JM 3100), an inhibitor of infectious cell entry. *Antimicrob Agents Chemother.* **40**:750-754

Deeks S., Brum S., and Xu Y., 2000, ABT378/ritonavir (ABT378/r) suppresses HIV RNA to <400 copies per ml in 84% of PI experienced patients at 4 weeks. VII Conference on Retroviruses and Opportunistic Infections. San Francisco, CA

De Bethune M., Andries K., and Ludovici D., 2001, TMC120 (R147681), a next-generation NNRTI, has potent *in vitro* activity against NNRTI-resistant HIV variants. Program and abstracts of the 8th Conference on Retroviruses and Opportunistic Infections Abstract 304. Chicago, Illinois

De la Monte S.M., Gabuzda D.H., Ho D.D., Brown R.J., Hedley W.E., Schooley R.T., Hirsch M.S., and Bhan A.K., 1988, Peripheral neuropathy in the acquired immunodeficiency syndrome. *Annals of Neurology* **23**:485-492

Erickson J., Gulnik S., and Suvorov L., 2001, A femtomolar HIV-1 protease inhibitor with subnanomolar activity against multidrug resistant HIV-1 strains. Program and abstracts of the 8th Conference on Retroviruses and Opportunistic Infections Abstract 12. Chicago, Illinois

Fischl M.A., Richman D.D., Causey D.M., Grieco M.H., Bryson Y., Mildvan D., Laskin O.L., Groopman J.E., Volberding P.A., Schooley R.T., et al., 1989, Prolonged zidovudine therapy in patients with AIDS and advanced AIDS-related complex. AZT Collaborative Working Group. *JAMA* **262**:2405-2410

Fischl M.A., Parker C.B., Pettinelli C., Wulfsohn M., Hirsch M.S., Collier A.C., Antoniskis D., Ho M., Richman D.D., Fuchs E., et al., 1990, A randomized controlled trial of a reduced daily dose of zidovudine in patients with the acquired immunodeficiency syndrome. The AIDS Clinical Trials Group. *New England Journal of Medicine* **323**:1009-1014

Fischl M.A., Richman D.D., Hansen N., Collier A.C., Carey J.T., Para M.F., Hardy W.D., Dolin R., Powderly W.G., Allan J.D., et al., 1990, The safety and efficacy of zidovudine (AZT) in the treatment of subjects with mildly symptomatic human immunodeficiency virus type 1 (HIV) infection. A double-blind, placebo-controlled trial. The AIDS Clinical Trials Group [see comments]. *Annals of Internal Medicine* **112**:727-737

Fischl M.A., Stanley K., Collier A.C., Arduino J.M., Stein D.S., Feinberg J.E., Allan J.D., Goldsmith J.C., and Powderly W.G., 1995, Combination and monotherapy with zidovudine and zalcitabine in patients with advanced HIV disease. The NIAID AIDS Clinical Trials Group [see comments]. *Annals of Internal Medicine* **122**:24-32

Fischl M.A., 1993, Treatment of HIV disease in 1993/1994. [Review]. *Aids Clinical Review* **94**:167-187

Fischl M.A., Olson R.M., Follansbee S.E., Lalezari J.P., Henry D.H., Frame P.T., Remick S.C., Salgo M.P., Lin A.H., Nauss K.C., et al., 1993, Zalcitabine compared with zidovudine in patients with advanced HIV-1 infection who received previous zidovudine therapy [see comments]. *Annals of Internal Medicine* **118**:762-769

Gatell M.L., Arnaiz, J.A. et al., 2000, A randomized study comparing continued indinavir (800 mg tid) vs switching to indinavir/ritonavir (800/100 mg bid) in HIV patients having achieved viral load suppression with indinavir plus 2 nucleoside analogues: the BID Efficacy and Safety Trial (BEST). XIII International AIDS Conference. Durban, South Africa

Gupta S., Gilbert R., and Brady A., 2000, CD4 cell counts in adults with newly diagnosed HIV infection. results of surveillance in England and Wales, 1990-98. CD4 Surveillance Scheme Advisory Group. *AIDS* **14**:853-861

Hammer S.M., 2002, Increasing choices for HIV therapy. *N Engl J Med* **346**:2022-2023

Hammer S., Squires K., and Hughes M., 1997, A controlled trial of two nucleoside analogues plus indinavir in persons with human immunodeficiency virus infection and CD4 cell counts of 200 per cubic millimeter or less. *N Engl J Med* **337**:725-733

Haas D., Zala C., and Schrader S., 2001, Once-daily atazanavir plus saquinavir favorably affects total cholesterol (TC) and fasting triglyceride (TG) profiles in patients failing prior

PI therapy (trial AI424-009, wk 24). Program and abstracts of the 41st Interscience Conference on Antimicrobial Agents and Chemotherapy. Chicago, Illinois Abstract LB-16

Hernandez J., Amador L., and Amantea M., 2000, Short-course monotherapy with AG1549, a novel nonnucleoside reverse transcriptase inhibitor (NNRTI), in antiretroviral naive patients. Program and abstracts of the 7th Conference on Retroviruses and Opportunistic Infections Abstract 669. San Francisco, CA

Hetherington S., Hughes A., and Mostellar M., 2002, Genetic variations in HLA-B region and hypersensitivity reactions to abacavir. *The Lancet.* **359**:1121-1122

Hoetelmans R., Reijers M., and Weverling G., 1998, The effect of plasma drug concentrations on HIV-1 clearance rate during quadruple drug therapy. *AIDS* **12**:F111-F115

Hoggard P.G., Kewn S., Barry M.G., Khoo S.H., and Back D.J, 1997, Effects of drugs on 2',3'-dideoxy-2',3'-didehydrothymidine phosphorylation *in vitro. Antimicrobial Agents and Chemotherapy* **41**:1231-1236

Hogg R., Yip B., and Wood E, 2001, Diminished effectiveness of antiretroviral therapy among patients initiating therapy with CD4+ cell counts below 200/mm3. VIII Conference on Retroviruses and Opportunistic Infections. VIII Conference on Retroviruses and Opportunistic Infections Abstract 342. Chicago, IL

Jeffrey S., Logue K., and Garber S., 2000, New second generation NNRTIs: improved resistance and cross-resistance profiles. Program and abstracts of the XIII International AIDS Conference. Program and abstracts of the XIII International AIDS Conference Abstract TuPpA1145. Durban, South Africa

Kaplan J.H., and Karon, J, 2001, Late initiation of antiretroviral therapy (at CD4+ lymphocyte count <200 cells/L) is associated with increased risk of death. VIII Conference on Retroviruses and Opportunistic Infections. Chicaogo IL

Kinloch-de Loes S., Hirschel B.J., Hoen B., Cooper D.A,. Tindall B., Carr A., Saurat J.H., Clumeck N., Lazzarin A., Mathiesen L., et al., 1995, A controlled trial of zidovudine in primary human immunodeficiency virus infection. *N Engl J Med* **333**:408-413

Lagakos S.W., 1993 Surrogate markers in AIDS clinical trials: a conceptual basis, validation, and uncertainties. *Clinical Infectious Diseases.* **11**:22-25

Larder B.A. Kellam P., and Kemp S.D., 1993, Convergent combination therapy can select viable multidrug-resistant HIV-1 *in vitro. Nature* **365**:451-453

Moyle G.P, and Opravil, M., 2000, The SPICE study: 48-week activity of combinations of saquinavir soft gelatin and nelfinavir with and without nucleoside analogues. Study of Protease Inhibitor Combinations in Europe. *J Aquir Immune Defic Syndr.* **23**:128-137

Mellors J.W., Rinaldo C.R., Gutpa P., White R.M., Todd J.A., and Kingsley L.A., 1996, Prognosis of HIV-1 infection predicted by the quantity of virus in plasma. *Science* **272**:1167-1170

Nelson M., Staszewski S., and Morales-Ramirez J., 2001, Successful virologic suppression with efavirenz in HIV-infected patients with low baseline CD4 cell counts: post hoc results from Study 006. X European Conference of Clinical Microbiology and Infectious Diseases. Stockholm

Opravil M., Lederberger B., and Furrer H., 2001, Clinical benefit of early initiation of HAART in patients with asymptomatic HIV infection and CD4 counts >350/mm3. VIII Conference on Retroviruses and Opportunistic Infections. Chicago, IL

Palella F.J, Delaney K.M., Moorman A.C., Loveless M.O., Fuhrer J., Satten G.A., Aschman D.J., and Holmberg S.D., 1998, Declining morbidity and mortality among patients with advanced human immunodeficiency virus infection. HIV Outpatient Study Investigators [see comments]. *N Engl J Med.* **38**:853-860

Phillips A., Staszewski S., and Weber R., 2000, Viral load changes in response to antiretroviral therapy according to the baseline CD4 lymphocyte count and viral load. *AIDS*. **14**:S3

Ragni M.V., Amato D.A,. LoFaro M.L., DeGruttola V., van der Horst, C., Eyster M.E., Kessler C.M., Gjerset G.F., Ho M., Parenti D.M., et al., 1995, Randomized study of didanosine monotherapy and combination therapy with zidovudine in hemophilic and nonhemophilic subjects with asymptomatic human immunodeficiency virus-1 infection. AIDS Clinical Trial Groups. *Blood* **85**:2337-2346

Randomised trial of addition of lamivudine or lamivudine plus loviride to zidovudine-containing regimens for patients with HIV-1 infection: the CAESAR trial, 1997, *The Lancet* **349**:1413-1421

Ruane P., Mendonca J., and Timerman A., 2001, Kaletra vs. nelfinavir in antiretroviral-naive subjects: week 60 comparison in a phase III, blinded, randomized clinical trial. Program and abstracts of The 1st IAS Conference on HIV Pathogenesis and Treatment Abstract 6. Buenos Aires, Argentina

Sha B., Proia L., and Kessler H., 2000, Adverse effects associated with use of nevirapine in HIV postexposure prophylaxis for 2 health care workers. *JAMA* **248**:2723

Staszewski S., Morales-Ramirez J., and Tashima K., 1999, Efavirenz plus zidovudine and lamivudine, efavirenz plus indinavir, and indinavir plus zidovudine and lamivudine in the treatment of HIV-1 infection in adults. Study 006 Team. *N Engl J Med.* **341**:1865-1873

Staszewski S., Keiser P., and Giorgi J., 1999, Comparison of antiviral response with abacavir/Combivir to indinavir/Combivir in therapy-naive adults at 48 weeks (CNA3005) Abstract 505. XXXIX Interscience Conference on Antimicrobial Agents and Chemotherapy. San Francisco, CA

Sterling T., Chaisson R., and Bartlett J, 2001, CD4+ lymphocyte level is better than HIV-1 plasma viral load in determining when to initiate HAART. VIII Conference on Retroviruses and Opportunistic Infections Abstract 519. Chicago IL

Walmsley S., Bernstein B., King M., Arribas J., Beall G., Ruane P., Johnson M., Johnson D., Lalonde R., Japour A., Brun S., and Sun E., 2002, Lopinavir-ritonavir versus nelfinavir for the initial treatment of HIV infection. *N Engl J Med.* **346**:2039-2046.

Wood R., and Team T., 2000, Sustained efficacy of nevirapine (NVP) in combination with two nucleosides in advanced treatment-naive HIV infected patients with high viral loads: a B1090 substudy Abstract WeOrB604. XIII International AIDS Conference. Durban, South Africa

Zhang L., Ramratnam B., and Tenner-Racz K., 1999, Quantifying residual HIV-1 replication in patients receiving combination antiretroviral therapy. *N Engl J Med.* **340**:1605-1613

Chapter 1.8

EVALUATION OF CURRENT STRATEGIES TO INHIBIT HIV ENTRY, INTEGRATION AND MATURATION

J. D. Reeves[1] , S. D. Barr[1] and S. Pöhlmann[2]

[1]*Department of Microbiology, University of Pennsylvania, 301B Johnson Pavilion, 3610 Hamilton Walk, Philadelphia, PA 19104, USA;* [2]*Institute for Clinical and Molecular Virology and Nikolaus-Fiebiger-Center for Molecular Medicine, University Erlangen-Nürnberg, Glückstraße 6, 91054 Erlangen, Germany*

Abstract: Human immunodeficiency virus type 1 (HIV-1) infection continues to be a massive global health crisis, particularly in developing countries. With no effective vaccine and no prospect for a cure in the foreseeable future, antiretroviral treatment is the only option at hand to combat HIV infection. Current HIV therapeutics target the viral enzymes reverse transcriptase and protease. The use of a combination of these drugs, termed highly active antiretroviral therapy (HAART), can efficiently reduce viral load in infected patients. Despite the success of HAART in reducing HIV related morbidity and mortality, HAART cannot eradicate virus in infected patients and might not confer life long suppression of HIV replication. In fact, due to ongoing HIV replication, drug resistant viruses frequently arise in treated patients and such viruses are increasingly transmitted between individuals. These observations, together with the considerable side effects of some HAART regimens, underline that current therapeutics need to be improved and that new antiviral agents with novel modes of action that are effective against current drug resistant viruses need to be sought. The replicative cycle of HIV affords multiple opportunities for therapeutic intervention. Entry of HIV into a cell, integration of the viral genome into the host cell chromosome and the generation of mature infectious progeny virions ("maturation") are promising targets for inhibitors. Here, we will discuss how HIV accomplishes entry, integration and maturation and which strategies are being pursued to inhibit these processes.

E. Bogner and A. Holzenburg (eds.), New Concepts of Antiviral Therapy, 213–254.

1.   INTRODUCTION

Human immunodeficiency virus (HIV) continues to be a significant global cause of mortality. More than 3 million deaths were attributed to acquired immune deficiency syndrome (AIDS) in 2003 alone (UNAIDS, 2004). In the same year, there was a record number of new HIV infections, estimated at around 5 million (UNAIDS, 2004). With no prospects for a vaccine or a cure in the foreseeable future, current therapy for HIV infection relies on the use of antiretroviral agents to reduce viral load, provide immunological benefit, delay disease progression and thus extend the life expectancy of infected individuals.

1.1 Current antiretroviral Therapy

There are currently twenty antiretroviral drugs approved for HIV therapy. These include agents that inhibit the activity of the viral enzymes reverse transcriptase (RT) and protease (Figure 1) or prevent the entry of HIV into cells (Figure 1 & 2; these are also discussed in chapter 1.7).

The first antiretroviral agent to be approved was the nucleoside analog zidovudine (AZT) that is incorporated by RT into the nascent polynucleotide chain during reverse transcription of viral RNA into a DNA copy (Figure 1). Subsequently, six other nucleoside reverse transcriptase inhibitors (NRTIs) and one nucleotide reverse transcriptase inhibitor (NtRTI) have been approved, along with three non-nucleoside reverse transcriptase inhibitors (nNRTIs), that bind and inactivate RT (reviewed in Balzarini, 2004; Ruane and DeJesus, 2004; Sharma *et al.*, 2004). These compounds are collectively referred to as RT inhibitors.

The second major class of antiretroviral agents target the protease enzyme directly to inhibit protease cleavage of viral Gag-Pol ($Pr160^{GagPol}$) and Gag ($Pr55^{Gag}$) polyprotein substrates and thus disrupt a late stage of the viral life cycle referred to as maturation (see Figure 1 and below). There are currently eight approved protease inhibitors (PIs) for antiretroviral therapy (reviewed in (Rodriguez-Barrios and Gago, 2004; Wynn *et al.*, 2004).

Enfuvirtide (Fuzeon/T-20) is presently the only approved member of a new class of antiretrovirals agents referred to as entry inhibitors (EIs) (reviewed in Kilby and Eron, 2003; Pöhlmann and Reeves, 2005; Reeves and Piefer, 2005). Enfuvirtide specifically targets the transmembrane subunit (gp41) of the envelope protein (Env) of HIV to block fusion of viral and cellular membranes, thus preventing entry of HIV into target cells (Greenberg *et al.*, 2004; Pöhlmann and Reeves, 2005; Reeves and Piefer, 2005).

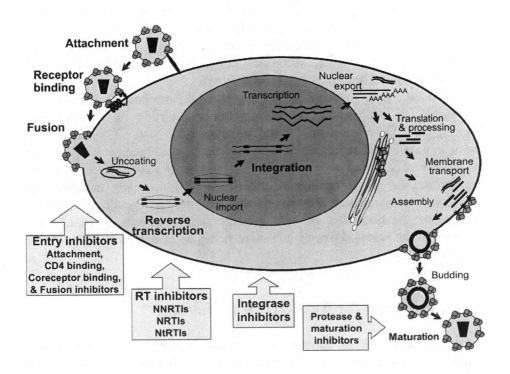

Figure 1: The life cycle of HIV and targets for inhibition. Steps of the HIV life cycle targeted by approved or investigational antiretrovirals arc labclcd in bold text. Classes of antiretroviral agents targeting these steps are boxed. Figure adapted from Reeves and Piefer (Reeves and Piefer, 2005). RT, reverse transcriptase; NRTI, nucleoside reverse transcriptase inhibitor; NNRTI, non-nucleoside reverse transcriptase inhibitor; NtRTI, nucleotide reverse transcriptase inhibitor.

Antiretroviral therapy (ART) typically employs a combination of protease and/or RT inhibitors. Modern ART regimens, referred to as "highly active antiretroviral therapy" (HAART), can successfully reduce viral load in infected individuals to below the level of detection for a number of years (Gulick *et al.*, 1997; Hammer *et al.*, 1997; Hicks *et al.*, 2004) and the advent of HAART has considerably improved the prognosis for HIV infected individuals. However, HAART may ultimately not be "highly active" and failing regimens are associated with the development of drug resistant virus, often contributed to by lack of compliance, due to difficult regimens and significant side effects associated with some inhibitors (Hammer and Pedneault, 2000). Furthermore, although there are nineteen approved RT and protease inhibitors, resistance to some inhibitors can confer cross-resistance to other members within the same class reducing the number of effective treatment options for some individuals.

Individuals failing current RT and protease inhibitor based regimens have the option of adding the fusion inhibitor enfuvirtide to an optimized background combination of RT and protease inhibitors. Enfuvirtide can result in a sustained reduction in viral load for a prolonged period (Lalezari *et al.*, 2003a; Lalezari *et al.*, 2003b; Lazzarin *et al.*, 2003). However, the requirement for administration of enfuvirtide by twice-daily subcutaneous injection, combined with the cost of this inhibitor, has resulted in limited acceptance. Additionally, as with all other antiretrovirals, resistant viral variants can be selected (reviewed in Greenberg and Cammack, 2004; Miller and Hazuda, 2004).

1.2 Prospects for Antiretroviral Therapy

Viruses resistant to current antiretrovirals are becoming increasingly common and drug resistant variants account for an increasing number of new infections (Little *et al.*, 2002; Wegner *et al.*, 2000; Weinstock *et al.*, 2004). Thus some individuals are left with a limited number of effective treatment options. These issues, along with the significant toxicities associated with some antiretrovirals, underscore the requirement for new antiretroviral agents that are active against current drug resistant strains as well as the necessity for new agents with reduced toxicities. Indeed, a number of investigational agents with novel mechanisms of action against current drug targets, or that block distinct stages of the viral life cycle, are under evaluation (Reeves and Piefer, 2005).

Antiretrovirals with novel mechanisms of action that are currently in clinical trials include inhibitors of viral entry, integration and maturation, as discussed below. The long-term side effects of these agents remain to be established, and experience with approved antiretrovirals would indicate that monotherapy directed against any target is likely to rapidly select for resistant viruses. However, addition of these investigational agents to an optimized background antiretroviral regimen may provide clinical benefit in patients with multi-drug-resistant viruses. Furthermore, combination therapy employing inhibitors directed against distinct targets, or with distinct modes of action against a single target, together with current antiretrovirals, will likely offer improved prospects for containment of viral replication in first line therapy.

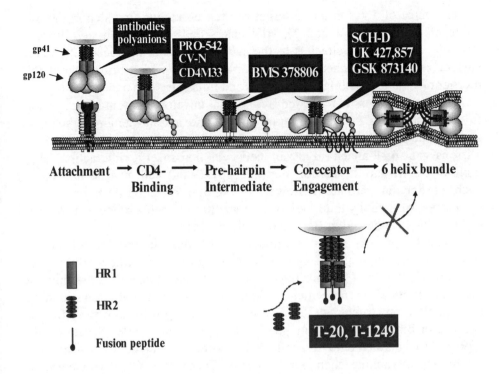

Figure 2: HIV entry and its inhibition. Binding of gp120 to heparan sulfates or lectins on the cell surface promotes HIV attachment to target cells. These interactions can augment infection efficiency but are dispensable for entry. The interaction of gp120 with CD4 initiates a series of interactions which are indispensable for viral entry. Binding of gp120 to CD4 induces conformational changes in Env that lead to the formation/exposure of a coreceptor binding site and trigger the exposure of structures in gp41 involved in membrane fusion. Subsequent binding of gp120 to coreceptor triggers further conformational changes in gp41 that promote the merger between the viral and the cellular membrane. Polyanions inhibit gp120 engagement of heparan sulfates, while antibodies prevent gp120 binding to lectins. CD4M33 and PRO-542 target the CD4 binding site of gp120 and block binding to cellular CD4. CV-N recognizes high-mannose carbohydrates on gp120 and inhibits gp120 engagement of CD4 and coreceptor, however, the precise mechanism of action of this agent remains to be determined. BMS-378806 targets the CD4 binding site in gp120 and arrests gp120 in a rigid conformation, thereby preventing the exposure of epitopes in gp41 involved in membrane fusion. Binding of gp120 to CCR5 can be inhibited by the CCR5 antagonists SCH-D, UK-427,857 and GSK873140. The peptide inhibitors T-20 (enfuvirtide/fuzeon) and T1249 mimic the HR2 region in gp41 and bind to HR1, thereby preventing the back-folding of HR2 onto HR1, which is required for membrane fusion. HR, helical region.

2.   ENTRY

The entry of a virion into a target cell represents the first step in the life cycle of HIV (Figure 1 and 2). HIV enters cells via a multi-step process whereby a virion first attaches to the surface of a cell, interacts with a cell surface receptor, and then a coreceptor molecule, which brings about the merger of viral and cellular membranes and thus entry into a target cell.

The entry process is mediated by the viral envelope protein (Env), which is comprised of a surface subunit, gp120, and a transmembrane subunit, gp41, which assemble as trimers of heterodimers on the virion surface. Virions can first attach to target cells via non-specific interactions that include electrostatic attraction between gp120 and cell surface molecules such as heparan sulfate (reviewed in Ugolini *et al.*, 1999), or by the interaction of carbohydrate moieties on gp120 with cell surface lectins such as DC-SIGN (reviewed in Baribaud *et al.*, 2001a).

The entry process proper is initiated by gp120 binding to the cell surface receptor CD4. The requirement of CD4 binding for entry largely governs the tropism of HIV for CD4-positive T-cells and macrophages. CD4 binding induces structural rearrangements within Env, that include the repositioning of variable loop structures (V1/V2 and V3) within gp120, to expose/form a coreceptor-binding site within gp120 (Kwong *et al.*, 1998; Wyatt *et al.*, 1998). Env can then bind to a cell surface coreceptor molecule, typically the seven transmembrane chemokine receptor CCR5 or CXCR4 (reviewed in Berger *et al.*, 1999; Reeves and Doms, 2003). Some primary HIV-2 isolates can interact directly with a coreceptor molecule to infect CD4-negative cells, whereas primary HIV-1 isolates are usually strictly CD4-dependent (Bhattacharya *et al.*, 2003; Reeves *et al.*, 1999).

Amino acid residues within the V3 loop of gp120 can determine whether an Env interacts with CCR5 or CXCR4, thus, in addition to CD4, V3 also governs cell tropism (Choe *et al.*, 1996; Cocchi *et al.*, 1996; Hartley *et al.*, 2005; Speck *et al.*, 1997; Wu *et al.*, 1996). Furthermore, the coreceptor specificity of Env influences HIV transmission as well as pathogenesis. Viruses that use CCR5 as a coreceptor (R5 tropic) are transmitted between individuals and those who fail to express functional CCR5 molecules, due to a deletion in both CCR5 alleles (Δ32*ccr5* homozygotes), are highly resistant to infection (Dean *et al.*, 1996; Huang *et al.*, 1996; Liu *et al.*, 1996; Samson *et al.*, 1996). In addition, the acquisition of viruses that use CXCR4 as a coreceptor (X4 tropic), either in addition to CCR5 (R5/X4 tropic) or instead of CCR5, is associated with, but is not a prerequisite for, disease progression (Connor *et al.*, 1997; Moyle *et al.*, 2005).

Coreceptor binding induces considerable conformational changes within gp41. The ectodomain of gp41 contains an amino-terminal hydrophobic fusion peptide and two heptad repeat regions, HR1 and HR2. Coreceptor binding is thought to result in insertion of fusion peptides into the membrane of the target cell. Then three HR2 regions, from individual gp41 subunits, are thought to fold to interact with grooves formed between three associated HR1 subunits, forming a highly stable six-helix bundle structure (Chan *et al.*, 1997; Weissenhorn *et al.*, 1997). This brings the amino and carboxy-terminal regions of the ectodomain of gp41 into proximity and, in consequence, the target cell membrane and viral membrane are brought into apposition. Six-helix bundle formation also pulls a fusion pore in the target cell membrane, and pore formation by a number of Env trimers is thought to cooperatively bring about fusion of viral and cellular lipid membranes (Kuhmann *et al.*, 2000; Melikyan *et al.*, 2000).

Each step of the HIV entry pathway presents a viable target for antiretroviral intervention. Indeed, inhibitors targeting attachment, CD4 binding, coreceptor binding, as well as fusion are currently in development, as described below.

2.1 Entry Inhbitiors

HIV entry into target cells involves a variety of sequential interactions between the viral envelope glycoprotein and cellular factors. Some of these interactions are essential for entry, others boost entry efficiency but are ultimately dispensable for infection, however, all are potential targets for therapeutic intervention. While most inhibitors directed against HIV Env face the same problems as the antiretrovirals used in current HAART – i.e. high variability of HIV and rapid outgrowth of resistant viruses – agents directed against invariant cellular factors involved in HIV entry might have a more sustained effect. Since HIV Env mediated membrane fusion is initiated at the cell surface (Smith and Helenius, 2004), this process can be efficiently targeted by non-membrane permeable inhibitors like antibodies or certain peptidic agents directed against structures intimately involved in the merger between viral and cellular membranes. In contrast, similar compounds might be less or non-effective against viruses which fuse with the membrane of endosomal vesicles, like influenza or vesicular stomatitis virus (Smith and Helenius, 2004). Thus, HIV entry is vulnerable to compounds with different chemical properties and modes of action. However, only inhibitors that are effective in the low nanomolar range *in vitro* and that are ideally orally bioavailable and reach effective concentrations systemically, with a reasonable half-life and minimal side effects, are likely to be successful in the clinic. Nevertheless, compounds that do not fit some of these criteria, for example those lacking oral bioavailability or with unwanted side effects upon systemic administration, might still be beneficial when applied topically within a microbicide formulation, aimed at preventing HIV transmission at mucosal sites (Shattock and Solomon, 2004; Shattock and Moore, 2003).

Current strategies to inhibit HIV entry target viral attachment to cells, Env binding to the cellular receptor CD4 and the coreceptors CCR5 and CXCR4, and transient structures in Env involved in membrane fusion. All of these approaches hold promise and one has already resulted in a drug approved for use in humans (enfuvirtide) (Greenberg and Cammack, 2004; Kilby and Eron, 2003). Inhibitors targeting the viral coreceptor CCR5 are currently in advanced stages of clinical development and might within the foreseeable future complement current HAART. Another promising potential for entry inhibitors is that a combination regimen of inhibitors to different entry targets might allow for sustained containment of viruses resistant to current inhibitors. Synergistic HIV inhibition *in vitro* by agents blocking different stages of the entry process indicates that this goal might indeed be attainable (Tremblay, 2004). Here, we will introduce compounds that target major steps of the entry process and will discuss their prospects as therapeutics – and the prospect of HIV to acquire resistance against these agents.

2.1.1 Inhibitors of viral attachment to cells

For the purpose of this review we will define viral attachment to cells as interactions between proteins inserted in the viral membrane and cellular factors other than CD4 and coreceptor, which mediate tethering of viral particles to the cell surface. Although these interactions are not essential for viral entry, they can dramatically increase infection efficiency and are thus targets for inhibitors (Ugolini *et al.*, 1999).

Polyanions inhibit binding of HIV Env to cellular heparan sulfates
Binding of HIV to cells, via electrostatic interactions or engagement of cellular lectins (see below), can not only increase infection of the cells to which the virus particles are attached (infection in cis), but in some cases can also augment entry into adjacent cells, in a process termed infection in trans. Both cis and trans infection can be mediated by interactions of positively charged residues in gp120 with negatively charged heparan sulfates on the surface of various cell types, especially the endothelium of blood vessels (Bobardt *et al.*, 2003; Gallay, 2004). Thus, the interaction of viral particles with endothelial cells has the potential to modulate infection within the

circulatory system. Furthermore, binding of virus to heparan sulfates on brain microsvascular endothelial cells followed by transcytosis of viral particles might also promote HIV infection of the brain (Argyris *et al.*, 2003; Bobardt *et al.*, 2004b; Liu *et al.*, 2002). The interaction of HIV with cellular heparan sulfates can be inhibited by anionic polymers that are negatively charged at neutral pH (Shattock and Moore, 2003). These compounds can be employed within a microbicide formulation to prevent HIV transmission (Shattock and Moore, 2003). The precise mechanism of action of polyanionic inhibitors is unclear, however, it has been suggested that they mainly target the positively charged V3 loop in Env and inhibit binding to coreceptor (Vives *et al.*, 2005), which requires V3 function (Hartley *et al.*, 2005). In this regard it is of note that V3 regions of R5-tropic viruses usually exhibit a lower positive charge than V3 regions of X4-tropic viruses and consequently polyanions inhibit the latter viruses more potently (Hartley *et al.*, 2005). Interestingly, HIV can acquire resistance to polyanionic inhibitors and resistant viruses exhibit multiple changes in gp120 (Bobardt *et al.*, 2004a), confirming that Env is indeed the major target of these compounds. Resistant variants are still capable of infecting cells that do not express heparan sulfate containing proteoglycans with high efficiency, but exhibit markedly reduced ability to infect cells that express these structures (Bobardt *et al.*, 2004a). Engagement of heparan sulfate harboring structures can thus impact HIV cell tropism and infectivity, validating this interaction as a target for inhibition.

Carbohydrate dependent binding of HIV Env to DC-SIGN on dendritic cells: A target for microbicides?

In addition to heparan sulfates, cellular lectins can also augment the infectivity of HIV. The lectins dendritic cell specific intercellular adhesion molecule grabbing non-integrin (DC-SIGN) and DC-SIGN related (DC-SIGNR, also termed L-SIGN, for liver SIGN) promote HIV-1, HIV-2 and SIV infection and might play an important role in viral dissemination (reviewed in Baribaud *et al.*, 2002a; van Kooyk and Geijtenbeek, 2003) . Thus, it has been postulated that DC-SIGN on submucosal dendritic cells can capture sexually transmitted HIV and that virus loaded cells could then migrate into lymphoid tissue where the virus could be transferred to adjoining T-cells – a process that might boost HIV spread in and between infected individuals (Geijtenbeek *et al.*, 2000). In contrast, expression of DC-SIGNR on endothelial cells lining liver and lymph node sinusoids as well as in placenta (Bashirova *et al.*, 2001; Pöhlmann *et al.*, 2001) might endow these cells with the ability to capture blood borne HIV and transfer it to transmigrating T-cells or adjacent macrophages. Moreover, DC-SIGNR might promote HIV infection of liver sinusoidal endothelial cells (Steffan *et al.*, 1992) and these cells might constantly release new virions into the blood stream. Both DC-SIGN and DC-SIGNR bind to high-mannose glycans in Env (Feinberg *et al.*, 2001; Guo *et al.*, 2004; Lin *et al.*, 2003a) and, upon expression on certain cell lines, facilitate infection in cis and in trans (Trumpfheller *et al.*, 2003; Wu *et al.*, 2004). Augmentation of HIV infectivity by DC-SIGN expressing cell lines might involve internalization and intracellular transport of virus particles (Kwon *et al.*, 2002; McDonald *et al.*, 2003). However, recent studies on a widely used cell culture system for DC-SIGN function revealed that these observations might have been confounded by productive infection of the lectin expressing cells in a CD4 and coreceptor dependent manner (Nobile *et al.*, 2005). Similarly, dendritic cell mediated HIV infection in trans, which has been mainly attributed to DC-SIGN function by some laboratories, probably involves direct infection of these cells and might also be promoted by HIV engagement of other lectins (Nobile *et al.*, 2005; Turville *et al.*, 2002). While these studies call a major role of DC-SIGN in HIV infection into question, two recent observations underline that DC-SIGN is likely important for HIV spread. First, inhibitors of DC-SIGN reduce HIV transmission in a mucosal explant model (Hu *et al.*, 2004), suggesting that a microbicide formulation should contain DC-SIGN inhibitors to be effective. Second, and perhaps more relevant, polymorphisms in the DC-SIGN gene can modulate the risk of HIV-1 infection (Liu *et al.*, 2004), at least upon certain routes of transmission (Martin *et al.*, 2004b) and it has been suggested that reduced DC-SIGN expression might hamper viral dissemination. Therefore, inhibition of DC-SIGN in HIV infected individuals might also be associated with a therapeutic benefit. Moreover, DC-SIGN and DC-SIGNR interact with a variety of viral and non viral pathogens (van Kooyk and Geijtenbeek, 2003), among them Ebolavirus (Alvarez *et al.*, 2002; Simmons *et al.*, 2003), hepatitis C virus (Gardner *et al.*, 2003; Lozach *et al.*, 2003), SARS associated coronavirus (Jeffers *et al.*, 2004; Marzi *et al.*, 2004; Yang *et al.*, 2004), Mycobacterium (Geijtenbeek *et al.*, 2003; Maeda *et al.*, 2003) and Leishmania (Colmenares *et al.*, 2002), suggesting that inhibitors of these lectins might be useful for treatment of a variety of infectious diseases.

Several antibodies have been described that potently block binding of ligands to DC-SIGN and DC-SIGNR (Baribaud *et al.*, 2002b; Wu *et al.*, 2002) and these reagents could be employed to inhibit sexual transmission of HIV. However, the *in vivo* potency of these antibodies still needs to be assessed in an experimental model such as SIV infection of macaques. Macaque and human DC-SIGN are equally adept in transmitting immuno-eficiency viruses (Baribaud *et al.*, 2001b; Geijtenbeek *et al.*, 2001) and both are expressed in the genital mucosa (Jameson *et al.*, 2002), underlining that experiments to block DC-SIGN in macaques would yield valuable information on the validity of this target in humans. Since DC-SIGN and DC-SIGNR recognize carbohydrates on the surface of ligands, it will be difficult to generate conventional small molecule inhibitors. However, glycodendritic structures that present multiple mannose residues prevent ligand binding to DC-SIGN and could be employed as microbicides (Lasala *et al.*, 2003; Rojo and Delgado, 2004). Finally, HIV engagement of DC-SIGN could be inhibited by compounds that bind glycans in the HIV-Env and such inhibitors are described below.

Cyanovirin-N binds mannose residues on HIV Env and inhibits infection.
Cyanovirin, a 101 amino acid protein derived from the bacterium *Nostoc ellipsosporum*, recognizes terminal di- and tri-mannose residues on high-mannose glycans (Barrientos and Gronenborn, 2005; Shenoy *et al.*, 2002) present on the HIV Env and on the glycoproteins of other viruses and can inhibit infection (Dey *et al.*, 2000). CV-N can interfere with gp120 binding to both CD4 and coreceptor (Dey *et al.*, 2000), however, the precise mechanism of inhibition needs to be determined. The structure of CV-N has been solved (Bewley *et al.*, 1998) and two carbohydrate binding sites with different affinities have been identified (Shenoy *et al.*, 2002), however, only the high affinity binding site is required for inhibition of HIV infection (Chang and Bewley, 2002). CV-N is present in solution as a monomer or a domain swapped dimer and both exhibit comparable antiviral activity (Barrientos *et al.*, 2004). When applied topically as a microbicide, CV-N can inhibit infection of macaques by a simian/human immunodeficiency virus hybrid (SHIV) upon vaginal (Tsai *et al.*, 2004) and rectal (Tsai *et al.*, 2003) challenge without eliciting major side effects, indicating the CV-N is a promising candidate compound for a microbicide formulation targeting HIV. Compounds with a similar mode of action might, at least in theory, be suitable for HIV-therapy. However, HIV can acquire resistance against such inhibitors by partially removing its carbohydrate shield (Balzarini *et al.*, 2004). Such variants are at least partially resistant to compounds that target glycans, like CV-N or the antibody 2G12 (directed against a carbohydrate epitope in HIV Env) (Balzarini *et al.*, 2004), and these variants are likely to exhibit enhanced neutralization sensitivity.

2.1.2 Inhibitors of gp120 binding to the CD4 receptor

The interaction of gp120 with cellular attachment factors can promote infection but is ultimately dispensable for viral entry. In contrast, binding of gp120 to CD4 is essential for HIV-1 entry into target cells. The structures of CD4 (Ryu *et al.*, 1990; Wang *et al.*, 1990; Wu *et al.*, 1997), gp120 (Chen *et al.*, 2005a; Chen *et al.*, 2005b), and gp120 bound to CD4 and a neutralizing antibody fragment (Kwong *et al.*, 1998) have been determined on the atomic level and the interface between CD4 and gp120 is therefore well characterized. Thus, 22 amino acids in domain 1 (D1) of CD4, particularly those located in the CDR2-like loop, contact 26 residues in gp120, located in a cavity formed at the interface between the inner and outer domain and the bridging sheet of gp120 (Kwong *et al.*, 1998). Despite, the wealth of information on the structures involved in gp120 binding to CD4, the screening for small molecule inhibitors targeting this interaction has thus far been unsuccessful. Antibodies that target the gp120 binding site in CD4 and block HIV infection are available, however, concerns have been raised that systemic administration of such antibodies might cause immunodeficiency (Shattock and Moore, 2003). Nevertheless, a single dose of a humanized antibody, that targets a different domain in CD4 (domain 2 (D2)) to block HIV entry, reduced viral load in infected individuals without eliciting major side effects (Kuritzkes *et al.*, 2004), indicating that this agent warrants further evaluation. Structure based approaches have been employed to inhibit the interaction between CD4 and gp120 and are discussed below.

CD4M33: A mini-CD4 protein that inhibits HIV infection

Several peptides that mimic portions of the CD4-Env interaction site have been generated and shown to inhibit HIV infection (Choi *et al.*, 2001; Ferrer and Harrison, 1999). Moreover, small proteins, that harbor the gp120 binding site of CD4 have also been generated and shown to have potent antiviral effects. Thus, 31 amino acids comprising the CRD2-like loop of CD4 were originally transferred onto a structural scaffold derived from the scorpion toxin charybdotoxin (Vita *et al.*, 1999), and the protein was subsequently optimized for efficient gp120 binding and antiviral activity and its structure determined (Huang *et al.*, 2005; Martin *et al.*, 2003). The most active compound generated, CD4M33, inhibited primary HIV isolates in the nanomolar range and induced structural changes in Env comparable to those triggered by binding to cellular CD4 (Huang *et al.*, 2005; Martin *et al.*, 2003). Although this compound has the general disadvantages of peptide inhibitors, such as low bioavailability and antigenicity, CD4M33 and derivatives thereof could be used as microbicides and, in combination with gp120, within vaccine formulations, since CD4M33 bound gp120 exposes otherwise hidden epitopes which may elicit broadly neutralizing antibodies (Huang *et al.*, 2005; Martin *et al.*, 2003).

PRO 542: A tetravalent CD4-immunoglobulin fusion protein with potent antiviral effects

Expression of the gp120 binding site of CD4 in a heterologous context is also the basis of the inhibitory activity of PRO 542, an immunoglobulin fusion protein, in which the variable regions of both the heavy and light chains are replaced by the D1D2 regions of human CD4 (Allaway *et al.*, 1995). PRO 542 is tetravalent and is most likely capable of cross linking Env timers on the surface of virions or infected cells (Zhu *et al.*, 2001), explaining its superior antiviral activity compared to previously investigated divalent compounds. The compound neutralizes primary and laboratory adapted HIV-1 strains independently of coreceptor usage (Gauduin *et al.*, 1996; Trkola *et al.*, 1998; Trkola *et al.*, 1995), protects SCID (severe combined immunodeficiency) mice harboring human peripheral blood mononuclear cells against challenge with primary HIV-1 isolates (Gauduin *et al.*, 1998), inhibits HIV-1 infection in a mucosal explant model (Hu *et al.*, 2004) and reduces viral load in HIV-1 infected individuals upon a single administration (Jacobson *et al.*, 2004; Jacobson *et al.*, 2000). Since no appreciable side effects were observed in PRO 542 treated individuals the compound merits further testing and could ultimately be employed for salvage therapy of patients with late stage AIDS or within a microbicide formulation.

BMS-378806 binds to the CD4 binding cavity in gp120 and blocks HIV infection

BMS-378806 (BMS-806) is a small molecule (407 Da) that potently inhibits subtype B HIV-1 strains but has no effect against HIV-2 and SIV (Lin *et al.*, 2003b). BMS-806 binds to gp120 and was originally shown to inhibit the interaction with CD4, suggesting that BMS-806 might target the CD4 binding site in gp120 (Lin *et al.*, 2003b). Two further observations supported such a mode of action. First, BMS-806 was shown to inhibit CD4-independent viruses (viral variants that do no require CD4 binding for infection but usually infect cells with higher efficiency in the presence of CD4) only when CD4 was expressed on target cells. Second, viruses that acquired resistance against the compound exhibited alterations in the CD4 binding site in gp120 (Guo *et al.*, 2003; Lin *et al.*, 2003b). Thus, it was suggested that BMS-806 blocks HIV-1 entry by preventing gp120 binding to CD4. However, subsequent studies indicate that BMS-806 might exert its inhibitory activity in a different fashion (Madani *et al.*, 2004; Si *et al.*, 2004). Thus, no inhibitory effect of BMS-806 on gp120 binding to CD4 (or coreceptor) was observed and CD4-independent viruses were equally sensitive to BMS-806 inhibition in the context of target cells expressing receptor and coreceptor or coreceptor alone (Madani *et al.*, 2004; Si *et al.*, 2004). These reports proposed that BMS-806 targets the recessed CD4 binding site and, instead of directly inhibiting CD4 binding to gp120, blocks structural rearrangements in Env that are normally induced by CD4 binding, as BMS-806 prevented the exposure of a structure in gp41 involved in the membrane fusion reaction (Si *et al.*, 2004). Structural analysis of gp120 in the absence of ligand indicates that BMS-806 might stabilize the unliganded conformation of Env and supports the finding that BMS-806 might act by inhibiting conformational changes required for fusion (Chen *et al.*, 2005b). In either case, BMS-806 is orally bioavailable and is highly effective at least against subtype B viruses (Lin *et al.*, 2003b), therefore compounds such as BMS-806 warrant further evaluation.

2.1.3 Inhibitors of gp120 binding to the CCR5 and CXCR4 coreceptors

CCR5 is currently the most promising new target for inhibitors of HIV entry and is an attractive target for several reasons. First, viruses transmitted between individuals via the sexual route employ CCR5 (Connor *et al.*, 1997). Second, individuals with two defective copies of the *ccr5* gene (Δ32 *ccr5* homozygotes) are healthy and third, these individuals are highly resistant to HIV-1 infection (Garred *et al.*, 1997; Huang *et al.*, 1996; Meyer *et al.*, 1997; Michael *et al.*, 1997). Several efforts are therefore underway to block HIV usage of CCR5 and some of these inhibitors are currently being tested in phase II/III clinical trials and might complement current HAART in the nearer future.

In a considerable percentage of infected individuals, viral variants arise that can use CXCR4 alone or in conjunction with CCR5 to infect cells, and the emergence of such variants is associated with progression to immunodeficiency (Connor *et al.*, 1997). Thus, compounds that block CXCR4 use by HIV are also under evaluation (reviewed in detail in (De Clercq, 2003; Schols, 2004)). Several such agents, usually positively charged compounds that specifically interact with the negatively charged surface of CXCR4, have been identified and shown to inhibit the spread of X4-tropic viruses in culture. However, CXCR4 and its natural ligand SDF are critical for hematopoiesis, cardiac function and cerebellar development (Nagasawa *et al.*, 1996; Zou *et al.*, 1998), therefore inhibition of CXCR4 may be associated with unwanted side effects. Indeed, development of the CXCR4 antagonist AMD3100 was halted due to side effects, as well as the lack of oral absorption. AMD070, a follow up CXCR4 antagonist with oral bioavailability, is currently in clinical development and was generally well tolerated in Phase I trials (Schols *et al.*, 2003; Stone *et al.*, 2004). As with AMD3100, CXCR4 antagonism by AMD070 results in a mobilization of white blood cells from the bone marrow (Stone *et al.*, 2004), for which the long-term consequences are unknown. Therefore, agents that block CXCR4 usage by HIV without interfering with the natural function of CXCR4 would be desirable.

Early strategies to inhibit HIV Env engagement of CCR5 relied on modifications of natural CCR5 ligands. Some of these compounds exhibited robust HIV inhibitory activity, but their poor oral bioavailability coupled with their capacity to induce receptor signaling likely impedes their usage as therapeutics (Pierson *et al.*, 2004; Pöhlmann and Reeves, 2005). In contrast, several small molecule CCR5 inhibitors have been generated that block Env-CCR5 interactions to inhibit infection and some of these exhibit favorable pharmacokinetic profiles. Three of these are currently poised to enter into Phase III clinical trials and are discussed below.

UK-427,857, SCH-D and GW873140

UK-427,857 (Maraviroc) is a small molecule CCR5 antagonist that inhibits primary HIV isolates and recombinant viruses harboring *env* genes from RT- and protease-inhibitor susceptible and resistant viruses in the low nanomolar range (MacCartney *et al.*, 2003; Westby *et al.*, 2003). This compound is orally bioavailable and no serious side effects were observed in treated patients following short-term administration (Russell *et al.*, 2003). Monotherapy with UK-427,857 significantly reduced viral load in infected individuals (average 1.6 log10 decline) (van der Ryst *et al.*, 2004). Similarly, SCH-D is an orally bioavailable, potent inhibitor of R5-tropic viruses that is well tolerated and upon monotherapy can also significantly reduce viral load in infected patients (average 1.62 log10 decline) (Schurmann *et al.*, 2004).

The spirodiketopiperazine GW873140 inhibits infection of R5-tropic HIV-1 primary isolates of different clades in the subnanomolar range and acts synergistic in conjunction with approved RT- and protease-inhibitors (Demarest *et al.*, 2004b). GW873140 is orally bioavailable, generally well tolerated and monotherapy with GW873140 can reduce viral load (average of 1.66 log10 decline) (Demarest *et al.*, 2004a; Lalezari *et al.*, 2004). Furthermore, GW873140 exhibits prolonged CCR5 occupancy *in vitro* and *in vivo* explaining sustained antiviral effects days following treatment cessation (Demarest *et al.*, 2004a; Demarest *et al.*, 2004c; Sparks *et al.*, 2005; Watson *et al.*, 2005).

GW873140 functions as a receptor antagonist (Watson *et al.*, 2005), similar to other well characterized small molecule CCR5 inhibitors, including SCH-D and UK-427,857. In contrast to these compounds however, GW873140 does not interfere with binding of the natural CCR5 ligands MIP-1α and RANTES (Watson *et al.*, 2005), indicating that GW873140 interacts with CCR5 differentially compared to SCH-D, UK-427,857 and other CCR5 antagonist, including TAK-779 and SCH-C. All of these compounds are believed to inhibit HIV use of CCR5 by exerting allosteric effects, rather than steric hindrance, on the receptor (Watson *et al.*, 2005).

Thus, binding of these compounds likely induces conformational changes in CCR5 that are incompatible with gp120 recognition of CCR5.

In the light of encouraging results from Phase I/II trials with UK-427,857, SCH-D and GW873140, all three are now poised to enter Phase III trials. The potential consequences of *in vivo* resistance to these inhibitors however, raise novels concerns, as discussed below.

Resistance against small molecule coreceptor inhibitors

As with other antiretroviral agents, viruses can acquire resistance to coreceptor inhibitors. Several resistance mechanisms are possible, however the virus might not always escape by the most obvious route. Thus, a switch from CCR5 to CXCR4 usage and vice versa has been observed when one coreceptor was blocked (Este *et al.*, 1999; Mosier *et al.*, 1999). Also, culture of viruses in cells that express CXCR4 in the presence of a CXCR4 inhibitor can lead to selection of resistant viruses that use this coreceptor in the presence of inhibitor (de Vreese *et al.*, 1996; Kanbara *et al.*, 2001; Schols *et al.*, 1998). Both findings are not unexpected given the variability in HIV Env and a certain degree of flexibility in gp120 interactions with coreceptor. Strikingly, however, in an experimental setting in which both CCR5 and CXCR4 were available, a R5-tropic HIV-1 isolate chose to adapt to CCR5 usage in the presence of drug rather than adapting to utilize CXCR4 (Trkola *et al.*, 2002). Resistance was associated with changes in Env that reduced affinity for CCR5 and diminished entry into CCR5 expressing cell lines, but presumably allowed CCR5 engagement in the presence of drug (Kuhmann *et al.*, 2004; Trkola *et al.*, 2002). These observations indicate that a coreceptor switch, for which as little as one to three substitutions in the V3 loop can be sufficient (Pastore *et al.*, 2004; Pöhlmann *et al.*, 2004; Shimizu *et al.*, 1999; Shioda *et al.*, 1994), might, at least under some circumstances, be associated with a considerable disadvantage to the virus. In such cases, the emergence of viruses that engage coreceptor in the presence of drug can be the consequence. If these results reflect the situation in HIV-1 infected patients treated with CCR5 inhibitors, the emergence of X4-tropic viruses might not be a general phenomenon. Nevertheless, HIV escape from CCR5 inhibitors by adaptation to CXCR4 usage, which is associated with disease progression, remains a major concern The use of CCR5 inhibitors will therefore require careful evaluation of relevant clinical parameters, including determination of the coreceptor tropism of patient derived viruses in order to maximize the potential therapeutic benefit of these inhibitors.

2.1.4 Peptide inhibitors targeting transient structures in gp41

Binding of HIV-1 Env to CD4 is thought to induce conformational changes in gp120 that allow engagement of a coreceptor and also lead to the exposure of epitopes in gp41 which are involved in driving membrane fusion. Coreceptor binding then induces further conformational re-arrangements in gp41 that facilitate fusion of the viral and the host cell membrane. As described above, the latter conformational changes in gp41 involves association of two helical regions, HR1 and HR2, which fold back onto each other to drive the formation of the six helix bundle, a structure intimately associated with membrane fusion. This process is conserved between class I fusion proteins of different viruses and can sometimes be inhibited by peptides that mimic either HR1 or HR2, with HR2 derived peptides often being more effective (Eckert and Kim, 2001). Enfuvirtide (T-20/Fuzeon) is a 36 amino acid peptide comprising sequences derived from HR2 and can efficiently block HIV-1 infection *in vitro* and *in vivo* (Kilby *et al.*, 1998). Use of enfuvirtide for therapy of individuals with multi drug resistant virus was approved after the demonstration that enfuvirtide treatment does not elicit major unwanted side effects and that administration of the compound in combination with an optimized RT- and protease-inhibitor regimen is more effective than treatment with RT- and protease-inhibitors alone . Enfuvirtide is the first entry inhibitor approved for therapy of HIV-1 infection and serves as proof of principle that inhibition of HIV entry is a promising new avenue for HIV drug development.

Despite the efficacy of enfuvirtide, HIV-1 variants resistant to enfuvirtide can develop readily in cell culture as well as in treated HIV patients, resulting in treatment failure (Greenberg and Cammack, 2004; Miller and Hazuda, 2004; Rimsky *et al.*, 1998; Wei *et al.*, 2002). Mutations are mainly localized to HR1 (Marcelin *et al.*, 2004; Rimsky *et al.*, 1998; Wei *et al.*, 2002), the target of enfuvirtide, and often alter a conserved GIV motif, required for optimal fusogenic activity of gp41 (Kinomoto *et al.*, 2005; Reeves *et al.*, 2005). Enfuvirtide resistance can result in slower Env fusion rates which can increase Env susceptibility to a subset of neutralizing antibodies (Reeves *et al.*, 2005) and, in general, resistant viruses exhibit reduced fitness *in vitro* (Lu *et al.*, 2004; Reeves *et al.*, 2005). It remains to be determined, however, if enfuvirtide resistant viruses are less pathogenic. One case of evolution of an enfuvirtide dependent virus has also been reported (Baldwin *et al.*, 2004), highlighting the complex interplay of this compound with the fusion machinery in gp41.

Enfuvirtide resistant viruses can be inhibited by T-1249, a related second generation compound that exhibits more potent antiviral activity and is effective against enfuvirtide resistant viruses *in vitro* and *in vivo* (Eron *et al.*, 2004; Lalezari *et al.*, 2005; Menzo *et al.*, 2004; Reeves *et al.*, 2005). Despite, the encouraging antiviral activity of T-1249, clinical trials to further evaluate the compound have been halted (Martin-Carbonero, 2004). A major disadvantage of enfuvirtide and T-1249 is the requirement for administration by twice daily intramuscular injection. This issue could be improved or resolved by methods to increase the circulatory half-life of these peptides or by devising improved delivery strategies for these drugs. Furthermore, the search for orally bioavailable small molecules that inhibit the membrane fusion machinery in gp41 is underway. Two such compounds have recently been described as potential leads (Jiang *et al.*, 2004).

2.2 Considerations for Antiretrovirals Targeting Entry

The inhibition of HIV entry into cells raises novel considerations specific to this class of antiretrovirals. All entry inhibitors target the Env protein, either directly or indirectly, and Env is the most variable HIV protein. Thus, it is perhaps not surprising that drug naïve viruses with divergent Env proteins can exhibit considerable variation in susceptibility to certain entry inhibitors *in vitro* (Derdeyn *et al.*, 2000; Derdeyn *et al.*, 2001; Labrosse *et al.*, 2003), whereas differences in susceptibility to RT and protease inhibitors are comparatively modest (Parkin *et al.*, 2004).

Factors that can contribute to marked differences in coreceptor antagonist susceptibility include variation in the affinity with which an Env protein binds a coreceptor molecule and differences in cell surface coreceptor expression levels (Reeves *et al.*, 2002; Reeves *et al.*, 2004), which can vary between targets cells and individuals (Lee *et al.*, 1999). Thus, low Env-coreceptor affinity or low coreceptor expression is associated with enhanced susceptibility to coreceptor antagonists (Reeves *et al.*, 2002; Reeves *et al.*, 2004). Less intuitively perhaps, these factors also affect susceptibility to the fusion inhibitor enfuvirtide (Reeves *et al.*, 2002; Reeves *et al.*, 2004). Mechanistically, a reduction in Env-coreceptor affinity or coreceptor levels can result in a slower rate of membrane fusion, which in consequence extends exposure of the temporal target for enfuvirtide (Reeves *et al.*, 2002; Reeves *et al.*, 2004). These factors likely explain synergistic inhibition of HIV by coreceptor antagonists and enfuvirtide *in vitro* (Tremblay, 2004; Tremblay *et al.*, 2002; Tremblay *et al.*, 2000). Thus, coreceptor antagonists will act to block Env-coreceptor binding and also, in effect, reduce the number of cell surface coreceptors available for Env interaction, thereby delaying fusion and enhancing enfuvirtide susceptibility. The use of a combination of these entry inhibitors for antiretroviral therapy, with or without RT and/or protease inhibitors, remains to be evaluated, as does the impact of Env-coreceptor affinity and coreceptor levels on entry inhibitor potency *in vivo*.

In addition to extensive variability, HIV Env exhibits considerable plasticity and is able to accommodate mutations that allow rapid adaptation to selective pressure. Thus the potential for HIV to readily acquire resistance to entry inhibitors was a concern. Indeed, HIV can rapidly escape from enfuvirtide inhibition both *in vitro* and *in vivo* (Greenberg and Cammack, 2004; Miller and Hazuda, 2004; Rimsky *et al.*, 1998; Wei *et al.*, 2002). Nevertheless, enfuvirtide therapy can remain effective for a prolonged time (Greenberg *et al.*, 2004; Lazzarin *et al.*, 2003), validating entry as a target for antiretrovirals. Furthermore, a number of reports indicate that it has been surprisingly difficult to generate viruses resistant to certain CCR5 antagonists *in vitro* (Trkola *et al.*, 2002; Westby *et al.*, 2004), further supporting a role for entry inhibitors in antiretroviral regimens. Mutations conferring relative resistance to enfuvirtide as well as CCR5 antagonists (in the absence of a coreceptor switch) can result in reduced viral infectivity and fitness *in vitro* (Lu *et al.*, 2004; Reeves *et al.*, 2005; Trkola *et al.*, 2002), and some enfuvirtide resistance mutations can confer enhanced susceptibility to a subset of neutralizing antibodies (Reeves *et al.*, 2005), thus drug resistance viruses might be less pathogenic and their emergence might still be associated with a clinical benefit.

A potential mechanism of escape from inhibitors that target CD4 binding is for viruses to adapt to utilize CCR5 or CXCR4 directly for infection. CD4-independent infection is usually less efficient than infection via CD4 and CD4-independent viruses are usually more susceptible to neutralizing antibodies (Bhattacharya *et al.*, 2003; Edwards *et al.*, 2001; Hoffman *et al.*, 1999; Reeves *et al.*, 1999), thus this mechanism of escape may also be associated with reduced fitness *in vivo*. However, CD4-independence does have the potential to broaden HIV cell tropism (Bhattacharya *et al.*, 2003; Reeves *et al.*, 1999; Willey *et al.*, 2003).

As discussed above, a poignant concern for antiretroviral therapy with CCR5 antagonists is the potential for virus to escape inhibition by switching to use CXCR4 as a coreceptor. While CXCR4 utilizing viruses are associated with pathogenesis, it still remains to be established whether their emergence is a cause or effect of disease progression. Thus, the clinical efficacy of CCR5 antagonists will require careful evaluation. Another factor specific to the utilization of coreceptor inhibitors is the requirement for determination of viral coreceptor specificity prior to treatment. Thus, a CCR5 antagonist will not be active against CXCR4 utilizing viruses and vice versa. Furthermore, the utilization of a CCR5 antagonist in individuals harboring a mixed population of R5, X4 and dual-tropic viruses might select for outgrowth of X4 and/or dual-tropic viruses. Again, any potential impact of this scenario on disease progression remains to be determined. The use of a combination of coreceptor antagonist that target both CCR5 and CXCR4 would alleviate these concerns, however there remains the potential for virus to escape these inhibitors by adapting to utilize drug bound coreceptor (as discussed above) or by utilizing one of a number of alternative coreceptors that can mediate infection *in vitro* (Berger *et al.*, 1999).

3.   INTEGRATION

3.1 The HIV Integration Reaction

Following entry of HIV into its target cell the viral RNA genome is converted into a double-stranded DNA molecule. In the cytoplasm, the DNA molecule associates with several viral and host proteins to form a multimeric complex that bridges both ends of the linear DNA molecule in what is referred to as a "preintegration complex" (PIC). Soon after completion of DNA synthesis, the viral DNA ends are primed by the enzyme integrase in a process called 3' processing. 3' processing involves cleavage of the terminal two nucleotides immediately 3' of a conserved CA dinucleotide motif of each LTR, resulting in two recessed 3'-hydroxyl groups (Bushman *et al.*, 1990; Craigie *et al.*, 1990; Katz *et al.*, 1990; Katzman *et al.*, 1989; Sherman and Fyfe, 1990). After docking with a host chromosome, the viral DNA undergoes a strand transfer reaction that involves a nucleophilic attack by each of the two recessed 3'-hydroxyl groups on a 5'-phosphate of the target DNA. For HIV, the points of joining of each end of the viral DNA are staggered by five base pairs across the major groove of the target DNA helix. After joining, the intervening five base pairs are melted, yielding gaps at the junctions of proviral and target DNA. The integration reaction is completed when the two protruding 5' proviral nucleotides are trimmed and the gaps repaired, most likely by host repair enzymes (Daniel *et al.*, 2004; Yoder and Bushman, 2000).

3.2 Integrase Structure and Function

HIV integrase is a critical component for the integration reaction. Integrase is encoded by the 3' end of the pol gene and is produced as a result of protease-mediated cleavage of the gag-pol precursor. The integrase enzyme is 288 amino acids long with a molecular weight of 32kDa. Residues 1-50 comprise the amino-terminal domain (NTD), 50-212 the catalytic core domain (CCD) and 212-288 the carboxy-terminal domain (CTD). The full-length structure of integrase complexed with viral DNA has eluded scientists for many years due to solubility difficulties with the complex. In the interim, many groups have used X-ray diffraction or solution NMR and solved the structures of individual integrase domains, the CCD complexed with the NTD and the CCD complexed with the CTD and two domain fragments (Bujacz *et al.*, 1995; Cai *et al.*, 1997; Chen *et al.*, 2000a; Chen *et al.*, 2000b; Dyda *et al.*, 1994; Eijkelenboom *et al.*, 1997; Goldgur *et al.*, 1998; Greenwald *et al.*, 1999; Lodi *et al.*, 1995; Maignan *et al.*, 1998; Reeves *et al.*, 2005; Wang *et al.*, 2001).

The NTD is characterized by a conserved two histidine and two cysteine motif, $HX_{3-7}HX_{23-32}CX_2C$ (referred to as the HHCC motif), that binds zinc (Burke *et al.*, 1992; Bushman *et al.*, 1993 1993; Cai *et al.*, 1997; Eijkelenboom *et al.*, 1997). This HHCC motif is essential for 3' processing and strand transfer activity *in vitro* and has been shown to promote tetramerization of integrase protomers and to enhance activity (Cannon *et al.*, 1994; Ellison *et al.*, 1995; Engelman *et al.*, 1995 1995; Zheng *et al.*, 1996). The CCD contains a D,DX35E motif formed by the catalytic triad D64, D116 and E152 embedded in a protein fold that is highly conserved among polynucleotide phosphotransferase enzymes (Bujacz *et al.*, 1995; Dyda *et al.*, 1994; Goldgur *et al.*, 1998; Greenwald *et al.*, 1999; Maignan *et al.*, 1998). Two monomeric core domains associate to form a dimer in solution, with each monomer being structurally similar to RNaseH, MuA transposase and RuvC resolvase, (Chiu and Davies, 2004). The D64 and the D116 residues form a coordination complex with a divalent metal (Mg^{2+} or Mn^{2+}) once the integrase binds to its DNA substrate (Gao *et al.*, 2004; Grobler *et al.*, 2002; Marchand *et al.*, 2003). The CCD requires the NTD and the CTD in a dimeric complex in order to maintain its 3' processing and strand transfer activities and mutation in any one of the three conserved D64, D116 or E152 residues abolishes this activity. The CTD sequence is less conserved and the overall structure resembles the Src-homology 3 (SH3) domain (Eijkelenboom *et al.*, 1995; Lodi *et al.*, 1995). The CTD binds DNA non-specifically and is required for 3' processing and strand transfer activities.

3.3 Integrase as a Drug Target

It is well known that HIV therapy is much more effective when combinations of drugs are used instead of single drug regimens, which has accelerated the search for additional anti-HIV drug targets. Currently there are no known human homologs of HIV integrase, making this enzyme a very attractive drug target. There are a number of criteria that are adhered to in screening integrase inhibitors. The inhibitor must have low cell toxicity and be specific for the integration step and not any of the other steps in the HIV lifecycle. Cells treated with the inhibitor should show an accumulation of 2-LTR circles, which is a by-product of unproductive integration, and cells should have a decreased number of integrated proviruses. Selection of drug-resistant viruses should also be shown to be a result of mutations solely in the integrase enzyme. This is usually verified by testing the inhibitor against recombinant integrase bearing the same mutations identified in the drug-resistant virus. Unfortunately, numerous drugs that inhibit integrase fail these criteria, especially in cell culture (see (Li *et al.*, 2004; Pommier *et al.*, 2000; Pommier and Neamati, 1999) for examples). Several assays exist for assessing integrase inhibition (reviewed in (Butler *et al.*, 2001; Hansen *et al.*, 1999; Witvrouw *et al.*, 2004)). The classical assays use LTR mimics to evaluate the 3' processing and strand transfer activities of integrase in the presence of inhibitors (Craigie *et al.*, 1991; Sherman and Fyfe, 1990). To date, the only class of integrase inhibitors that meet all of the above criteria is the diketo acids and their related naphthyridines. Another class of inhibitors called pyranodipyrimidines is currently under study and is hoped to provide an alternative to the diketo acid inhibitor family. Several other integrase inhibitors are currently under investigation and will advance our knowledge of integrase function and will ultimately help to define far superior integrase inhibitors in the future. These inhibitors are far too numerous to discuss here and so readers are referred to several reviews on these inhibitors (Johnson *et al.*, 2004; Pommier *et al.*, 2005; Pommier *et al.*, 2000; Pommier and Neamati, 1999; Pommier *et al.*, 1997).

3.4 Integrase Inhibitors

Diketo Acids

The most advanced integrase inhibitors reported to date are the diketo acids and their derivatives discovered by Merck Research Laboratories (Hazuda *et al.*, 2000) and Shionogi & Co. Ltd (Goldgur *et al.*, 1999). The diketo acids were the first reported integrase inhibitors that showed high specificity for the integration reaction and antiviral activity in cells (Goldgur *et al.*, 1999; Hazuda *et al.*, 2000). The diketo moiety is usually flanked by an acidic group or an equivalent group such as a carboxyl, tetrazole or triazole group, and an aromatic group. Several substitutions of the aromatic group have been studied and shown to be critical for activity (Marchand *et al.*, 2003; Marchand *et al.*, 2002; Pais *et al.*, 2002; Wai and al., 2000). The diketo acids selectively recognize a particular conformation of the integrase active site only after it assembles on the viral DNA ends. Once bound, the diketo acids compete with the target DNA and inhibit strand transfer, most likely by sequestering the active site metal (Mg^{2+}) (Espeseth *et al.*, 2000; Grobler *et al.*, 2002; Pommier *et al.*, 2005). Merck increased enthusiasm in the field when they showed that naphthyridines (derivatives of diketo acids) have potent antiviral activity against HIV-1, HIV-2 and simian immuno-deficiency virus (SIV) with no cross-resistance to drugs that target other aspects of the viral lifecycle (Hazuda *et al.*, 2000; Pluymers *et al.*, 2002). At that time, L-870,810 was the most promising diketo acid, having the most potent anti-viral activity and entered clinical trials. However, L-870,810 studies were recently halted due to liver and kidney toxicity observed in dogs. Currently, Merck is developing the integrase inhibitors L-870,812 and Compound B. L-870,812 has been tested in rhesus macaques and has been shown to suppress viremia and chronic infections caused by SIV (Hazuda *et al.*, 2004).

Detailed analyses of the integrase gene of viruses that have become resistant to the Merck diketo acids (L-708,906 and L731,988) have revealed several mutations located near the metal coordinating residues of the D,DX35E motif in the CCD subunit of the enzyme. Specificity of these drugs was further supported by the finding that the resistant viruses still maintained sensitivity to inhibitors of reverse transcriptase, protease and viral entry (Fikkert *et al.*, 2003; Hazuda *et al.*, 2000). L-708,906-resistant viruses were also shown to exhibit cross-resistance to the diketo analogue S-1360, but they remained fully susceptible to the pyranodipyrimidine inhibitor V-165 (discussed later).

The diketo acid 1-(5-chloroindole-3-yl)-3-hydroxy-3(2H-tetrazol-5-yl)-propenone, otherwise know as 5CITEP, from Shionogi & Co. Ltd made a breakthrough when they co-crystallized 5CITEP with the CCD of HIV integrase and showed it to be in close association with the conserved D,DX35E motif (Goldgur *et al.*, 1999). 5CITEP is active against both 3' processing and strand transfer, thereby distinguishing this diketo derivative from the Merck derivatives (Marchand *et al.*, 2003; Marchand *et al.*, 2002; Pais *et al.*, 2002). In addition, molecular docking and dynamics simulation studies suggest that that the Merck inhibitor L-731,988 and 5CITEP bind to integrase in a different way than do the diketo acids, likely involving metal chelation differences contributed by the aromatic and acidic groups of the diketo moiety (Keseru and Kolossvary, 2001; Marchand *et al.*, 2003).

Shionogi & Co. Ltd developed a more potent derivative of 5CITEP called S-1360 (Billich, 2003). S-1360 retains the diketo functionality but contains a triazole instead of the tetrazole group of 5CITEP. Numerous mutations arising from the selection of resistant virus in the presence of S-1360 appear to be in the vicinity of the highly conserved D,DX35E motif of the CCD.

Cross-resistance to the diketo acid L-708,906 was observed with these integrase mutants, but they remained fully susceptible to V-165 (Fikkert et al., 2004). S-1360 has recently entered phase II clinical trials.

Pyranodipyrimidine

5H-pyrano[2,3-d:-6,5-d']dipyrimidines (PDPs) is a second class of inhibitors showing promise as a new integrase inhibitor. 5-(4-nitrophenyl)-2,8-dithiol-4,6-dihydroxy-5H-pyrano[2,3-d:-6,5-d']dipyrimidine (referred to as V-165) has been shown to inhibit integrase activities in enzymatic assays, although inhibition of reverse transcriptase activities has also been observed (Pannecouque et al., 2002). In contrast to the diketo acids, V-165 was shown to have inhibitory effects against both the 3' processing and strand transfer activities in enzymatic assays, however in cell culture the anti-HIV activity of V-165 appears to correlate with inhibition of the strand transfer activity during integration. The mechanism of this inhibition is likely attributed to inhibition of integrase-DNA complex formation (Pannecouque et al., 2002). Interestingly, V-165 remained fully effective against viruses resistant to the diketo inhibitors and inhibitors of viral entry and reverse transcription (Fikkert et al., 2004; Fikkert et al., 2003; Pannecouque et al., 2002). Recently some mutations in the RT and env genes of resistant viruses that altered viral phenotype was reported (Cold Spring Harbor Retrovirus Meeting May 2005) and therefore further studies on the characterization of V-165-resistant HIV strains are required to verify that V-165 specifically targets integrase.

3.5 Considerations for Antiretrovirals Targeting Integration

The absence of a host equivalent to integrase greatly increases the therapeutic index of integrase inhibitors. However, caution must be taken since integrase shares mechanistic and structural similarities with various recombinases, RNases and integrases (Chiu and Davies, 2004; Rice and Baker, 2001; Shaw-Reid et al., 2003). Such similarities help explain the finding that diketo acids inhibit the V(D)J RAG1/2 recombinases, albeit at about a 20 fold higher concentration than that needed to inhibit integrase (Melek et al., 2002).

Other targets of the integration reaction that may lead to the discovery of new inhibitors include the 3' end processing of the viral DNA, multimerization of the integrase complex, assembly of the PIC and targeting of the PIC to chromosomes. Such studies will benefit tremendously once the crystal structure of full-length integrase complexed with DNA is solved.

4. MATURATION

The last step in the life cycle of HIV is referred to as "maturation" (Figure 1; reviewed in (Bukrinskaya, 2004; Vogt, 1996)). Immature HIV virions undergo morphologic changes that include condensation of the viral capsid protein (CA) to form a cone shaped core structure that is associated with mature, infectious virions. Maturation occurs during and following viral egress and is coordinated by the viral protease enzyme. Protease cleaves the Gag precursor polyprotein (Pr55^{Gag}) into the individual protein and peptide subunits, matrix (MA), CA, spacer peptide 1 (SP1), nucleocapsid (NC), spacer peptide 2 (SP2) and p6.

4.1 Maturation Inhibitors

Protease inhibitors, a core constituent of antiretroviral therapy, act by directly targeting the protease enzyme to block enzymatic activity and thus viral maturation. Issues that include resistance to current protease inhibitors and drug toxicity are driving the development of new inhibitors that target the protease enzyme either directly or indirectly (Rodriguez-Barrios and Gago, 2004; Wynn *et al.*, 2004). Indeed, a derivative of betulinic acid (3-O-betulinic acid, referred to as PA-457, DSB or YK-FH312) (Zhou *et al.*, 2004a), that is under clinical development by V. I. Technologies (Vitech; formally Panacos), acts to block HIV maturation via a novel mechanism of action and is a representative of a new class of antiretroviral agents referred to as maturation inhibitors (Kanamoto *et al.*, 2001; Li *et al.*, 2003; Zhou *et al.*, 2004b).

PA-457 is active against diverse primary HIV isolates *in vitro*, including viruses resistant to approved protease and RT inhibitors, with IC$_{50}$s in the low nanomolar range (Li *et al.*, 2003). PA-457 blocks virion maturation by inhibiting protease cleavage between the junction of CA and SP1, the last step in the processing of Pr55^{Gag} (Li *et al.*, 2003; Zhou *et al.*, 2004b). Processing of CA-SP1 into CA and SP1 is a prerequisite for condensation of CA into a mature viral core, thus virions produced in the presence of PA-457 have defective core structures and are not infectious. Passage of virus in the presence of PA-457 *in vitro* selects for mutations at the CA-SP1 junction (Li *et al.*, 2003). Mutations within this region are associated with resistance to PA-457, but also correlate with reduced viral fitness (Li *et al.*, 2003; Liang *et al.*, 2002; Zhou *et al.*, 2004a; Zhou *et al.*, 2004b). The precise mechanism of action of PA-457 is under investigation.

PA-457 is orally bioavailable, has favorable pharmacokinetics and was well tolerated in Phase I trials (Martin *et al.*, 2005a; Martin *et al.*, 2004a). Phase I/II trials of a single dose in HIV-infected individuals demonstrated that PA-457 could reduce viral load up to approximately 0.5 log (Martin *et al.*, 2005b). Phase II trials are underway and PA-457 has been granted fast-track review status by the FDA. A distinct mode of action from approved protease inhibitors means that PA-457 will likely be active against protease inhibitor resistant viruses *in vivo* as well as viruses resistant to other currently approved antiretroviral agents (Li *et al.*, 2003; Martin *et al.*, 2005b). Additionally, escape from PA-457 *in vivo* will likely come at a cost to viral fitness.

5. CONCLUSIONS

Despite significant advances in antiretroviral therapy over the past few years, an increasing number of individuals are harboring multi drug resistant viruses and have little options for effective therapy. Thus there is a pressing need for new antiretroviral agents that are active against viruses resistant to current drugs. Indeed, new inhibitors to current drug targets as well as inhibitors to new drug targets are in various stages of development, fueled by advances in our understanding of the viral life cycle. The life cycle of HIV presents numerous potential targets for intervention, and, as reviewed here, inhibitors to new targets that are furthest along in development include agents that interfere with various steps of the entry process, compounds that inhibit the integration of HIV into the host cell genome and an agent that prevents the formation of mature infectious virions. Distinct modes of action from approved antiretrovirals mean that these inhibitors will likely be effective against viruses resistant to current drugs. Furthermore, the use of a combination of inhibitors directed against distinct targets in first line therapy holds promise for enhanced viral containment. Entry inhibitors are also being developed as candidate microbicides and hold promise for the prevention of HIV transmission.

In summary, the development of novel entry, integration and maturation inhibitors will complement current antiretroviral therapy and future HAART regimens that attack HIV from various angles will likely offer better prospects for sustained inhibition of viral replication.

ACKNOWLEDGEMENTS

JDR is supported by NIH grant AI 058701 and amfAR fellowship 106437-34-RFGN. SB is supported by Alberta Heritage Foundation for Medical Research and Natural Sciences and Engineering Research Council of Canada. SP is supported by SFB 466. We thank Frederic Bushman for helpful comments and criticisms. We also thank F. Neipel for continuous instructions on the use of the Endnote program.

REFERENCES

Allaway, G. P., Davis-Bruno, K. L., Beaudry, G. A., Garcia, E. B., Wong, E. L., Ryder, A. M., Hasel, K. W., Gauduin, M. C., Koup, R. A., McDougal, J. S., and *et al.*, 1995, Expression and characterization of CD4-IgG2, a novel heterotetramer that neutralizes primary HIV type 1 isolates. *AIDS Res Hum Retroviruses* **11**: 533-9.

Alvarez, C. P., Lasala, F., Carrillo, J., Muniz, O., Corbi, A. L., and Delgado, R., 2002, C-type lectins DC-SIGN and L-SIGN mediate cellular entry by Ebola virus in cis and in trans. *J Virol* **76**: 6841-4.

Argyris, E. G., Acheampong, E., Nunnari, G., Mukhtar, M., Williams, K. J., and Pomerantz, R. J., 2003, Human immunodeficiency virus type 1 enters primary human brain micro-vascular endothelial cells by a mechanism involving cell surface proteoglycans independent of lipid rafts. *J Virol* **77**: 12140-51.

Baldwin, C. E., Sanders, R. W., Deng, Y., Jurriaans, S., Lange, J. M., Lu, M., and Berkhout, B., 2004, Emergence of a drug-dependent human immunodeficiency virus type 1 variant during therapy with the T20 fusion inhibitor. *J Virol* **78**: 12428-37.

Balzarini, J., 2004, Current status of the non-nucleoside reverse transcriptase inhibitors of human immunodeficiency virus type 1. *Curr Top Med Chem* **4**: 921-944.

Balzarini, J., Van Laethem, K., Hatse, S., Vermeire, K., De Clercq, E., Peumans, W., Van Damme, E., Vandamme, A. M., Bolmstedt, A., and Schols, D., 2004, Profile of resistance of human immunodeficiency virus to mannose-specific plant lectins. *J Virol* **78**: 10617-27.

Baribaud, F., Pöhlmann, S., and Doms, R. W., 2001a, The role of DC-SIGN and DC-SIGNR in HIV and SIV attachment, infection, and transmission. *Virology* **286**: 1-6.

Baribaud, F, Doms, R.W., and Pöhlmann, S. 2002a, The role of DC-SIGN and DC-SIGNR in HIV and Ebola virus infection: can potential therapeutics block virus transmission and dissemination? *Expert Opin Ther Targets.* **6**: 423-31.

Baribaud, F., Pöhlmann, S., Leslie, G., Mortari, F., and Doms, R. W., 2002b, Quantitative expression and virus transmission analysis of DC-SIGN on monocyte-derived dendritic cells. *J Virol* **76**: 9135-42.

Baribaud, F., Pöhlmann, S., Sparwasser, T., Kimata, M. T., Choi, Y. K., Haggarty, B. S., Ahmad, N., Macfarlan, T., Edwards, T. G., Leslie, G. J., Arnason, J., Reinhart, T. A., Kimata, J. T., Littman, D. R., Hoxie, J. A., and Doms, R. W., 2001b, Functional and antigenic characterization of human, rhesus macaque, pigtailed macaque, and murine DC-SIGN. *J Virol* **75**: 10281-9.

Barrientos, L. G., and Gronenborn, A. M., 2005, The highly specific carbohydrate-binding protein cyanovirin-N: structure, anti-HIV/Ebola activity and possibilities for therapy. *Mini Rev Med Chem* **5**: 21-31.

240 *J.D. Reeves, S.D. Barr and S. Pöhlmann*

Barrientos, L. G., Lasala, F., Delgado, R., Sanchez, A., and Gronenborn, A. M., 2004, Flipping the switch from monomeric to dimeric CV-N has little effect on antiviral activity. *Structure* (Camb) **12**: 1799-807.
Bashirova, A. A., Geijtenbeek, T. B., van Duijnhoven, G. C., van Vliet, S. J., Eilering, J. B., Martin, M. P., Wu, L., Martin, T. D., Viebig, N., Knolle, P. A., KewalRamani, V. N., van Kooyk, Y., and Carrington, M., 2001, A dendritic cell-specific intercellular adhesion molecule 3-grabbing nonintegrin (DC-SIGN)-related protein is highly expressed on human liver sinusoidal endothelial cells and promotes HIV-1 infection. *J Exp Med* **193**: 671-8.
Berger, E. A., Murphy, P. M., and Farber, J. M., 1999, Chemokine receptors as HIV-1 coreceptors: roles in viral entry, tropism, and disease. *Annu Rev Immunol* **17**: 657-700.
Bewley, C. A., Gustafson, K. R., Boyd, M. R., Covell, D. G., Bax, A., Clore, G. M., and Gronenborn, A. M., 1998, Solution structure of cyanovirin-N, a potent HIV-inactivating protein. *Nat Struct Biol* **5**: 571-8.
Bhattacharya, J., Peters, P. J., and Clapham, P. R., 2003, CD4-independent infection of HIV and SIV: implications for envelope conformation and cell tropism in vivo. *Aids* **17** (Suppl 4): S35-43.
Billich, A., 2003, S-1360 Shionogi-GlaxoSmithKline. *Curr Opin Investig Drugs* **4**: 206-9.
Bobardt, M. D., Armand-Ugon, M., Clotet, I., Zhang, Z., David, G., Este, J. A., and Gallay, P. A., 2004a, Effect of polyanion-resistance on HIV-1 infection. *Virology* **325**: 389-98.
Bobardt, M. D., Salmon, P., Wang, L., Esko, J. D., Gabuzda, D., Fiala, M., Trono, D., Van der Schueren, B., David, G., and Gallay, P. A., 2004b, Contribution of proteoglycans to human immunodeficiency virus type 1 brain invasion. *J Virol* **78**: 6567-84.
Bobardt, M. D., Saphire, A. C., Hung, H. C., Yu, X., Van der Schueren, B., Zhang, Z., David, G., and Gallay, P. A., 2003, Syndecan captures, protects, and transmits HIV to T lymphocytes. *Immunity* **18**: 27-39.
Bujacz, G., Jaskolski, M., Alexandratos, J., Wlodawer, A., Merkel, G., Katz, R. A., and Skalka, A. M., 1995, High-resolution Structure of the Catalytic Domain of Avain Sarcoma Virus Integrase. *J. Mol. Biol.* **253**: 333-346.
Bukrinskaya, A. G., 2004, HIV-1 assembly and maturation. *Arch Virol* **149**: 1067-1082.
Burke, C. J., Sanyal, G., Bruner, M. W., Ryan, J. A., LaFemina, R. L., Robbins, H. L., Zeft, A. S., Middaugh, C. R., and Cordingley, M. G., 1992, Structural Implications of Spectroscopic Characterization of a Putative Zinc Finger Peptide from HIV-1 Integrase. *J. Biol. Chem.* **267**: 9639-9644.
Bushman, F. D., Engelman, A., Palmer, I., Wingfield, P., and Craigie, R., 1993, Domains of the integrase protein of human immunodeficiency virus type 1 responsible for poly-nucleotidyl transfer and zinc binding. *Proc. Natl. Acad. Sci. USA* **90**: 3428-3432.
Bushman, F. D., Fujiwara, T., and Craigie, R., 1990, Retroviral DNA integration directed by HIV integration protein in vitro. *Science* **249**: 1555-1558.
Butler, S., Hansen, M., and Bushman, F. D., 2001, A Quantitative Assay for HIV cDNA Integration In vivo. *Nat. Med.* **7**: 631-634.
Cai, M., Zheng, R., Caffrey, M., Craigie, R., Clore, G. M., and Gronenborn, A. M., 1997, Solution structure of the N-terminal zinc binding domain of HIV-1 integrase. *Nat. Struct. Biol.* **4**: 567-577.
Cannon, P. M., Wilson, W., Byles, E., Kingsman, S. M., and Kingsman, A. J., 1994, Human immunodeficiency virus type 1 integrase: Effect on viral replication of mutations at highly conserved residues. *J. Virol.* **68**: 4768-4775.
Chan, D. C., Fass, D., Berger, J. M., and Kim, P. S., 1997, Core structure of gp41 from the HIV envelope glycoprotein. *Cell* **89**: 263-273.

Chang, L. C., and Bewley, C. A., 2002, Potent inhibition of HIV-1 fusion by cyanovirin-N requires only a single high affinity carbohydrate binding site: characterization of low affinity carbohydrate binding site knockout mutants. *J Mol Biol* **318**: 1-8.

Chen, B., Vogan, E. M., Gong, H., Skehel, J. J., Wiley, D. C., and Harrison, S. C., 2005a, Determining the structure of an unliganded and fully glycosylated SIV gp120 envelope glycoprotein. *Structure* (Camb) **13**: 197-211.

Chen, B., Vogan, E. M., Gong, H., Skehel, J. J., Wiley, D. C., and Harrison, S. C., 2005b, Structure of an unliganded simian immunodeficiency virus gp120 core. *Nature* **433**: 834-41.

Chen, J. C.-H., Krucinski, J., Miercke, L. J. W., Finer-Moore, J. S., Tang, A. H., Leavitt, A. D., and Stroud, R. M., 2000a, Crystal structure of the HIV-1 integrase catalytic core and C-terminal domains: A model for viral DNA binding. *Proc. Natl. Acad. Sci. USA* **97**: 8233-8238.

Chen, Z., Yan, Y., Munshi, S., Li, Y., Zugay-Murphy, J., Xu, B., Witmer, M., Felock, P., Wolfe, A., Sardana, V., Emini, E. A., Hazuda, D., and Kuo, L. C., 2000b, X-ray structure of simian immunodeficiency virus integrase containing the core and C-terminal domain (residues 50-293)–an initial glance of the viral DNA binding platform. *J. Mol. Biol.* **296**: 521-533.

Chiu, T. K., and Davies, D. R., 2004, Structure and function of HIV-1 integrase. *Curr Top Med Chem* **4**: 965-77.

Choe, H., Farzan, M., Sun, Y., Sullivan, N., Rollins, B., Ponath, P. D., Wu, L., Mackay, C. R., LaRosa, G., Newman, W., Gerard, N., Gerard, C., and Sodroski, J., 1996, The beta-chemokine receptors CCR3 and CCR5 facilitate infection by primary HIV-1 isolates. *Cell* **85**: 1135-1148.

Choi, Y. H., Rho, W. S., Kim, N. D., Park, S. J., Shin, D. H., Kim, J. W., Im, S. H., Won, H. S., Lee, C. W., Chae, C. B., and Sung, Y. C., 2001, Short peptides with induced beta-turn inhibit the interaction between HIV-1 gp120 and CD4. *J Med Chem* **44**: 1356-63.

Cocchi, F., DeVico, A. L., Garzino-Demo, A., Cara, A., Gallo, R. C., and Lusso, P., 1996, The V3 domain of the HIV-1 gp120 envelope glycoprotein is critical for chemokine-mediated blockade of infection. *Nat Med* **2**: 1244-1247.

Colmenares, M., Puig-Kroger, A., Pello, O. M., Corbi, A. L., and Rivas, L., 2002, Dendritic cell (DC)-specific intercellular adhesion molecule 3 (ICAM-3)-grabbing nonintegrin (DC-SIGN, CD209), a C-type surface lectin in human DCs, is a receptor for Leishmania amastigotes. *J Biol Chem* **277**: 36766-9.

Connor, R. I., Sheridan, K. E., Ceradini, D., Choe, S., and Landau, N. R., 1997, Change in coreceptor use coreceptor use correlates with disease progression in HIV-1 infected individuals. *J Exp Med* **185**: 621-628.

Craigie, R., Fujiwara, T., and Bushman, F., 1990, The IN protein of Moloney murine leukemia virus processes the viral DNA ends and accomplishes their integration in vitro. *Cell* **62**: 829-837.

Craigie, R., Mizuuchi, K., Bushman, F. D., and Engelman, A., 1991, A rapid in vitro assay for HIV DNA integration. *Nuc. Acids Res.* **19**: 2729-2734.

Daniel, R., Greger, J. G., Katz, R. A., Taganov, K. D., Wu, X., Kappes, J. C., and Skalka, A. M., 2004, Evidence that stable retroviral transduction and cell survival following DNA integration depend on components of the nonhomologous end joining repair pathway. *J Virol* **78**: 8573-81.

De Clercq, E., 2003, The bicyclam AMD3100 story. *Nat Rev Drug Discov* **2**: 581-587.

de Vreese, K., Kofler-Mongold, V., Leutgeb, C., Weber, V., Vermeire, K., Schacht, S., Anne, J., de Clercq, E., Datema, R., and Werner, G., 1996, The molecular target of bicyclams, potent inhibitors of human immunodeficiency virus replication. *J Virol* **70**: 689-96.

Dean, M., Carrington, M., Winkler, C., Huttley, G. A., Smith, M. W., Allikmets, R., Goedert, J. J., Buchbinder, S. P., Vittinghoff, E., Gomperts, E., Donfield, S., Vlahov, D., Kaslow, R., Saah, A., Rinaldo, C., Detels, R., and O'Brien, S. J., 1996, Genetic restriction of HIV-1 infection and progression to AIDS by a deletion allele of the CKR5 structural gene. Hemophilia Growth and Development Study, Multicenter AIDS Cohort Study, Multicenter Hemophilia Cohort Study, San Francisco City Cohort, ALIVE Study. *Science* **273**: 1856-1862.

Demarest, J., Adkinson, K., Sparks, S., Shachoy-Clark, A., Schell, K., Reddy, S., Fang, L., O'Mara, K., Shibayama, S., and Piscitelli, S. (2004a). 11th Conference on Retroviruses and Opportunistic Infections.

Demarest, J., Shibayama, S., Ferris, R., Vavro, C., St Clair, M., and Boone, L. (2004b). 15th International AIDS Conference, Bangkok, Thailand.

Demarest, J., Sparks, S., Watson, C., McDanal, C., Kenakin, T., and Shibayama, S. (2004c). Interscience Conference on Antimicrobial Agents and Chemotherapy.

Derdeyn, C. A., Decker, J. M., Sfakianos, J. N., Wu, X., O'Brien, W. A., Ratner, L., Kappes, J. C., Shaw, G. M., and Hunter, E., 2000, Sensitivity of human immunodeficiency virus type 1 to the fusion inhibitor T-20 is modulated by coreceptor specificity defined by the V3 loop of gp120. *J Virol* **74**, 8358-67.

Derdeyn, C. A., Decker, J. M., Sfakianos, J. N., Zhang, Z., O'Brien, W. A., Ratner, L., Shaw, G. M., and Hunter, E., 2001, Sensitivity of human immunodeficiency virus type 1 to fusion inhibitors targeted to the gp41 first heptad repeat involves distinct regions of gp41 and is consistently modulated by gp120 interactions with the coreceptor. *J Virol* **75**: 8605-14.

Dey, B., Lerner, D. L., Lusso, P., Boyd, M. R., Elder, J. H., and Berger, E. A., 2000, Multiple antiviral activities of cyanovirin-N: blocking of human immunodeficiency virus type 1 gp120 interaction with CD4 and coreceptor and inhibition of diverse enveloped viruses. *J Virol* **74**: 4562-9.

Dyda, F., Hickman, A. B., Jenkins, T. M., Engelman, A., Craigie, R., and Davies, D. R., 1994, Crystal Structure of the Catalytic Domain of HIV-1 Integrase: Similarity to Other Polynucleotidyl Transferases. *Science* **266**: 1981-1986.

Eckert, D. M., and Kim, P. S., 2001, Mechanisms of viral membrane fusion and its inhibition. *Annu Rev Biochem* **70**: 777-810.

Edwards, T. G., Hoffman, T. L., Baribaud, F., Wyss, S., LaBranche, C. C., Romano, J., Adkinson, J., Sharron, M., Hoxie, J. A., and Doms, R. W., 2001, Relationships between CD4 independence, neutralization sensitivity, and exposure of a CD4-induced epitope in a human immunodeficiency virus type 1 envelope protein. *J Virol* **75**: 5230-5239.

Eijkelenboom, A. P. A. M., Puras Lutzke, R. A., Boelens, R., Plasterk, R. H. A., Kaptein, R., and Hard, K., 1995, The DNA binding domain of HIV-1 integrase has an SH3-like fold. *Nature Struc. Biol.* **2**: 807-810.

Eijkelenboom, A. P. A. M., van den Ent, F. M. I., Vos, A., Doreleijers, J. F., Hard, K., Tullius, T., Plasterk, R. H. A., Kaptein, R., and Boelens, R., 1997, The solution structure of the amino-terminal HHCC domain of HIV-2 integrase; a three-helix bundle stabilized by zinc. *Cur. Biol.* **1**: 739-746.

Ellison, V., Gerton, J., Vincent, K. A., and Brown, P. O., 1995, An essential interaction between distinct domains of HIV-1 integrase mediates assembly of the active multimer. *J. Biol. Chem.* **270**: 3320-3326.

Engelman, A., Englund, G., Orenstein, J. M., Martin, M. A., and Craigie, R., 1995, Multiple effects of mutations in human immunodeficiency virus type 1 integrase on viral replication. *J. Virol.* **69**: 2729-2736.

Eron, J. J., Gulick, R. M., Bartlett, J. A., Merigan, T., Arduino, R., Kilby, J. M., Yangco, B., Diers, A., Drobnes, C., DeMasi, R., Greenberg, M., Melby, T., Raskino, C., Rusnak, P., Zhang, Y., Spence, R., and Miralles, G. D., 2004, Short-term safety and antiretroviral activity of T-1249, a second-generation fusion inhibitor of HIV. *J Infect Dis* **189**: 1075-83.

Espeseth, A. S., Felock, P., Wolfe, A., Witmer, M., Grobler, J., Anthony, N., Egbertson, M., Melamed, J. Y., Young, S., Hamill, T., Cole, J. L., and Hazuda, D. J., 2000, HIV-1 integrase inhibitors that compete with the target DNA substrate define a unique strand transfer conformation for integrase. *Proc Natl Acad Sci U S A* **97**: 11244-9.

Este, J. A., Cabrera, C., Blanco, J., Gutierrez, A., Bridger, G., Henson, G., Clotet, B., Schols, D., and De Clercq, E., 1999, Shift of clinical human immunodeficiency virus type 1 isolates from X4 to R5 and prevention of emergence of the syncytium-inducing phenotype by blockade of CXCR4. *J Virol* **73**: 5577-85.

Feinberg, H., Mitchell, D. A., Drickamer, K., and Weis, W. I., 2001, Structural basis for selective recognition of oligosaccharides by DC-SIGN and DC-SIGNR. *Science* **294**: 2163-6.

Ferrer, M., and Harrison, S. C., 1999, Peptide ligands to human immunodeficiency virus type 1 gp120 identified from phage display libraries. *J Virol* **73**: 5795-802.

Fikkert, V., Hombrouck, A., Van Remoortel, B., De Maeyer, M., Pannecouque, C., De Clercq, E., Debyser, Z., and Witvrouw, M., 2004, Multiple mutations in human immunodeficiency virus-1 integrase confer resistance to the clinical trial drug S-1360. *Aids* **18**: 2019-28.

Fikkert, V., Van Maele, B., Vercammen, J., Hantson, A., Van Remoortel, B., Michiels, M., Gurnari, C., Pannecouque, C., De Maeyer, M., Engelborghs, Y., De Clercq, E., Debyser, Z., and Witvrouw, M., 2003, Development of resistance against diketo derivatives of human immunodeficiency virus type 1 by progressive accumulation of integrase mutations. *J Virol* **77**: 11459-70.

Gallay, P., 2004, Syndecans and HIV-1 pathogenesis. *Microbes Infect* **6(6)**, 617-22.

Gao, K., Wong, S., and Bushman, F., 2004, Metal binding by the D,DX35E motif of human immunodeficiency virus type 1 integrase: selective rescue of Cys substitutions by Mn2+ in vitro. *J Virol* **78**: 6715-22.

Gardner, J. P., Durso, R. J., Arrigale, R. R., Donovan, G. P., Maddon, P. J., Dragic, T., and Olson, W. C., 2003, L-SIGN (CD 209L) is a liver-specific capture receptor for hepatitis C virus. *Proc Natl Acad Sci U S A* **100**: 4498-503.

Garred, P., Eugen-Olsen, J., Iversen, A. K., Benfield, T. L., Svejgaard, A., and Hofmann, B., 1997, Dual effect of CCR5 delta 32 gene deletion in HIV-1-infected patients. Copenhagen AIDS Study Group. *Lancet* **349**: 1884.

Gauduin, M. C., Allaway, G. P., Maddon, P. J., Barbas, C. F., 3rd, Burton, D. R., and Koup, R. A., 1996, Effective ex vivo neutralization of human immunodeficiency virus type 1 in plasma by recombinant immunoglobulin molecules. *J Virol* **70**: 2586-92.

Gauduin, M. C., Allaway, G. P., Olson, W. C., Weir, R., Maddon, P. J., and Koup, R. A., 1998, CD4-immunoglobulin G2 protects Hu-PBL-SCID mice against challenge by primary human immunodeficiency virus type 1 isolates. *J Virol* **72**: 3475-8.

Geijtenbeek, T. B., Koopman, G., van Duijnhoven, G. C., van Vliet, S. J., van Schijndel, A. C., Engering, A., Heeney, J. L., and van Kooyk, Y., 2001, Rhesus macaque and chimpanzee DC-SIGN act as HIV/SIV gp120 trans-receptors, similar to human DC-SIGN. *Immunol Lett* **79**: 101-7.

Geijtenbeek, T. B., Kwon, D. S., Torensma, R., van Vliet, S. J., van Duijnhoven, G. C., Middel, J., Cornelissen, I. L., Nottet, H. S., KewalRamani, V. N., Littman, D. R., Figdor, C. G., and van Kooyk, Y., 2000, DC-SIGN, a dendritic cell-specific HIV-1-binding protein that enhances trans-infection of T cells. *Cell* **100**: 587-97.

Geijtenbeek, T. B., Van Vliet, S. J., Koppel, E. A., Sanchez-Hernandez, M., Vandenbroucke-Grauls, C. M., Appelmelk, B., and Van Kooyk, Y., 2003, Mycobacteria target DC-SIGN to suppress dendritic cell function. *J Exp Med* **197**: 7-17.

Goldgur, Y., Craigie, R., Cohen, G. H., Fujiwara, T., Yoshinaga, T., Fujishita, T., Sugimoto, H., Endo, T., Murai, H., and Davies, D. R., 1999, Structure of the HIV-1 integrase catalytic domain complexed with an inhibitor: A platform for antiviral drug design. *Proc. Natl. Acad. Sci. USA* **96**: 13040-13043.

Goldgur, Y., Dyda, F., Hickman, A. B., Jenkins, T. M., Craigie, R., and Davies, D. R., 1998, Three new structures of the core domain of HIV-1 integrase: An active site that binds magnesium. *Proc. Natl. Acad. Sci. USA* **95**: 9150-9154.

Greenberg, M., Cammack, N., Salgo, M., and Smiley, L., 2004, HIV fusion and its inhibition in antiretroviral therapy. *Rev Med Virol* **14**: 321-337.

Greenberg, M. L., and Cammack, N., 2004, Resistance to enfuvirtide, the first HIV fusion inhibitor. *J Antimicrob Chemother.* **54**: 333-400.

Greenwald, J., Le, V., Butler, S. L., Bushman, F. D., and Choe, S., 1999, Mobility of an HIV-1 Integrase Active Site Loop is Correlated with Catalytic Activity. *Biochemistry* **38**: 8892-8898.

Grobler, J. A., Stillmock, K., Hu, B., Witmer, M., Felock, P., Espeseth, A. S., Wolfe, A., Egbertson, M., Bourgeois, M., Melamed, J., Wai, J. S., Young, S., Vacca, J., and Hazuda, D. J., 2002, Diketo acid inhibitor mechanism and HIV-1 integrase: implications for metal binding in the active site of phosphotransferase enzymes. *Proc Natl Acad Sci U S A* **99**: 6661-6666.

Gulick, R. M., Mellors, J. W., Havlir, D., Eron, J. J., Gonzalez, C., McMahon, D., Richman, D. D., Valentine, F. T., Jonas, L., Meibohm, A., Emini, E. A., and Chodakewitz, J. A., 1997, Treatment with indinavir, zidovudine, and lamivudine in adults with human immunodeficiency virus infection and prior antiretroviral therapy. *N Engl J Med* **337**: 734-739.

Guo, Q., Ho, H. T., Dicker, I., Fan, L., Zhou, N., Friborg, J., Wang, T., McAuliffe, B. V., Wang, H. G., Rose, R. E., Fang, H., Scarnati, H. T., Langley, D. R., Meanwell, N. A., Abraham, R., Colonno, R. J., and Lin, P. F., 2003, Biochemical and genetic characterizations of a novel human immunodeficiency virus type 1 inhibitor that blocks gp120-CD4 interactions. *J Virol* **77**: 10528-36.

Guo, Y., Feinberg, H., Conroy, E., Mitchell, D. A., Alvarez, R., Blixt, O., Taylor, M. E., Weis, W. I., and Drickamer, K., 2004, Structural basis for distinct ligand-binding and targeting properties of the receptors DC-SIGN and DC-SIGNR. *Nat Struct Mol Biol* **11**: 591-8.

Hammer, S. M., and Pedneault, L., 2000, Antiretroviral resistance testing comes of age. *Antivir Ther* **5**: 23-26.

Hammer, S. M., Squires, K. E., Hughes, M. D., Grimes, J. M., Demeter, L. M., Currier, J. S., Eron, J. J., Jr., Feinberg, J. E., Balfour, H. H., Jr., Deyton, L. R., Chodakewitz, J. A., and Fischl, M. A., 1997, A controlled trial of two nucleoside analogues plus indinavir in persons with human immunodeficiency virus infection and CD4 cell counts of 200 per cubic millimeter or less. AIDS Clinical Trials Group 320 Study Team. *N Engl J Med* **337**: 725-733.

Hansen, M. S. T., Smith, G. J. I., Kafri, T., Molteni, V., Siegel, J. S., and Bushman, F. D., 1999, Integration complexes derived from HIV vectors for rapid assays in vitro. *Nat. Biotech.* **17**: 578-582.

Hartley, O., Klasse, P. J., Sattentau, Q. J., and Moore, J. P., 2005, V3: HIV's switch-hitter. *AIDS Res Hum Retroviruses* **21**: 171-89.

Hazuda, D. J., Felock, P., Witmer, M., Wolfe, A., Stillmock, K., Grobler, J. A., Espeseth, A., Gabryelski, L., Schleif, W., Blau, C., and Miller, M. D., 2000, Inhibitors of Strand Transfer That Prevent Integration and Inhibit HIV-1 Replication in Cells. *Science* **287**: 646-650.

Hazuda, D. J., Young, S. D., Guare, J. P., Anthony, N. J., Gomez, R. P., Wai, J. S., Vacca, J. P., Handt, L., Motzel, S. L., Klein, H. J., Dornadula, G., Danovich, R. M., Witmer, M. V., Wilson, K. A., Tussey, L., Schleif, W. A., Gabryelski, L. S., Jin, L., Miller, M. D., Casimiro, D. R., Emini, E. A., and Shiver, J. W., 2004, Integrase inhibitors and cellular immunity suppress retroviral replication in rhesus macaques. *Science* **305**: 528-32.

Hicks, C., King, M. S., Gulick, R. M., White, A. C., Jr., Eron, J. J., Jr., Kessler, H. A., Benson, C., King, K. R., Murphy, R. L., and Brun, S. C., 2004, Long-term safety and durable antiretroviral activity of lopinavir/ritonavir in treatment-naive patients: 4 year follow-up study. *Aids* **18**: 775-779.

Hoffman, T. L., LaBranche, C. C., Zhang, W., Canziani, G., Robinson, J., Chaiken, I., Hoxie, J. A., and Doms, R. W., 1999, Stable exposure of the coreceptor-binding site in a CD4-independent HIV-1 envelope protein. *Proc Natl Acad Sci USA* **96**: 6359-6364.

Hu, Q., Frank, I., Williams, V., Santos, J. J., Watts, P., Griffin, G. E., Moore, J. P., Pope, M., and Shattock, R. J., 2004, Blockade of attachment and fusion receptors inhibits HIV-1 infection of human cervical tissue. *J Exp Med* **199**: 1065-75.

Huang, C. C., Stricher, F., Martin, L., Decker, J. M., Majeed, S., Barthe, P., Hendrickson, W. A., Robinson, J., Roumestand, C., Sodroski, J., Wyatt, R., Shaw, G. M., Vita, C., and Kwong, P. D., 2005, Scorpion-toxin mimics of CD4 in complex with human immunodeficiency virus gp120 crystal structures, molecular mimicry, and neutralization breadth. *Structure (Camb)* **13**: 755-68.

Huang, Y., Paxton, W. A., Wolinsky, S. M., Neumann, A. U., Zhang, L., He, T., Kang, S., Ceradini, D., Jin, Z., Yazdanbakhsh, K., Kunstman, K., Erickson, D., Dragon, E., Landau, N. R., Phair, J., Ho, D. D., and Koup, R. A., 1996, The role of a mutant CCR5 allele in HIV-1 transmission and disease progression. *Nat Med* **2**: 1240-1243.

Jacobson, J. M., Israel, R. J., Lowy, I., Ostrow, N. A., Vassilatos, L. S., Barish, M., Tran, D. N., Sullivan, B. M., Ketas, T. J., O'Neill, T. J., Nagashima, K. A., Huang, W., Petropoulos, C. J., Moore, J. P., Maddon, P. J., and Olson, W. C., 2004, Treatment of advanced human immunodeficiency virus type 1 disease with the viral entry inhibitor PRO 542. *Antimicrob Agents Chemother* **48**: 423-9.

Jacobson, J. M., Lowy, I., Fletcher, C. V., O'Neill, T. J., Tran, D. N., Ketas, T. J., Trkola, A., Klotman, M. E., Maddon, P. J., Olson, W. C., and Israel, R. J., 2000, Single-dose safety, pharmacology, and antiviral activity of the human immunodeficiency virus (HIV) type 1 entry inhibitor PRO 542 in HIV-infected adults. *J Infect Dis* **182**: 326-9.

Jameson, B., Baribaud, F., Pöhlmann, S., Ghavimi, D., Mortari, F., Doms, R. W., and Iwasaki, A., 2002, Expression of DC-SIGN by dendritic cells of intestinal and genital mucosae in humans and rhesus macaques. *J Virol* **76**: 1866-75.

Jeffers, S. A., Tusell, S. M., Gillim-Ross, L., Hemmila, E. M., Achenbach, J. E., Babcock, G. J., Thomas, W. D., Jr., Thackray, L. B., Young, M. D., Mason, R. J., Ambrosino, D. M., Wentworth, D. E., Demartini, J. C., and Holmes, K. V., 2004, CD209L (L-SIGN) is a receptor for severe acute respiratory syndrome coronavirus. *Proc Natl Acad Sci U S A* **101**: 15748-53.

Jiang, S., Lu, H., Liu, S., Zhao, Q., He, Y., and Debnath, A. K., 2004, N-substituted pyrrole derivatives as novel human immunodeficiency virus type 1 entry inhibitors that interfere with the gp41 six-helix bundle formation and block virus fusion. *Antimicrob Agents Chemother* **48**: 4349-59.

Johnson, A. A., Marchand, C., and Pommier, Y., 2004, HIV-1 integrase inhibitors: a decade of research and two drugs in clinical trial. Curr Top Med Chem **4**: 1059-77.

Kanamoto, T., Kashiwada, Y., Kanbara, K., Gotoh, K., Yoshimori, M., Goto, T., Sano, K., and Nakashima, H., 2001, Anti-human immunodeficiency virus activity of YK-FH312 (a betulinic acid derivative), a novel compound blocking viral maturation. *Antimicrob Agents Chemother* **45**: 1225-1230.

Kanbara, K., Sato, S., Tanuma, J., Tamamura, H., Gotoh, K., Yoshimori, M., Kanamoto, T., Kitano, M., Fujii, N., and Nakashima, H., 2001, Biological and genetic characterization of a human immunodeficiency virus strain resistant to CXCR4 antagonist T134. *AIDS Res Hum Retroviruses* **17**: 615-22.

Katz, R. A., Merkel, G., Kulkosky, J., Leis, J., and Skalka, A. M., 1990, The avian retroviral IN protein is both necessary and sufficient for integrative recombination in vitro. *Cell* **63**: 87-95.

Katzman, M., Katz, R. A., Skalka, A. M., and Leis, J., 1989, The avian retroviral integration protein cleaves the terminal sequences of linear viral DNA at the in vivo sites of integration. *J. Virol.* **63**: 5319-5327.

Keseru, G. M., and Kolossvary, I., 2001, Fully flexible low-mode docking: application to induced fit in HIV integrase. *J Am Chem Soc* **123**: 12708-9.

Kilby, J. M., and Eron, J. J., 2003, Novel therapies based on mechanisms of HIV-1 cell entry. *N Engl J Med* **348**: 2228-38.

Kilby, J. M., Hopkins, S., Venetta, T. M., DiMassimo, B., Cloud, G. A., Lee, J. Y., Alldredge, L., Hunter, E., Lambert, D., Bolognesi, D., Matthews, T., Johnson, M. R., Nowak, M. A., Shaw, G. M., and Saag, M. S., 1998, Potent suppression of HIV-1 replication in humans by T-20, a peptide inhibitor of gp41-mediated virus entry. *Nat Med* **4**: 1302-7.

Kinomoto, M., Yokoyama, M., Sato, H., Kojima, A., Kurata, T., Ikuta, K., Sata, T., and Tokunaga, K., 2005, Amino acid 36 in the human immunodeficiency virus type 1 gp41 ectodomain controls fusogenic activity: implications for the molecular mechanism of viral escape from a fusion inhibitor. *J Virol* **79**: 5996-6004.

Kuhmann, S. E., Platt, E. J., Kozak, S. L., and Kabat, D., 2000, Cooperation of multiple CCR5 coreceptors is required for infections by human immunodeficiency virus type 1. *J Virol* **74**(15), 7005-7015.

Kuhmann, S. E., Pugach, P., Kunstman, K. J., Taylor, J., Stanfield, R. L., Snyder, A., Strizki, J. M., Riley, J., Baroudy, B. M., Wilson, I. A., Korber, B. T., Wolinsky, S. M., and Moore, J. P., 2004, Genetic and phenotypic analyses of human immunodeficiency virus type 1 escape from a small-molecule CCR5 inhibitor. *J Virol* **78**: 2790-807.

Kuritzkes, D. R., Jacobson, J., Powderly, W. G., Godofsky, E., DeJesus, E., Haas, F., Reimann, K. A., Larson, J. L., Yarbough, P. O., Curt, V., and Shanahan, W. R., Jr., 2004, Antiretroviral activity of the anti-CD4 monoclonal antibody TNX-355 in patients infected with HIV type 1. *J Infect Dis* **189**: 286-91.

Kwon, D. S., Gregorio, G., Bitton, N., Hendrickson, W. A., and Littman, D. R., 2002, DC-SIGN-mediated internalization of HIV is required for trans-enhancement of T cell infection. *Immunity* **16**: 135-44.

Kwong, P. D., Wyatt, R., Robinson, J., Sweet, R. W., Sodroski, J., and Hendrickson, W. A., 1998, Structure of an HIV gp120 envelope glycoprotein in complex with the CD4 receptor and a neutralizing human antibody. *Nature* **393**: 648-659.

Labrosse, B., Labernardiere, J. L., Dam, E., Trouplin, V., Skrabal, K., Clavel, F., and Mammano, F., 2003, Baseline susceptibility of primary human immunodeficiency virus type 1 to entry inhibitors. *J Virol* **77**: 1610-3.

Lalezari, J. P., Bellos, N. C., Sathasivam, K., Richmond, G. J., Cohen, C. J., Myers, R. A., Jr., Henry, D. H., Raskino, C., Melby, T., Murchison, H., Zhang, Y., Spence, R., Greenberg, M. L., Demasi, R. A., and Miralles, G. D., 2005, T-1249 retains potent antiretroviral activity in patients who had experienced virological failure while on an enfuvirtide-containing treatment regimen. *J Infect Dis* **191**: 1155-63.

Lalezari, J. P., Eron, J. J., Carlson, M., Cohen, C., DeJesus, E., Arduino, R. C., Gallant, J. E., Volberding, P., Murphy, R. L., Valentine, F., Nelson, E. L., Sista, P. R., Dusek, A., and Kilby, J. M., 2003a, A phase II clinical study of the long-term safety and antiviral activity of enfuvirtide-based antiretroviral therapy. *Aids* **17**: 691-698.

Lalezari, J. P., Henry, K., O'Hearn, M., Montaner, J. S., Piliero, P. J., Trottier, B., Walmsley, S., Cohen, C., Kuritzkes, D. R., Eron, J. J., Jr., Chung, J., DeMasi, R., Donatacci, L., Drobnes, C., Delehanty, J., and Salgo, M., 2003b, Enfuvirtide, an HIV-1 fusion inhibitor, for drug-resistant HIV infection in North and South America. *N Engl J Med* **348**: 2175-2185.

Lalezari, J. P., Thompson, M., Kumar, P., Piliero, P. J., Davey, R., Murtaugh, T., Patterson, K., Shachoy-Clark, A., Adkinson, K., Demarest, J., Sparks, S., Fang, L., Lou, Y., Berrey, M., and Piscitelli, S., 2004, *Interscience Conference on Antimicrobial Agents and Chemotherapy*. Washingtion, DC, USA.

Lasala, F., Arce, E., Otero, J. R., Rojo, J., and Delgado, R., 2003, Mannosyl glycodendritic structure inhibits DC-SIGN-mediated Ebola virus infection in cis and in trans. *Antimicrob Agents Chemother* **47**: 3970-2.

Lazzarin, A., Clotet, B., Cooper, D., Reynes, J., Arasteh, K., Nelson, M., Katlama, C., Stellbrink, H. J., Delfraissy, J. F., Lange, J., Huson, L., DeMasi, R., Wat, C., Delehanty, J., Drobnes, C., and Salgo, M., 2003, Efficacy of enfuvirtide in patients infected with drug-resistant HIV-1 in Europe and Australia. *N Engl J Med* **348**: 2186-2195.

Lee, B., Sharron, M., Montaner, L. J., Weissman, D., and Doms, R. W., 1999, Quantification of CD4, CCR5, and CXCR4 levels on lymphocyte subsets, dendritic cells, and differentially conditioned monocyte-derived macrophages. *Proc Natl Acad Sci USA* **96**: 5215-5220.

Li, F., Goila-Gaur, R., Salzwedel, K., Kilgore, N. R., Reddick, M., Matallana, C., Castillo, A., Zoumplis, D., Martin, D. E., Orenstein, J. M., Allaway, G. P., Freed, E. O., and Wild, C. T., 2003, PA-457: a potent HIV inhibitor that disrupts core condensation by targeting a late step in Gag processing. *Proc Natl Acad Sci U S A* **100**: 13555-13560.

Li, W.-X., Li, H., Lu, R., Li, F., Dus, M., Atkinson, P., Brydon, E. W. A., Johnson, K. L., Garcia-Sastre, A., Ball, L. A., Palese, P., and Ding, S.-W., 2004, Interferon antagonist proteins of influenza and vaccinia viruses are suppressors of RNA silencing. *Proc. Natl. Acad. Sci. U S A* **101**: 1350-1355.

Liang, C., Hu, J., Russell, R. S., Roldan, A., Kleiman, L., and Wainberg, M. A., 2002, Characterization of a putative alpha-helix across the capsid-SP1 boundary that is critical for the multimerization of human immunodeficiency virus type 1 gag. *J Virol* **76**: 11729-11737.

Lin, G., Simmons, G., Pöhlmann, S., Baribaud, F., Ni, H., Leslie, G. J., Haggarty, B. S., Bates, P., Weissman, D., Hoxie, J. A., and Doms, R. W., 2003a, Differential N-linked glycosylation of human immunodeficiency virus and Ebola virus envelope glycoproteins modulates interactions with DC-SIGN and DC-SIGNR. *J Virol* **77**: 1337-46.

Lin, P. F., Blair, W., Wang, T., Spicer, T., Guo, Q., Zhou, N., Gong, Y. F., Wang, H. G., Rose, R., Yamanaka, G., Robinson, B., Li, C. B., Fridell, R., Deminie, C., Demers, G., Yang, Z., Zadjura, L., Meanwell, N., and Colonno, R., 2003b, A small molecule HIV-1 inhibitor that targets the HIV-1 envelope and inhibits CD4 receptor binding. *Proc Natl Acad Sci U S A* **100**: 11013-8.

Little, S. J., Holte, S., Routy, J. P., Daar, E. S., Markowitz, M., Collier, A. C., Koup, R. A., Mellors, J. W., Connick, E., Conway, B., Kilby, M., Wang, L., Whitcomb, J. M., Hellmann, N. S., and Richman, D. D., 2002, Antiretroviral-drug resistance among patients recently infected with HIV. *N Engl J Med* **347**: 385-394.

Liu, H., Hwangbo, Y., Holte, S., Lee, J., Wang, C., Kaupp, N., Zhu, H., Celum, C., Corey, L., McElrath, M. J., and Zhu, T., 2004, Analysis of genetic polymorphisms in CCR5, CCR2, stromal cell-derived factor-1, RANTES, and dendritic cell-specific intercellular adhesion molecule-3-grabbing nonintegrin in seronegative individuals repeatedly exposed to HIV-1. *J Infect Dis* **190**: 1055-8.

Liu, N. Q., Lossinsky, A. S., Popik, W., Li, X., Gujuluva, C., Kriederman, B., Roberts, J., Pushkarsky, T., Bukrinsky, M., Witte, M., Weinand, M., and Fiala, M., 2002, Human immunodeficiency virus type 1 enters brain microvascular endothelia by macropinocytosis dependent on lipid rafts and the mitogen-activated protein kinase signaling pathway. *J Virol* **76**: 6689-700.

Liu, R., Paxton, W. A., Choe, S., Ceradini, D., Martin, S. R., Horuk, R., MacDonald, M. E., Stuhlmann, H., Koup, R. A., and Landau, N. R., 1996, Homozygous defect in HIV-1 coreceptor accounts for resistance of some multiply-exposed individuals to HIV-1 infection. *Cell* **86**: 367-377.

Lodi, P. J., Ernst, J. A., Kuszewski, J., Hickman, A. B., Engelman, A., Craigie, R., Clore, G. M., and Gronenborn, A. M., 1995, Solution Structure of the DNA Binding Domain of HIV-1 Integrase. *Biochemistry* **34**: 9826-9833.

Lozach, P. Y., Lortat-Jacob, H., de Lacroix de Lavalette, A., Staropoli, I., Foung, S., Amara, A., Houles, C., Fieschi, F., Schwartz, O., Virelizier, J. L., Arenzana-Seisdedos, F., and Altmeyer, R., 2003, DC-SIGN and L-SIGN are high affinity binding receptors for hepatitis C virus glycoprotein E2. *J Biol Chem* **278**: 20358-66.

Lu, J., Sista, P., Giguel, F., Greenberg, M., and Kuritzkes, D. R., 2004, Relative replicative fitness of human immunodeficiency virus type 1 mutants resistant to enfuvirtide (T-20). *J Virol* **78**: 4628-37.

MacCartney, M. J., Dorr, P. K., and Smith-Burchnel, C. (2003). *43rd Intersience Conference on Antimicrobial Agents and Chemotherapy.* Chicago, IL, USA.

Madani, N., Perdigoto, A. L., Srinivasan, K., Cox, J. M., Chruma, J. J., LaLonde, J., Head, M., Smith, A. B., 3rd, and Sodroski, J. G., 2004, Localized changes in the gp120 envelope glycoprotein confer resistance to human immunodeficiency virus entry inhibitors BMS-806 and #155. *J Virol* **78**: 3742-52.

Maeda, N., Nigou, J., Herrmann, J. L., Jackson, M., Amara, A., Lagrange, P. H., Puzo, G., Gicquel, B., and Neyrolles, O., 2003, The cell surface receptor DC-SIGN discriminates between Mycobacterium species through selective recognition of the mannose caps on lipoarabinomannan. *J Biol Chem* **278**: 5513-6.

Maignan, S., Guilloteau, J. P., Zhou-Liu, Q., Clement-Mella, C., and Mikol, V., 1998, Crystal Structures of the Catalytic Domain of HIV-1 Integrase Free and Complexed with its Metal Cofactor: High Level of Similarity of the Active Site with Other Viral Integrases. *J. Mol. Biol.* **282**: 359-368.

Marcelin, A. G., Reynes, J., Yerly, S., Ktorza, N., Segondy, M., Piot, J. C., Delfraissy, J. F., Kaiser, L., Perrin, L., Katlama, C., and Calvez, V., 2004, Characterization of genotypic determinants in HR-1 and HR-2 gp41 domains in individuals with persistent HIV viraemia under T-20. *Aids* **18**: 1340-2.

Marchand, C., Johnson, A. A., Karki, R. G., Pais, G. C., Zhang, X., Cowansage, K., Patel, T. A., Nicklaus, M. C., Burke, T. R., Jr., and Pommier, Y., 2003, Metal-dependent inhibition of HIV-1 integrase by beta-diketo acids and resistance of the soluble double-mutant (F185K/C280S). *Mol Pharmacol* **64**: 600-9.

Marchand, C., Zhang, X., Pais, G. C., Cowansage, K., Neamati, N., Burke, T. R., Jr., and Pommier, Y., 2002, Structural determinants for HIV-1 integrase inhibition by beta-diketo acids. *J Biol Chem* **277**: 12596-603.

Martin, D., Ballow, C., Doto, J., Blum, R., Wild, C., and Allaway, G., 2005a, *12th Conference on retroviruses and opportunistic infections*, Boston, MA, USA.

Martin, D., Jacobson, J., Schurmann, D., Osswald, E., Doto, J., Wild, C., and Allaway, G. (2005b). 12th Conference on retroviruses and opportunistic infections, Boston, MA, USA.

Martin, D. E., Smith, P., Wild, C. T., and Allaway, G. P., 2004a, *XV International AIDS Conference*, Bangkok, Thailand.

Martin, L., Stricher, F., Misse, D., Sironi, F., Pugniere, M., Barthe, P., Prado-Gotor, R., Freulon, I., Magne, X., Roumestand, C., Menez, A., Lusso, P., Veas, F., and Vita, C., 2003, Rational design of a CD4 mimic that inhibits HIV-1 entry and exposes cryptic neutralization epitopes. *Nat Biotechnol* **21**:, 71-6.

Martin, M. P., Lederman, M. M., Hutcheson, H. B., Goedert, J. J., Nelson, G. W., van Kooyk, Y., Detels, R., Buchbinder, S., Hoots, K., Vlahov, D., O'Brien, S. J., and Carrington, M., 2004b, Association of DC-SIGN promoter polymorphism with increased risk for parenteral, but not mucosal, acquisition of human immunodeficiency virus type 1 infection. *J Virol* **78**: 14053-6.

Martin-Carbonero, L., 2004, Discontinuation of the clinical development of fusion inhibitor T-1249. *AIDS Rev* **6**: 61.

Marzi, A., Gramberg, T., Simmons, G., Moller, P., Rennekamp, A. J., Krumbiegel, M., Geier, M., Eisemann, J., Turza, N., Saunier, B., Steinkasserer, A., Becker, S., Bates, P., Hofmann, H., and Pöhlmann, S., 2004, DC-SIGN and DC-SIGNR interact with the glycoprotein of Marburg virus and the S protein of severe acute respiratory syndrome coronavirus. *J Virol* **78**: 12090-5.

McDonald, D., Wu, L., Bohks, S. M., KewalRamani, V. N., Unutmaz, D., and Hope, T. J., 2003, Recruitment of HIV and its receptors to dendritic cell-T cell junctions. *Science* **300**: 1295-7.

Melek, M., Jones, J. M., O'Dea, M. H., Pais, G., Burke, T. R., Jr., Pommier, Y., Neamati, N., and Gellert, M., 2002, Effect of HIV integrase inhibitors on the RAG1/2 recombinase. *Proc Natl Acad Sci U S A* **99**: 134-7.

Melikyan, G. B., Markosyan, R. M., Hemmati, H., Delmedico, M. K., Lambert, D. M., and Cohen, F. S., 2000, Evidence that the transition of HIV-1 gp41 into a six-helix bundle, not the bundle configuration, induces membrane fusion. *J Cell Biol* **151**: 413-423.

Menzo, S., Castagna, A., Monachetti, A., Hasson, H., Danise, A., Carini, E., Bagnarelli, P., Lazzarin, A., and Clementi, M., 2004, Resistance and replicative capacity of HIV-1 strains selected in vivo by long-term enfuvirtide treatment. *New Microbiol* **27**: 51-61.

Meyer, L., Magierowska, M., Hubert, J. B., Rouzioux, C., Deveau, C., Sanson, F., Debre, P., Delfraissy, J. F., and Theodorou, I., 1997, Early protective effect of CCR-5 delta 32 heterozygosity on HIV-1 disease progression: relationship with viral load. The SEROCO Study Group. *Aids* **11**: F73-8.

Michael, N. L., Chang, G., Louie, L. G., Mascola, J. R., Dondero, D., Birx, D. L., and Sheppard, H. W., 1997, The role of viral phenotype and CCR-5 gene defects in HIV-1 transmission and disease progression. *Nat Med* **3**: 338-40.

Miller, M. D., and Hazuda, D. J., 2004, HIV resistance to the fusion inhibitor enfuvirtide: mechanisms and clinical implications. *Drug Resist Updat* **7**: 89-95.

Mosier, D. E., Picchio, G. R., Gulizia, R. J., Sabbe, R., Poignard, P., Picard, L., Offord, R. E., Thompson, D. A., and Wilken, J., 1999, Highly potent RANTES analogues either prevent CCR5-using human immunodeficiency virus type 1 infection in vivo or rapidly select for CXCR4-using variants. *J Virol* **73**: 3544-50.

Moyle, G. J., Wildfire, A., Mandalia, S., Mayer, H., Goodrich, J., Whitcomb, J., and Gazzard, B. G., 2005, Epidemiology and predictive factors for chemokine receptor use in HIV-1 infection. *J Infect Dis* **191**: 866-872.

Nagasawa, T., Hirota, S., Tachibana, K., Takakura, N., Nishikawa, S., Kitamura, Y., Yoshida, N., Kikutani, H., and Kishimoto, T., 1996, Defects of B-cell lymphopoiesis and bone-marrow myelopoiesis in mice lacking the CXC chemokine PBSF/SDF-1. *Nature* **382**: 635-8.

Nobile, C., Petit, C., Moris, A., Skrabal, K., Abastado, J. P., Mammano, F., and Schwartz, O., 2005, Covert human immunodeficiency virus replication in dendritic cells and in DC-SIGN-expressing cells promotes long-term transmission to lymphocytes. *J Virol* **79**: 5386-99.

Pais, G. C., Zhang, X., Marchand, C., Neamati, N., Cowansage, K., Svarovskaia, E. S., Pathak, V. K., Tang, Y., Nicklaus, M., Pommier, Y., and Burke, T. R., Jr., 2002, Structure activity of 3-aryl-1,3-diketo-containing compounds as HIV-1 integrase inhibitors. *J Med Chem* **45**: 3184-94.

Pannecouque, C., Pluymers, W., Van Maele, B., Tetz, V., Cherepanov, P., De Clercq, E., Witvrouw, M., and Debyser, Z., 2002, New class of HIV integrase inhibitors that block viral replication in cell culture. *Curr Biol* **12**: 1169-77.

Parkin, N. T., Hellmann, N. S., Whitcomb, J. M., Kiss, L., Chappey, C., and Petropoulos, C. J., 2004, Natural variation of drug susceptibility in wild-type human immunodeficiency virus type 1. *Antimicrob Agents Chemother* **48**: 437-443.

Pastore, C., Ramos, A., and Mosier, D. E., 2004, Intrinsic obstacles to human immunodeficiency virus type 1 coreceptor switching. J Virol **78**: 7565-74.

Pierson, T. C., Doms, R. W., and Pöhlmann, S., 2004, Prospects of HIV-1 entry inhibitors as novel therapeutics. *Rev Med Virol* **14**: 255-70.

Pluymers, W., Pais, G., Van Maele, B., Pannecouque, C., Fikkert, V., Burke, T. R., Jr., De Clercq, E., Witvrouw, M., Neamati, N., and Debyser, Z., 2002, Inhibition of human immunodeficiency virus type 1 integration by diketo derivatives. *Antimicrob Agents Chemother* **46**: 3292-7.

Pöhlmann, S., Davis, C., Meister, S., Leslie, G. J., Otto, C., Reeves, J. D., Puffer, B. A., Papkalla, A., Krumbiegel, M., Marzi, A., Lorenz, S., Munch, J., Doms, R. W., and Kirchhoff, F., 2004, Amino acid 324 in the simian immunodeficiency virus SIVmac V3 loop can confer CD4 independence and modulate the interaction with CCR5 and alternative coreceptors. *J Virol* **78**: 3223-32.

Pöhlmann, S., and Reeves, J. D., 2005, Cellular entry of HIV: Evaluation of therapeutic targets. *Curr Pharm Des*, In press.

Pöhlmann, S., Soilleux, E. J., Baribaud, F., Leslie, G. J., Morris, L. S., Trowsdale, J., Lee, B., Coleman, N., and Doms, R. W., 2001, DC-SIGNR, a DC-SIGN homologue expressed in endothelial cells, binds to human and simian immunodeficiency viruses and activates infection in trans. *Proc Natl Acad Sci U S A* **98**: 2670-5.

Pommier, Y., Johnson, A. A., and Marchand, C., 2005, Integrase inhibitors to treat HIV/AIDS. *Nat Rev Drug Discov* **4**: 236-48.

Pommier, Y., Marchand, C., and Neamati, N., 2000, Retroviral integrase inhibitors year 2000: update and perspectives. *Antiviral Res* **47**: 139-48.

Pommier, Y., and Neamati, N., 1999, Inhibitors of human immunodeficiency virus integrase. *Adv Virus Res* **52**: 427-58.

Pommier, Y., Pilon, A. A., Bajaj, K., Mazumder, A., and Neamati, N., 1997, HIV-1 integrase as a target for antiviral drugs. *Antiviral Chem. and Chemother.* **8**: 463-483.

Reeves, J. D., and Doms, R. W. (2003). The Role of chemokine receptors in HIV infection of host cells. In *"Handbook of cell signalling"* (R. A. Bradshaw, and E. A. Dennis, Eds.), Vol. 1, pp. 191-196. 3 vols. Academic Press, New York.

Reeves, J. D., Gallo, S. A., Ahmad, N., Miamidian, J. L., Harvey, P. E., Sharron, M., Pöhlmann, S., Sfakianos, J. N., Derdeyn, C. A., Blumenthal, R., Hunter, E., and Doms, R. W., 2002, Sensitivity of HIV-1 to entry inhibitors correlates with envelope/coreceptor affinity, receptor density, and fusion kinetics. *Proc Natl Acad Sci U S A* **99**: 16249-54.

Reeves, J. D., Hibbitts, S., Simmons, G., McKnight, A., Azevedo-Pereira, J. M., Moniz-Pereira, J., and Clapham, P. R., 1999, Primary human immunodeficiency virus type 2 (HIV-2) isolates infect CD4-negative cells via CCR5 and CXCR4: comparison with HIV-1 and simian immunodeficiency virus and relevance to cell tropism in vivo. *J Virol* **73**: 7795-804.

Reeves, J. D., Lee, F. H., Miamidian, J. L., Jabara, C. B., Juntilla, M. M., and Doms, R. W., 2005, Enfuvirtide resistance mutations: impact on human immunodeficiency virus envelope function, entry inhibitor sensitivity, and virus neutralization. *J Virol* **79**: 4991-9.

Reeves, J. D., Miamidian, J. L., Biscone, M. J., Lee, F. H., Ahmad, N., Pierson, T. C., and Doms, R. W., 2004, Impact of mutations in the coreceptor binding site on human immunodeficiency virus type 1 fusion, infection, and entry inhibitor sensitivity. *J Virol* **78**: 5476-85.

Reeves, J. D., and Piefer, A. J., 2005, Emerging drug targets for antiretroviral therapy. *Drugs*, In press.

Rice, P. A., and Baker, T. A., 2001, Comparative architecture of transposase and integrase complexes. *Nat Struct Biol* **8**: 302-7.

Rimsky, L. T., Shugars, D. C., and Matthews, T. J., 1998, Determinants of human immunodeficiency virus type 1 resistance to gp41-derived inhibitory peptides. *J Virol* **72**: 986-93.

Rodriguez-Barrios, F., and Gago, F., 2004, HIV protease inhibition: limited recent progress and advances in understanding current pitfalls. *Curr Top Med Chem* **4**: 991-1007.

Rojo, J., and Delgado, R., 2004, Glycodendritic structures: promising new antiviral drugs. *J Antimicrob Chemother* **54**: 579-81.

Ruane, P. J., and DeJesus, E., 2004, New Nucleoside/Nucleotide Backbone Options: A Review of Recent Studies. *J Acquir Immune Defic Syndr* **37**: S21-S29.

Russell, D., Bakhtyari, A., and Jazrawi, R. P. 2003, *43rd Conference on Antimicrobial Agents and Chemotherapy*. Chicago, IL, USA.

Ryu, S. E., Kwong, P. D., Truneh, A., Porter, T. G., Arthos, J., Rosenberg, M., Dai, X. P., Xuong, N. H., Axel, R., Sweet, R. W., and et al., 1990, Crystal structure of an HIV-binding recombinant fragment of human CD4. *Nature* **348**: 419-26.

Samson, M., Libert, F., Doranz, B. J., Rucker, J., Liesnard, C., Farber, C. M., Saragosti, S., Lapoumeroulie, C., Cognaux, J., Forceille, C., Muyldermans, G., Verhofstede, C., Burtonboy, G., Georges, M., Imai, T., Rana, S., Yi, Y., Smyth, R. J., Collman, R. G., Doms, R. W., Vassart, G., and Parmentier, M., 1996, Resistance to HIV-1 infection in caucasian individuals bearing mutant alleles of the CCR-5 chemokine receptor gene. *Nature* **382**: 722-725.

Schols, D., 2004, HIV co-receptors as targets for antiviral therapy. *Curr Top Med Chem* **4**: 883-893.

Schols, D., Claes, S., Hatse, S., Princen, K., Vermeire, K., De Clercq, E., Skerlj, R., Bridger, G., and Calandra, G. (2003). *10th Conference on Retroviruses and Opportunistic Infections*, Boston, MA.

Schols, D., Este, J. A., Cabrera, C., and De Clercq, E., 1998, T-cell-line-tropic human immunodeficiency virus type 1 that is made resistant to stromal cell-derived factor 1alpha contains mutations in the envelope gp120 but does not show a switch in coreceptor use. *J Virol* **72**: 4032-7.

Schurmann, D., Rouzier, R., Nougarede, R., Reynes, J., Fatkenheuer, G., Raffi, F., Michelet, C., Tarral, A., Hoffmann, C., Kiunke, J., Sprenger, H., vanLier, J., Sansone, A., Jackson, M., and Laughlin, M., 2004, SCH D: Antiviral activity of a CCR5 receptor antagonist. *11th Conference on Retroviruses and Opportunistic Infections,* San Francisco, CA, USA.

Sharma, P. L., Nurpeisov, V., Hernandez-Santiago, B., Beltran, T., and Schinazi, R. F., 2004, Nucleoside inhibitors of human immunodeficiency virus type 1 reverse transcriptase. *Curr Top Med Chem* **4**(9), 895-919.

Shattock, R., and Solomon, S., 2004, Microbicides--aids to safer sex. *Lancet* **363**: 1002-3.

Shattock, R. J., and Moore, J. P., 2003, Inhibiting sexual transmission of HIV-1 infection. *Nat Rev Microbiol* **1**(1), 25-34.

Shaw-Reid, C. A., Munshi, V., Graham, P., Wolfe, A., Witmer, M., Danzeisen, R., Olsen, D. B., Carroll, S. S., Embrey, M., Wai, J. S., Miller, M. D., Cole, J. L., and Hazuda, D. J., 2003, Inhibition of HIV-1 ribonuclease H by a novel diketo acid, 4-[5-(benzoylamino)thien-2-yl]-2,4-dioxobutanoic acid. *J Biol Chem* **278**: 2777-80.

Shenoy, S. R., Barrientos, L. G., Ratner, D. M., O'Keefe, B. R., Seeberger, P. H., Gronenborn, A. M., and Boyd, M. R., 2002, Multisite and multivalent binding between cyanovirin-N and branched oligomannosides: calorimetric and NMR characterization. *Chem Biol* **9**: 1109-18.

Sherman, P. A., and Fyfe, J. A., 1990, Human immunodeficiency virus integration protein expressed in Escherichia coli possesses selective DNA cleaving activity. *Proc. Natl. Acad. Sci. USA* **87**: 5119-5123.

Shimizu, N., Haraguchi, Y., Takeuchi, Y., Soda, Y., Kanbe, K., and Hoshino, H., 1999, Changes in and discrepancies between cell tropisms and coreceptor uses of human immunodeficiency virus type 1 induced by single point mutations at the V3 tip of the env protein. *Virology* **259**: 324-33.

Shioda, T., Oka, S., Ida, S., Nokihara, K., Toriyoshi, H., Mori, S., Takebe, Y., Kimura, S., Shimada, K., Nagai, Y., and *et al.*, 1994, A naturally occurring single basic amino acid substitution in the V3 region of the human immunodeficiency virus type 1 env protein alters the cellular host range and antigenic structure of the virus. *J Virol* **68**: 7689-96.

Si, Z., Madani, N., Cox, J. M., Chruma, J. J., Klein, J. C., Schon, A., Phan, N., Wang, L., Biorn, A. C., Cocklin, S., Chaiken, I., Freire, E., Smith, A. B., 3rd, and Sodroski, J. G., 2004, Small-molecule inhibitors of HIV-1 entry block receptor-induced conformational changes in the viral envelope glycoproteins. *Proc Natl Acad Sci* U S A **101**: 5036-41.

Simmons, G., Reeves, J. D., Grogan, C. C., Vandenberghe, L. H., Baribaud, F., Whitbeck, J. C., Burke, E., Buchmeier, M. J., Soilleux, E. J., Riley, J. L., Doms, R. W., Bates, P., and Pöhlmann, S., 2003, DC-SIGN and DC-SIGNR bind ebola glycoproteins and enhance infection of macrophages and endothelial cells. *Virology* **305**: 115-23.

Smith, A. E., and Helenius, A., 2004, How viruses enter animal cells. *Science* **304**: 237-42.

Sparks, S., Adkison, K., Shachoy-Clark, A., Piscitelli, S., and Demarest, J. 2005, *12th Conference on Retroviruses and Opportunistic Infections*, Boston, MA, USA.

Speck, R. F., Wehrly, K., Platt, E. J., Atchison, R. E., Charo, I. F., Kabat, D., Chesebro, B., and Goldsmith, M. A., 1997, Selective employment of chemokine receptors as human immunodeficiency virus type 1 coreceptors determined by individual amino acids within the envelope V3 loop. *J Virol* **71**: 7136-7139.

Steffan, A. M., Lafon, M. E., Gendrault, J. L., Schweitzer, C., Royer, C., Jaeck, D., Arnaud, J. P., Schmitt, M. P., Aubertin, A. M., and Kirn, A., 1992, Primary cultures of endothelial cells from the human liver sinusoid are permissive for human immunodeficiency virus type 1. *Proc Natl Acad Sci U S A* **89**: 1582-6.

Stone, N., Dunaway, S., Flexner, C., Calandra, G., Wiggins, I., Conley, J., Snyder, S., Tierney, C., and Hendrix, C. W. 2004, *XV International AIDS Conference*, Bangkok, Thailand.

Tremblay, C., 2004, Effects of HIV-1 entry inhibitors in combination. Curr Pharm Des **10**: 1861-5.

Tremblay, C. L., Giguel, F., Kollmann, C., Guan, Y., Chou, T. C., Baroudy, B. M., and Hirsch, M. S., 2002, Anti-human immunodeficiency virus interactions of SCH-C (SCH 351125), a CCR5 antagonist, with other antiretroviral agents in vitro. *Antimicrob Agents Chemother* **46**: 1336-9.

Tremblay, C. L., Kollmann, C., Giguel, F., Chou, T. C., and Hirsch, M. S., 2000, Strong in vitro synergy between the fusion inhibitor T-20 and the CXCR4 blocker AMD-3100. *J Acquir Immune Defic Syndr* **25**: 99-102.

Trkola, A., Ketas, T., Kewalramani, V. N., Endorf, F., Binley, J. M., Katinger, H., Robinson, J., Littman, D. R., and Moore, J. P., 1998, Neutralization sensitivity of human immunodeficiency virus type 1 primary isolates to antibodies and CD4-based reagents is independent of coreceptor usage. *J Virol* **72**: 1876-85.

Trkola, A., Kuhmann, S. E., Strizki, J. M., Maxwell, E., Ketas, T., Morgan, T., Pugach, P., Xu, S., Wojcik, L., Tagat, J., Palani, A., Shapiro, S., Clader, J. W., McCombie, S., Reyes, G. R., Baroudy, B. M., and Moore, J. P., 2002, HIV-1 escape from a small molecule, CCR5-specific entry inhibitor does not involve CXCR4 use. *Proc Natl Acad Sci U S A* **99**: 395-400.

Trkola, A., Pomales, A. B., Yuan, H., Korber, B., Maddon, P. J., Allaway, G. P., Katinger, H., Barbas, C. F., 3rd, Burton, D. R., Ho, D. D., and *et al.*, 1995, Cross-clade neutralization of primary isolates of human immunodeficiency virus type 1 by human monoclonal antibodies and tetrameric CD4-IgG. *J Virol* **69**: 6609-17.

Trumpfheller, C., Park, C. G., Finke, J., Steinman, R. M., and Granelli-Piperno, A., 2003, Cell type-dependent retention and transmission of HIV-1 by DC-SIGN. *Int Immunol* **15**: 289-98.

Tsai, C. C., Emau, P., Jiang, Y., Agy, M. B., Shattock, R. J., Schmidt, A., Morton, W. R., Gustafson, K. R., and Boyd, M. R., 2004, Cyanovirin-N inhibits AIDS virus infections in vaginal transmission models. *AIDS Res Hum Retroviruses* **20**: 11-8.

Tsai, C. C., Emau, P., Jiang, Y., Tian, B., Morton, W. R., Gustafson, K. R., and Boyd, M. R., 2003, Cyanovirin-N gel as a topical microbicide prevents rectal transmission of SHIV89.6P in macaques. *AIDS Res Hum Retroviruses* **19**: 535-41.

Turville, S. G., Cameron, P. U., Handley, A., Lin, G., Pöhlmann, S., Doms, R. W., and Cunningham, A. L., 2002, Diversity of receptors binding HIV on dendritic cell subsets. *Nat Immunol* **3**(10), 975-83.

Ugolini, S., Mondor, I., and Sattentau, Q. J., 1999, HIV-1 attachment: another look. *Trends Microbiol* **7**: 144-149.

UNAIDS (2004). 2004 Report on the global AIDS epidemic: Executive Summary. *UNAIDS.* June.

van der Ryst, E., Rosario, M. C., Poland, W., Felstead, S., Jenkins, T. M., and Sullivan, J. F., 2004, *15th International AIDS Conference,* Bangkok, Thailand.

van Kooyk, Y., and Geijtenbeek, T. B., 2003, DC-SIGN: escape mechanism for pathogens. *Nat Rev Immunol* **3**: 697-709.

Vita, C., Drakopoulou, E., Vizzavona, J., Rochette, S., Martin, L., Menez, A., Roumestand, C., Yang, Y. S., Ylisastigui, L., Benjouad, A., and Gluckman, J. C., 1999, Rational engineering of a miniprotein that reproduces the core of the CD4 site interacting with HIV-1 envelope glycoprotein. *Proc Natl Acad Sci U S A* **96**: 13091-6.

Vives, R. R., Imberty, A., Sattentau, Q. J., and Lortat-Jacob, H., 2005, Heparan sulfate targets the HIV-1 envelope glycoprotein GP120 coreceptor binding site. *J Biol Chem.* **280**:21353-7.

Vogt, V. M., 1996, Proteolytic processing and particle maturation. *Curr Top Microbiol Immunol* **214**: 95-131.

Wai, J. S., and al., e., 2000, 4-Aryl-2,4-diosobutanoic acid inhibitors of HIV-1 integrase and viral replication in cells. *J. Med. Chem.* **43**: 4923-4926.

Wang, J. H., Yan, Y. W., Garrett, T. P., Liu, J. H., Rodgers, D. W., Garlick, R. L., Tarr, G. E., Husain, Y., Reinherz, E. L., and Harrison, S. C., 1990, Atomic structure of a fragment of human CD4 containing two immunoglobulin-like domains. *Nature* **348**: 411-8.

Wang, J. Y., Ling, H., Yang, W., and Craigie, R., 2001, Structure of a two-domain fragment of HIV-1 integrase: implications for domain organization in the intact protein. *EMBO J* **20**: 7333-7343.

Watson, C., Jenkinson, S., Kazmierski, W., and Kenakin, T., 2005, The CCR5 receptor-based mechanism of action of 873140, a potent allosteric noncompetitive HIV entry inhibitor. *Mol Pharmacol* **67**: 1268-82.

Wegner, S. A., Brodine, S. K., Mascola, J. R., Tasker, S. A., Shaffer, R. A., Starkey, M. J., Barile, A., Martin, G. J., Aronson, N., Emmons, W. W., Stephan, K., Bloor, S., Vingerhoets, J., Hertogs, K., and Larder, B., 2000, Prevalence of genotypic and phenotypic resistance to anti-retroviral drugs in a cohort of therapy-naive HIV-1 infected US military personnel. *Aids* **14**: 1009-1015.

Wei, X., Decker, J. M., Liu, H., Zhang, Z., Arani, R. B., Kilby, J. M., Saag, M. S., Wu, X., Shaw, G. M., and Kappes, J. C., 2002, Emergence of resistant human immunodeficiency virus type 1 in patients receiving fusion inhibitor (T-20) monotherapy. *Antimicrob Agents Chemother* **46**: 1896-905.

Weinstock, H. S., Zaidi, I., Heneine, W., Bennett, D., Garcia-Lerma, J. G., Douglas, J. M., Jr., LaLota, M., Dickinson, G., Schwarcz, S., Torian, L., Wendell, D., Paul, S., Goza, G. A., Ruiz, J., Boyett, B., and Kaplan, J. E., 2004, The epidemiology of antiretroviral drug resistance among drug-naive HIV-1-infected persons in 10 US cities. *J Infect Dis* **189**: 2174-2180.

Weissenhorn, W., Dessen, A., Harrison, S. C., Skehel, J. J., and Wiley, D. C., 1997, Atomic structure of the ectodomain from HIV-1 gp41. *Nature* **387**: 426-430.

Westby, M., Napier, C., Mansfield, R., Collins, D., Huang, W., Hellmann, N. S., Lie, Y., and Perros, M. (2003). *12th International HIV Drug Resistance Workshop.* Los Cabos, Mexico.

Westby, M., Smith-Burchnell, C., Mori, J., Lewis, M., Mansfield, R., Whitcomb, J., Petropoulos, C., and Perros, M., 2004, In vitro escape of R5 primary isolates from the CCR5 antagonist, UK-427,857, is difficult and involves continued use of the CCR5 receptor. *Antiviral Therapy* **9**: 10.

Chapter 1.9

MANAGING ANTIRETROVIRAL RESISTANCE

B. SCHMIDT, M. TSCHOCHNER, H. WALTER, and K. KORN
Institute of Clinical and Molecular Virology, German National Reference Centre for Retroviruses, Schlossgarten 4, D-91054 Erlangen, Germany

Abstract: Human immunodeficiency virus type 1 (HIV-1) was identified as causative agent of the acquired immune deficiency syndrome (AIDS) in 1983. Since then, 20 antiretroviral drugs inhibiting different steps of the viral life cycle have been approved for the treatment of HIV-infected individuals. Combinations of these drugs enable a sustained suppression of viral replication in the majority of treated subjects. Therapy failure frequently occurs with suboptimal drug concentrations promoting the development of drug-resistant viral strains. The increased capacity of a virus to replicate in the presence of antiretroviral drugs, based on mutations in the respective genes, can be detected by phenotypic or genotypic analysis. Retro- and prospective studies confirm that viral replication can be suppressed more effectively when antiretroviral therapy is adjusted to the individual resistance profile. Therefore, drug resistance testing is recommended for all cases of therapy failure; and for newly infected individuals as well, because current epidemiological data show a 10% risk for the transmission of drug-resistant viruses in this group. This review will focus on three aspects: (i) the development of drug resistance interpretation systems and the move towards a consensus, (ii) the impact of pharmacokinetics in drug applications and the design of new antiretrovirals, and (iii) the role of viral fitness in HIV-infected individuals harboring multidrug-resistant viruses. Thus, HIV infection has become a treatable chronic disease, which requires all our efforts to maintain the current truce, until a cure will eventually be found.

E. Bogner and A. Holzenburg (eds.), New Concepts of Antiviral Therapy, 255–279.

1. INTRODUCTION

The causative agent of the acquired immune deficiency syndrome (AIDS) was first identified in 1983, when a T-lymphotropic retrovirus was recovered from the lymph node of a French patient with persistent generalized lymphadenopathy often preceding the development of full blown AIDS (Barré-Sinoussi *et al.*, 1983). It only took two more years to discover that viral infectivity and cytopathic effects could be blocked by 3'-azido-3'-deoxythymidine, a thymidine analogue, which inhibited the viral reverse transcriptase at concentrations not toxic to eukaryotic cells (Mitsuya *et al.*, 1985).

The euphoria of having a drug to treat HIV-infected patients, however, was soon curbed by the finding that the clinical improvement did not last longer than six months. The therapeutic failure could eventually be tracked down to the development of drug-resistant viruses, which were able to replicate in the presence of this antiretroviral drug (Larder *et al.*, 1989). Subsequent research led to the discovery of other reverse transcriptase inhibitors (DDC, DDI, D4T, 3TC). These nucleoside analogues also followed the principle of competitive chain termination (Table 1). Again, the effects of antiretroviral therapy were soon compromised by the emergence of drug-resistant viruses.

In 1993, the concept of combination therapy was developed when zidovudine and lamivudine were co-administered in cell culture. Notably, susceptibility to zidovudine was restored by development of resistance against lamivudine (Tisdale *et al.*, 1993; Boucher *et al.*, 1993). This finding was the first evidence that mutations conferring resistance to a certain drug can be resensitized by others. Unfortunately, mutations were subsequently identified which mediate dual resistance against zidovudine and lamivudine (Schinazi *et al.*, 2000).

A break-through in antiretroviral therapy was achieved in 1994, when protease inhibitors became available (Craig *et al.*, 1991; Kageyama *et al.*, 1994). These drugs block the function of the viral protease and thus the cleavage of the *gag-pol* precursor protein (Flexner, 1998). The resulting particles were described with irregular morphology, delayed maturation, and severely affected infectivity (Wiegers *et al.*, 1998). The introduction of protease inhibitors as antiretroviral drugs enabled a combination therapy attacking HIV at different points of the viral life cycle. This highly active antiretroviral therapy (HAART) achieved for the first time a sustained suppression of viral replication in the majority of HIV-infected patients, which led to a dramatic reduction in HIV morbidity and mortality (Mocroft *et al.*, 1998; Palella *et al.*, 1998).

The virus's capacity to adapt to all adversities – based on the low fidelity of the HIV-1 reverse transcriptase – gave rise to viral variants resistant to protease inhibitors. Unfortunately, the first-generation protease inhibitors (IDV, SQV, RTV, NFV) were characterized by a broad cross-resistance (Race *et al.*, 1999). The phenomenon that patients' viruses were resistant against drugs they had not previously been exposed to, was due to the structure-based drug design: all inhibitors were constructed to fit into the active pocket of the viral protease.

In contrast, the second-generation protease inhibitors (APV, LPV, ATV) are characterized by a lower degree of cross-resistance (Schmidt *et al.*, 2000a, Schnell *et al.*, 2003). Furthermore, response to protease inhibitor therapy was substantially improved by a more favorable pharmacokinetics: the blocking of cytochrome P450 by baby-dosed RTV smoothens the profile of high peak and low trough levels of most protease inhibitors, preventing suboptimal drug concentrations at dosing intervals (Moyle and Back, 2001).

The arsenal of antiretroviral drugs has been enlarged by three other drug classes. Non-nucleoside inhibitors of the reverse transcriptase (NVP, DLV, EFV) effectively suppress viral replication by acting as allosteric inhibitors of the enzyme (De Clercq, 1991). Again, resistance as well as broad cross-resistance has been described for this group of drugs (Miller *et al.*, 2001).

Within the group of reverse transcriptase inhibitors, tenofovir, an acyclic nucleotide analogue, has become available. The oral prodrug, tenofovir disoproxil fumarate, is rapidly converted into the active drug after cell membrane penetration (Naesens *et al.*, 1998). This drug promised to be active against viruses with multiple resistance against nucleoside analogues; unfortunately, cross-resistance proved to be larger than previously suspected (Wolf *et al.*, 2003).

As a most recent development, inhibitors of the viral fusion process have been introduced into the antiviral therapy. Peptides such as enfuvirtide (also known as T-20) mimic the heptad region of gp41, thus blocking fusion of the viral particle with the cell membrane (reviewed in Greenberg *et al.*, 2004).

Table 1. FDA-approved antiretroviral drugs.

Groups	Substance	Abbreviation	Trade name
NRTI	Zidovudine	ZDV	Retrovir®
	Lamivudine	3TC	Epivir®
		ZDV/3TC	Combivir®
	Didanosine	DDI	Videx®
	Zalcitabine	DDC	Hivid®
	Stavudine	D4T	Zerit®
	Abacavir	ABC	Ziagen®
		AZT/3TC/ABC	Trizivir®
	Emtricitabine	FTC	Emtriva®
NtRTI	Tenofovir	TDF	Viread®
NNRTI	Nevirapin	NVP	Viramune®
	Delavirdin	DLV	Rescriptor®
	Efavirenz	EFV	Sustiva®
PI	Ritonavir	RTV	Norvir®
	Saquinavir	SQV	Invirase® / Fortovase®
	Indinavir	IDV	Crixivan®
	Nelfinavir	NFV	Viracept®
	Amprenavir	APV	Agenerase®
	Fosamprenavir		Telzir®, Lexiva®
	Lopinavir/r*	LPV/ r*	Kaletra®
	Atazanavir	ATV	Reyataz®
Fusion inhibitors	Enfuvirtide	T-20	Fuzeon®

RT reverse transcriptase, NRTI nucleoside RT inhibitors, NtRTI nucleotide RT inhibitors, NNRTI non-nucleoside RT inhibitors, PI protease inhibitors.

2.   GENO-/PHENOTYPIC RESISTANCE TESTING

2.1   Methods of resistance testing

In principle, there are two methods to test for HIV-1 drug resistance: phenotyping, which is characterized by the determination of viral replication in the presence of increasing drug concentrations, and genotyping, which analyses drug resistance-associated mutations in the genes coding for the respective viral enzymes (reviewed in: Schmidt *et al.*, 2000a).

Functional testing of HIV-1 drug resistance was first performed on peripheral blood mononuclear cells (PBMC) (Japour *et al.*, 1993). After isolation of the patient's virus within 2-6 weeks, PBMC were inoculated with a standardized amount of virus and cultivated in the presence of antiretroviral drugs using p24 antigen as read-out. This assay was not only time-consuming, but implicated the risk of selecting against drug-resistance associated mutations during prolonged passaging in cell culture. Therefore, recombinant virus assays have been introduced which shorten the time period for generating the input virus considerably. The first kind of these assays was based on homologous recombination of the HIV-1 reverse transcriptase amplified from the patient's virus and a correspondingly deleted provirus, which could effectively be propagated on cell lines (Kellam and Larder, 1994). To further reduce the time necessary for homologous recombination (7-10 days), subsequently developed assays made use of ligating the patient-derived part into the viral backbone (Walter *et al.*, 1999, Petropoulos *et al.*, 2000; Jármy *et al.*, 2001). The use of reporter cell lines improved the standardization of the read-out systems.

Despite all modifications, functional resistance assays still require a great deal of technical know-how and will therefore remain restricted to specialized laboratories. In contrast, determination of drug resistance-associated mutations is now offered as routine diagnostic service. Commercial as well as in-house systems are available which amplify the genes for the protease and reverse transcriptase from patient's plasma. Subsequently, the nucleotide sequence of the amplified fragment is determined and then compared to a prototypic HIV-1 strain, e.g. pNL4-3 or HxB2. Drug resistance-associated mutations are identified using tables (Schinazi *et al.*, 2000) or electronic websites (http://www.iasusa.org, http://resdb.lanl.gov/Resist_DB/default.htm, http://hivdb.stanford.edu/).

Although genotyping makes life much easier, phenotyping will remain the gold standard of resistance testing, since it allows the determination of resistance profiles of new drugs ahead of all *in vivo* experience. In addition, phenotyping can identify unexpected effects of new combinations of mutations for drugs already in clinical use (Mueller *et al.*, 2004).

2.2 Retro- and prospective studies

Clinicians soon realized that resistance testing was a useful tool to figure out which antiretroviral drugs were still active in patients after treatment failure. This experience was documented in a number of retrospective studies, which were re-evaluated in a meta-analysis confirming that resistance testing significantly improved response to antiretroviral therapy in HIV-1 infected patients (DeGruttola *et al.*, 2000).

These data were confirmed in several prospective studies. In the majority of studies, change of treatment based on the results of genotypic resistance testing was evaluated vs. standard-of-care, *i.e.* selection of a new treatment regimen based only on the information about treatment history and the course of viral load and CD4 cell counts. Endpoints were changes in viral load (expressed either as differences between baseline and 3 – 12 months after treatment change or as percentage of patients reaching undetectable viral load after different time intervals). Significant advantages in favour of genotyping were found in the following trials: VIRADAPT (Durant *et al.*, 1999; Clevenbergh *et al.*, 2000), GART (Baxter *et al.*, 2000), HAVANA (Tural *et al.*, 2002), and ARGENTA (Cingolani *et al.*, 2002). Clinical studies evaluating phenotypic resistance testing either detected only a moderate or no additional benefit to standard-of-care: VIRA3001 (Cohen *et al.*, 2002), NARVAL (Meynard *et al.*, 2002) and CCTG 575 (Haubrich *et al.*, 2005).

This discrepancy of phenotypic to genotypic resistance testing may in part be due to the cut-off problem of phenotyping, *i.e.* the determination of the fold reduction in drug susceptibility which separates active from inactive drugs. Such "clinical" cut-offs are difficult to determine in the era of antiretroviral combination therapy and are available only for a minority of antiretroviral drugs. A much easier task is to determine "technical" cut-offs, defined as the variability within repeated determinations of a wildtype reference virus, or "biological" cut-offs representing the variability of drug susceptibilities in viruses obtained from a larger number of drug-naïve HIV-1 infected patients.

Unfortunately, clinical cut-offs can be very low for some drugs and thus overlap with technical cut-offs. This is particularly true for some drugs which are characterized by an overall narrow range of resistance, in particular the dideoxynucleoside analogues. For example, data have been published indicating that a subtle reduction in D4T susceptibility of 1.4-fold was already associated with failure to achieve significant viral load reduction after eight weeks of D4T monotherapy (Shulman *et al.*, 2002). On the other hand, there are also drugs for which the clinical cut-off is substantially higher than the technical cut-off, which may lead to inadequate restriction of the use of potentially active drugs. Recently published data on one of the prospective studies comparing phenotypic resistance testing to standard-of-care suggest that use of inappropriate cut-off values may be responsible for the lack of an overall benefit for patients in the phenotyping arm, although at least a subgroup of more heavily pretreated patients had better virological response if treatment was changed according to the results of the phenotypic resistance test (Haubrich *et al.*, 2005).

Although the proven benefit of antiretroviral resistance testing is mostly limited to short-term improvements of virologic, but not immunologic response (Panidou *et al.*, 2004), some pragmatism has been incorporated into current European and international guidelines for the clinical use of HIV-1 drug resistance testing (Hirsch *et al.*, 2003; Vandamme *et al.*, 2004). The widespread recommendation of testing after treatment failure was motivated by the opportunity to select the optimal antiretroviral therapy for each individual patient, which additionally proved to be cost-effective (Weinstein *et al.*, 2001; Corzillius *et al.*, 2004).

3. CURRENT EPIDEMIOLOGICAL SITUATION

3.1 Resistance data in treated patients

Despite of a large number of publications about HIV drug resistance, systematic analyses of the frequency of resistance in treated patients are quite rare. Richman and colleagues studied a random sample of about 1800 patients, who were supposed to be representative for more than 200.000 HIV-infected adults in the USA receiving medical care in early 1996, *i.e.* at the beginning of the HAART era (Richman *et al.*, 2004). 61% of these patients had a viral load > 500 copies/ml and a drug resistance test available. Among these, 76% harboured HIV with resistance to at least one antiretroviral drug in a phenotypic assay. Assuming that those with a viral load < 500 copies/ml do not have resistant viruses, the overall prevalence of drug resistance would be 46.5% in this patient population.

Others found similar rates of resistance in the range of 70 – 80% in patients with treatment failure. Thus, in the CAPTURE study, which included genotypes from more than 1200 patients with treatment failure from 12 European countries, 80% had at least one resistance associated mutation (Van de Vijver *et al.*, 2005). Concerning individual mutations, those conferring resistance to nucleoside analogues were the most common in this study with an overall rate of 69%, followed by mutations associated with NNRTI resistance (41%) and protease inhibitor resistance (36%). Half of the patients showed resistance to at least two classes of antiretroviral drugs and one in six patients (17%) harboured viruses with resistance mutations for three classes of antiretroviral drugs. In an analysis of our own phenotypic data from 1998 to 2004, focusing on resistance to individual drugs rather than on the percentage of patients with resistant viruses, we found that for each of 17 drugs tested, between 35% and 55% of viruses exhibited resistance, with the exception of lopinavir, where only 27% of isolates showed resistance (Fig. 1).

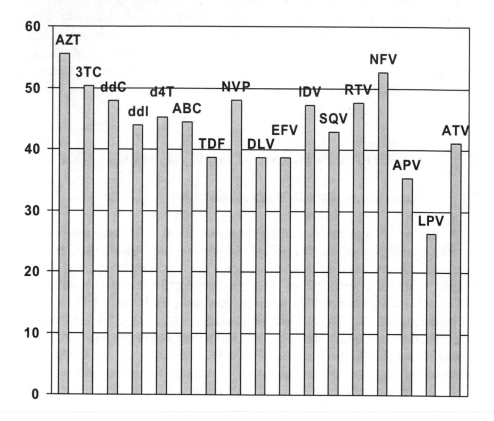

Figure 1. Percentage of resistance to individual drugs among samples submitted to the German National Reference Center for Retroviruses from 1998 to 2004 (n = 804).

Larger differences exist when looking at the overall rate of patients with evidence of drug resistant viruses. This is exemplified by looking at recently published data from the United Kingdom (The UK Collaborative Group on HIV Drug Resistance and UK CHIC Study Group, 2005). In this study, the cumulative risk of developing at least one resistance associated mutation was only 27% over a 6-year period, which is substantially lower than in the study from the United States with 46.5% (Richman *et al.*, 2004). The main reason for this discrepancy is that the UK data focus exclusively on patients who started treatment with a potent three- or four-drug combination therapy, whereas the US analysis also includes subjects who have received previous monotherapy or dual therapy. Thus, the major difference is in the rate of patients with treatment failure (38% vs. 63%), whereas the rate of drug resistance among those patients with treatment failure is comparable (71% vs. 76%).

Whereas these studies already show high rates of resistance in patients experiencing treatment failure, longitudinal analyses suggest that the cross-sectional approach may lead to a substantial underestimation of the magnitude of HIV drug resistance (Harrigan *et al.*, 2005). Looking at more than 1700 patients for whom at least 3 genotypic resistance tests were available over the period from 1996 to 2004, the detection rate of many mutations, particularly those associated with nucleoside analogue resistance, was much higher when all genotypes were taken into account compared to only the most recent genotype. Conversely, whereas almost 54% of patients had no major resistance mutation in their most recent genotype, this proportion declined to 28.3% if all available genotypes were considered. Thus, especially in patients with a long treatment history and frequent changes of drug regimen, combined current and historical genotypic data (if available) and the complete treatment history should be considered when trying to select a new treatment regimen.

3.2 Resistance data in untreated patients

Published information about the prevalence of drug resistant viruses in untreated patients is available for many European countries as well the USA (reviewed in: Vandamme *et al.*, 2004). The percentages of untreated patients reported to carry resistant viruses differ greatly between less than 2% and more than 50%. However, many of the published studies are small and often not representative. Furthermore, differences in the selection of patients, in the types of drug resistance tests used and in the criteria to define what is called a resistant virus account for a large part of the differences.

If one focuses on studies with comparable patient groups (for example recently infected patients) and defined criteria for resistance (the IAS mutation list (Johnson *et al.*, 2003) is most frequently used), the results become more comparable. Thus, in recently infected drug-naïve patients, at least one in ten harbours a virus with one or more mutations from the IAS list. This is true for the USA (Grant *et al.*, 2002; Little *et al.*, 2002) as well as for Europe (Deschamps *et al.*, 2005; Wensing *et al.*, 2005).

If newly diagnosed untreated patients with chronic infection are considered, the rates of patients with resistant viruses tend to be lower. Thus, in the French national survey (Deschamps *et al.*, 2005) in 2001/2002, the rate of viruses with at least one resistance mutation was 14% in newly infected patients versus 6% in chronically infected patients. Similarly, the respective figures for the retrospective CATCH study analyzing patients from 19 European countries between 1996 and 2002 are 13.5% vs. 8.7%.

Concerning trends over time, resistance to nucleoside analogues is decreasing in most studies, whereas resistance to NNRTI and to a lesser extent also to PI is increasing. In studies from Europe, the overall percentage of resistance in untreated patients seems to be rather stable, whereas increasing rates are observed in the USA. This difference may be due to the fact that among newly diagnosed and also recently infected patients in Europe, the proportion of patients with viruses of non-B subtypes has been increasing to about 30%. Resistance rates in these viruses originating mainly from developing countries without broad access to antiretroviral therapy are usually lower than for subtype B viruses (e.g. 4.8% vs. 12.9% in the CATCH study).

However, also in this patient group, the percentage of viruses with resistance mutations is increasing over time. Thus, we have to anticipate a further increase of transmitted HIV drug resistance in the future. This is especially true since transmitted resistant viruses have a tendency to persist for long periods even in the absence of drug pressure (Brenner *et al.*, 2002; Little *et al.*, 2002), whereas in patients with treatment failure, resistant viruses are usually replaced by susceptible viruses in the plasma population after treatment is stopped.

A further issue of concern is the recent description of transmission of a multidrug-resistant virus leading to rapid progression to AIDS in the newly infected individual (Markowitz *et al.*, 2005). Although multidrug resistance is far less common in transmitted resistant viruses than after treatment failure and although the importance of viral factors compared to host factors for the rapid disease progression remains to be determined, the epidemiology of HIV drug resistance in newly infected patients must be carefully followed.

4. GENOTYPIC DRUG RESISTANCE INTERPRETATION SYSTEMS

With the introduction of two commercial systems for genotypic drug resistance testing, the technical quality of HIV drug resistance testing has substantially improved (Korn *et al.*, 2003). The remaining technical issue is the detection of resistant minority species, which may promote therapy failure. If present in less than 20 – 30% of the viral population, resistant minorities are likely to be overlooked by conventional sequencing techniques. New technologies such as real-time based amplifications of mutations may help to overcome this problem at least for selected mutations (Metzner *et al.*, 2003).

The major challenge of genotyping lies in the interpretation of test results. There are many ways how drug resistance-associated mutations can influence each other. Mutations have been described in resistant viruses which restore the susceptibility to certain drugs such as ZDV (Larder *et al.*, 1995; Schinazi *et al.*, 2000) or confer hypersusceptibility to other drugs, *e.g.* APV (Ziermann *et al.*) or EFV (Shulman *et al.*, 2001). Conversely, other mutations can counteract these effects again (Kemp *et al.*, 1998).

The high number of mutations in clinical samples and the different types of possible interactions between them make predictions of phenotypic resistance or clinical response to therapy highly dependent on the expertise of the clinician and/or virologist. Thus, expert advice has been shown to be an independent predictor of virologic therapy response (Tural *et al.*, 2002).

Knowledge about resistance is based on several sources: (i) the development of mutations when wild-type virus is exposed to a certain drug in cell culture, (ii) corresponding pairs of genotypic and phenotypic resistance, and (iii) the influence of drug resistance-associated mutations on the therapy outcome. The information about drug resistance is summarized in so-called rules-based interpretation systems, which are composed of more or less sophisticated rules predicting susceptibility or different degrees of resistance from the mutation profile. Several of these interpretation systems can be accessed online, providing convenient sequence input and immediate prediction output (reviewed in: Schmidt *et al.*, 2002b).

Another group of interpretation systems are database-driven systems. One of these (Virtual phenotype™) is based on a commercial relational database of more than 30,000 corresponding genotype-phenotype pairs. A query sequence is compared to all sequences in the database and the phenotypes of those sequences that share a certain mutational profile with the query sequence are averaged and make up the "virtual phenotype" of the query sequence. Another database-driven system uses a bioinformatic approach to either classify the query sequence as susceptible or resistant according to its path through a decision tree, in which certain drug resistance-associated mutations have key positions (Beerenwinkel *et al.*, 2002), or calculate resistance factors using support vector machines as machine-learning technique (Beerenwinkel *et al.*, 2001).

When the outputs of frequently accessed interpretation systems were compared with each other, different degrees of consensus were found. The discrepancies between the interpretation systems were based on different interpretations of resistance against inhibitors of the reverse transcriptase (Puchhammer-Stockl *et al.*, 2002; Kijak *et al.*, 2003; Ravela *et al.*, 2003; Stürmer *et al.*, 2003). However, a retrospective comparison of 11 interpretation systems revealed no major differences with respect to the prediction of viral load decreases and CD4+ cell increases (De Luca *et al.*, 2003).

The prediction of virological and immunological therapy response is the major challenge for current drug resistance interpretation systems, because several of these systems were originally designed to predict phenotype from genotype. In a retrospective comparison of 9 different interpretation systems, the percentages of correctly predicted viral load increases or decreases ranged between 73% and 80% for the online-accessible systems (M. Helm *et al.*, unpublished data).

Clinical validation is crucial for all systems. Since the differences between the systems are relatively small, it will be difficult to perform prospective clinical studies to compare the predictive value of one system to another. Instead, it may be advisable to evaluate the algorithms for all drugs retrospectively in large clinical databases. This approach could help to define those rules or other means of interpretation which will perform extraordinarily well in comparison to others.

Towards a consensus algorithm, one has to keep in mind that drug resistance is a developing field, which will make regular updates inevitable. Notably, not only new drugs will appear, but also the philosophy of combination therapy will change. The interpretation systems have already modified their predictions of protease inhibitor resistance in response to the boosting with baby-dosed RTV. Drugs may soon become available with longer half-life, which will affect pharmacokinetics and pharmacodynamics more favorably. New viral and cellular targets such as RNAse H, integrase, and coreceptors will open new fields of both antiretroviral therapy and resistance.

5. INFLUENCE OF DRUG LEVELS ON VIROLOGIC AND IMMUNOLOGIC THERAPY RESPONSE

In retrospective databases, the comparison of interpretation systems is compromised by the lack of information on drug levels. If no therapeutic drug levels are achieved either because of non-adherence to therapy or because of insufficient resorption or pharmacological interactions, none of the systems can reliably predict therapy success or failure.

Adherence has been identified as major factor for optimal viral load decrease (Arnsten *et al.*, 2001; Van Vaerenbergh *et al.*, 2002) and in particular, for the immunologic therapy response (Cingolani *et al.*, 2002). Various methods of monitoring adherence have been compared: medication event monitoring system (MEMS) caps as electronic pill counts, pharmacy refill data, questionnaires and diaries, assessment of adherence by the primary physician and specialist nurses, and therapeutic drug monitoring (Hugen *et al.*, 2002).

The most reliable parameter for adherence seems to be the determination of drug levels, which also allows to test for complex pharmacokinetic problems such as insufficient absorption or metabolism (Back *et al.*, 2002, Gerber and Acosta, 2003). The most frequent method used is high-pressure liquid chromatography combined with tandem mass spectrometry (Kurowski *et al.*, 1999). High individual variability of plasma drug levels has been reported (Boffito *et al.*, 2003).

Retrospective studies show a significant correlation of drug levels and virologic therapy response (Durant *et al.*, 2000; Baxter *et al.*, 2002; Van Rossum *et al.*, 2002). Prospective studies, however, yielded discrepant results. While lower viral loads were observed in HIV-infected patients with regular drug monitoring (Fletcher *et al.*, 2002; Burger *et al.*, 2003), two other studies could not confirm this effect at least for the short-term virologic response (Bossi *et al.*, 2002; Clevenbergh *et al.*, 2002). More prospective data are needed to decide whether drug monitoring should be included into the regular monitoring of subjects receiving antiretroviral therapy.

The prediction accuracy of genotypic interpretation systems can be improved considerably by combining the drug levels at the end of each dosing interval (= trough levels) with the resistance data, which is expressed as the phenotypic or genotypic inhibitory quotient (IQ) (Kempf et al., 2001; Marcelin et al., 2003). Since low trough levels promote the development of drug resistance, pharmacokinetics is one of the most challenging fields of future antiretroviral therapy.

6.   VIRAL FITNESS

In 1998, Perrin and Telenti analyzed the virological and immunological outcome of antiretroviral therapy in HIV-1-infected patients. Four groups could be distinguished: patients with a decrease in viral load concomitant with an increase in CD4+ cells (40%), those with an increase in viral load accompanied by a decrease in CD4+ cells (15%), a rare group with a decrease in both viral load and CD4+ cells (5%), and patients with an increase in viral load, but also an increase of CD4+ cells (40%).

Since the latter group comprises patients who benefit from antiretroviral therapy despite a rebound of viral load, reduced viral fitness of drug-resistant viruses was discussed as reason for this phenomenon. This was based on data showing that protease inhibitor-resistant viruses developed mutations at gag cleavage sites to re-adjust the substrate (= gag-pol precursor) to the enzyme (= protease) (Doyon et al., 1996; Zhang et al., 1997). Removal of these cleavage site mutations led to a decrease in viral growth, confirming their role in viral fitness.

The term 'viral fitness' has been used and interpreted in many different ways. In its most comprehensive way, fitness defines the replicative adaptability of an organism to its environment (Quinones-Mateu and Arts, 2002). With HIV, all steps of the viral life cycle are included from the infection of a target cell until the production of progeny viruses. It starts with the entry of the viral particle (infectivity), reverse transcription, integration of the provirus, generation of viral mRNAs, cleavage of the polyprotein precursor, budding and maturation.

The 'competitive replicative ability' has first been described using vesicular stomatitis virus in growth competition assays (Holland et al., 1991). Since then, assays have been published with measure enzyme catalytic activities of the HIV-1 protease (Nijhuis et al., 1999) and reverse transcriptase (Brenner et al., 2002) as well as the replicative capacity of either primary isolates or pseudotyped recombinant viruses (Table 2). All systems have inherent advantages and draw-backs.

Coinfection/competition assays are closest to the *in vivo* situation, in which wild-type and resistant viral strains compete with each other in the presence and absence of drug pressure. Reduced protease activity was associated with a reduction in viral replication capacity, resulting in an evolution of novel variants with compensatory mutations, which displayed increased protease activity, but not increased resistance (Nijhuis *et al.*, 1999).

The one-replication cycle assay was originally developed to test for HIV-1 drug resistance in the context of recombinant viruses (Petropoulos *et al.*, 2000). A modified version was used to evaluate the relative replicative capacity of viruses in patients before and after treatment interruptions (Deeks *et al.*, 2001). Concomitant with an increase in viral load and decrease of T helper cells, resistant viruses lost their protease inhibitor resistance and also regained viral fitness within a short period of time. Viral growth kinetics were mostly used to test for fitness in the context of viral evolution (Maeda *et al.*, 1998). Importantly, the authors found a difference in fitness when antiretroviral drugs were added to the cell culture, indicating that viral fitness can be different in the presence and in the absence of drug.

Both coinfection/competition assays and viral growth kinetics can be performed with primary isolates, whereas the one-replication assay uses recombinant viruses containing patient-derived *pol* genes. The latter assay can only determine the relative replicative capacity of the patient-derived insert in the context of the vector backbone. Mutations in the *pol* gene certainly contribute substantially, but may not be sufficient for the level of viral fitness displayed by the primary isolate (Simon *et al.*, 2003).

The effect of single or combined mutations on viral fitness can only be assessed with great difficulty (reviewed in: Quinones-Mateu *et al.*, 2002). Results are given as orders of mutations with relative increases or decreases of viral fitness, but not as absolute values. Due to different assay systems, the results cannot be compared with each other and are sometimes contradictory, which may also be due to different viral backbones.

Nonetheless, it will certainly be valuable to collect and summarize all information on viral fitness within a database, which will be able to characterize mutations and combination of mutations with respect to replication capacity. It may even be possible to identify patterns that result in particularly unfit viruses, which may be exploited therapeutically when options for a suppressive therapy are no longer available (Clementi, 2004).

Table 2. Methods of viral fitness testing.

Method	Authors[1]	Cells	Advantages	Disadvantages	Read-out	Length
Coinfection/ competition assay	Nijhuis *et al.*, 1999	PBMC	- Testing of primary isolates possible - minor differences in fitness can be detected - internal control (dual infection)	- Passaging of virus can lead to loss of resistance-associated mutations	Sequencing, HTA[3], real-time based detection of mutations[3]	Up to 6 wks
One-replication assay	Deeks *et al.*, 2001	Cell line	- Rapid performance - good reproducibility	- Incomplete viral replication cycle of recombinant pseudotyped viruses	Luciferase	2d[2]
Viral growth kinetics	Maeda *et al.*, 1998	PBMC or cell line	- Simple performance - Complete viral replication cycle - testing of primary isolates possible	- High variability - only major differences in fitness can be detected	p24, RT activity	15d and longer

[1] only one of the prototypic assays is included

[2] after generation of recombinant virus

[3] used in recently published assays (Quinones-Mateu *et al.*, 2000; Weber *et al.*, 2003)

CONCLUSIONS

The therapeutic options for HIV-1 infected patients have greatly improved. With 20 antirctroviral drugs attacking the virus at different steps of the viral life cycle, HIV infection has become a treatable, although currently not curable chronic disease. Future options include (i) the development of new drugs with activity against cellular coreceptors, the viral RNAse H and the viral integrase, (ii) the introduction of drugs with more favourable pharmacokinetics and less side-effects, and (iii) resistance-guided treatment choices.

As long as antiretroviral therapy as well as diagnostic and therapeutic monitoring are available, HIV-1 replication can be suppressed in the majority of infected individuals. If not, impaired fitness of drug-resistant viral strains seems to reduce viral pathogenicity to an extent the immune system can cope with.

Therefore, current efforts should focus on bringing antiretroviral therapy to all HIV-infected individuals, in particular in Africa. Since every new drug is just a few years ahead of the virus before resistant viral strains will compromise the drug's therapeutic effect again, pharmaceutical companies are encouraged to continue their research and development. It is most important to ensure the survival of HIV-infected subjects until someone will eventually find a cure.

ACKNOWLEDGEMENTS

We would like to thank all colleagues who worked alongside with us in generating and publishing information about drug resistance and making this knowledge publicly accessible. We thank all pharmaceutical companies for their commitment in developing new antiretroviral drugs and providing them for resistance testing. We are also indebted to all HIV-infected individuals whose numerous data about drug resistance have contributed to optimize the antiretroviral therapy strategies for future patients.

We would like to acknowledge the continuous support by the Robert Koch Institute, Berlin (National Reference Centre for Retroviruses), and the Federal Ministry of Education and Research (HIV Competence Network, AZ 01 KI 0211).

NOTES

Drug resistance is a rapidly evolving field, which makes it increasingly difficult to write comprehensive reviews without omitting important information published by appreciated colleagues. This article was written to give an overview of current therapeutic strategies in the light of HIV-1 drug resistance. It does not claim to be exhaustive nor does it imply that data not mentioned here are not as valuable as those presented.

REFERENCES

Arnsten, J.H., Demas, P.A., Farzadegan, H., Grant, R.W., Gourevitch, M.N., Chang, C.J., Buono, D., Eckholdt, H., Howard, A.A., and Schoenbaum, E.E., 2001, Antiretroviral therapy adherence and viral suppression in HIV-infected drug users: comparison of self-report and electronic monitoring. *Clin. Infect. Dis.* **33**:1417-1423

Back, D., Gatti, G., Fletcher, C., Garaffo, R., Haubrich, R., Hoetelmans, R., Kurowski, M., Luber, A., Merry, C., and Perno, C.F., 2002, Therapeutic drug monitoring in HIV infection: current status and future directions. *AIDS* **16 (Suppl. 1)**:S5-S37

Barré-Sinoussi, F., Chermann, J.C., Rey, F., Nugeyre, M.T., Chamaret, S., Gruest, J., Dauguet, C., Axler-Blin, C., Vezinet-Brun, F., Rouzioux, C., Rozenbaum, W., and Montagnier, L., 1983, Isolation of a T-lymphotropic retrovirus from a patient at risk for acquired immune deficiency syndrome (AIDS). *Science* **220**:868-871

Baxter, J., Mayers, D., Wentworth, D., Neaton, J.D., Hoover, M.L., Winters, M.A., Mannheimer, S.B., Thompson, M.A., Abrams, D.I., Brizz, B.J., Ioannidis, J.P., and Merigan, T.C., 2000, A randomized study of antiretroviral management based on plasma genotypic antiretroviral testing in patients failing therapy. CPCRA 046 Study Team for the Terry Beirn Community Programs for Clinical Research on AIDS. *AIDS* **14**:83-93

Baxter, J.D., Merigan, T.C., Wentworth, D.N., Neaton, J.D., Hoover, M.L., Hoetelmans, R.M., Piscitelli, S.C., Verbiest, W.H., and Mayers, D.L.; CPCRA 046 Study Team for the Terry Beirn Community Programs for Clinical Research on AIDS, 2002, Both baseline HIV-1 drug resistance and antiretroviral drug levels are associated with short-term virologic response to salvage therapy. *AIDS* **16**:1131-1138

Beerenwinkel, N., Schmidt, B., Walter, H., Kaiser, R., Lengauer, T., Hoffmann, D., Korn, K., and Selbig, J., 2001, Geno2pheno: interpreting genotypic HIV drug resistance tests. *IEEE Intellig. Syst.* **16**:35-41

Beerenwinkel, N., Schmidt, B., Walter, H., Kaiser, R., Lengauer, T., Hoffmann, D., Korn, K., and Selbig, J., 2002, Diversity and complexity of HIV-1 drug resistance: a bioinformatics approach to predicting phenotype from genotype. *Proc. Natl. Acad. Sci. USA.* **99**:8271-8276

Boffito, M., Back, D.J., Hoggard, P.G., Caci, A., Bonora, S., Raiteri, R., Sinicco, A., Reynolds, H., Khoo, S., and Di Perri, G., 2003, Intra-individual variability in lopinavir plasma trough concentrations supports therapeutic drug monitoring. *AIDS* **17**:1107-1108

Bossi, P., Peytavin, G., Ait-Mohand, H., Delaugerre, C., Ktorza, N., Paris, L., Bonmarchand, M., Cacace, R., David, D.J., Simon, A., Lamotte, C., Marcelin, A.G., Calvez, V., Bricaire, F., Costagliola, D., and Katlama, C., 2004, GENOPHAR: a randomized study of plasma drug measurements in association with genotypic resistance testing and expert advice to optimize therapy in patients failing antiretroviral therapy. *HIV Med.* **5:**352-359

Boucher, C.A., Cammack, N., Schipper, P., Schuurman, R., Rouse, P., Wainberg, M.A., and Cameron, J.M., 1993, High-level resistance to (-) enantiomeric 2'-deoxy-3'-thiacytidine in vitro is due to one amino acid substitution in the catalytic site of human immunodeficiency virus type 1 reverse transcriptase. *Antimicrob. Agents Chemother.* **37:**2231-2234

Brenner, B.G., Routy, J.P., Petrella, M., Moisi, D., Oliveira, M., Detorio, M., Spira, B., Essabag, V., Conway, B., Lalonde, R., Sekaly, R.P., and Wainberg, M.A., 2002, Persistence and fitness of multidrug-resistant human immunodeficiency virus type 1 acquired in primary infection. *J. Virol.* **76:**1753-1761

Burger, D., Hugen, P., Reiss, P., Gyssens, I., Schneider, M., Kroon, F., Schreij, G., Brinkman, K., Richter, C., Prins, J., Aarnoutse, R., and Lange, J., ATHENA Cohort Study Group, 2003, Therapeutic drug monitoring of nelfinavir and indinavir in treatment-naive HIV-1-infected individuals. *AIDS* **17:**1157-1165

Cingolani, A., Antinori, A., Rizzo, M. G. Murri, R., Ammassari, A., Baldini, F., Di Giambenedetto, S., Cauda, R., and De Luca, A, 2002, Usefulness of monitoring HIV drug resistance and adherence in individuals failing highly active antiretroviral therapy: a randomized study (ARGENTA). *AIDS* **16:**369-379

Clementi, M., 2004, Can modulation of viral fitness represent a target for anti-HIV-1 strategies? *New Microbiol.* **27:**207-214

Clevenbergh, P., Durant, J., Halfon, P., del Giudice, P., Mondain, V., Montagne, N., Schapiro, J.M., Boucher, C.A., and Dellamonica, P., 2000, Persisting long-term benefit of genotype-guided treatment for HIV-infected patients failing HAART. The Viradapt study: week 48 follow-up. *Antivir. Ther.* **5:**65-70

Clevenbergh, P., Garraffo, R., Durant, J., and Dellamonica, P., 2002, PharmAdapt: a randomised prospective study to evaluate the benefit of therapeutic monitoring of protease inhibitors: 12 week results. *AIDS* **16:**2311-2315

Cohen, C.J., Hunt, S., Sension, M., Farthing, C., Conant, M., Jacobson, S., Nadler, J., Verbiest, W., Hertogs, K., Ames, M., Rinehart, A.R., and Graham, N.M.; VIRA3001 Study Team, 2002, A randomized trial assessing the impact of phenotypic resistance testing on antiretroviral therapy. *AIDS* **16:**579-588

Corzillius, M., Muhlberger, N., Sroczynski, G., Jaeger, H., Wasem, J., and Siebert, U., 2004, Cost effectiveness analysis of routine use of genotypic antiretroviral resistance testing after failure of antiretroviral treatment for HIV. *Antivir. Ther.* **9:**27-36

Craig, J.C., Duncan, I.B., Hockley, D., Grief, C., Roberts, N.A., and Mills, J.S., 1991, Antiviral properties of Ro 31-8959, an inhibitor of human immunodeficiency virus (HIV) proteinase. *Antiviral Res.* **16:**295-305

De Clercq, E., 1991, Chemotherapy of the acquired immune deficiency syndrome (AIDS): non-nucleoside inhibitors of the human immunodeficiency virus type 1 reverse transcriptase. *Int. J. Immunopharmacol.* **13 (Suppl 1):**83-89

Deeks, S.G., Wrin, T., Liegler, T., Hoh, R., Hayden, M., Barbour, J.D., Hellmann, N.S., Petropoulos, C.J., McCune, J.M., Hellerstein, M.K., and Grant, R.M., 2001, Virologic and immunologic consequences of discontinuing combination antiretroviral-drug therapy in HIV-infected patients with detectable viremia. *N. Engl. J. Med.* **344:**472-480

De Luca, A., Cingolani, A., Di Giambenedetto, S., Trotta, M.P., Baldini, F., Rizzo, M.G., Bertoli, A., Liuzzi, G., Narciso, P., Murri, R., Ammassari, A., Perno, C.F., and Antinori, A., 2003, Variable prediction of antiretroviral treatment outcome by different systems for interpreting genotypic human immunodeficiency virus type 1 drug resistance. *J. Infect. Dis.* **187**:1934-1943

DeGruttola, V., Dix, L., D'Aquila, R., Holder, D., Phillips, A., Ait-Khaled, M., Baxter, J., Clevenbergh, P., Hammer, S., Harrigan, R., Katzenstein, D., Lanier, R., Miller, M., Para, M., Yerly, S., Zolopa, A., Murray, J., Patick, A., Miller, V., Castillo, S., Pedneault, L., and Mellors, J., 2000, The relation between baseline HIV drug resistance and response to antiretroviral therapy: re-analysis of retrospective and prospective studies using a standardized data analysis plan. *Antivir. Ther.* **5**:41-48

Descamps, D., Chaix, M.L., Andre, P., Brodard, V., Cottalorda, J., Deveau, C., Harzic, M., Imgrand, D., Izopet, J., Kohli, E., Masquelier, B., Mouajjah, S., Palmer, P., Pellegrin, I., Plantier, J.C., Poggi, C., Rogez, S., Ruffault, A., Schneider, V., Signori-Schmuck, A., Tamalet, C., Wirden, M., Rouzioux, C., Brun-Vezinet, F., Meyer, L., and Costagliola, D., 2005, French national sentinel survey of antiretroviral drug resistance in patients with HIV-1 primary infection and in antiretroviral-naïve chronically infected patients in 2001-2002. *J. Acquir. Immmune Defic. Syndr.* **38**:545-552

Doyon, L., Croteau, G., Thibeault, D., Poulin, F., Pilote, L., Lamarre, D., 1996, Second locus involved in human immunodeficiency virus type 1 resistance to protease inhibitors. *J. Virol.* **70**:3763-3769

Durant, J., Clevenbergh, P., Halfon, P., Delgiudice, P., Porsin, S., Simonet, P., Montagne, N., Boucher, C.A., Schapiro, J.M., and Dellamonica, P., 1999. Drug resistance genotyping in HIV-1 therapy: the VIRADAPT randomized controlled trial. *Lancet* **353**:2195-2199

Durant, J., Clevenbergh, P., Garraffo, R., Halfon, P., Icard, S., Del Giudice, P., Montagne, N., Schapiro, J. M., and Dellamonica, P., 2000, Importance of protease inhibitor plasma levels in HIV-infected patients treated with genotypic-guided therapy: pharmacological data from the Viradapt Study. *AIDS* **14**:1333-1339

Fletcher, C.V., Anderson, P.L., Kakuda, T.N., Schacker, T.W., Henry, K., Groß, C.R., and Brandage, R.C., 2002, Concentration-controlled compared with conventional antiretroviral therapy for human immunodeficiency-virus infection. *AIDS* **16**:551-560

Flexner, C., 1998, HIV-protease inhibitors. *N. Engl. J. Med.* **338**:1281-1292

Gerber, J.G., and Acosta, E.P., 2003, Therapeutic drug monitoring in the treatment of HIV-infection. *J. Clin. Virol.* **27**:117-128

Grant, R.M., Hecht, F.M., Warmerdam, M., Liu, L., Liegler, T., Petropoulos, C.J., Hellmann, N.S., Chesney, M., Busch, M.P, and Kahn, J.O., 2002, Time trends in primary HIV-1 drug resistance among recently infected persons. *JAMA* **288**:181-188

Greenberg, M., Cammack, N., Salgo, M., and Smiley, L., 2004, HIV fusion and its inhibition in antiretroviral therapy. *Rev. Med. Virol.* **14**:321-337

Harrigan, P.R., Wynhoven, B., Brumme, Z.L., Brumme, C.J., Sattha, B., Major, J.C., de la Rosa, R., and Montaner, J.S.G., 2005, HIV-1 drug resistance: degree of underestimation by a cross-sectional versus a longitudinal testing approach. *J. Infect. Dis.* **191**:1325-1330

Haubrich, R.H., Kemper, C.A., Hellmann, N.S., Keiser, P.H., Witt, M.D., Tilles, J.G., Forthal, D.N., Leedom, J., Leibowitz, M., McCutchan, J.A., and Richman, D.D., California Collaborative Treatment Group, 2005, A randomized, prospective study of phenotype susceptibility testing versus standard of care to manage antiretroviral therapy: CCTG 575. *AIDS.* **19**:295-302

Hirsch, M.S., Brun-Vezinet, F., Clotet, B., Conway, B., Kuritzkes, D.R., D'Aquila, R.T., Demeter, L.M., Hammer, S.M., Johnson, V.A., Loveday, C., Mellors, J.W., Jacobsen, D.M., and Richman, D.D., 2003, Antiretroviral drug resistance testing in adults infected with human immunodeficiency virus type 1: 2003 recommendations of an International AIDS Society-USA panel. *Clin. Infect. Dis.* **37**:113-128

Holland, J.J., de la Torre, J.C., Clarke, D.K., and Duarte, E., 1991, Quantitation of relative fitness and great adaptability of clonal populations of RNA viruses. *J. Virol.* **65**:2960-2967

Hugen, P.W.H., Langebeek, N., Burger, D.M., Zomer, B., van Leusen, R., Schuurman, R., Koopmans, P.P., and Hekster, Y.A., 2002, Assessment of adherence to HIV protease inhibitors: comparison and combination of various methods, including MEMS (electronic monitoring), patient and nurse report, and therapeutic drug monitoring. *J. Acquir. Immmune Defic. Syndr.* **30**:324-334

Japour, A.J., Mayers, D.L., Johnson, V.A., Kuritzkes, D.R., Beckett, L.A., Arduino, J.M., Lane, J., Black, R.J., Reichelderfer, P.S., D'Aquila, R.T., and Crumpacker, C.S., The RV-43 Study Group and The AIDS Clinical Trials Group Virology Committee Resistance Working Group, 1993, Standardized peripheral blood mononuclear cell culture assay for determination of drug susceptibilities of clinical human immunodeficiency virus type 1 isolates. The RV-43 Study Group, the AIDS Clinical Trials Group Virology Committee Resistance Working Group. *Antimicrob. Agents Chemother.* **37**:1095-1101

Jármy, G., Heinkelein, M., Weissbrich, B., Jassoy, C., and Rethwilm, A., 2001, Phenotypic analysis of the sensitivity of HIV-1 to inhibitors of the reverse transcriptase, protease, and integrase using a self-inactivating virus vector system. *J. Med. Virol.* **64**:223-231

Johnson, V.A., Brun-Vezinet, F., Clotet, B., Conway, B., D'Aquila, R.T., Demeter, L.M., Kuritzkes, D.R., Pillay, D., Schapiro, J.M., Telenti, A., and Richman, D.D., International AIDS Society-USA Drug Resistance Mutations Group, 2003, Drug resistance mutations in HIV-1. *Top HIV Med.* **11**:215-221

Kageyama, S., Anderson, B.D., Hoesterey, B.L., Hayashi, H., Kiso, Y., Flora, K.P., and Mitsuya, H., 1994, Protein binding of human immunodeficiency virus protease inhibitor KNI-272 and alteration of its in vitro antiretroviral activity in the presence of high concentrations of proteins. *Antimicrob. Agents Chemother.* **38**:1107-1111

Kellam, P., Larder, B.A., 1994, Recombinant virus assay: a rapid, phenotypic assay for assessment of drug susceptibility of human immunodeficiency virus type 1 isolates. *Antimicrob. Agents Chemother.* **38**:23-30

Kemp, S.D., Shi, C., Bloor, S., Harrigan, P.R., Mellors, J.W., and Larder, B.A., 1998, A novel polymorphism at codon 333 of human immunodeficiency virus type 1 reverse transcriptase can facilitate dual resistance to zidovudine and L-2',3'-dideoxy-3'-thiacytidine. *J. Virol.* **72**:5093-5098

Kempf, D., Isaacson, J., King, M.S., Brun, S.C., Xu, Y., Real, K., Bernstein, B.M., Japour, A.J., Sun, E., and Rode, R.A., 2001, Identification of genotypic changes in HIV protease that correlate with reduced susceptibility to the protease inhibitor lopinavir among viral isolates from protease inhibitor-experienced patients. *J. Virol.* **75**:7462-7469

Kijak, G.H., Rubio, A.E., Pampuro, S.E., Zala, C., Cahn, P., Galli, R., Montaner, J.S., and Salomon, H., 2003, Discrepant results in the interpretation of HIV-1 drug-resistance genotypic data among widely used algorithms. *HIV Medicine* **4**:72-78

Korn, K., Reil, H., Walter, H., and Schmidt, B., 2003, Quality control trial for human immunodeficiency virus type 1 drug resistance testing using clinical samples reveals problems with detecting minority species and interpretation of test results. *J. Clin. Microbiol.* **41**:3559-3565

Kurowski, M., Müller, M., Donath, F., Mrozikiewicz, M., and Möcklinghoff, C., 1999, Single daily doses of saquinavir achieves HIV inhibitory concentrations when combined with baby-dose ritonavir. *Eur. J. Med. Res.* **4:**101-104

Larder, B.A., Darby, G., and Richman, D.D., 1989, HIV with reduced sensitivity to zidovudine (AZT) isolated during prolonged therapy. *Science* **243:**1731-1734

Larder, B., Kemp, S., and Harrigan, R., 1995, Potential mechanism of sustained antiretroviral efficacy of AZT-3TC combination therapy. *Science* **269:**696-699

Little, S.J., Holte, S., Routy, J.P. Daar, E.S., Markowitz, M., Collier, A.C., Koup, R.A., Mellors, J.W., Connick, E., Conway, B., Kilby, M., Wang, L., Whitcomb, J.M., Hellmann, N.S., and Richman, D.D., 2002, Antiretroviral-drug resistance among patients recently infected with HIV. *N Engl J Med* **347:**385-394

Maeda, Y., Venzon, D.J., Mitsuya, H., 1998, Altered drug sensitivity, fitness, and evolution of human immunodeficiency virus type 1 with pol gene mutations conferring multi-dideoxynucleoside resistance. *J. Infect. Dis.* **177:**1207-1213

Marcelin, A.G., Lamotte, C., Delauguerre, C., Ktorza, N., Ait Mohand, H., Cacace, R., Bonmarchand, M., Wirden, M., Simon, A., Bossi, P., Bricaire, F., Costagliola, D., Katlama, C., Peytavin, G., Calvez, V, and the Genophar Study Group., 2003, Genotypic inhibitory quotient as predictor of virological response to ritonavir-amprenavir in human immunodeficiency virus type 1 protease inhibitor-experienced patients. *Antimicrob. Agents Chemother.* **47:**594-600

Markowitz, M., Mohri, H., Mehandru, S., Shet, A., Berry, L., Kalyanaraman, R., Kim, A., Chung, C., Jean-Pierre, P., Horowitz, A., La Mar, M., Wrin, T., Parkin, N., Poles, M., Petropoulos, C., Mullen, M., Boden, D., and Ho, D.D., 2005, Infection with multidrug resistant, dual-tropic HIV-1 and rapid progression to AIDS: a case report. *Lancet* **365:**1031-1038

Metzner, K.J., Bonhoeffer, S., Fischer, M., Karanicolas, R., Allers, K., Joos, B., Weber, R., Hirschel, B., Kostrikis, L.G., and Gunthard, H.F.; The Swiss HIV Cohort Study, 2003, Emergence of minor populations of human immunodeficiency virus type 1 carrying the M184V and L90M mutations in subjects undergoing structured treatment interruptions. *J. Infect. Dis.* **188:**1433-1343

Meynard, J., Vray, M., Morand-Joubert, L., Race, E., Descamps, D., Peytavin, G., Matheron, S., Lamotte, C., Guiramand, S., Costagliola, D., Brun-Vezinet, F., Clavel, F., and Girard, P. M.; Narval Trial Group, 2002, Phenotypic or genotypic resistance testing for choosing antiretroviral therapy after treatment failure: a randomized trial. *AIDS* **16:**727-736

Miller, V., and Larder, B.A., 2001, Mutational patterns in the HIV genome and cross-resistance following nucleoside and nucleotide analogue drug exposure. *Antivir. Ther.* **6 (Suppl 3):**25-44

Mitsuya, H., Weinhold, K.J., Furman, P.A., St Clair, M.H., Lehrman, S.N., Gallo, R.C., Bolognesi, D., Barry, D.W., and Broder, S., 1985, 3'-Azido-3'-deoxythymidine (BW A509U): an antiviral agent that inhibits the infectivity and cytopathic effect of human T-lymphotropic virus type III/lymphadenopathy-associated virus in vitro. *Proc. Natl. Acad. Sci. U.S.A.* **82:**7096-7100

Mocroft, A., Vella, S., Benfield, T.L., Chiesi, A., Miller, V., Gargalianos, P., d'Arminio Manforte, A., Yust, I., Bruun, J.N., Phillips, A.N., and Lundgren, J.D., 1998, Changing patterns of mortality across Europe in patients infected with HIV-1. *Lancet* **352:**1725-1730

Moyle, G.J., and Back, D., 2001, Principles and practice of HIV-protease inhibitor pharmaco-enhancement. *HIV Med.* **2:**105-113

Mueller, S.M., Daeumer, M, Kaiser, R., Walter, H., Colonno, R., and Korn, K, 2004, Susceptibility to saquinavir and atazanavir in highly protease inhibitor (PI) resistant HIV-1 is caused by lopinavir-induced drug resistance mutation L76V. *Antivir. Ther.* **9**:S44.

Naesens, L., Bischofberger, N., Augustijns, P., Annaert, P., Van den Mooter, G., Arimilli, M.N., Kim, C.U., and De Clercq, E., 1998, Antiretroviral efficacy and pharmacokinetics of oral bis(isopropyloxycarbonyloxymethyl)-9-(2-phosphonylmethoxypropyl)adenine in mice. *Antimicrob. Agents Chemother.* **42**:1568-1573

Nijhuis, M., Schuurman, R., de Jong, D., Erickson, J., Gustchina, E., Albert, J., Schipper, P., Gulnik, S., and Boucher, C.A., 1999, Increased fitness of drug resistant HIV-1 protease as a result of acquisition of compensatory mutations during suboptimal therapy. *AIDS* **13**:2349-2359

Palella, F.J. Jr., Delaney, K.M., Moorman, A.C., Loveless, M.O., Fuhrer, J., Satten, G.A., Aschman, D.J., and Holmberg, S.D., 1998, Declining morbidity and mortality among patients with advanced human immunodeficiency virus infection. HIV Outpatient Study Investigators. *N. Engl. J. Med.* **338**:853-860

Panidou, E.T., Trikalinos, T.A., Ioannidis, J.P., 2004, Limited benefit of antiretroviral resistance testing in treatment-experienced patients: a meta-analysis. *AIDS* **18**:2153-2161.

Perrin, L., and Telenti, A., 1998, HIV treatment failure: testing for HIV resistance in clinical practice. *Science* **280**:1871-1873

Petropoulos, C.J., Parkin, N.T., Limoli, K.L., Lie, Y.S., Wrin, T., Huang, W., Tian, H., Smith, D., Winslow, G.A., Capon, D.J., and Whitcomb, J.M., 2000, A novel phenotypic drug susceptibility assay for human immunodeficiency virus type 1. *Antimicrob. Agents Chemother.* **44**:920-928

Puchhammer-Stockl, E., Steininger, C., Geringer, E., and Heinz, F.X., 2002, Comparison of virtual phenotype and HIV-SEQ program (Stanford) interpretation for predicting drug resistance of HIV strains. *HIV Med.* **3**:200-206

Quinones-Mateu, M.E., Ball, S.C., Marozsan, A.J., Torre, V.S., Albright, J.L., Vanham, G., van Der Groen, G., Colebunders, R.L., and Arts, E.J., 2000, A dual infection/competition assay shows a correlation between ex vivo human immunodeficiency virus type 1 fitness and disease progression. *J. Virol.* **74**:9222-9233

Quinones-Mateu, M. E., and Arts, E. J., 2002, Fitness of drug resistant HIV-1: methodology and clinical implications. *Drug Resist. Updat.* **5**:224-233

Race, E., Dam, E., Obry, V., Paulous, S., and Clavel, F., 1999, Analysis of HIV cross-resistance to protease inhibitors using a rapid single-cycle recombinant virus assay for patients failing on combination therapies. *AIDS* **13**:2061-2081

Ravela, J., Betts, B.J., Brun-Vezinet, F., Vandamme, A.M., Descamps, D., van Laethem, K., Smith, K., Schapiro, J.M., Winslow, D.L., Reid, C., and Shafer, R.W., 2003, HIV-1 protease and reverse transcriptase mutation patterns responsible for discordances between genotypic drug resistance interpretation systems. *J. Acquir. Immune Defic. Syndr.* **33**:8-14

Richman, D.D., Morton, S.C., Wrin, T., Hellmann, N., Berry, S., Shapiro, M.F., and Bozzette, S.A., 2004, The prevalence of antiretroviral drug resistance in the United States. *AIDS* **18**:1393-1401

Schinazi, R.F., Larder, B., and Mellors, J.W., 2000, Mutations in retroviral genes associated with drug resistance: 2000-2001 update. *Int. Antivir. News* **8**:65-92

Schmidt, B., Korn, K., Moschik, B., Paatz, C., Uberla, K., and Walter, H., 2000a, Low level of cross-resistance to amprenavir (141W94) in samples from patients pretreated with other protease inhibitors. *Antimicrob. Agents Chemother.* **44**:3213-3216.

Schmidt, B., Walter, H., Moschik, B., Paatz, C., van Vaerenbergh, K., Vandamme, A.-M., Schmitt, M., Harrer, T., Überla, K., and Korn, K., 2000b, Simple algorithm derived from a geno-/phenotypic database to predict HIV-1 protease inhibitor resistance. *AIDS* **14**:1731-1738

Schmidt, B., Korn, K., and Walter, H., 2002a. Technologies for measuring HIV-1 drug resistance. *HIV Clin. Trials* **3**:227-236

Schmidt, B., Walter, H., Zeitler, N., and Korn, K., 2002b. Genotypic drug resistance interpretation systems – the cutting edge of antiretroviral therapy. *AIDS Reviews* **4**:148-156. Erratum in: *AIDS Reviews* 2003, **5**:63

Schnell, T., Schmidt, B., Moschik, G., Thein, C., Paatz, C., Korn, K., and Walter, H., 2003, Distinct cross-resistance profiles of the new protease inhibitors amprenavir, lopinavir, and atazanavir in a panel of clinical samples. *AIDS* **17**:1258-1261

Shulman, N.S., Hughes, M.D., Winters, M.A., Shafer, R.W., Zolopa, A.R., Hellmann, N.S., Bates, M., Whitcomb, J.M., and Katzenstein, D.A., 2002, Subtle decreases in stavudine phenotypic susceptibility predict poor virologic response to stavudine monotherapy in zidovudine-experienced patients. *J. Acquir. Immune Defic. Syndr.* **31**:121-127

Shulman, N., Zolopa, A.R., Passaro, D., Shafer, R.W., Huang, W., Katzenstein, D., Israelski, D.M., Hellmann, N., Petropoulos, C., and Whitcomb, J., 2001, Phenotypic hyper-susceptibility to non-nucleoside reverse transcriptase inhibitors in treatment-experienced HIV-infected patients: impact on virological response to efavirenz-based therapy. *AIDS* **15**:1125-1132

Simon, V., Padte, N., Murray, D., Vanderhoeven, J., Wrin, T., Parkin, N., Di Mascio, M., and Markowitz, M., 2003, Infectivity and replication capacity of drug-resistant human immunodeficiency virus type 1 variants isolated during primary infection. *J. Virol.* **77**:7736-7745

Sturmer, M., Doerr, H.W., Staszewski, S., and Preiser, W., 2003, Comparison of nine resistance interpretation systems for HIV-1 genotyping. *Antivir. Ther.* **8**:239-244

The UK Collaborative Group on HIV Drug Resistance and UK CHIC Study Group, 2005, Long term probability of detection of HIV-1 drug resistance after starting antiretroviral therapy in routine clinical practice. *AIDS* **19**:487-494

Tisdale, M., Kemp, S.D., Parry, N.R., Larder, B.A., 1993, Rapid in vitro selection of human immunodeficiency virus type 1 resistant to 3'-thiacytidine inhibitors due to a mutation in the YMDD region of reverse transcriptase. *Proc. Natl. Acad. Sci. U. S. A.* **90**:5653-5656

Tural, C., Ruiz, L., Holtzer, C., Schapiro, J., Viciana, P., Gonzalez, J., Domingo, P., Boucher, C., Rey-Joly, C., and Clotet, B.; Havana Study Group., 2002, Clinical utility of HIV-1 genotyping and expert advice: the Havana trial. *AIDS* **16**:209-218

Vandamme, A.M., Sonnerborg, A., Ait-Khaled, M., Albert, J., Asjo, B., Bacheler, L., Banhegyi, D., Boucher, C., Brun-Vezinet, F., Camacho, R., Clevenbergh, P., Clumeck, N., Dedes, N., De Luca, A., Doerr, H.W., Faudon, J.L., Gatti, G., Gerstoft, J., Hall, W.W., Hatzakis, A., Hellmann, N., Horban, A., Lundgren, J.D., Kempf, D., Miller, M., Miller, V., Myers, T.W., Nielsen, C., Opravil, M., Palmisano, L., Perno, C.F., Phillips, A., Pillay, D., Pumarola, T., Ruiz, L., Salminen, M., Schapiro, J., Schmidt, B., Schmit, J.C., Schuurman, R., Shulse, E., Soriano, V., Staszewski, S., Vella, S., Youle, M., Ziermann, R., and Perrin, L., 2004, Updated European recommendations for the clinical use of HIV drug resistance testing. *Antivir. Ther.* **9**:829-848

Van de Vijver, D.A.M.C., Wensing, A.M.J., Asjo, B., Bruckova, M., Jorgensen, L.B., Horban, A., Linka, M., Lazanas, M., Loveday, C., MacRae, E., Nielsen, C., Paraskevis, D., Poljak, M., Puchhammer-Stöckl, E., Ruiz, L., Schmit, J.C., Stanczak, G., Stanojevic, M., Vandamme, A.M., Vercauteren, J., and Boucher, C.A.B., on behalf of the SPREAD Programme, Frequency of antiretroviral drug resistance associated mutations in sequences from treated individuals across Europe: the CAPTURE study. 3rd European HIV Drug Resistance Conference, Athens 2005 (abstract no. 2).

Van Rossum, A.M., Berghoeff, A.S., Fraaij, P.L., Hugen, P.W., Hartwig, N.G., Geelen, S.P., Wolfs, T.F., Weemaes, C.M., De Groot, R., and Burger, D.M., 2002, Therapeutic drug monitoring of indinavir and nelfinavir to assess adherence to therapy in human immunodeficiency virus-infected children. *Pediatr. Infect. Dis. J.* **21**:743-747

Van Vaerenbergh, K., Harrer, T., Schmit, J.-C., Carbonez, A., Fontaine, E., Kurowski, M., Grünke, M., Löw, P., Rascu, A., Schmidt, B., Schmitt, M., Thoelen, I., Walter, H., van Laethem, K., van Ranst, M., Desmyter, J., de Clercq, E., and Vandamme, A.-M., 2002, Initiation of HAART in drug-naïve HIV type 1 patients prevents viral breakthrough for a median period of 35.5 months in 60 % of patients. *AIDS Res. Hum. Retrovir.* **18**:419-426

Walter, H., Schmidt, B., Korn, K., Vandamme, A.M., Harrer, T., and Uberla, K., 1999, Rapid, phenotypic HIV-1 drug sensitivity assay for protease and reverse transcriptase inhibitors. *J. Clin. Virol.* **13**:71-80

Weber, J., Rangel, H.R., Chakraborty, B., Tadele, M., Martinez, M.A., Martinez-Picado, J., Marotta, M.L., Mirza, M., Ruiz, L., Clotet, B., Wrin, T., Petropoulos, C.J., Quinones-Mateu, M.E., 2003, A novel TaqMan real-time PCR assay to estimate ex vivo human immunodeficiency virus type 1 fitness in the era of multi-target (pol and env) antiretroviral therapy. *J. Gen. Virol.* **84**:2217-2228

Weinstein, M.C., Goldie, S.J., Losina, E., Cohen, C.J., Baxter, J.D., Zhang, H., Kimmel, A.D., and Freedberg, K.A., 2001, Use of genotypic resistance testing to guide hiv therapy: clinical impact and cost-effectiveness. *Ann. Intern Med.* **134**:440-450

Wensing, A.M.J, van de Vijver, D.A., Angarano, G., Åsjö, B., Balotta, C., Boeri, E., Camacho, R., Chaix, M.-L., Costagliola, D., De Luca, A., Derdelinckx, I., Grossman, Z., Hamouda, O., Hatzakis, A., Hemmer, R., Hoepelman, A., Horban, A., Korn, K., Kücherer, C., Leitner, T., Loveday, C., MacRae, E., Maljkovic, I., de Mendoza, C., Meyer, L., Nielsen, C., Op de Coul, E.L., Ormaasen, V., Paraskevis, D., Perrin, L., Puchhammer-Stöckl, E., Ruiz, L., Salminen, M., Schmit, J.-C., Schneider, F., Schuurman, R., Soriano, V., Stanczak, G., Stanojevic, M., Vandamme, A.-M., Van Laethem, K., Violin, M., Wilbe, K., Yerly, S., Zazzi, M., and Boucher, C.A. on behalf of the SPREAD Programme, 2005, Prevalence of drug-resistant HIV-1 variants in untreated individuals in Europe: implications for clinical management. *J. Infect. Dis.* **192**:958-966

Wiegers, K., Rutter, G., Kottler, H., Tessmer, U., Hohenberg, H., and Krausslich, H.G., 1998, Sequential steps in human immunodeficiency virus particle maturation revealed by alterations of individual Gag polyprotein cleavage sites. *J. Virol.* **72**:2846-2854

Wolf, K., Walter, H., Beerenwinkel, N., Keulen, W., Kaiser, R., Hoffmann, D., Lengauer, T., Selbig, J., Vandamme, A.M., Korn, K., and Schmidt, B., 2003, Tenofovir resistance and resensitization. *Antimicrob Agents Chemother.* **47**:3478-3484

Zhang, Y.M., Imamichi, H., Imamichi, T., Lane, H.C., Falloon, J., Vasudevachari, M.B., and Salzman, N.P., 1997, Drug resistance during indinavir therapy is caused by mutations in the protease gene and in its Gag substrate cleavage sites. *J. Virol.* **71**:6662-6670

Ziermann, R., Limoli, K., Das, K., Arnold, E., Petropoulos, C.J., and Parkin, N.T., 2000, A mutation in HIV type 1 protease, N88S, that causes in vitro hypersensitivity to amprenavir. *J. Virol.* **74**:4414-4419

2. Concepts of therapy for DNA viruses

Chapter 2.1

SELECTIVE INHIBITORS OF THE REPLICATION OF POXVIRUSES

J. NEYTS[1] and E. DE CLERCQ[1]

[1]*Rega Institute for Medical Research, University of Leuven, Minderbroedersstraat 10, 3000 Leuven, Belgium*

Abstract: In case of an inadvertent epidemic with the smallpox virus that could not immediately (or not only) be controlled by vaccination, it will be of utmost importance to have effective antiviral drugs at hand for treatment and/or short-term prophylaxis. Most advanced as a therapeutic or early prophylactic modality is cidofovir, a compound that has been licensed (as Vistide®) for the intravenous treatment of CMV retinitis in AIDS patients. Cidofovir is active in vitro against all (ortho)poxviruses studied so far; is effective in various relevant animal models, including cynomolgus monkeys infected with the monkeypox virus, and has shown efficacy in the clinical setting for the treatment of infections with molluscum contagiosum and orf. Cidofovir should also allow to treat severe complications of vaccination, as may occur in immunodeficient patients. Recently a selective non-nucleoside inhibitor of orthopoxvirus replication (ST-246) was reported that targets a major envelope protein that is involved in viral maturation and the production of infectious extracellular virus. ST-246 exhibits potent activity in mouse models of poxvirus infection. Also, the Abl-family tyrosine kinases and the ErbB-1 kinase were shown to play a crucial role in the replication cycle of poxviruses. Hence inhibitors of these kinases, turned out to be selective inhibitors of poxvirus replication, both in cell culture and in experimentally infected animals.

1.   INTRODUCTION

Variola virus, the causative agent of smallpox, is highly transmissible by the aerosol route from infected to susceptible persons (Mahy *et al.*, 2003). The last case of naturally occurring smallpox was in 1977 in Somalia; the last laboratory infection occurred in 1978 in Birmingham (UK). The World

E. Bogner and A. Holzenburg (eds.), New Concepts of Antiviral Therapy, 283–307.

Health Organization (WHO) announced the global eradication of smallpox in 1980. The discontinuation of vaccination against smallpox has rendered most humans vulnerable to the virus. Virtually all children and many adults, are now fully susceptible to smallpox. Human infection with monkeypox occurs sporadically in parts of Western and Central Africa. In 1996 and 1997 an important outbreak of monkeypox occurred in humans in the Democratic Republic of Congo (Fleischauer *et al.*, 2005; Heymann *et al.*, 1998; Hutin *et al.*, 2001; Learned *et al.*, 2005). More recently there has been an introduction of monkeypox in the USA with human cases (Sejvar *et al.*, 2004). It is thus important to have selective inhibitors of poxviruses at hand that can be used in the wake of a possible bioterrorist attack with variola virus, for the treatment of monkeypox in regions where the virus is endemic or in case of an epidemic with monkeypox in other locations. Such antivirals are also needed for the treatment of complications that are associated with the use of the existing live smallpox vaccine. Furthermore, selective inhibitors of the replication of poxviruses may also be of interest for the treatment of molluscum contagiosum and orf, particularly in immunosuppressed patients. It has been suggested that an ideal poxvirus compound should (i) be active against vaccinia, monkeypox and variola virus; (ii) be active orally for ease of administration; (iii) have a long intracellular half-life so that dosing could be infrequent; (iv) be stable for long periods under adverse storage conditions; (v) be inexpensive, so that large amounts could be stockpiled, and (vi) have a tolerable safety profile also for children and immunocompromised individuals (Kern *et al.*, 2003).

2.   TARGETS FOR ANTIVIRAL TREATMENT AGAINST POXVIRUSES

Poxviruses are the largest viruses, having the largest viral genome and encoding for the largest number of specific viral proteins, that could be envisaged as targets for antiviral intervention. Several of such virus-encoded enzymes and factors are packaged in the infectious virion and are directly involved in the synthesis and modification of mRNA (e.g. RNA polymerases and an RNA polymerase-associated protein (RAP94), capping and methylation enzymes (RNA triphosphatase, guanylyltransferase, methyl-transferase, poly A polymerase) (Moss, 2001). Many viral proteins are involved in processes that are required for pox virus replication, such as viral entry, uncoating, viral gene expression, DNA replication, viral trafficking, virion assembly, maturation and release. Likewise, from recent work it has become evident that also certain cellular factors are involved in these processes and that these can be specifically targeted to inhibit the replication of the virus (Reeves *et al.*, 2005; Yang *et al.*, 2005b). Molecules that interfere with the host cell nucleoside/nucleotide metabolism have also been shown to inhibit the replication of poxviruses; these include inhibitors of the inosine monophosphate dehydrogenase (IMP-DH), S-adenosylhomocysteine (SAH) hydrolase, thymidylate synthase, CTP synthetase and OMP-decarboxylates inhibitors.

3. VIRAL TARGETS

3.1 INHIBITORS OF THE VIRAL DNA POLYMERASE

3.1.1 Cidofovir; analogues and prodrugs

In vitro efficacy of cidofovir and related analogues

The antiviral activity spectrum of the acyclic nucleoside phosphonate analogue (*S*)-1-(3-hydroxy-2-phosphonylmethoxypropyl) cytosine (HPMPC, cidofovir; Fig. 1) encompasses all DNA viruses, in particular papillomaviruses, polyomaviruses, adenoviruses, herpesviruses and poxviruses (De Clercq *et al.*, 1987; Naesens *et al.*, 1997). Cidofovir has shown *in vitro* activity against all poxviruses against which the compound has been evaluated, including vaccinia virus, cowpox virus, ectromelia (mousepox) virus, variola, monkeypox virus, camelpox virus, orf (sheep pox) and molluscum contagiosum (De Clercq, 2002). Poxviruses have an obvious tropism for skin and mucosa, for which reason it is important to study the efficacy of cidofovir (and other agents with anti-poxvirus activity) in skin organotypic raft cultures. Organotypic raft cultures of epithelial cells allow the reconstitution of a skin equivalent that can be easily infected with different viruses with cutaneous tropism such as vaccinia and orf. Cidofovir was shown to effectively inhibit poxvirus replication in these organotypic raft cultures (Andrei *et al.*, 2003).

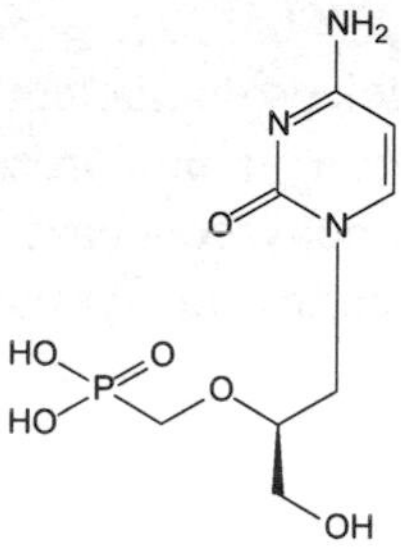

Figure 1: Structure of Cidofovir.

Mechanism of action.

Cidofovir is presumably taken up into cells by fluid phase endocytosis (Connelly *et al.*, 1993). In comparison with nucleoside analogues, cidofovir (and other acyclic nucleoside phosphonate analogues) have the capability to bypass the first phosphorylation step, normally carried out by a cellular or virus-encoded nucleoside kinase, because the molecule carries already a phosphate-mimick. Thus cidofovir needs only two phosphorylations (instead of three phosphorylations for nucleoside analogues) to be converted to the active metabolite. Pyrimidine nucleoside monophosphate (PNMP) kinase converts cidofovir to its monophosphate (HPMPCp), which is then further phosphorylated by nucleoside diphosphate (NDP) kinase, pyruvate kinase, or creatine kinase to cidofovir diphosphate, the antivirally active metabolite (Cihlar and Chen, 1996). Cidofovir is endowed with a long–lasting antiviral activity, both *in vitro*, in experimental animal models and in the clinical setting (De Clercq, 2001). The long-lasting antiviral action of cidofovir can be attributed to the long half-life of its mono- and diphosphate metabolites, but in particular to the cidofovir-phophate-choline metabolite. The latter may serve as the intracellular depot form of HPMPC, since its intracellular half-life is extremely long (48 h) (Ho *et al.*, 1992; Connelly *et al.*, 1993).

The purified vaccinia polymerase was shown to use the 5'-diphosphate of cidofovir as a dCTP mimic and to incorporate the nucleotide analogue into the growing DNA strand (opposite of a G). The next deoxynucleoside monophosphate is still efficiently added to cidofovir-terminated primers but the cidofovir + 1 reaction products are poor substrates for further DNA synthesis, which thus result in a slowing down of the rate of primer extension. Incorporated cidofovir can be excised from the primer 3'-terminus by the 3'-to-5' proofreading exonuclease activity of the polymerase. However, DNA that contains cidofovir at the penultimate 3' position is completely resistant to the exonuclease activity of the enzyme which may thus result in error-prone viral DNA synthesis (Magee *et al.*, 2005).

Resistance to cidofovir

Cidofovir-resistant variants of camelpox, cowpox, monkeypox, and vaccinia viruses (8 to 27-fold reduced susceptibility) were obtained by serial *in vitro* passage of the viruses in the presence of the compound. The cidofovir-resistant cowpox virus DNA polymerase was 8.5-fold less sensitive to inhibition by cidofovir diphosphate than the wild-type polymerase. In intranasally cowpox virus-infected BALB/c mice, however, the cidofovir-resistant virus proved 80-fold less virulent than the wild-type (WT) strain. However, doses of cidofovir that protected mice against infection with a WT cowpox virus, did not suffice to protect mice against an infection with the resistant virus (Smee *et al.*, 2002). A cidofovir-resistant vaccinia mutant that was also generated in cell culture was completely attenuated for virulence [at 10^7 PFU per mouse] in normal BALB/c mice and in SCID mice. In mouse cells the virus replicated less efficiently than the WT virus. A WT virus that was passaged in cell culture for 15 passage, in the absence of cidofovir, was, in BALB/c mice, also 100-fold less virulent than WT virus. The authors therefore concluded that the lack of virulence of the resistant virus in mice was partly explained by its reduced ability to replicate in mouse cells and by an attenuation occurring as a result of extensive cell culturing (Smee *et al.*, 2005).

Efficacy in small animal models for poxvirus infections

In 1993, we were the first to report on the successful use of cidofovir in the prevention and therapy of a lethal vaccinia virus infection in immunosuppressed (SCID) mice (Neyts and De Clercq, 1993). These findings were then corroborated by others (Smee *et al.*, 2001a,b). Even when given as a single dose of 100 mg/kg at 7 days before vaccinia virus infection, cidofovir was able to delay virus-induced mortality by 6 days, and a single dose of 100 mg/kg of cidofovir on the day before the infection delayed virus-induced mortality by about 20 days. Moreover, even if the start of treatment with cidofovir (25 mg/kg per day, for 5 consecutive days) was delayed until day 6 post-infection, virus mortality was markedly delayed (Neyts and De Clercq, 1993). Cidofovir, when given subcutaneously as one dosis of 100 mg/kg on day 0, 2 or 4 after infection, was found to be able to protect mice (90–100% survival) that had been exposed to the cowpox virus by aerosol or by the intranasal route (Bray *et al.*, 2000). When cidofovir was given as an aerosol (0.5–5 mg/kg) to mice that had been infected via aerosol with cowpox, the compound was always more effective than 25 mg/kg of the compound given subcutaneously (Bray *et al.*, 2002). Even if administered as a single intranasal dose (at 10, 20 or 40 mg/kg) at 24 h after intranasal challenge with cowpox virus, cidofovir conferred up to 100% protection against mortality (Smee *et al.*, 2000).

Accidental infection with vaccinia can occur in immunocompetent patients, but complications of vaccination are most often seen either in immunodeficient patients or in patients with eczema or other forms of atopic dermatitis (Engler *et al.*, 2002). To study whether systemically administrated cidofovir would also be effective for the treatment of disseminated vaccinia, a mouse model was elaborated for this purpose (Neyts *et al.*, 2004). Athymic-nude (nu/nu) mice (which suffer from a severe deficiency in cell-mediated immunity) that have been inoculated intracutaneously with vaccinia virus, develop typical vaccinia lesions at the site of virus ino- culation. By two weeks p.i., the infection disseminates to other parts of the skin. Systemically administered cidofovir was found to cause complete, or nearly complete, healing of disseminated vaccinia lesions in this model (Neyts *et al.*, 2004).

Even infrequent dosing of cidofovir (1, 2, or 3 times/week) resulted in an improvement or healing of the lesions and markedly delayed virus-induced mortality. Following cessation of therapy, however, the virus recurred, not a surprising observation in a severely immunocompromised host. Also topical treatment with cidofovir, was able to completely prevent the animals against virus-induced morbidity and mortality, provided, however, that treatment was initiated within the first 2 to 3 days post-infection, thus before systemic spread of the virus (Neyts *et al.*, 2004).

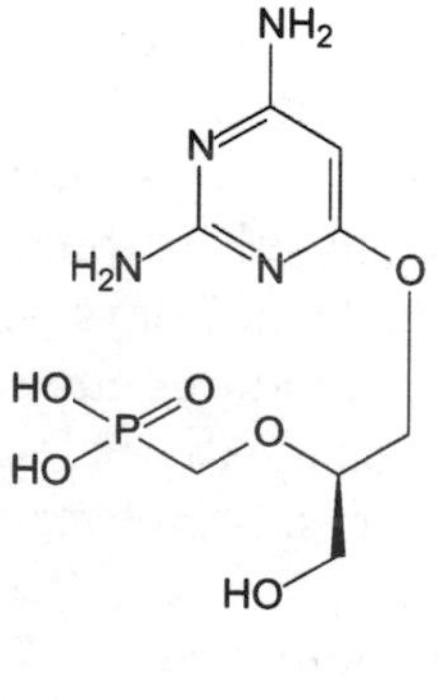

Figure 2: Structure of (S)-HPMPO-DAPy.

The cidofovir analogue HPMPO-DAPy (Fig. 2) proved equipotent to cidofovir *in vitro* and in the mouse vaccinia pox tail lesion model (Leyssen *et al.*, 2005). Also in the cutaneous vaccinia model in athymic-nude mice,

HPMPO-DAPy proved highly effective, even when treatment was started as late as day 15 post infection (Leyssen *et al.*, 2005a).

Lipid conjugate prodrugs of cidofovir

Cidofovir penetrates only poorly and slowly into cells and is virtually not taken up by the oral route (Cundy *et al.*, 1999). In attempts to circumvent these problems, several prodrugs of cidofovir have been synthesized such as its cyclic form (cHPMPC; Fig. 3) (Bischofberger *et al.*, 1994) and a series of lipid conjugates.

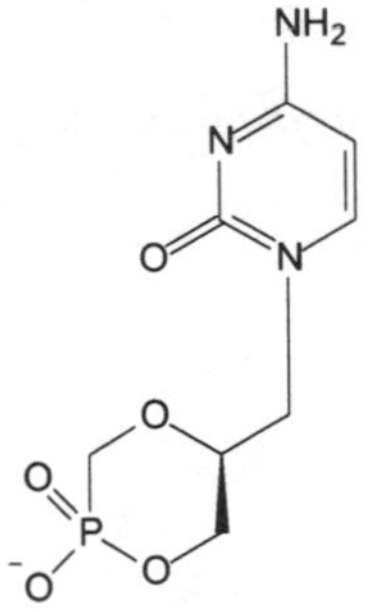

Figure 3: Structure of cHPMPC.

The latter were synthesized by covalently coupling cidofovir to an alkoxyalkanol such as hexadecylpropanediol (HDP-CDV; Fig. 4) or octadecylethanediol (ODE-CDV; Fig. 4) to form an ether lipid CDV conjugate (Hostetler *et al.*, 1997; Kern *et al.*, 2002; Beadle *et al.*, 2002). The rationale behind the synthesis of such conjugates is that the conjugate will mimic lysophosphatidylcholine (LPC) and thus use the natural LPC uptake pathway in the small intestine. Such conjugates should thus be readily absorbed intact from the small intestine and distributed to tissues via plasma and/or lymph thus achieving high oral availability (Painter and Hostetler, 2004). The concentrations of drug achieved in the liver, spleen and lungs of mice were indeed considerably higher after the oral administration of HDP-CDV or ODE-CDV than after the administration of CDV. The lipid conjugates of cidofovir accumulate less efficiently than cidofovir in the proximal tubules of the kidney which may also be important to avoid the nephrotoxicity that is associated with the use of cidofovir. Linking cidofovir onto a lipid tail such as HDP enhanced the *in vitro* potency against several poxviruses (including variola virus) by one to three orders of magnitude (Kern *et al.*, 2002, Quenelle *et al.*, 2004, Keith *et al.*, 2004). HDP-CDV also showed markedly increased oral bioavailability (93% as compared to 0.6%

for cidofovir itself) and provided 100% protection against aerosolized cowpox virus infection in mice when administered orally at 5, 10 or 20 mg/kg, once daily for 5 days (Winegarden *et al.*, 2002).

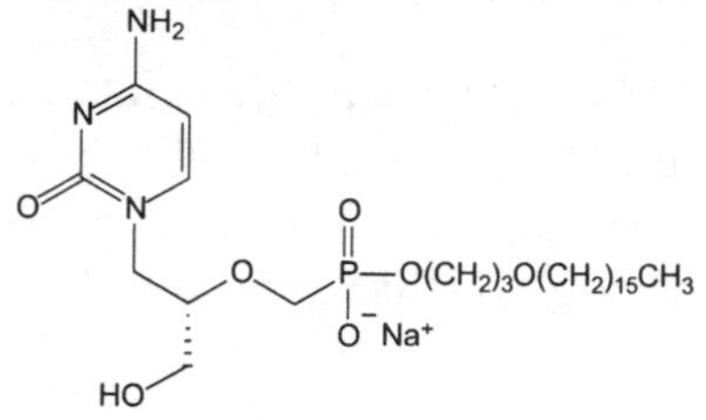

Hexadecyloxypropyl cidofovir (HDP-CDV)

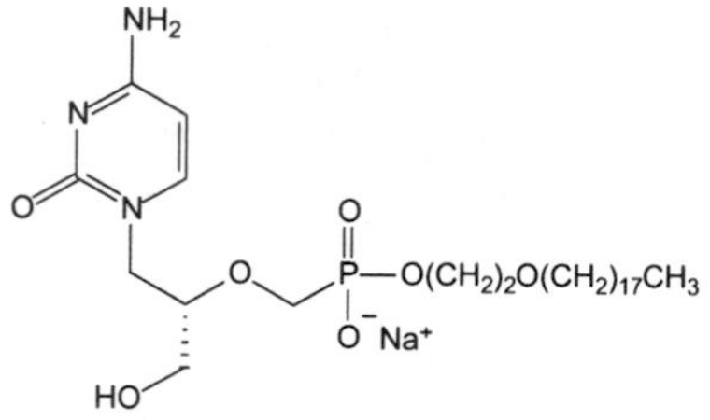

Octadecyloxyethyl cidofovir (ODE-CDV)

Figure 4: Structures of alkoxyalkanols HDP-CDV and ODE-CDV.

The efficacy of these cidofovir analogues was also evaluated in mice that had been infected intranasally with ectromelia (mouse pox) or vaccinia. HDP-CDV or ODE-CDV given orally proved as effective as cidofovir given parenterally. It should be noted that however that there was little or no reduction of viral titer in the lungs (Quenelle *et al.*, 2004).

Efficacy of cidofovir and HPMPO-DAPy in a lethal monkeypox model

The effectiveness of antiviral treatment (with either cidofovir or HPMPO–DAPy) was compared to post exposure smallpox vaccination in a lethal intratracheal monkeypox infection model in cynomolgus monkeys (*Macaca fascicularis*). Monkeypox virus (MPXV) causes in cynomolgus monkeys a disease that resembles in many ways that of human smallpox. Antiviral treatment initiated 24 hours after lethal intratracheal MPXV infection, with either cidofovir or HPMPO-DAPy, using various systemic treatment regimens, resulted in a significantly reduced mortality and reduced numbers of cutaneous monkeypox lesions. This efficacy contrasted with the lack of effect of vaccination (at 24 hr post infection) with a standard human dose of a currently recommended smallpox vaccine (Elstree-RIVM). When antiviral therapy was terminated at day 13 post infection all surviving animals had mounted virus-specific serum antibodies and antiviral T-lymphocytes (Stittelaar *et al.*, 2005).

Clinical use of cidofovir for the treatment of poxvirus infections

Cidofovir has been licensed (as Vistide®) for the intravenous treatment of CMV retinitis in AIDS patients, but it also has therapeutic potential, upon either systemic or topical administration, for the treatment of various other infections; such as with herpesviruses, polyomaviruses, papillomaviruses, adenoviruses and poxviruses, as reviewed previously (De Clercq, 2002, 2003; Snoeck and De Clercq 2002). Cidofovir is available as an aqueous solution of 375 mg/5 ml, intended for intravenous infusion at a dose of (maximally) 5 mg/kg [in the treatment of CMV retinitis, once weekly for the first 2 weeks, and thereafter once every other week]. In humans, cidofovir has so far been used only in the treatment of two types of poxvirus infections, namely molluscum contagiosum and orf (ecthyma infectiosum); in molluscum contagiosum, by both the parenteral (intravenous) and local route (Meadows, 1997; Davies *et al.*, 1999; Zabawski and Cockerell, 1999; Ibarra *et al.*, 2000; Toro *et al.*, 2000) and in orf, by topical application (Geerinck *et al.*, 2001). In all these cases, cidofovir proved highly effective in curbing the infection and the therewith associated symptoms; in the case of orf, the ecthyma lesion completely disappeared following topical application. Obviously, this has not been, and could not be, proven for smallpox, as the disease has been officially declared eradicated before cidofovir was discovered.

3.1.2 Nucleoside analogues that presumably target the viral polymerase

Of a series of 2-, 6-, and 8-alkylated adenosine analogues, 8-methyladenosine (Fig. 5) emerged as a potent and selective inhibitor of vaccinia virus. In addition, the compound proved, not unexpectedly, active against a cidofovir-resistant vaccinia virus strain. In NMRI mice that had been inoculated intravenously with vaccinia, the compound significantly reduced the number of pox tail lesions and, when applied topically, also inhibited the development of cutaneous vaccinia lesions following intracutaneous inoculation of athymic-nude mice. (Van Aerschot *et al.*, 1993; Leyssen *et al.*, 2005b).

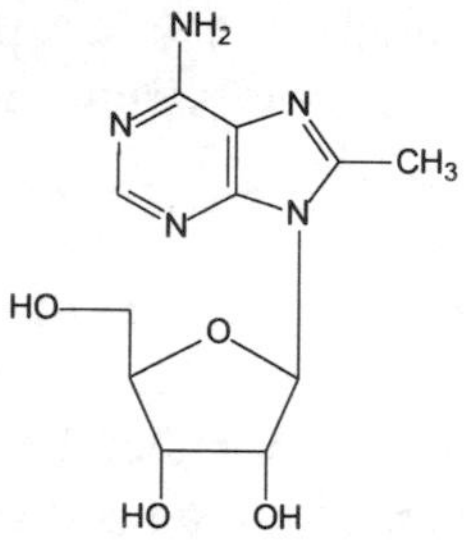

Figure 5: Structure of 8-Methyladenosine.

Of particular interest is the nucleoside analogue 2-amino-7-[(1,3-dihydroxy-2-propoxy)methyl]purine (or S2242; Fig. 6). This compound is a potent and selective inhibitor of virtually all herpesviruses and is an efficient inhibitor of vaccinia virus replication (Neyts and De Clercq, 2001). Although the mode of anti-vaccinia virus activity of S2242 has not been established, it can be surmised that the compound is phosphorylated intracellularly to its triphosphate (Neyts *et al.*, 1998) before it blocks viral DNA synthesis. The diacetate ester of S2242, an oral prodrug form, was shown to be highly protective against vaccinia virus infection in both immunocompetent and SCID mice (Neyts and De Clercq, 2001). A compound that could be given orally would certainly provide an advantage over compounds that must be administered intravenously, particularly when many people would have to be treated in an epidemic situation.

The protective effect of S2242 and its prodrug was confirmed in mice that had been lethally infected, intranasally, with cowpox virus (Smee *et al.*, 2002a).

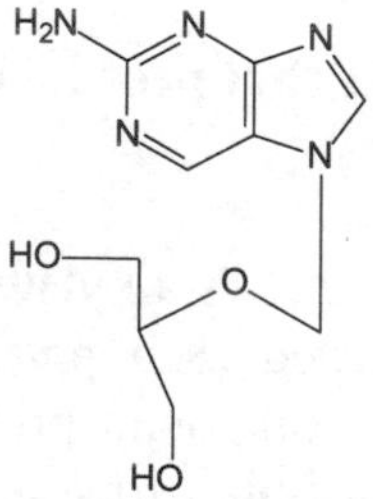

Figure 6: Structure of 2-amino-7-[(1,3-dihydroxy-2-propoxy)methyl]purine (S2242).

A well known inhibitor of herpesvirus replication, i.e., 5'-iodo-2'-deoxyuridine (IDU; Fig. 7) (Herpid®, Stoxil®, Idoxene®, Virudox®,) also inhibits vaccinia virus replication in cell cultures and is able to markedly delay vaccinia virus-induced mortality in SCID mice, even when treatment is postponed until 2 or 4 days after infection (Neyts and De Clercq, 2002). The protective activity of IDU in a non-lethal vaccinia pox tail lesion model has already been reported in the mid seventies (De Clercq *et al.*, 1975). Other compounds, such as trifluorothymidine (TFT) and arabinofuranosyl cytosine (Ara-C) that were also reported to cause protection in the vaccinia tail lesion model (De Clercq *et al.*, 1975) did not prove effective in the lethal vaccinia virus infection model in SCID mice (Neyts and De Clercq, 2002).

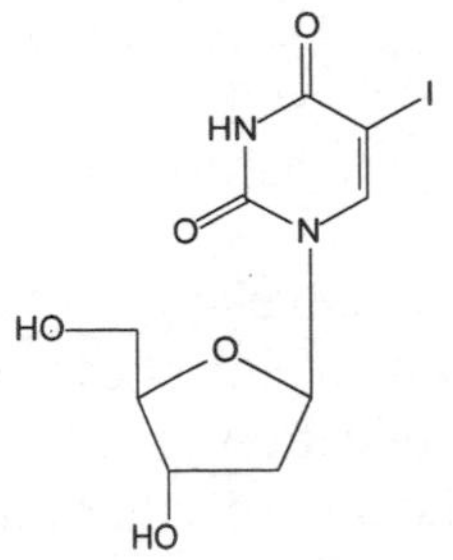

Figure 7: Structure of 5'-iodo-2'-deoxyuridine (IDU).

The use of IDU for the treatment of herpesvirus infections is restricted to topical use because the compound was found to be too toxic for intravenous use (Alford and Whitley, 1976); hence systemic use of IDU for the treatment of poxvirus infections would not be advisable. Although the mechanism of antiviral activity of 8-methyladenosine, S2242 and IDU has not been elucidated, it may be reasonable to assume that the 5'-triphosphate metabolite of these compounds inhibits the viral DNA polymerase.

3.2 Viral Maturation, the F13L gene as an antiviral target

ST-246 or 4-trifluoromethyl-N-(3,3a,4,4a,5,5a,6,6a-octahydro-1,3-dioxo-4,6-ethenocycloprop[f]isoindol-2(1H)-yl)-benzamide was recently reported to be a selective inhibitor of the replication of orthopoxviruses including the vaccinia, monkeypox, camelpox, cowpox, ectromelia, and variola viruses (Yang *et al.*, 2005a; Fig. 8). The compound is equally effective against wild-type cowpox virus and a cidofovir-resistant virus variant thereof. ST-246

294   *J. Neyts and E. De Clercq*

targets the cowpox virus V061 gene, which is homologous to the vaccinia virus F13L gene (encoding for the p37).

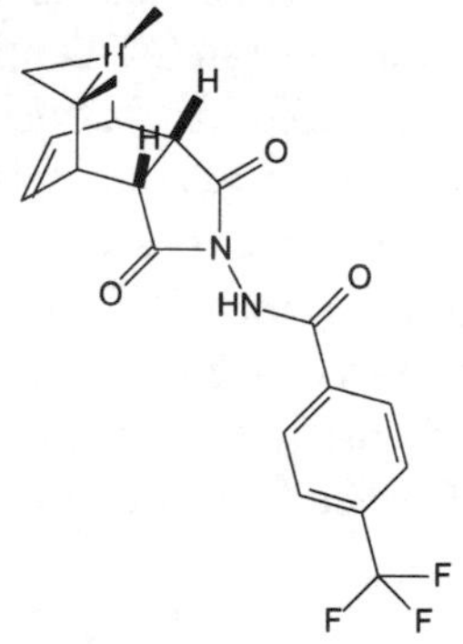

Figure 8: Structure of ST-246.

The latter encodes a major envelope protein that is involved in viral maturation and the production of infectious extracellular virus particles. Hence, ST-246 reduces the production of extracellular virus, whereas it has little or no effect on the production of intracellular virus. The single amino acid change detected in the resistant virus was reintroduced in the WT virus and resulted again in a virus with a ST-246-resistant phenotype. In mice, oral administration of the compound (at a dose of 50 mg/kg) was well tolerated an resulted in a Cmax that was 4000-fold higher than the *in vitro* 50% effective concentration. This was corroborated by the pronounced activity of the compound in various mouse models for poxvirus infections. ST-246 protected A/Ncr mice against a lethal intranasal challenge with as much as 40.000 x LD_{50} of ectromelia; titers of infectious virus in various organs of these mice were below the limit of detection at day 8 post infection. ST-246, when given orally, also efficiently reduced vaccinia virus-induced pox tail lesions in NMRI mice (Yang *et al.*, 2005a).

3.3   Thiosemicarbazones

The thiosemicarbazones (tuberculostatic agents), were found almost 60 years ago to be active against vaccinia virus in cell culture and in vaccinia virus-infected mice (Domagk *et al.*, 1946; Bauer, 1955). Later, the thiosemicarbazone derivative methisazone (Marboran, N-methylisatin 3-thiosemicarbozone; Fig. 9) (Bauer *et al*, 1963) was shown to be effective in the prophylaxis of smallpox and also proved effective in the treatment of complications of smallpox vaccination, i.e. vaccinia gangrenosa and eczema vaccinatum (Fenner and White, 1970).

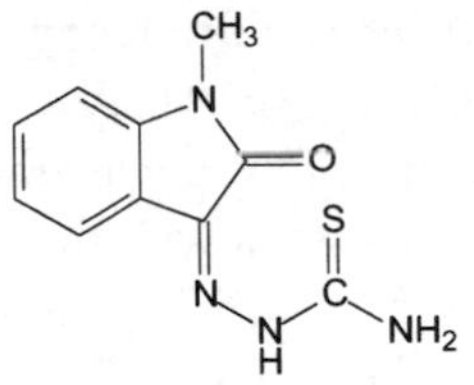

Figure 9: Structure of N-methylisatin 3-thiosemicarbazone.

In a double-blind field trial, however, others were unable to confirm the prophylactic activity of methisazone against smallpox (Heiner *et al.*, 1971). Following successful implementation of the smallpox vaccine, the use of methisazone was not further pursued. Almost 35 years later, a combinatorial approach was employed to generate variants of the isatin-beta-thiosemicarbazone scaffold. N-aminomethyl-isatin-beta-thiosemicarbazones (Fig. 10) with a markedly increased *in vitro* antiviral activity against vaccinia virus and coxpox virus were discovered. The mechanism of antiviral activity of this class of compounds has not been unraveled, but it can be assumed that the resistance profile is likely to be different from that of other classes of anti-pox virus agents (Pirrung *et al.*, 2005).

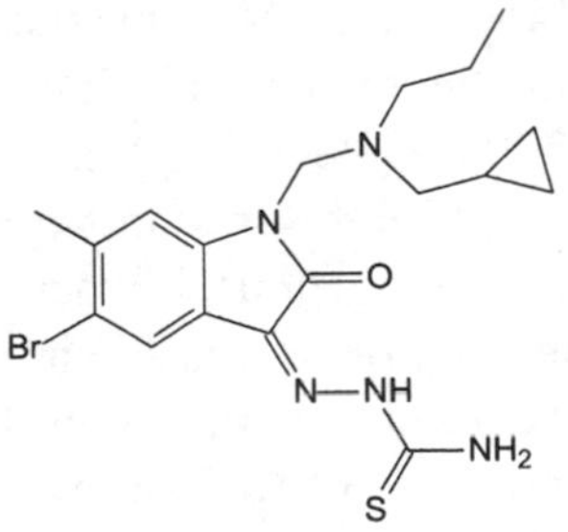

Figure 10: Example of a N-aminomethyl-isatin-beta-thiosemicarbazone.

4. HOST CELL TARGETS

4.1 Abl-family tyrosine kinases

The viral genome is packaged individually in intracellular mature virions (IMVs). Some of these IMV obtain a second membrane and become intracellularly-enveloped virions (IEV). IMVs are released from the cell by cytolysis and are believed to be rapidly recognized and inactivated by the immune system. Before cytolysis, IEV may travel to the periphery of the host cell by means of a kinesin-microtubule transport system (Smith *et al.*, 2003). This particle (IEV), after fusion with the host cell membrane becomes a cell associated enveloped virion (CEV), after having left behind one of the outer envelopes. Unlike IMV, CEV and extracellular-enveloped virions (EEV) evade the immune system and result in spread of the virus (Smith *et al.*, 2002; Smith *et al.*, 2003).

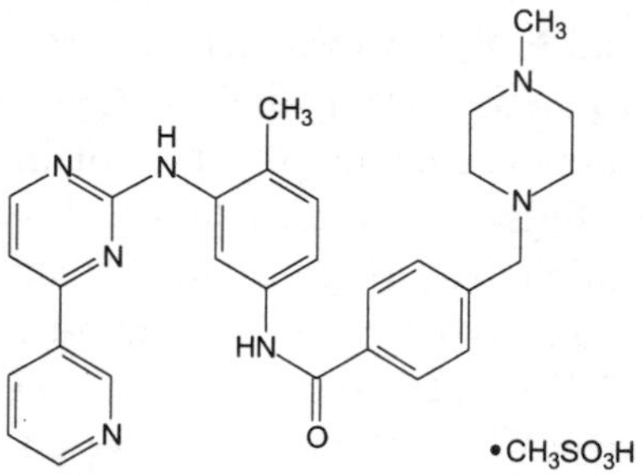

Figure 11: Structure of Gleevec (Imatinib mesylate).

The intracellular mobility of EEV is mediated by Scr- and Abl-family kinases. In addition, also the release of CEV is mediated by Abl-family kinases (Reeves *et al.*, 2005). Gleevec, [2-phenyl pyrimidine (or Imatinib mesylate; Fig. 11) is a potent inhibitor of this tyrosine kinase and is used in the treatment of chronic myelogenous leukaemia] efficiently inhibits this process. C57/B6 mice that had been infected intraperitoneally with the WR strain of vaccinia and that were treated with 100 mg/kg/day of Gleevec survived the infection, whereas 50 to 75% of the untreated mice died. In addition Gleevec resulted in a 4-log reduction in viral load in ovaries of intected animals (Reeves *et al.*, 2005).

4.2   Inhibitors of the ErbB-1 kinase

Small molecule inhibitors of the ErB-1 kinases, in particular CI-1033 (Fig. 12) and related 4-anilinoquinazolines, were shown to exhibit anti-poxvirus activity *in vitro* in Vero and BSC-40 cells that had been infected with the variola strain Solaimen (Yang *et al.*, 2005b) The smallpox virus encodes a growth factor (smallpox growth factor; SPGF) that targets ErbB-1. As a result, this kinase phosphorylates the tyrosine residues of certain cellular factors, which in turn facilitates viral replication.

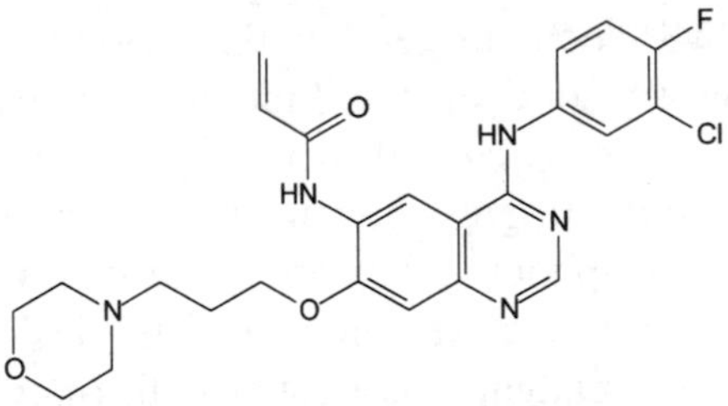

Figure 12: Structure of CI-1033.

As for ST-246 and the Abl-family kinase inhibitors, the compound primarily inhibits the secondary spread of the virus. The efficacy of CI-1033 was monitored in B6 mice that had been inoculated intranasally with a lethal dose of the vaccinia virus WR. Treatment with the compound at a dose of 50 mg/kg resulted, when using a variety of treatment schedules, in a delay or the prevention of virus-induced mortality. In conjunction with a single dose of a monoclonal antibody (targeting IMV), CI-1033 resulted in an almost complete clearance of the virus from the lungs of infected mice by the eighth day after infection (Yang *et al.*, 2005b).

4.3   Inhibitors of nucleoside metabolism

Inhibitors of IMP dehydrogenase

IMP dehydrogenase converts IMP to xanthine 5'-monophospohate (XMP), which is a crucial step in the biosynthesis of the purine mono-nucleotides GMP, GDP, GTP, dGDP and dGTP. Inhibition of IMP dehydrogenase leads to a depletion of GMP, GDP, dGDP, GTP and dGTP pools and, hence, inhibition of both RNA and DNA synthesis.

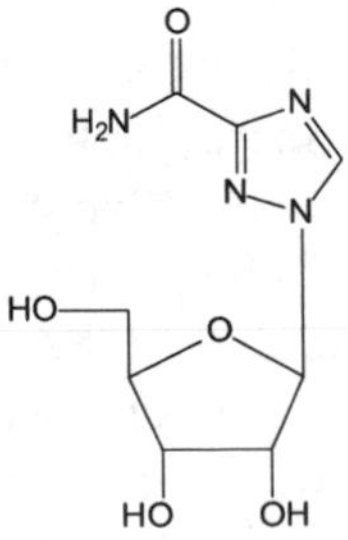

Figure 13: Structure of Ribavirin.

Ribavirin (virazole, 1-beta-ribofuranosyl-1,2,4-triazole-3-carboxamide; Fig. 13) inhibits IMP dehydrogenase through its 5'-monophosphate metabolite. Ribavirin was one of the first compounds shown to inhibit vaccinia virus replication *in vitro* and *in vivo* (Sidwell *et al.*, 1972, 1973). Ribavirin was used to treat progressive vaccinia in a patient with metastatic melanoma and chronic lymphocytic leukemia that was inadvertently given a vaccinia melanoma oncolysate vaccination. The lesion started to heal apparently only when the patient was treated with ribavirin (Kesson *et al.*, 1997). Recently, treatment of cowpox virus respiratory infection in mice with a combination of cidofovir and ribavirin was reported (Smee *et al.*, 2000).

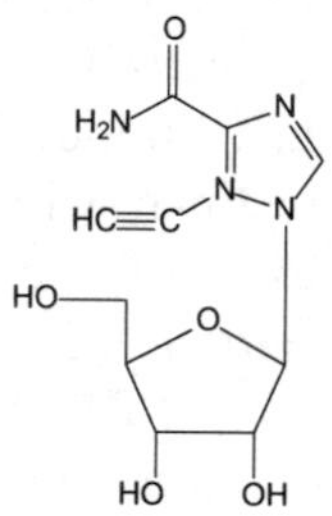

Figure 14: Structure of 5-ethynyl-1-β-D-ribofuranosylimidazole-4-carboxamide (EICAR).

EICAR, (5-ethynyl-1-β-D-ribofuranosylimidazole-4-carboxamide; Fig. 14) a 5-ethynyl derivative of ribavirin is significantly more potent than ribavirin; it inhibits vaccinia virus replication *in vitro* with a 50% inhibitory concentration of 0.2 µg/ml. Importantly, EICAR was found to inhibit vaccinia virus-induced pox tail lesion formation in mice at doses that were not toxic to the host (De Clercq *et al.*, 1991a).

Inhibitors of the S-adenosyl homocysteine hydrolase

S-adenosyl homocysteine (SAH) is a product/inhibitor of the SAM (S-adenosylmethionine)-dependent methyltransferase reactions; and it must thus been removed by SAH hydrolase (that cleaves SAH into homocysteine and adenosine) to allow an efficient methylation. If this hydrolysis is suppressed by SAH hydrolase inhibitors, SAH accumulates and negatively affects the methyltransfer reactions. A wide variety of carbocyclic adenosine analogues that are potent inhibitors of SAH hydrolase have been found to selectively inhibit vaccinia virus replication *in vitro*.

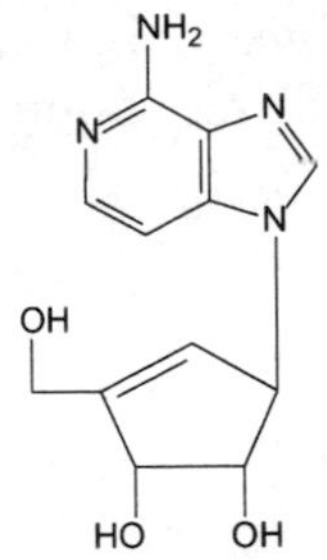

Figure 15: Structure of 3-Deazaneplanocin A.

The replication of vaccinia, and other viruses that are inhibited by SAH hydrolase inhibitors, such as 3-deazaneplanocin A (Fig. 15), strongly depends on methylations for 5'-cap formation. Poxviruses encode for their own methyltransferase (mRNA capping enzyme). SAM-dependent methyltransferases play an important role in the 5'-cap formation and, hence, the maturation of vaccinia mRNA (Borchardt *et al.*, 1984) Viruses, in their replicative cycle, are apparently more sensitive to the action of the SAH hydrolase inhibitors than uninfected non-dividing cells (De Clercq *et al.*, 1990). Vaccinia virus-induced pox tail lesion formation in mice was inhibited by treatment with 3-deazaneplanocin A (Tseng *et al.*, 1989).

OMP decarboxylase and CTP synthetase inhibitors

OMP decarboxylase inhibitors prevent the conversion of OMP to UMP and thus lead to an inhibition of UTP and CTP pools. CTP synthetase inhibitors block the conversion of UTP to CTP and hence deplete CTP pools. As can be deduced from their target of action, inhibitors of both OMP decarboxylase and CTP synthetase should suppress RNA synthesis, which may, in non-dividing cells, result in a distinct antiviral effect. Pyrazofurin, a prototype OMP decarboxylase inhibitor, is a potent inhibitor of vaccinia virus replication in cell culture. (Descamps and De Clercq, 1978). Among the CTP synthetase inhibitors, cyclopentyl cytosine (C-Cyd, carbodine) and cyclopentenyl cytosine (Ce-Cyd) result in stationary (non-dividing) cells in potent *in vitro* anti-vaccinia virus activity (De Clercq *et al.*, 1991b).

Thymidylate synthase inhibitors

Thymidylate synthase converts dUMP to dTMP and inhibition of the enzyme causes depletion of dTTP pools that are required for efficient viral (and cellular) DNA synthesis. Several inhibitors of thymidylate synthase

(TS) were shown to elicit anti-vaccinia virus activity *in vitro* including 5-trifluoromethyl-dUrd, 5-nitro-dUrd, 5-formyl-dUrd, 5-ethynyl-dUrd and 5-amino-dUrd (De Clercq, 1980).

5.   CONCLUSION

The use of highly potent and selective antiviral therapy against smallpox may offer an important alternative to vaccination for short-term prophylaxis against smallpox should the smallpox vaccine not be available or provide insufficient protection (either prophylactically or therapeutically). In cynomolgus monkeys that had been infected with the monkeypox virus, post-exposure antiviral therapy (with the acyclic nucleoside phosphonates), but not post-exposure vaccination, protected the animals against morbidity and mortality (Stittelaar *et al.*, 2005). Antiviral therapy may also be a supplement to vaccination, should the vaccine not completely prevent the viral infection, or should vaccination by itself lead to severe complications (i.e., in immunodeficient patients). A consequence of using an antiviral drug concomitantly with the smallpox vaccine could be a reduced efficacy of the vaccine, as the drug will inhibit the replication of the vaccine virus. Antiviral therapy, in the context for smallpox, may also be important in reducing patient-to-patient transmission.

Cidofovir, a compound that has been licensed (as Vistide®) for the treatment of CMV retinitis in AIDS patients, shows potent activity against infections with various poxviruses in relevant animal models, whether administered intravenously, subcutaneously, topically, or intranasally (aerosolized). Of interest, even if cidofovir is highly effective against experimental DNA virus infections, several of these viruses are in *in vitro* less sensitive to the antiviral activity of cidofovir than the variola virus. There is thus compelling evidence to assume that cidofovir should be efficacious in the therapy and short-term prophylaxis of smallpox and the complications of smallpox vaccination. Cidofovir has already been used, with success, in humans, against molluscum contagiosum and orf, which further underscores the potential of this compound for treatment of infections with poxviruses at large. If huge numbers of individuals would need to be treated (either prophylactically or therapeutically), an oral prodrug form or an aerosolized formulation of an antiviral drug would of course mean a significant improvement. Oral prodrugs of cidofovir have been developed that show promising antiviral activity in various animal models and also aerosolized cidofovir has proved to be effective in such cases.

Until recently most compounds that had been reported to inhibit the *in vitro* (and sometimes also the *in vivo*) replication of poxviruses were either nucleoside or nucleotide analogues that target the viral polymerase or nucleoside analogues that interfere with the synthetic or metabolic pathways of natural nucleoside/nucleotides and, hence, inhibit viral replication by a rather non-specific mechanism. Since the mortality rate, associated with smallpox infection, can be as high as 30 to 40%, the use of such non-selective, and often relatively toxic compounds, may be aimed at reducing viral replication during the most acute phase of the infection. Recently, several novel strategies have been reported to inhibit in a more selective way the replication of poxviruses. These include a molecule (ST-246) that inhibits viral maturation and the production of extracellular virus (Yang *et al.*, 2005a). Also recently inhibitors of the cellular ErbB-1 kinase, (an enzyme that is activated by growth factors encoded by poxviruses), were reported to selectively disrupt viral replication (Yang *et al.*, 2005b). Moreover, the cellular Abl kinase family was shown to be a good target for anti-poxvirus therapy. Abl kinases are needed for efficient egress of the virus from the host cell. Blocking this kinase with Gleevec, a drug used for the treatment of chronic myelogenous leukemia, resulted in a marked antiviral effect. Thousands of compounds that target ErB-1 and Abl-kinases have in recent years been synthesized in anti-cancer programs. Such compounds could be further studied for their potential to inhibit the replication of poxviruses (Fauci and Challberg, 2005). Compounds that target viral factors are also less likely to induce virus-drug resistance. On the other hand, a compound such as cidofovir does not readily generate drug-resistant viruses, and drug-resistant viruses are apparently less virulent than the wild-type virus.

ACKNOWLEDGEMENTS

The authors appreciate the fine editorial help of Mrs. Inge Aerts. Supported by EU grant (Contract no.: 022639) "RiViGene - Genomic inventory, forensic markers, and assessment of potential therapeutic and vaccine targets for viruses relevant in biological crime and terrorism" and NIAID/NIH grant (1UC1AI062540-01) "Novel Inhibitors of Poxvirus Infections".

REFERENCES

Alford, C.A. Jr, and Whitley, R.J., 1976, Treatment of infections due to Herpesvirus in humans: a critical review of the state of the art. *J. Infect. Dis.* **133**: Suppl: A101-108.

Anonymous. From the Centers for Disease Control and Prevention, 1998, Human monkeypox-Kasai Oriental, Democratic Republic of Congo, February 1996–October 1997. *JAMA* **279**: 189-190.

Bauer, D.J., 1955, The antiviral and synergistic actions of isathin thiosemicarbazone and certain phenoxypyrimidines in vaccinia infection in mice. *Br. J. Exp. Pathol.* **36**: 105-114.

Bauer, D.J., St. Vincent, L., Kempe, C.H., and Downie, A.W., 1963, Prophylactic treatment of smallpox contacts with *N*-methylisatin-thiosemicarbazone. *Lancet* **35**: 494-496.

Beadle, J.R., Hartline, C., Aldern, K.A., Rodriguez, N., Harden, E., Kern, E.R., and Hostetler, K.Y., 2002, Alkoxyalkyl esters of cidofovir and cyclic cidofovir exhibit multiple-log enhancement of antiviral activity against cytomegalovirus and herpesvirus replication in vitro. *Antimicrob. Agents Chemother.* **46**: 2381-2386.

Bischofberger, N., Hitchcock, M.J., Chen, M.S., Barkhimer, D.B., Cundy, K.C., Kent, K.M., Lacy, S.A., Lee, W.A., Li, Z.H., Mendel, D.B., Smee, D.F., and Smith, J.L., 1994, 1-[(((S)-2-hydroxy-2-oxo-1,4,2-dioxaphosphorinan-5-yl)methyl] cytosine, an intracellular prodrug for (S)-1-(3-hydroxy-2-phosphonylmethoxypropyl)cytosine with improved therapeutic index in vivo. *Antimicrob. Agents Chemother.* **38**: 2387-2391.

Borchardt, R.T., Keller, B.T., and Patel-Thombre, U., 1984, Neplanocin, A. A potent inhibitor of *S*-adenosylhomocysteine hydrolase and of vaccine virus multiplication in mouse L929 cells. *J. Biol. Chem.* **259**: 4353-4358.

Bray, M., Martinez, M., Kefauver, D., West, M., and Roy, C., 2002, Treatment of aerosolized cowpox virus infection in mice with aerosolized cidofovir. *Antiviral Res.* **54**: 129-142.

Bray, M., Martinez, M., Smee, D.F., Kefauver, D., Thompson, E., and Huggins, J.W., 2000, Cidofovir protects mice against lethal aerosol or intranasal cowpox virus challenge. *J. Infect. Dis.* **181**: 10-19.

Buller, R.M., Owens, G., Schriewer, J., Melman, L., Beadle, J.R., and Hostetler, K.Y., 2004, Efficacy of oral active ether lipid analogs of cidofovir in a lethal mousepox model. *Virology* **318**: 474-481.

Cihlar, T., and Chen, M.S., 1996, Identification of enzymes catalyzing two-step phosphorylation of cidofovir and the effect of cytomegalovirus infection on their activities in host cells. *Mol. Pharmacol.* **50**: 1502-1510.

Connelly, M.C., Robbins, B.L., and Fridland, A., 1993, Mechanism of uptake of the phosphonate analog (S)-1-(3-hydroxy-2-phosphonylmethoxypropyl)cytosine (HPMPC) in Vero cells. *Biochem. Pharmacol.* **46**: 1053-1057.

Cundy, K.C., 1999, Clinical pharmacokinetics of the antiviral nucleotide analogues cidofovir and adefovir. *Clin. Pharmacokinet.* **36**: 127-143.

Davies, E.G., Thrasher, A., Lacey, K., and Harper, J., 1999, Topical cidofovir for severe molluscum contagiosum. *Lancet* **353**: 2042.

De Clercq, E., 1980, Antiviral and antitumor activities of 5-substituted 2'-deoxyuridines. *Methods Find. Exp. Clin. Pharmacol.* **2**: 253-267.

De Clercq, E., 2001, Vaccinia virus inhibitors as paradigm for the chemotherapy of poxvirus infections. *Clin. Microbiol. Rev.* **14**: 382-397.

De Clercq, E., 2002, Cidofovir in the treatment of poxvirus infections. *Antiviral Res.* **55**: 1-13.

De Clercq, E., 2003, Clinical potential of the acyclic nucleoside phosphonates cidofovir, adefovir, and tenofovir in treatment of DNA virus and retrovirus infections. *Clin. Microbiol. Rev.* **16**: 569-596.

De Clercq, E., Bernaerts, R., Shealy, Y.F., and Montgomery, J.A., 1990, Broad-spectrum antiviral activity of carbodine, the carbocyclic analogue of cytidine. *Biochem. Pharmacol.* **39:** 319-325.

De Clercq, E., Cools, M., Balzarini, J., Snoeck, R., Andrei, G., Hosoya, M., Shigeta, S., Ueda, T., Minakawa, N., and Matsuda, A., 1991a, Antiviral activities of 5-ethynyl-1-β-D-ribofuranosylimidazole-4-carboxamide and related compounds. *Antimicrob. Agents Chemother.* **35:** 679-684.

De Clercq, E., Luczak, M., Shugar, D., Torrence, P.F., Waters, J.A., and Witkop, B., 1975, Effect of cytosine arabinoside, iododeoxyuridine, ethyldeoxyuridine, thiocyanato-deoxyuridine, and ribavirin on tail lesion formation in mice infected with vaccinia virus. *Proc. Soc. Exp. Biol. Med.* **151:** 487-490.

De Clercq, E., Murase, J., and Marquez, V.E., 1991b, Broad-spectrum antiviral and cytocidal activity of cyclopentenylcytosine, a carbocyclic nucleoside targeted at CTP synthetase. *Biochem. Pharmacol.* **41:** 1821-1829.

De Clercq, E., Sakuma, T., Baba, M., Pauwels, R., Balzarini, J., Rosenberg, I., and Holý, A., 1987, Antiviral activity of phosphonylmethoxyalkyl derivatives of purine and pyrimidines. *Antiviral Res.* **8:** 261-272.

Descamps, J., and De Clercq, E., 1978, Broad-spectrum antiviral activity of pyrazofurin (pyrazomycin). In *Current Chemotherapy. Proceedings of the 10th International Congress of Chemotherapy, Zürich, Switzerland, September 18-23, 1977* (W. Siegenthaler and R. Lüthy, eds.), American Society for Microbiology, Washington, DC, 354 pp.

Domagk, G., Behnisch, R., Mietzch, F., and Schmidt, H., 1946, Über eineneue, gegen Tuberkelbazillen *in vitro* wirksame Verbindungsklasse. *Naturwissenschaften* **10:** 315.

Engler, R.J., Kenner, J., and Leung, D.Y., 2002, Smallpox vaccination: risk considerations for patients with atopic dermatitis. *J. Allergy Clin. Immunol.* **110:** 357-365.

Enserink, M., 2002, Smallpox vaccines. New cache eases shortage worries. *Science* **296:** 25-27.

Fauci, A.S., and Challberg, M.D., 2005, Host-based antipoxvirus therapeutic strategies: turning the tables. *J Clin Invest.* **115:** 231-233.

Fenner, F.J., and White, D.O., 1970, *Medical Virology* (390 p.). Academic Press, New York, p. 190.

Fleischauer, A.T., Kile, J.C., Davidson, M., Fischer, M., Karem, K.L., Teclaw, R., Messersmith, H., Pontones, P., Beard, B.A., Braden, Z.H., Cono, J., Sejvar, J.J., Khan, A.S., Damon, I., and Kuehnert, M.J., 2005, Evaluation of human-to-human transmission of monkeypox from infected patients to health care workers. *Clin. Infect. Dis.* **40:** 689-694.

Geerinck, K., Lukito, G., Snoeck, R., De Vos, R., De Clercq, E., Vanrenterghem, Y., Degreef, H., and Maes, B., 2001, A case of human orf in an immunocompromised patient treated successfully with cidofovir cream. *J. Med. Virol.* **64:** 543-549.

Heiner, G.G., Fatima, N., Russell, P.K., Haase, A.T., Ahmad, N., Mohammed,N., Thomas, D.B., Mack, T.M., Khan, M.M., Knatterud,G.L., Anthony, R.L., and McCrumb Jr., F.R., 1971, Field trials of methisazoneas a prophylactic agent against smallpox. *Am. J. Epidemiol.* **94:** 435-449.

Heymann, D.L., Szczeniowski, M., and Esteves, K., 1998, Re-emergence of monkeypox in Africa: a review of the past six years. *Br. Med. Bull.* **54:** 693-702.

Ho, H.T., Woods, K.L., Bronson, J.J., De Boeck, H., Martin, J.C., and Hitchcock, M.J., 1992, Intracellular metabolism of the antiherpes agent (S)-1-[3-hydroxy-2-(phosphonyl-methoxy)propyl]cytosine. *Mol. Pharmacol.* **41:** 197-202.

Hostetler, K.Y., Beadle, J.R., Kini, G.D., Gardner, M.F., Wright, K.N., Wu, T.H., and Korba, B.A., 1997, Enhanced oral absorption and antiviral activity of 1-O-octadecyl-sn-glycero-3-phospho-acyclovir and related compounds in hepatitis B virus infection, in vitro. *Biochem. Pharmacol.* **53**: 1815-1822.

Hutin, Y.J., Williams, R.J., Malfait, P., Pebody, R., Loparev, V.N., Ropp, S.L., Rodriguez, M., Knight, J.C., Tshioko, F.K., Khan, A.S., Szczeniowski, M.V., and Esposito, J.J., 2001, Outbreak of human monkeypox, Democratic Republic of Congo, 1996 to 1997. *Emerg. Infect. Dis.* **7**: 434-438.

Ibarra, V., Blanco, J.R., Oteo, J.A., and Rosel, L., 2000, Efficacy of cidofovir in the treatment of recalcitrant molluscum contagiosum in an AIDS patient. *Acta Derm. Venereol.* **80**: 315-316.

Keith, K.A., Wan, W.B., Ciesla, S.L., Beadle, J.R., Hostetler, K.Y., and Kern, E.R., 2004, Inhibitory activity of alkoxyalkyl and alkyl esters of cidofovir and cyclic cidofovir against orthopoxvirus replication *in vitro. Antimicrob. Agents Chemother.* **48**: 1869-1871.

Kern, E.R., 2003, *In vitro* activity of potential anti-poxvirus agents. *Antiviral Res.* **57**: 35-40.

Kern, E.R., Hartline, C., Harden, E., Keith, K., Rodriguez, N., Beadle, J.R., and Hostetler, K.Y., 2002, Enhanced inhibition of orthopoxvirus replication *in vitro* by alkoxyalkyl esters of cidofovir and cyclic cidofovir. *Antimicrob. Agents Chemother.* **46**: 991-995.

Kesson, A.M., Ferguson, J.K., Rawlinson, W.D., and Cunningham, A.L., 1997, Progressive vaccinia treated with ribavirin and vaccinia immune globulin. *Clin. Infect. Dis.* **25**: 911-914.

Learned, L.A., Reynolds, M.G., Wassa, D.W., Li, Y., Olson, V.A., Karem, K., Stempora, L.L., Braden, Z.H., Kline, R., Likos, A., Libama, F., Moudzeo, H., Bolanda, J.D., Tarangonia, P., Boumandoki, P., Formenty, P., Harvey, J.M., and Damon, I.K., 2005, Extended interhuman transmission of monkeypox in a hospital community in the Republic of the Congo, 2003. *Am. J. Trop. Med. Hyg.* **73**: 428-434.

Levy, H.B., and Lvovsky, E., 1978, Topical treatment of vaccinia virus infection with an interferon inducer in rabbits. *J. Infect. Dis.* **137**: 78-81.

Leyssen, P., Andrei, G., Snoeck, R., De Clercq, E., Holý, A., and Neyts, J., 2005a, Submitted for publication, (S)-HPMPO-DAPy is a potent inhibitor of orthopoxvirus replication.

Leyssen, P., Vliegen, I., De Clercq, E., Van Aerschot, A., and Neyts, J., 2005b, Submitted for publication, Selective inhibition of vaccinia virus replication in vitro and in mouse models.

Magee, W.C., Hostetler, K.Y., and Evans, D.H., 2005, Mechanism of inhibition of vaccinia virus DNA polymerase by cidofovir diphosphate. *Antimicrob. Agents Chemother.* **49**: 3153-3162.

Mahy, B.W., 2003, An overview of the use of a viral pathogen as a bioterrorism agent: why smallpox? *Antiviral Res.* **57**: 1-5.

Meadows, K.P., 1997, Resolution of recalcitrant molluscum contagiosum virus lesions in human immunodeficiency virus-infected patients treated with cidofovir. *Arch. Dermatol.* **133**: 987-990.

Moss, B., 2001, Poxviridae: the viruses and their replication. In *Fields Virology, 4th ed.* (D.M. Knipe and P.P. Howley, eds.), Lippincott Williams & Wilkins, pp. 2849-2883.

Mukinda, V.B., Mwema, G., Kilundu, M., Heymann, D.L., Khan, A.S., and Esposito, J.J., 1997, Re-emergence of human monkeypox in Zaire in 1996. Monkeypox Epidemiologic Working Group. *Lancet* **349**: 1449-1450.

Naesens, L., Snoeck, R., Andrei, G., Balzarini, J., Neyts, J., and De Clercq, E., 1997, HPMPC (cidofovir), PMEA (adefovir) and related acyclic nucleoside phosphonate analogues: a review of their pharmacology and clinical potential in the treatment of viral infections. *Antiviral Chem. Chemother.* **8:** 1-23.

Neyts, J., and De Clercq, E., 1993, Efficacy of (S)-1-(3-hydroxy-2-phosphonyl-methoxypropyl) cytosine for the treatment of lethal vaccinia virus infections in severe combined immune deficiency (SCID) mice. *J. Med. Virol.* **41:** 242-246.

Neyts, J., and De Clercq, E., 2001, Efficacy of 2-amino-7-[(1,3-dihydroxy-2-propoxy) methyl]-purine for the treatment of vaccinia (orthopox-) virus infections in mice. *Antimicrob. Agents Chemother.* **45:** 84-87.

Neyts, J., Balzarini, J., Andrei, G., Chaoyong, Z., Snoeck, R., Zimmerman, A., Mertens, T., Karlsson, A., and De Clercq, E., 1998, Intracellular metabolism of the N7-substituted acyclic nucleoside analog 2-amino-7-(1,3-dihydroxy-2-propoxymethyl)purine, a potent inhibitor of herpesvirus replication. *Mol. Pharmacol.* **53:** 157-165.

Neyts, J., Leyssen, P., Verbeken, E., and De Clercq, E., 2004, Efficacy of cidofovir in a murine model of disseminated progressive vaccinia. *Antimicrob. Agents Chemother.* **48:** 2267-2273.

Neyts, J., Verbeken, E., and De Clercq, E., 2002, Effect of 5-iodo-2'-deoxyuridine on vaccinia virus (orthopoxvirus) infections in mice. *Antimicrob. Agents Chemother.* **46:** 2842-2847.

Painter, G.R., and Hostetler, K.Y., 2004, Design and development of oral drugs for the prophylaxis and treatment of smallpox infection. *Trends Biotechnol.* **22:** 423-427.

Pirrung, M.C., Pansare, S.V., Sarma, K.D., Keith, K.A., and Kern, E.R., 2005, Combinatorial optimization of isatin-beta-thiosemicarbazones as anti-poxvirusagents. *J. Med. Chem.* **48:** 3045-3050.

Quenelle, D.C., Collins, D.J., Wan, W.B., Beadle, J.R., Hostetler, K.Y., and Kern, E.R., 2004, Oral treatment of cowpox and vaccinia virus infections inmice with ether lipid esters of cidofovir. *Antimicrob. Agents Chemother.* **48:** 404-412.

Redfield, R.R., Wright, D.C., James, W.D., Jones, T.S., Brown, C., and Burke, D.S., 1987, Disseminated vaccinia in a military recruit with human immunodeficiency virus (HIV) disease. *N. Engl. J. Med.* **316:** 673-676.

Reeves, P.M., Bommarius, B., Lebeis, S., McNulty, S., Christensen, J., Swimm, A., Chahroudi, A., Chavan, R., Feinberg, M.B., Veach, D., Bornmann, W., Sherman, M., and Kalman, D., 2005, Disabling poxvirus pathogenesis by inhibition of Abl-family tyrosine kinases. *Nature Med.* **11:** 731-739.

Sejvar, J.J., Chowdary, Y., Schomogyi, M., Stevens, J., Patel, J., Karem, K., Fischer, M., Kuehnert, M.J., Zaki, S.R., Paddock, C.D., Guarner, J., Shieh, W.J., Patton, J.L., Bernard, N., Li, Y., Olson, V.A., Kline, R.L., Loparev, V.N., Schmid, D.S., Beard, B., Regnery, R.R., and Damon, I.K., 2004, Human monkeypox infection: a family cluster in the midwestern United States. *J. Infect. Dis.* **190:** 1833-1840.

Sidwell, R.W., Allen, L.B., Khare, G.P., Huffman, J.H., Witkowski, J.T., Simon, L.N., and Robins, R.K., 1973, Effect of 1-beta-D-ribofuranosyl- 1,2,4-triazole-3-carboxamide (Virazole, ICN 1229) on herpes and vaccinia keratitis and encephalitis in laboratory animals. *Antimicrob. Agents Chemother.* **3:** 242-246.

Sidwell, R.W., Huffman, J.H., Khare, G.P., Allen, L.B., Witkowski, J.T., and Robins, R.K., 1972, Broad-spectrum antiviral activity of virazole: 1-beta-d-ribofuranosyl-1,2,4-triazole-3-carboxamide. *Science* **177:** 705-706.

Smee, D. F., Bailey, K. W., and Sidwell, R.W., 2000, Treatment of cowpox virus respiratory infections in mice with ribavirin as a single agent or followed sequentially by cidofovir. *Antiviral Chem. Chemother.* **11:** 303-309.

Smee, D.F., Bailey, K.W., and Sidwell, R.W., 2001b, Treatment of lethal vaccinia virus respiratory infections in mice with cidofovir. *Antiviral Chem. Chemother.* **12:** 71-76.

Smee, D.F., Bailey, K.W., and Sidwell, R.W., 2002a, Treatment of lethal cowpox virus respiratory infections in mice with 2-amino-7-[(1,3-dihydroxy-2-propoxy)methyl]purine and its orally active diacetate ester prodrug. *Antiviral Res.* **54:** 113-120.

Smee, D.F., Bailey, K.W., Wong, M., and Sidwell, R.W., 2000, Intranasal treatment of cowpox virus respiratory infections in mice with cidofovir. *Antiviral Res.* **47:** 171-177.

Smee, D.F., Bailey, K.W., Wong, M.H., and Sidwell, R.W., 2001a, Effects of cidofovir on the pathogenesis of a lethal vaccinia virus respiratory infection in mice. *Antiviral Res.* **52:** 55-62.

Smee, D.F., Sidwell, R.W., Kefauver, D., Bray, M., and Huggins, J.W., 2002, Characterization of wild-type and cidofovir-resistant strains of camelpox, cowpox, monkeypox, and vaccinia viruses. *Antimicrob. Agents Chemother.* **46:** 1329-1335.

Smee, D.F., Wandersee, M.K., Bailey, K.W., Hostetler, K.Y., Holy, A., and Sidwell, R.W., 2005, Characterization and treatment of cidofovir-resistant vaccinia (WR strain) virus infections in cell culture and in mice. *Antiviral Chem. Chemother.* **16:** 203-211.

Smith, G.L., Murphy, B.J., and Law, M., 2003, Vaccinia virus motility. *Ann. Rev. Microbiol.* **57:** 323-342.

Smith, G.L., Vanderplasschen, A., and Law, M., 2002, The formation and function of extracellular enveloped vaccinia virus. *J. Gen. Virol.* **83:** 2915-2931

Snoeck, R., and De Clercq, E., 2002, Role of cidofovir in the treatment of DNA virus infections, other than CMVinfections, in immunocompromised patients. *Curr. Opin. Investig. Drugs* **3:** 1561-1566.

Snoeck, R., Holy, A., Dewolf-Peeters, C., Van Den Oord, J., De Clercq, E., and Andrei, G., 2002, Antivaccinia activities of acyclic nucleoside phosphonate derivatives in epithelial cells and organotypic cultures. *Antimicrob. Agents Chemother.* **46:** 3356-3361.

Stittelaar, K.J., Neyts, J., Naesens, L., van Amerongen, G., van Lavieren, R.F., Holý, A., De Clercq, E., Niesters, B.G.M., Fries, E., Maas, C., Mulder, P.G.H., van der Zeijst, B.A., and Osterhaus, A.D.M.E., 2005, Antiviral treatment with acyclic nucleotides is more effective than smallpox vaccination upon lethal monkeypox virus infection. *Nature*, in press.

Toro, J.R., Wood, L.V., Patel, N.K., and Turner, M.L., 2000, Topical cidofovir. A novel treatment for recalcitrant molluscum contagiosum in children infected with human immunodeficiency virus 1. *Arch. Dermatol.* **136:** 983-985.

Tseng, C.K.H., Marquez, V.E., Fuller, R.W., Goldstein, B.M., Haines,D.R., McPherson, H., Parsons, J.L., Shannon, W.M., Arnett, G.,Hollingshead, M., and Driscoll, J.S., 1989, Synthesis of 3-deazaneplanocinA, a powerful inhibitor of *S*-adenosylhomocysteine hydrolase withpotent and selective *in vitro* and *in vivo* antiviral activities. *J. Med. Chem.* **32:** 1442-1446.

Van Aerschot, A., Mamos, P., Weyns, N.J., Ikeda, S., De Clercq, E., and Herdewijn, P., 1993, Antiviral activity of C-alkylated purine nucleosides obtained by cross-coupling with tetraalkyltin reagents. *J. Med. Chem.* **36:** 2938-2942.

Winegarden, K.L., Ciesla, S.L., Aldern, K.A., Beadle, J.R., and Hostetler, K.Y., 2002, Oral pharmacokinetics and preliminary toxicology of 1-*O*-hexadecyloxypropyl-cidofovir in mice. Abstracts of the 15th International Conference on Antiviral Research, Prague, Czech Republic, March 17-21, 2002. *Antiviral Res.* **53(105):** A67.

Xiong, X., Smith, J.L., and Chen, M.S., 1997, Effect of incorporation of cidofovir into DNA by human cytomegalovirus DNA polymerase on DNA elongation. *Antimicrob. Agents Chemother.* **41:** 594–599.

Yang, G., Pevear, D.C., Davies, M.H., Collett, M.S., Bailey, T., Rippen, S., Barone, L., Burns, C., Rhodes, G., Tohan, S., Huggins, J.W., Baker, R.O., Buller, R.L., Touchette, E., Waller, K., Schriewer, J., Neyts, J., De Clercq, E., Jones, K., Hruby, D., and Jordan, R., 2005a, An orally bioavailable antipoxvirus compound (ST-246) inhibits extracellular virus formation and protects mice from lethal orthopoxvirus challenge. *J. Virol.* **79:** 13139-13149.

Yang H, Kim SK, Kim M, Reche PA, Morehead TJ, Damon IK, Welsh RM, Reinherz EL., 2005b, Antiviral chemotherapy facilitates control of poxvirus infections through inhibition of cellular signal transduction. *J. Clin. Invest.* **115:** 379-387.

Zabawski, E.J., and Cockerell, C.J., 1999, Topical cidofovir for molluscum contagiosum in children. *Pediatr. Dermatol.* **16:** 414-415.

Chapter 2.2

MARIBAVIR: A PROMISING NEW ANTIHERPES THERAPEUTIC AGENT

K.K. BIRON[1]

[1]*Department of Clinical Virology, GlaxoSmithKline, Research Triangle Park, NC 27709 USA.*

Abstract: Maribavir (1263W94) is a potent and selective, orally bioavailable antiviral drug with a novel mechanism of action against CMV This benzimidazole riboside (1-(ß-L-ribofuranosyl)-2-isopropylamino-5,6-dichlorobenzimidazole) is 3- to 20-fold more potent than ganciclovir or cidofovir, and has not exhibited cross resistance to these drugs. Maribavir inhibits CMV replication by interfering with CMV DNA synthesis and disrupting viral capsid egress. Maribavir has demonstrated positive *in vivo* anti-CMV activity in semen and urine in HIV-infected subjects with asymptomatic CMV shedding treated for 28 days. It is currently being developed for CMV infections in transplant and HIV-infected subjects. The preclinical, pharmacokinetic, and clinical development of maribavir is reviewed.

1.   INTRODUCTION

Maribavir,   1-(ß-L-ribofuranosyl)-2-isopropylamino-5,6-dichlorobenz-imidazole, also known as BW1263W94, GW-1263, or benzimidavir (Fig. 1) is a potent antiviral agent that selectively inhibits two members of the human herpesvirus family: human cytomegalovirus (hCMV) and Epstein-Barr virus (EBV). Maribavir is the most advanced drug candidate in the benzimidazole nucleoside analog class; a series originally aimed at modifying the broad spectrum transcriptional inhibitor, DRB, into an anti-tumor agent (Townsend and Revankar 1970; Yankulov *et al.* 1995). The discovery of the selective anti-CMV activity of two of these early analogs, TCRB (2,5,6-trichloro-1-(beta-D-ribofuranosyl) benzimidazole), and its 2-bromo derivative (BDCRB) led to extensive chemical synthesis and SAR studies. This effort produced a compound series rich in antiviral potency, and diverse in mechanisms of

E. Bogner and A. Holzenburg (eds.), New Concepts of Antiviral Therapy, 309–336.

action (Townsend *et al.* 1995; Townsend *et al.* 1999; Chan *et al.* 2000; Biron *et al.* 2002; Evers *et al.* 2004). Ultimately two attractive clinical candidates, maribavir and GW275175X (Fig. 1), were progressed into clinical development. Interestingly, these two related analogs show different, yet novel, modes of antiviral action. The pre-clinical and clinical development of maribavir is the focus of this review. This promising therapeutic agent is currently (2005) in Phase II development, and will be progressed for CMV infections, including stem cell and solid organ transplant indications, congenital disease, and CMV disease in HIV-infected individuals (Lu and Thomas 2004).

Maribavir
GW1263W94

GW275175X

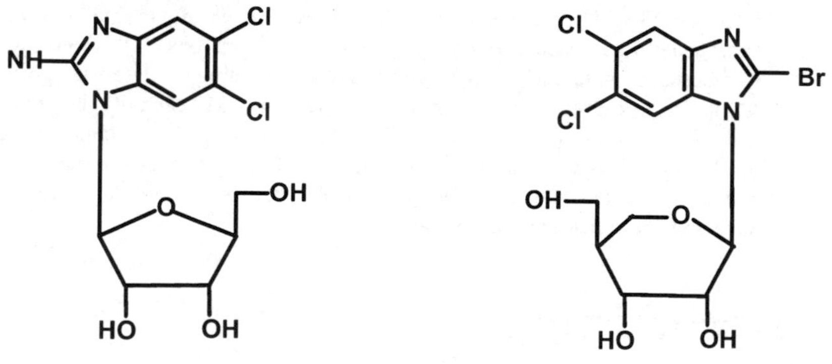

1-(ß-L-ribofuranosyl)-2-isopropylamino-5,6-dichlorobenzimidazole

1-(ß-D-ribopyranosyl)-2-bromo-5,6-dichloro-1*H*-benzimidazole

Figure 1. Structures of Maribavir and GW275175X

2. CMV AND EBV DISEASES: UNMET MEDICAL NEED

CMV and EBV are members of the beta and gamma classes of human herpesvirus, respectively. The herpesviruses share common features of virion structure, genome organization, and their replication cycles (Roizman 1996). Importantly, they have all evolved unique strategies for lifelong persistence in the host, with distinct disease consequences of reactivation

(Roizman 1996). Symptomatic CMV disease occurs almost exclusively in individuals with immature or compromised immune systems such as neonates, organ transplant recipients, and AIDS patients, following either a primary infection, or more commonly, viral reactivation. Cytomegalovirus-associated disease is responsible for a number of syndromes, including acute mononucleosis, retinitis, colitis, esophagitis, pneumonia, hepatitis, and meningoencephalitis (Vancikova and Dvorak 2001). Congenital CMV infection remains a concern worldwide despite the relatively low incidence (estimated 0.15% - 2.4%) because of the se-verity of the long-term sequelae (hearing loss and mental deficits) (Whitley 2004). CMV, as well as other viral and bacterial pathogens, has been implicated in inflammatory processes that subsequently play a role in the development of arteriosclerosis and its associated complications (Valantine 2004). It is noteworthy that the prevalence of CMV disease in HIV-infected patients, especially CMV retinitis, has dramatically decreased in developed countries after the introduction in the mid 90's of highly active antiretroviral therapy (HAART) (Springer and Weinberg 2004).

In the case of the gamma herpesvirus, EBV, primary infection is also subclinical in immunocompetent individuals, especially when acquired early in life. Acquisition of EBV during adolescence can result in infectious mononucleosis. Reactivation of EBV in immunocompromised individuals may manifest clinically as oral hairy leukoplakia (OHL), a lytic infection of epithelial cells of the tongue. EBV infection has been linked to oncogenic diseases: post-transplant lymphoproliferative disease, lymphoma (Burkitt's and AIDS-related), Hodgkin's disease, and nasopharyngeal carcinoma (Pagano 1999; Thorley-Lawson and Gross 2004). Recent epidemiologic studies have implicated EBV serologic and virologic status with numerous autoimmune diseases, including systemic lupus erythematous, (James *et al.* 2001; Chen *et al.* 2005; Parks *et al.* 2005) Sjogren's Syndrome (Perrot *et al.* 2003; Yamazaki *et al.* 2005) and multiple sclerosis (Sundstrom *et al.* 2004; Levin *et al.* 2005; Cepok *et al.* 2005). A causal relationship of EBV infection to these diseases has not been shown (Gross *et al.* 2005). The potential for antiviral treatments in EBV-mediated diseases may best be determined in the context of clinical trials. Several experimental approaches for treatment of EBV-related malignancies that may not respond to conventional antiviral drug treatment are currently being evaluated.

Based on serologic evidence, the prevalence of both CMV and EBV is high worldwide and is universally distributed among human populations. The incidence is somewhat higher and the rate of seroconversion more rapid in developing countries. However, even in developed countries, most people have been exposed to both CMV (>60%) and EBV (>90%) by mid-adulthood.

3.   CURRENT ANTIVIRAL THERAPIES FOR CMV & EBV

The latent reservoirs of the herpesviruses, like those of HIV, remain the ultimate goal of antiviral therapeutic research. These reservoirs are not responsive to the standard antiviral drugs that block active virus replication, which therefore can only modulate disease activity. A broad spectrum anti-herpes agent was partially achieved with the discovery of acyclovir (ACV) (Elion *et al.* 1977). This guanosine nucleoside analog and its L-valine ester prodrug valacyclovir, have demonstrated clinical efficacy against various diseases caused by HSV types 1 and 2, VZV, EBV (OHL) (Resnick *et al.* 1988; Walling *et al.* 2003), and as a prophylactic therapy against reactivated CMV disease in renal transplant patients (Balfour *et al.* 1989; Lowance *et al.* 1999). Although the safety profile of ACV has been excellent overall, potency is lacking for management of active CMV diseases.

The first line therapy for treatment of CMV disease is currently the related nucleoside analog ganciclovir (GCV), and its analogous amino acid ester prodrug, valganciclovir (Reusser 2001). In practice, valganciclovir has provided an advance in controlling CMV disease, especially when used as a prophylactic in high-risk solid organ transplant recipients (Taber *et al.* 2004; Cvetkovic and Wellington 2005; Hodson *et al.* 2005). Adverse side effects include leukopenia, thrombocytopenia, anemia, and bone marrow hypoplasia (Crumpacker 1996; ganciclovir package insert 2000). Alternate therapies include the nucleotide analog, cidofovir (CDV), and the pyrophosphate analog, foscarnet (PFA); however, their utility is limited due to the requirement for intravenous administration and toxicity. PFA treatment is associated with dose-limiting renal impairment (Jacobson 1992; foscarnet injection package insert 2000); while CDV can cause life-threatening nephotoxicity as well as bone marrow depletion (Safrin *et al.* 1999; cidofovir injection package insert, 1996). Formivirsen, a CMV retinitis drug, is the first antiviral antisense compound that works due to its high affinity and specificity for hCMV RNAs. Nevertheless, its usefulness is severely limited by the need for repeated intraocular injections. Clearly new safe, orally bioavailability drugs with novel mechanisms of action are needed for long-term prophylaxis and treatment of established CMV disease in immunocompromised patient populations.

All three of the approved systemic drugs ultimately act at the viral DNA polymerase to block viral DNA synthesis, and consequently, cross-resistance can occur (Erice 1999; Chou *et al.* 2003). With chronic use in immunocompromised patients, CMV resistance to GCV arises in the two genes responsible for the mechanism of action (MOA); the UL97 gene encoding the phosphotransferase or viral protein kinase that mono-phosphorylates GCV, and in the UL54 gene encoding the ultimate target, the viral DNA polymerase (Erice 1999; Chou 1999). CDV and PFA resistance also results from mutations in the CMV polymerase gene, and mutation-specific cross-resistance has been reported. Given the need for long-term prophylactic therapy in many immunocompromised patients, the development of viral resistance is a continuing problem.

There are no drugs with approved indications for the treatment of EBV diseases. Infectious mononucleosis is usually beyond the benefit of a standard replication inhibitor at diagnosis. ACV and valacyclovir have shown some benefit in the treatment of OHL (Resnick *et al.* 1988; Walling *et al.* 2003). Recent progress in understanding the immunology and patho-genesis of reactivating and oncogenic EBV in disease will facilitate the design of therapeutic strategies (Thorley-Lawson 2001; Okano 2003; Gross *et al.* 2005).

4.   SPECTRUM OF ANTIVIRAL ACTIVITY

Maribavir has an unusual and restricted spectrum of antiviral activity as defined thus far. It has demonstrated potent and specific activity against CMV and replicating EBV in various cell assay systems, but not against other human herpes viruses HSV 1 and 2, VZV, HHV 6, 7, and 8, nor against animal CMVs (Williams *et al.* 2003). Other viruses reported to be negative included hepatitis B, HIV-1, HPV and BVDV.

Maribavir inhibited viral replication of numerous laboratory strains and clinical isolates of human CMV in a concentration-dependent manner, with a range of IC_{50} values from 0.06-19.4 μM (McSharry *et al.* 2001b; Biron *et al.* 2002; Williams *et al.* 2003). The reported *in vitro* potency of maribavir varies according to assay methods, cell type and growth state of the cultures. This feature may reflect the non-essential nature of the key targets of maribavir in cell culture, and the complementing/compensating capabilities of the host cells. In general, maribavir's anti-CMV activity was up to ten-fold more potent than that of GCV in side-by-side assays.

The mechanism of action, although only partly understood, is distinct from those of the approved drugs, and thus maribavir is active against those strains resistant to GCV, CDV and PFA. This includes CMV strains with mutations in either or both the DNA pol gene and the UL97 gene (McSharry *et al.* 2001a; Biron *et al.* 2002).

Combination therapy in HIV (HAART) and in bacterial infections (trimethoprim-sulfamethoxazole) can increase efficacy and delay the emergence of resistance; often because the molecular sites of action for the combined agents are distinct. In transplant settings, patients are likely to be treated with multiple antimicrobials, along with immunosuppressive agents. The antiviral effect of maribavir in combination with the approved CMV drugs GCV, CDV and PFA has been found to be additive for GCV, and in the case of PFA and CDV, additive to synergistic (Evers *et al.* 2002; Selleseth *et al.* 2003). Importantly, no *in vitro* mechanism-based antagonism has been reported to date. As expected, maribavir did not interfere with the anti-HIV activity of representatives from the three major classes of approved antiretrovirals, nor did they reduce the anti-CMV action of maribavir *in vitro* (GlaxoSmithKline; data on file).

Animal models of disease pathogenicity can be useful in demonstrating antiviral efficacy and correlating outcome with plasma levels of a candidate drug. Murine and guinea pig herpes virus models were used in the early evaluations of ACV, GCV, CDV and PFA. Maribavir has no activity against the nonhuman strains of CMV, but was evaluated in two models in which severe combined immunodeficient (SCID) mice were implanted with human fetal tissue (Kern *et al.* 2004). In one model, human fetal retinal tissue was implanted into the anterior chamber of the SCID mouse eye, and in the second model, human fetal thymus and liver tissues were implanted under the kidney capsule. Established implants were infected with known quantities of CMV, and antiviral treatment by oral or intraperitoneal route was initiated and compared to placebo vehicle controls. In the retinal tissue model, maribavir reduced CMV replication about fourfold through 21 days post infection, and in the thymus/liver tissue model, maribavir was effective in inhibiting CMV replication by approximately 30- to 3,000-fold (Kern *et al.* 2004). These preclinical results were encouraging, although the infection system was too variable to provide a robust tool for efficacy-pharmacokinetic correlations.

Maribavir's activity against EBV has not been as extensively studied. The *in vitro* systems available for determining the antiviral activity of a compound against EBV are limited. Typically, immortalized cell lines carrying the episomal reservoir of EBV genome under basal transcriptional programs are induced to a lytic replication cycle by agents that trigger the B cell activation state, and the effects of antiviral treatment on EBV linear DNA molecules or structural proteins are measured. The activity of maribavir against EBV was assessed in two laboratories, using three assay methods. The reported potencies ranged from 0.15 to approximately 10 μM for maribavir, compared to 3.6 to 17 μM for the reference standard ACV (Zacny *et al.* 1999; Kern *et al.* 2004).

5.   MECHANISM OF MARIBAVIR'S ANTIVIRAL ACTION AND RESISTANCE

The discovery of the selectivity and potency of BDCRB and TCRB generated excitement early in the preclinical development program. However, the *in vivo* lability of the glycosidic linkage precluded their clinical development. This prompted extensive chemical exploration within the series, and several strategies were identified to stabilize the ribose-base linkage and still maintain antiviral potency (see Chapter 3: Drach, Townsend, and Bogner). Ultimately, the distinct MOAs exhibited within the benzimidazole series correlated with the stereochemistry of the ribose (D or L conformations), and with the nature of the substituents in the 2-position of the benzimidazole ring as illustrated in Figure 2 (Townsend *et al.* 1995; Townsend *et al.* 1999; Biron *et al.* 2002).

The β-D-ribose-containing compounds, exemplified by the original selective hits BDCRB and TCRB, blocked the maturational cleavage and subsequent packaging of newly-synthesized concatameric CMV DNA (Underwood *et al.* 1998). This MOA was mediated by inhibitor interaction with two of the proteins (UL56 and UL89) comprising the terminase complex, which is responsible for site-specific viral DNA cleavage and packaging (Underwood *et al.* 1998; Krosky *et al.* 1998; Bogner 2002) and through their interaction with the putative portal protein, pUL104 (Komazin *et al.* 2004; Dittmer *et al.* 2005). Disguising the pentose sugar of BDCRB in a six-membered pyranosyl ring structure generated the metabolically stable clinical candidate, compound GW275175X (Fig. 1), which retained the mechanism of action of BDCRB based on its cross-resistance profile and phenotype in infected cells (Underwood *et al.* 2004).

Enantiomeric Pairs

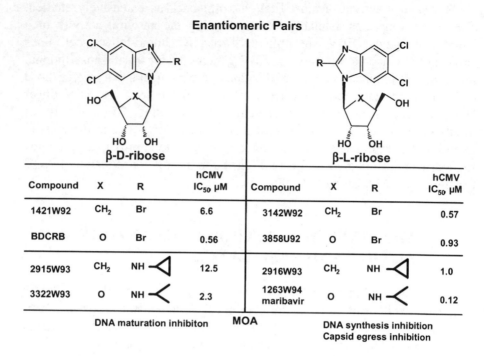

Compound	X	R	hCMV IC$_{50}$ μM	Compound	X	R	hCMV IC$_{50}$ μM
1421W92	CH$_2$	Br	6.6	3142W92	CH$_2$	Br	0.57
BDCRB	O	Br	0.56	3858U92	O	Br	0.93
2915W93	CH$_2$	NH—	12.5	2916W93	CH$_2$	NH—	1.0
3322W93	O	NH—	2.3	1263W94 maribavir	O	NH—	0.12

<table>
<tr><td>DNA maturation inhibiton</td><td>MOA</td><td>DNA synthesis inhibition
Capsid egress inhibition</td></tr>
</table>

Figure 2. Correlation of benzimidazole conformation and hCMV MOA

The more significant change of the β-D-ribose conformation to the "unnatural" L orientation was another approach designed to prevent the *in vivo* cleavage of the glycosidic bond and avoid toxicity, although the *in vitro* potency of the resulting β-L-BDCRB (3858U92) was also slightly reduced (Fig. 2) (Biron *et al.* 2002). The CMV potency of 3858U92 and of the carbocyclic ribose analog (77W94) was enhanced about 3-12-fold by substitution of the halogen in the 2-position of the ring with isopropyl- or cyclopropyl-amino substituents (Townsend *et al.* 1999). Unexpectedly, these few structural changes altered the scope of activity to include EBV, and totally altered the mechanisms of action away from those involving inhibition of CMV DNA maturation. The mechanisms of action are incompletely understood, in part because the precise functions of the target gene products in the virus life cycle and in disease pathogenesis are not fully defined. A summary of the current state of knowledge of maribavir's MOA follows.

5.1   Mechanism of CMV inhibition by maribavir

Maribavir exhibits a complex mode of action, affecting both early and late events of the CMV replication cycle. The phenotype of maribavir-treated CMV-infected cells showed variable reduction in the levels of viral DNA, and a late effect on the production of infectious virus (Biron *et al.* 2002; Krosky *et al.* 2003a). The relative effects of maribavir measured on these two stages of the life cycle varied somewhat according to the physiologic state of the cell, and perhaps cell type. However, the mechanism of interference with viral DNA synthesis was distinct from that of the approved drugs GCV and CDV. The mechanism was not mediated through a direct substrate analog inhibition of the CMV DNA polymerase. Maribavir was not phosphorylated in infected cells, nor did the compound itself, or any of its synthetic phosphorylated derivatives, inhibit the CMV DNA polymerase or the mammalian DNA polymerases tested (Biron *et al.* 2002).

This action of maribavir across temporally separated stages of the viral life cycle implicated several direct targets (an unlikely scenario), or a key target that served a regulatory role throughout the life cycle, effecting multiple substrates. The selection and characterization of drug-resistant virus was ultimately crucial to the identification of two CMV proteins involved in the novel mechanism of action

The first maribavir-resistant virus encoded a Leu397Arg amino acid substitution in pUL97, the hCMV protein kinase (PK), which conferred a 20- to 30-fold less sensitive phenotype. This resistant virus (2916rA, selected with the carbocyclic sugar analog of maribavir, 2916W93) remained susceptible to BDCRB and other approved anti-CMV drugs, including GCV. Moreover, this maribavir-resistant strain was fully competent for *in vitro* growth, in contrast to the growth-impaired phenotype of the UL97-knock out strains (RCΔ97) (Prichard *et al.* 1999). Supporting evidence for the pUL97 as the target was provided by studies with the expressed, purified pUL97 enzyme: the wild type pUL97-catalyzed phosphorylation of histone 2b was inhibited by maribavir ($IC_{50}=2\mu M$), while pUL97 with the Leu397Arg mutation was not ($IC_{50}>1000$ nM) (Biron *et al.* 2002). The inhibition of the pUL97 appeared selective; maribavir was inactive at 10 μM against the UL97 homolog in HSV-1, the pUL13 PK, and against several cell cycle kinases (GlaxoSmithKline; data on file). Studies of CMV replication in cells infected with the knockout of functional UL97 (RCΔ97) also indicated roles for this gene product in viral DNA synthesis, capsid maturation and egress

(Wolf *et al.* 2001; Krosky *et al.* 2003a). Thus, maribavir is the first selective antiviral that inhibits a viral-encoded protein kinase.

The CMV pUL97 protein kinase has the characteristics of an ideal target; it is conserved and essential for disease pathogenesis in the human host, and it regulates multiple viral events throughout the viral life cycle. The UL97 protein kinase is one of a family of serine-threonine protein kinases encoded by all of the human herpesviruses (Hanks *et al.* 1988; Smith and Smith 1989; Chee *et al.* 1989). The UL97 gene is highly conserved across CMV clinical strains, with nucleotide sequence identity ranging from 98.6% to 100%, and amino acid identity >99% in the 28 isolates evaluated (Lurain *et al.* 2001). The protein sequence is invariant at codon 397, the site of maribavir resistance. The analogous protein kinases in HSV, VZV and EBV have been shown to play a variety of regulatory roles in the viral replication cycles, including regulation of gene expression, and virion maturation (Purves *et al.* 1993; Reddy *et al.* 1998; Kato *et al.* 2001; Kenyon *et al.* 2001; Kawaguchi and Kato 2003). They appear dispensable for viral growth in cultured cells, but these herpes protein kinases are required for viral pathogenesis in animal models of disease (Coulter *et al.* 1993; Moffat *et al.* 1998; Michel and Mertens 2004).

Defining maribavir's action via inhibition of the pUL97 will require a better understanding of this protein kinase's role in CMV replication (identification and functional characterization of all its natural substrates) and pathogenesis. The UL97 is a virion component that carries a nuclear localization signal (NLS), and locates to the nucleus during the replication cycle (Michel *et al.* 1996; van Zeijl *et al.* 1997). The pUL97 is auto-phosphorylated at serine/threonine sites and using the synthetic substrate histone 2b, was shown to have a preference for the amino acid motif Ser-38 with peptides containing an Arg or Lys residue 5 positions downstream (P+5) from the serine (He *et al.* 1997; Michel *et al.* 1999; Baek *et al.* 2002) The pUL97 protein kinase possesses unusual biochemical capabilities, notably the ability to monophosphorylate the guanine nucleoside analogs ACV and GCV, yet not the normal nucleoside substrates of DNA replication (Sullivan *et al.* 1992; Michel *et al.* 1996; Talarico *et al.* 1999). This property is the reason that GCV is active against CMV, which does not encode a viral thymidine kinase (TK) analogous to the TK genes of HSV and VZV. Resistance to GCV in the clinical setting initially results from UL97 mutations encoding changes in the substrate binding domain (Fig. 3). Maribavir maintains activity against these UL97-altered strains resistant to GCV (Chou 2001; Biron *et al.* 2002), consistent with the observations that mutations affecting recognition of GCV do not impair the autophosphorylation of pUL97 (He *et al.* 1997; Michel *et al.* 1999). Baculovirus-expressed and purified GCV-resistant UL97 proteins also phosphorylate the synthetic substrate histone 2b with kinetics similar to those of wild-type enzyme (He *et al.* 1997).

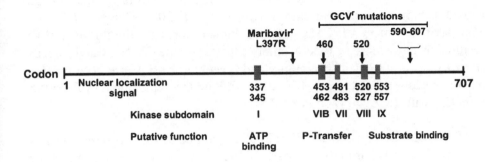

Figure 3. hCMV UL97 protein kinase

Maribavir's impact on hCMV replication, which somewhat mimics the RCΔ97 phenotype, must result from inadequate or improper phosphorylation of viral (host) substrates involved in viral DNA synthesis and capsid egress. Phosphorylation studies with maribavir have identified at least two candidate substrates. Regulation of the pUL44 phosphorylated states appears related to the reduction in viral DNA synthesis observed with maribavir treatment. The hCMV-encoded pUL44 is a phosphoprotein (ser298 and ser301) that is an essential component of the viral DNA replication machinery. The pUL44 associates with the viral DNA polymerase and serves as the processivity factor (Weiland *et al.* 1994). A direct interaction between the pUL97 and pUL44 has been reported, based on reciprocal immunoprecipitation from infected or co-transfected cells (Marschall *et al.* 2003; Krosky *et al.* 2003b). This interaction was linked to the co-localization of pUL44 and pUL97 in the CMV replication compartments (RCs) (Marschall *et al.* 2003). Multiple phosphorylated forms of the pUL44 are found in hCMV-infected cells, and several of these forms were absent in extracts of cells infected with RCΔ97, virus that is known to exhibit reduced viral DNA synthesis. Likewise, pUL44 forms were deficient in maribavir-treated cells infected with wild type virus (Krosky *et al.* 2003b). Thus an essential component of the hCMV DNA synthesis complex, the pUL44 polymerase processivity factor, is a pUL97 substrate whose functional co-localization in the RCs is dependent upon correct phosphorylation. This phosphorylation appears to be sensitive to inhibition by maribavir; the consequence is a reduction in viral DNA synthesis.

The second gene mediating laboratory-derived resistance to maribavir was the UL27 ORF, which may relate to the effects of this drug on late stage capsid maturation or nuclear egress (Komazin *et al.* 2003; Chou *et al.* 2004). Maribavir treatment of CMV-infected human foreskin fibroblasts (HFF) resulted in an increase of type A empty capsids or type C DNA filled capsids (Krosky *et al.* 2003a). These empty and precursor capsids also accumulated in the nucleus of infected HFF cells late in the replication cycle, after infection with RCΔ97 (Wolf *et al.* 2001).

The UL27 ORF encodes a protein of 608 amino acids. It is transcribed as an early-late gene, initially as part of longer transcript encompassing the ORFs UL26 through UL29 (Stamminger *et al.* 2002; Chou *et al.* 2004). Sequence database searches indicate limited homologies within various animal species of CMV, but no homologs are apparent in the other human herpesvirus or in mammalian cells (Chou *et al.* 2004). The function of pUL27 is unknown at this time, but is reported to be non-essential for growth *in vitro*. However, the UL27 homolog of murine CMV, (M27), also shown by mutagenesis studies to be nonessential for growth in culture, was required for disease virulence *in vivo* (Abenes *et al.* 2001). The pUL27 carries a NLS; deletional mutagenesis resulted in cytoplasmic retention and a one-half log reduction in viral titers (Chou *et al.* 2004).

The UL27 mutations conferring maribavir resistance were mapped to four codons that clustered in the region 335 to 415 (Fig. 4). The mutations in UL27 were shown to be necessary and sufficient to confer the maribavir-resistant phenotype in otherwise wild type strain AD169 or Towne (Komazin *et al.* 2003; Chou *et al.* 2004). However, the resistance level was weak compared to that conferred by the L397R in the pUL97 (3-5 fold vs 30-50 fold, respectively). The mutation at codon 415 is predicted to severely truncate the pUL27, and consistent with engineered UL27 null strains, this mutation abrogated the intranuclear localization of the pUL27 protein. The UL27 gene sequence was 96% conserved in the 16 clinical isolates studied and no changes were noted at the locations of the 4 UL27 resistance mutations (Chou *et al.* 2004).

Maribavir resistance mutations:

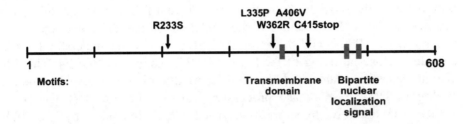

Figure 4. hCMV UL27 polypeptide

Viral encapsidation and nuclear egress involve the action of a number of viral gene products (Mettenleiter 2002), and the pUL97 clearly plays a role in the process. It is tempting to speculate that the pUL27 is a substrate for phosphorylation by pUL97. The pUL27 contains several proposed pUL97 substrate motifs (Baek *et al.* 2002; Chou *et al.* 2004). The elucidation of the role of pUL27 in the maribavir-resistant phenotype awaits further investigation of its function. Another potential substrate of the pUL97, the pp65 tegument protein, may also be involved in the normal virion particle maturation. This pp65 protein has been shown to interact with pUL97 based on reciprocal co-immunoprecipitation from infected cells, and direct phosphorylation of pp65 by the pUL97 (GlaxoSmithKline; data on file). Aggregates of pp65 accumulate in the presence of maribavir, or in cells infected with RCΔ97 (Prichard *et al.* 2005).

In conclusion, maribavir is the first antiviral drug that blocks viral growth by direct inhibition of a virally-encoded protein kinase. This inhibition appears selective, based on *in vitro* studies to date. The pUL97 protein kinase, by virtue of its regulatory functions, enables maribavir inhibition at multiple steps in viral replication. Moreover, genetic analyses indicates that both the UL97 and UL27 CMV genes that mediate maribavir resistance encode highly conserved products essential for pathogenesis in the host, thus providing ideal antiviral targets.

5.2   Mechanism of EBV inhibition by maribavir

Maribavir treatment of lytic EBV replication induced in Burkitt's lymphoma cells blocked both the appearance of linear forms of newly synthesized EBV DNA and the accumulation of the early antigen EA-D (Zacny *et al.* 1999). The basis for this effect is not yet understood. EBV encodes a close homolog of the hCMV UL97, designated BGLF4. This PK also autophosphorylates, and has been reported to directly phosphorylate the analogous DNA polymerase processivity factor, the EA-D (Chen *et al.* 2000). The level of hyperphosphorylated EA-D (encoded by the EBV BMRF1 gene) was reduced during viral reactivation in Akata cells by drug treatment, similar to the impact of maribavir treatment on the hCMV pUL44. Direct inhibition of the phosphorylation of EA-D has not been demonstrated, and maribavir did not block the phosphorylation of EA-D by EBV PK in transient co-expression assays with these two viral genes (Gershburg and Pagano 2002). Maribavir's mechanism of action against EBV requires further study.

5.3   Resistance

Early in the pre-clinical development of an antiviral agent, studies of laboratory-selected resistant virus aids in the identification of the viral target gene and provides a tool to study its function. Laboratory resistance genotypes may sometimes predict resistant genotypes that arise in the clinic, in association with loss of virologic response. In the case of maribavir, there is limited clinical experience to date and no reports of resistance (28-day maximum treatment exposure in study CMAA1003).

Maribavir resistance in laboratory strains selected thus far is mediated through mutations in the CMV UL97 PK gene (1 strain) or the UL27 gene (5 strains). Both genes are highly conserved, and based on herpes homolog studies, are essential for disease pathogenesis. The tolerance for mutations in UL97 and UL27 may be limited. The modest *in vitro* resistance mediated by mutations in the UL27 gene may not preclude clinical efficacy of the drug. As always, the germane, definitive characteristics of antiviral resistance await clinical experience.

6. PRECLINICAL TOXICOLOGY AND PHARMACOLOGY

Early in the pre-clinical evaluation, maribavir showed advantages in both pharmacokinetic and safety properties relative to the approved drugs for treatment of CMV diseases (Koszalka *et al.* 2002). *In vitro* cytotoxicity screens indicated that maribavir was not cytotoxic to rapidly dividing cells at antiviral concentrations. Importantly, maribavir was less inhibitory to the growth of human marrow progenitor cells compared to GCV. Also in contrast to GCV and CDV, maribavir was not genotoxic (Ames bacterial screen, *in vitro* mouse lymphoma assay below cytotoxic concentrations, rodent bone marrow micronucleus test *in vivo*). Maribavir did not evoke any significant responses in a broad pharmacological screen of central nervous system, cardiovascular, metabolic, or gastrointestinal activities (PanLabs PharmaScreen).

The pharmacokinetic profiles of maribavir in rodents and primates looked encouraging in a standard series of absorption, distribution, metabolism, and elimination (ADME) studies. Average oral bioavailability of maribavir was approximately 50% in monkeys and >90% in rats, and doses of 10 mg/kg resulting in plasma concentrations that were 24 to 114-fold higher than the mean IC$_{50}$ reported for 10 hCMV clinical isolates (Biron *et al.* 2002). Maribavir clearance was 1.8 L/kg/h in the rat, and 0.8 L/kg/h in the monkey. Drug clearance in both rats and monkeys was mediated primarily through biliary secretion, with metabolic and renal mechanisms contributing a minor role. The principal metabolite identified in rats and monkeys was the N-dealkylated 2-amino analog (5% and 8-14% of the plasma levels, respectively). Evidence of enterohepatic recirculation was observed in both species, complicating accurate calculations of elimination half-life.

Maribavir was able to penetrate the brain, cerebrospinal fluid, and vitreous humor of cynomolgus monkeys, but tissue levels were low, resulting in approximately 20%, 1-2% and <1%, of corresponding plasma concentrations. Limited penetration may be partly attributable to extensive protein binding. Maribavir was highly bound to plasma proteins, 84% and 88% in the monkey and rat, respectively. The principal binding site was serum albumin; however, based on *in vitro* binding kinetic studies, the binding was predicted to be readily reversible.

Comprehensive pre-clinical toxicology testing of maribivar has been completed with overall results establishing an acceptable safety profile (Koszalka *et al.* 2002). In acute oral toxicity studies in mice and rats, the maximum tolerated single doses were 250 and 1,000 mg/kg, respectively. These doses represented respectively at least 42 and 167 times the initial clinical dose tested in man (300 mg/day or 6.0 mg/kg using a 50 kg individual). In the acute intravenous toxicity studies, the maximum tolerated single dose was approximately 30 mg/kg in mice, and 75 mg/kg in rats.

One month sub-chronic treatment studies were conducted in the rat (100, 200, and 400 mg/kg/day) and monkey (20, 60 and 180 mg/kg/day) with no remarkable clinical signs in either species. The no-observable effect level of oral drug administered to cynomolgus monkeys for 30 consecutive days was 180mg/kg/day (plasma Cmax 8ug/mL) with mild reversible decreases in erythroid parameters and increases in reticulocytes seen at the two higher exposures.

Prolonged exposures to maribavir in the 6-month chronic rat (25, 100, and 400 mg/kg/day) and monkey (100, 200 and 500 mg/kg/day) study, and in the one-year chronic monkey study at the same doses, revealed mild, reversible treatment-related effects on WBC (elevations), RBC (anemias), and liver enzyme functions. A consistent treatment-related finding in both rats and monkeys at all doses was mild to moderate mucosal hyperplasia of the intestinal tract, which was partially reversed during the one-month recovery period, and was manifested clinically as altered stools or diarrhea at the higher doses (400mg/kg/day in the rats, 200 and 400 mg/kg/day in the monkeys.

Reproductive toxicity studies (oral fertility and embryo-fetal development studies in rats and rabbits) did not identify serious reproductive risks with maribavir. However, embryo-fetal survival *in utero* in rats was marginally reduced at all dosages when treatment commenced before fertilization and implantation, although development of surviving fetuses was not affected.

The generally favorable preclinical profile of maribavir suggests there may be advantages in safety over currently marketed CMV drugs, especially GCV and CDV, where multiple treatment–limiting toxicities were observed. While the lack of mutagenicity in preclinical studies is encouraging, the in-life carcinogenicity studies with maribavir have not yet been conducted. These test results are important to help define the long-term risk of the gut-tract hyperplasia observed in the 6 month (rat, monkey) and 1-year (monkey) chronic toxicity studies.

7. PHARMACOKINETICS AND METABOLISM IN MAN

Several phase I trials were conducted to investigate the safety, tolerability, and pharmacokinetic parameters of maribavir. Pharmacokinetic and safety parameters were consistent between the trials. In the first study, 13 healthy male subjects received single oral doses of maribavir at doses of 50, 100, 200, 400, 800 and 1600 mg, while in the second study, 17 HIV-infected male subjects were administered doses of 100, 400, 800 and 1,600 mg. (Wang *et al.* 2003). A subsequent study conducted in 16 healthy human volunteers evaluated the pharmacokinetics of 400 mg twice daily in the context of a drug interaction study. (Ma *et al.* 2005b). In both of the single oral dose escalation studies, dose proportional pharmacokinetics (Cmax or maximum concentration of the drug in plasma, and the area under the concentration-time curve, or AUC 0-∞) were demonstrated over the dose range tested; however, Cmax increased slightly less than proportionally to the dose, and AUC 0-∞ increased slightly more than proportionally to the dose. Maribavir was rapidly absorbed after oral administration, with peak concentrations of maribavir in plasma occurring 1 to 3 hours after dosing. The mean half-life in plasma was 3 to 5 hours, independent of the dose level. Clearance was metabolically mediated, and based on human liver microsomal assays, CYP3A4 is the major cytochrome P450 isozyme responsible for the metabolism. The N-dealkylated metabolite, 4469W94, (Koszalka *et al.* 2002) accounts for approximately 40% of the dose recovered in the urine (only 2% recovered as unchanged drug). The mean AUC 0-infinity ratio of the metabolite 4469W94 to maribavir (molar basis) was 15-20% in healthy male subjects, and 20-40% in HIV-infected male subjects. In the single dose escalation trial in HIV-infected subjects, consumption of a high-fat meal decreased the maribavir AUC 0- infinity and Cmax in plasma by approximately 30% (N=17).

The metabolism and clearance of maribavir in humans differed some-what from that observed in the rat and monkey models, where clearance was mediated primarily by biliary secretion of the parent drug and an N dealkylated metabolite. In humans, metabolic clearance and kidney excretion were the major mechanism. The extent of protein binding was also higher in humans, with >97% of drug in plasma bound to proteins (principally albumin), and the extent of binding appeared to be constant over the dose range studied (GlaxoSmithKline; data on file).

A metabolic study assessing the effects of maribavir on the activities of cytochrome P450 (CYP) 1A2, CYP2C9, CYP2C19, CYP2D6, CYP3A, N-acetyltransferase-2 (NAT-2), and xanthine oxidase has recently been completed. Activities of these enzymes were evaluated in a double-blind placebo-controlled trial in 20 healthy adults (16 receiving 400 mg maribavir twice daily for 10 days and 4 receiving placebo), utilizing a modified 5-drug phenotyping cocktail. No effect was shown on the activities of CYP1A2, CYP2C9, CYP3A, NAT-2, or xanthine oxidase; however, equivalence was not established for CYP2C19 or CYP2D6 where decreased activity was demonstrated (Ma *et al.* 2005a). The clinical significance of these observations is unknown, since other factors such as the dose and therapeutic window of concomitant medications, would need to be evaluated. Maribavir pharmacokinetic and safety parameters evaluated in this study were comparable to previous results.

Other metabolic studies are in progress, including a drug-drug interaction study between ketoconazole and maribavir, important since ketoconazole is an inducer of the liver P450 CYP3A4 enzyme which metabolizes maribavir. A study assessing maribavir pharmacokinetics in subjects with renal impairment is also in progress (2005).

8.   MARIBAVIR PROOF OF CONCEPT (POC) TRIAL

Proof of concept for the antiviral activity of maribavir in man was demonstrated within a dose-escalation trial in HIV-1 infected adult males with asymptomatic CMV shedding in the semen. Six oral dosage regimens (100, 200, or 400 mg three times a day, or 600, 900, or 1,200 mg twice a day), or a placebo were evaluated for 28 days. Asymptomatic CMV shedding was monitored in patients receiving the four lowest dose regimens. Maribavir exhibited *in vivo* anti-CMV activity in semen at all tested dosages, with mean reductions in semen CMV titers of 2.9 to 3.7 $\log_{10}$ on day 28 (Lalezari *et al.* 2002) (Fig. 5). These results were comparable to those seen in a similar trial with the approved dose (5mg/kg) of CDV (Lalezari *et al.* 1995). While the levels of infectious virus in semen were reduced in a dose-response fashion, the total viral DNA as quantified by PCR was less responsive to drug treatment (range 1.1-1.5 $\log_{10}$ for all cohorts). This result may be a reflection of the unusual mechanism of action of the drug;

maribavir treatment may allow incomplete virion packaging of viral DNA, and the non-infectious viral DNA would be detected. Between 17 and 50% of the patients in the various cohorts with positive urine cultures at start of study showed conversion to negative status for infectious virus at day 28. Only one subject, who received maribavir at 600 mg twice daily, had quantifiable CMV DNA in whole blood at baseline (4.81 $\log_{10}$ copies/ml). In this subject CMV DNA amounts decreased to below the limit of detection by PCR by day 28.

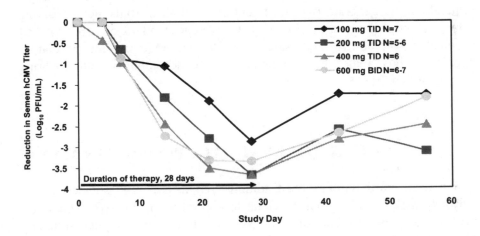

Figure 5. Maribavir POC: Phase 1 study CMAA1003

CMV viral load reductions were also reported in a pilot pharmacokinetic study evaluating 800 mg TID or 1200 mg BID of maribavir for 7 days in subjects with CMV retinitis (Hendrix *et al.* 2000). The purpose of this study was to determine whether the steady state concentrations of maribavir in vitreous fluids could impact CMV retinitis. CMV levels in the blood were reduced by 1.9, 0.26, and 1.1 $\log_{10}$ in the 3 subjects with CMV at baseline. Vitreous concentrations above the IC_{50} for CMV clinical isolates were achieved in 8 of 8 subjects treated at a dose of 1200 mg BID or 800 mg TID for 8 days (GlaxoSmithKline; data on file).

An additional POC trial in CMV seropositive allogenic bone marrow transplant patients is in progress, with results expected in early 2006. In this phase II double-blind, placebo-controlled trial, subjects will be treated prophylactically with increasing doses of maribavir for 12 weeks, and the incidence of CMV reactivation will be monitored, as well as pharmacokinetics, dose-response, safety and tolerability. The requirements for approval of a new a drug by the Food and Drug Administration are stringent and vary somewhat depending on the risk-risk analyses of the specific circumstances. As appropriate, some latitude is allowed when the unmet medical need is great and other treatment options limited. Nevertheless, an extensive number of studies are required to establish an acceptable preclinical profile and preliminary phase I studies are required to establish safety and tolerability in both normal volunteers and the target population. In general, two well-controlled clinical trials demonstrating effectiveness at least comparable to the current standard of care, with an acceptable tolerability and safety profile are required for regulatory approval. Thus, major hurdles remain in the development path to bring maribavir to market.

9.   SAFETY & TOLERABILITY PROFILE IN MAN

To date, maribavir has been administered to over 100 subjects in the various Phase I studies that cover up to a duration of 28 days. The safety profile was generally consistent between the trials and compiled summary information from the trials will be presented to provide the most comprehensive information (GlaxoSmithKline; data on file). Maribavir was generally well tolerated in these phase I trials; however, the vast majority of subjects had at least one adverse event during the study period. Taste disturbance, reported by 79% of maribavir recipients and 3/16 (19%) of placebo recipients, was the most frequently reported adverse event. Other adverse events reported by at least 2 subjects and considered to be drug-related included headache, dry mouth, fatigue, drowsiness/tiredness, light-headedness/dizziness, flatulence, nausea, bloating, diarrhea, stomach cramps, decreased appetite, vomiting, fever, rash, pruritis, discolored urine, and pyuria.

The high frequency of taste disturbance was considered a concern as it could negatively affect patient acceptance and adherence, although it was not considered a safety concern. The taste disturbance appeared to be dose dependent, generally occurring within minutes to up to 3 hours post dosing and lasting from 1 to several hours. The taste disturbance was described as altered, bad, terrible, strange, unusual, bitter, chemical, stale, and/or metallic.

None of the 79 subjects who reported taste disturbance discontinued study participation due to the taste disturbance, but 6 did rate the disturbance as severe. Secretion of drug or the metabolite 4469W94 into the salivary glands after systemic absorption was presumed to cause the taste disturbance.

Five subjects receiving maribavir prematurely discontinued the trial after 7-12 days due to a drug-related diffuse maculopapular rash of moderate (grade-2) intensity. Two of these subjects received 200 mg TID, and one each received 600, 900, and 1,200 mg BID. Four of these subjects had a history of allergic reaction to other drugs. The rash resolved quickly (within 1 to 4 days without sequelae) after discontinuation of the study drug.

Clinical laboratory values were uneventful, and no dose-related trends were observed. There were modest decreases in hemoglobin that were consistent with the phlebotomy requirements of the protocol and hemoglobin values increased to screening values by the 4-week post study visit. Additionally, some observed decreases in lymphocytes and increases in total protein were considered consistent with the HIV-infected status of the subjects. No issues relative to vital signs or electrocardiograms were observed

10. CONCLUSION

As briefly reviewed herein, wide-ranging basic scientific, pre-clinical, Phase I and metabolic studies in humans have been conducted to establish the overall profile of maribavir as a new antiviral agent. Information to date is encouraging and suggests that maribavir has the potential to present a significant advance over currently available CMV therapies. This novel benzimidazole riboside, if approved as an anti-CMV drug, will be the first selective protein kinase inhibitor among all marketed antivirals. If the on-going (2005) phase II dose-ranging study in allogeneic stem cell transplant patients is successful, the phase III studies required for drug registration could begin in 2006.

ACKNOWLEDGEMENTS

I want to acknowledge the pioneering work of John C. Drach and Leroy B. Townsend in the benzimidazole riboside compound series, without whose insightful work this promising class of antiviral agents would not have been pursued. I thank all the members of the GlaxoWellcome 1263W94 project team, my laboratory group (Sylvia Stanat, Mark Underwood, Christine Talarico, Michelle Davis) and colleagues Susan Daluge, George (Barney) Koszalka, Stanley Chamberlin, and Robert Harvey for their valuable contributions and dedication to the discovery and development of maribavir. I also thank Gloria Boone for editing and writing assistance.

REFERENCES

Cytovene-IV (ganciclovir sodium for intravenous infusion only) and Cytovene (ganciclovir capsules for oral administration only), package insert. Roche Laboratories, Inc. June 2000.

Foscavir (foscarnet injection) package insert. Astra Zeneca, Inc. June 2000.

Vistide (cidofovir injection) package insert. Gilead Sciences, Inc. September 1996.

Abenes, G., Lee, M., Haghjoo, E., Tong, T., Zhan, X. and Liu, F., 2001, Murine cytomegalovirus open reading frame M27 plays an important role in growth and virulence in mice. *J. Virol.* **75**: 1697-1707.

Baek, M C., Krosky, P. M., He, Z. and Coen, D.M., 2002, Specific phosphorylation of exogenous protein and peptide substrates by the human cytomegalovirus UL97 protein kinase. Importance of the P+5 position. *J. Biol .Chem.* **277**: 29593-29599.

Balfour, H. H., Jr., Chace, B.A., Stapleton, J.T., Simmons, R.L. and Fryd, D.S., 1989, A randomized, placebo-controlled trial of oral acyclovir for the prevention of cytomegalovirus disease in recipients of renal allografts. *N. Engl. J. Med.* **320**: 1381-1387.

Biron, K.K., Harvey, R.J., Chamberlain, S.C., Good, S.S., Smith, A.A., III, Davis, M.G., Talarico, C.L., Miller, W.H., Ferris, R., Dornsife, R.E., Stanat, S.C., Drach, J.C., Townsend, L.B. and Koszalka, G.W., 2002, Potent and selective inhibition of human cytomegalovirus replication by 1263W94, a benzimidazole L-riboside with a unique mode of action. *Antimicrob. Agents Chemother.* **46**: 2365-2372.

Bogner, E., 2002, Human cytomegalovirus terminase as a target for antiviral chemotherapy. *Rev. Med. Virol.* **12**: 115-127.

Cepok, S., Zhou, D., Srivastava, R., Nessler, S., Stei, S., Bussow, K., Sommer, N. and Hemmer, B., 2005, Identification of Epstein-Barr virus proteins as putative targets of the immune response in multiple sclerosis. *J. Clin. Invest* **115**: 1352-1360.

Chan, J.H., Chamberlain, S.D., Biron, K.K., Davis, M.G., Harvey, R.J., Selleseth, D.W., Dornsife, R.E., Dark, E H., Frick, L.W., Townsend, L B., Drach, J C. and Koszalka, G.W., 2000, Synthesis and evaluation of a series of 2'-deoxy analogues of the antiviral agent 5,6-dichloro-2-isopropylamino-1-(beta-L-ribofuranosyl)-1H-benzimidazole (1263W94). *Nucleosides Nucleotides Nucleic Acids* **19**: 101-123.

Chee, M.S., Lawrence, G.L. and Barrell, B.G., 1989, Alpha-, beta- and gammaherpesviruses encode a putative phosphotransferase. *J. Gen. Virol.* **70** (Pt 5): 1151-1160.

Chen, C J., Lin, K.H., Lin, S.C., Tsai, W.C., Yen, J.H., Chang, S.J., Lu, S.N. and Liu, H.W., 2005, High prevalence of immunoglobulin A antibody against Epstein-Barr virus capsid antigen in adult patients with lupus with disease flare: case control studies. *J. Rheumatol.* **32**: 44-47.

Chen, M.R., Chang, S.J., Huang, H. and Chen, J Y., 2000, A protein kinase activity associated with Epstein-Barr virus BGLF4 phosphorylates the viral early antigen EA-D *in vitro*. *J. Virol.* **74**: 3093-3104.

Chou, S., 1999, Antiviral drug resistance in human cytomegalovirus. *Transpl. Infect. Dis.* **1**: 105-114.

Chou, S., Lurain, N.S., Thompson, K.D., Miner, R.C. and Drew, W.L., 2003, Viral DNA polymerase mutations associated with drug resistance in human cytomegalovirus. *J. Infect. Dis.* **188**: 32-39.

Chou, S., Marousek, G.I., Senters, A.E., Davis, M.G. and Biron, K.K., 2004, Mutations in the human cytomegalovirus UL27 gene that confer resistance to maribavir. *J. Virol.* **78**: 7124-7130.

Chou, S.W., 2001,. Cytomegalovirus drug resistance and clinical implications. *Transpl. Infect. Dis.* **3** Suppl 2: 20-24.

Coulter, L.J., Moss, H.W., Lang, J. and McGeoch, D.J., 1993, A mutant of herpes simplex virus type 1 in which the UL13 protein kinase gene is disrupted. *J. Gen. Virol.* **74** (Pt 3): 387-395.

Crumpacker, C.S., 1996, Ganciclovir. *N. Engl. J. Med.* **335**: 721-729.

Cvetkovic, R.S. and Wellington, K., 2005, Valganciclovir: a review of its use in the management of CMV infection and disease in immunocompromised patients. *Drugs* **65**: 859-878.

Dittmer, A., Drach, J.C., Townsend, L.B., Fischer, A. and Bogner, E., 2005, Interaction of the putative HCMV portal protein pUL104 with the large terminase subunit pUL56 and its inhibition by benzimidazole-D-ribonucleosides. *J. Virol.* **79**: in press.

Elion, G.B., Furman, P.A., Fyfe, J.A., de Miranda, P., Beauchamp, L. and Schaeffer, H.J., 1977, Selectivity of action of an antiherpetic agent, 9-(2-hydroxyethoxymethyl) guanine. *Proc. Natl. Acad. Sci. U.S.A* **74**: 5716-5720.

Erice, A. 1999. Resistance of human cytomegalovirus to antiviral drugs. *Clin. Microbiol. Rev.* **12**: 286-297.

Evers, D.L., Komazin, G., Ptak, R.G., Shin, D., Emmer, B.T., Townsend, L.B. and Drach, J.C., 2004, Inhibition of human cytomegalovirus replication by benzimidazole nucleosides involves three distinct mechanisms. *Antimicrob. Agents Chemother.* **48**: 3918-3927.

Evers, D.L., Komazin, G., Shin, D., Hwang, D.D., Townsend, L.B. and Drach, J.C., 2002, Interactions among antiviral drugs acting late in the replication cycle of human cytomegalovirus. *Antiviral Res.* **56**: 61-72.

Gershburg, E. and Pagano, J.S., 2002, Phosphorylation of the Epstein-Barr virus (EBV) DNA polymerase processivity factor EA-D by the EBV-encoded protein kinase and effects of the L-riboside benzimidazole 1263W94. *J. Virol.* **76**: 998-1003.

Gross, A.J., Hochberg, D., Rand, W.M. and Thorley-Lawson, D.A., 2005, EBV and systemic lupus erythematosus: a new perspective. *J. Immunol.* **174**: 6599-6607.

Hanks, S.K., Quinn, A.M. and Hunter, T., 1988, The protein kinase family: conserved features and deduced phylogeny of the catalytic domains. *Science* **241**: 42-52.

He, Z., He, Y.S., Kim, Y., Chu, L., Ohmstede, C., Biron, K.K. and Coen, D.M., 1997, The human cytomegalovirus UL97 protein is a protein kinase that autophosphorylates on serines and threonines. *J. Virol.* **71**: 405-411.

Hendrix C., Kuppermann, B.D., Dunn J.P., Wang L.H., Starns E., Snowden B.W., Hamzeh F. and Wire M.B., 2000, A phase I trial to evaluate the ocular (intravitreal) penetration of 1263W94 after multiple dose, oral administration in AIDS patients with CMV retinitis. 40th ICAAC.

Hodson, E.M., Jones, C.A., Webster, A.C., Strippoli, G.F., Barclay, P.G., Kable, K., Vimalachandra, D. and Craig, J.C., 2005, Antiviral medications to prevent cytomegalovirus disease and early death in recipients of solid-organ transplants: a systematic review of randomised controlled trials. *Lancet* **365**: 2105-2115.

332   *K. K. Biron*

Jacobson, M.A., 1992, Review of the toxicities of foscarnet. *J. Acquir. Immune. Defic. Syndr.* **5** Suppl 1: S11-S17.

James, J.A., Neas, B. R., Moser, K.L., Hall, T., Bruner, G.R., Sestak, A.L. and Harley, J.B., 2001, Systemic lupus erythematosus in adults is associated with previous Epstein-Barr virus exposure. *Arthritis Rheum.* **44**: 1122-1126.

Kato, K., Kawaguchi, Y., Tanaka, M., Igarashi, M., Yokoyama, A., Matsuda, G., Kanamori, M., Nakajima, K., Nishimura, Y., Shimojima, M., Phung, H. T., Takahashi, E. and Hirai, K., 2001, Epstein-Barr virus-encoded protein kinase BGLF4 mediates hyper-phosphorylation of cellular elongation factor 1delta (EF-1delta): EF-1delta is universally modified by conserved protein kinases of herpesviruses in mammalian cells. *J. Gen. Virol.* **82**: 1457-1463.

Kawaguchi, Y. and Kato, K., 2003, Protein kinases conserved in herpesviruses potentially share a function mimicking the cellular protein kinase cdc2. *Rev. Med. Virol.* **13**: 331-340.

Kenyon, T.K., Lynch, J., Hay, J., Ruyechan, W. and Grose, C., 2001, Varicella-zoster virus ORF47 protein serine kinase: characterization of a cloned, biologically active phosphor-transferase and two viral substrates, ORF62 and ORF63. *J. Virol.* **75**: 8854-8858.

Kern, E.R., Hartline, C.B., Rybak, R.J., Drach, J.C., Townsend, L.B., Biron, K.K. and Bidanset, D.J., 2004, Activities of benzimidazole D- and L-ribonucleosides in animal models of cytomegalovirus infections. *Antimicrob. Agents Chemother.* **48**: 1749-1755.

Komazin, G., Ptak, R.G., Emmer, B.T., Townsend, L.B. and Drach, J.C., 2003, Resistance of human cytomegalovirus to the benzimidazole L-ribonucleoside maribavir maps to UL27. *J. Virol.* **77**: 11499-11506.

Komazin, G., Townsend, L.B. and Drach, J.C., 2004, Role of a mutation in human cytomegalovirus gene UL104 in resistance to benzimidazole ribonucleosides. *J. Virol.* **78**: 710-715.

Koszalka, G.W., Johnson, N.W., Good, S.S., Boyd, L., Chamberlain, S.C., Townsend, L.B., Drach, J.C. and Biron, K.K., 2002, Pre-clinical and toxicology studies of 1263W94, a potent and selective inhibitor of human cytomegalovirus replication. *Antimicrob. Agents Chemother.* **46**: 2373-2380.

Krosky, P.M., Baek, M.C. and Coen, D.M., 2003a, The human cytomegalovirus UL97 protein kinase, an antiviral drug target, is required at the stage of nuclear egress. *J. Virol.* **77**: 905-914.

Krosky, P. M., Baek, M. C., Jahng, W. J., Barrera, I., Harvey, R. J., Biron, K. K., Coen, D. M., and Sethna, P. B. 2003b. The human cytomegalovirus UL44 protein is a substrate for the UL97 protein kinase. *J. Virol.* **77**: 7720-7727.

Krosky, P.M., Underwood, M.R., Turk, S.R., Feng, K.W., Jain, R.K., Ptak, R.G., Westerman, A.C., Biron, K.K., Townsend, L.B. and Drach, J.C., 1998, Resistance of human cytomegalovirus to benzimidazole ribonucleosides maps to two open reading frames: UL89 and UL56. *J. Virol.* **72**: 4721-4728.

Lalezari, J.P., Aberg, J.A., Wang, L. H., Wire, M.B., Miner, R., Snowden, W., Talarico, C. L., Shaw, S., Jacobson, M.A. and Drew, W.L., 2002, Phase I dose escalation trial evaluating the pharmacokinetics, anti-human cytomegalovirus (HCMV) activity, and safety of 1263W94 in human immunodeficiency virus-infected men with asymptomatic HCMV shedding. *Antimicrob. Agents Chemother.* **46**: 2969-2976.

Lalezari, J. P., Drew, W. L., Glutzer, E., James, C., Miner, D., Flaherty, J., Fisher, P. E., Cundy, K., Hannigan, J., and Martin, J. C. 1995. (S)-1-[3-hydroxy-2-(phosphonyl-methoxy)propyl]cytosine (cidofovir): results of a phase I/II study of a novel antiviral nucleotide analogue. *J. Infect. Dis.* **171**: 788-796.

Levin, L.I., Munger, K.L., Rubertone, M., Peck, C.A., Lennette, E.T., Spiegelman, D. and Ascherio, A., 2005, Temporal relationship between elevation of Epstein-Barr virus antibody titers and initial onset of neurological symptoms in multiple sclerosis. *JAMA* **293**: 2496-2500.

Lowance, D., Neumayer, H.H., Legendre, C.M., Squifflet, J.P., Kovarik, J., Brennan, P.J., Norman, D., Mendez, R., Keating, M.R., Coggon, G.L., Crisp, A. and Lee, I.C., 1999, Valacyclovir for the prevention of cytomegalovirus disease after renal transplantation. International Valacyclovir Cytomegalovirus Prophylaxis Transplantation Study Group. *N. Engl. J. Med.* **340**: 1462-1470.

Lu, H. and Thomas, S., 2004, Maribavir (ViroPharma). *Curr. Opin. Investig. Drugs* **5**: 898-906.

Lurain, N.S., Weinberg, A., Crumpacker, C.S. and Chou, S., 2001, Sequencing of cytomegalovirus UL97 gene for genotypic antiviral resistance testing. *Antimicrob.Agents Chemother.* **45**: 2775-2780.

Ma, J.D., Nafziger, A.N., Villano, S.A., Gaedigk, A. and Victory, J., 2005a, Effect of the anti-cytomegalovirus drug maribavir on the activities of cytochrome P450 (CYP) 1A2, 2C9, 2C19, 2D6, 3A, N-acetyltransferase-2, and xanthine oxidase as Assessed by the Modified Cooperstown 5+1 Drug Cocktail. 2005, American Society for Clinical Pharmacology and Therapeutics (ASCPT) Annual meeting in Orlando, Florida , Abstract PI-43.

Ma, J.D., Nafziger, A.N., Villano, S.A., Victory, J. and Bertino, J.S., 2005b, Single and multiple dose pharmacokinetics (PK) of maribavir (MB) in healthy adults. 2005 American Society for Clinical Pharmacology and Therapeutics (ASCPT) Annual meeting in Orlando, Florida.

Marschall, M., Freitag, M., Suchy, P., Romaker, D., Kupfer, R., Hanke, M. and Stamminger, T., 2003, The protein kinase pUL97 of human cytomegalovirus interacts with and phosphorylates the DNA polymerase processivity factor pUL44. *Virology* **311**: 60-71.

McSharry, J.J., McDonough, A., Olson, B., Hallenberger, S., Reefschlaeger, J., Bender, W. and Drusano, G.L., 2001a, Susceptibilities of human cytomegalovirus clinical isolates to BAY38-4766, BAY43-9695, and ganciclovir. *Antimicrob. Agents Chemother.* **45**: 2925-2927.

McSharry, J.J., McDonough, A., Olson, B., Talarico, C., Davis, M. and Biron, K.K., 2001b, Inhibition of ganciclovir-susceptible and -resistant human cytomegalovirus clinical isolates by the benzimidazole L-riboside 1263W94. *Clin. Diagn. Lab Immunol.* **8**: 1279-1281.

Mettenleiter, T.C., 2002, Herpesvirus assembly and egress. *J. Virol.* **76**: 1537-1547.

Michel, D., Kramer, S., Hohn, S., Schaarschmidt, P., Wunderlich, K. and Mertens, T., 1999, Amino acids of conserved kinase motifs of cytomegalovirus protein UL97 are essential for autophosphorylation. *J. Virol.* **73**: 8898-8901.

Michel, D. and Mertens, T., 2004, The UL97 protein kinase of human cytomegalovirus and homologues in other herpesviruses: impact on virus and host. *Biochim. Biophys. Acta* **1697**: 169-180.

Michel, D., Pavic, I., Zimmermann, A., Haupt, E., Wunderlich, K., Heuschmid, M. and Mertens, T., 1996, The UL97 gene product of human cytomegalovirus is an early-late protein with a nuclear localization but is not a nucleoside kinase. *J. Virol.* **70**: 6340-6346.

Moffat, J.F., Zerboni, L., Sommer, M.H., Heineman, T.C., Cohen, J.I., Kaneshima, H. and Arvin, A.M., 1998, The ORF47 and ORF66 putative protein kinases of varicella-zoster virus determine tropism for human T cells and skin in the SCID-hu mouse. *Proc. Natl. Acad. Sci. U.S.A* **95**: 11969-11974.

Okano, M., 2003, The evolving therapeutic approaches for Epstein-Barr virus infection in immunocompetent and immunocompromised individuals. *Curr. Drug Targets. Immune. Endocr. Metabol. Disord.* **3**: 137-142.

Pagano, J.S., 1999, Epstein-Barr virus: the first human tumor virus and its role in cancer. *Proc. Assoc. Am. Physicians* **111**: 573-580.

Parks, C.G., Cooper, G.S., Hudson, L.L., Dooley, M.A., Treadwell, E.L., St Clair, E.W., Gilkeson, G.S. and Pandey, J.P., 2005, Association of Epstein-Barr virus with systemic lupus erythematosus: effect modification by race, age, and cytotoxic T lymphocyte-associated antigen 4 genotype. *Arthritis Rheum.* **52**: 1148-1159.

Perrot, S., Calvez, V., Escande, J.P., Dupin, N. and Marcelin, A.G., 2003, Prevalences of herpesviruses DNA sequences in salivary gland biopsies from primary and secondary Sjogren's syndrome using degenerated consensus PCR primers. *J. Clin. Virol.* **28**: 165-168.

Prichard, M.N., Gao, N., Jairath, S., Mulamba, G., Krosky, P., Coen, D.M., Parker, B.O. and Pari, G.S., 1999, A recombinant human cytomegalovirus with a large deletion in UL97 has a severe replication deficiency. *J. Virol.* **73**: 5663-5670.

Prichard, M.N., Hartline, C.B., Britt, W.J. and Kern, E.R., 2005, Maribavir induces the formation of tegument aggregates in cells infected with human cytomegalovirus Abstracts of the 18th International Conference on Antiviral Research. Barcelona, Spain, April 11-14, 2005. *Antiviral Res* **65**: A27-108.

Purves, F.C., Ogle, W.O. and Roizman, B., 1993, Processing of the herpes simplex virus regulatory protein alpha 22 mediated by the UL13 protein kinase determines the accumulation of a subset of alpha and gamma mRNAs and proteins in infected cells. *Proc. Natl. Acad. Sci. U.S.A* **90**: 6701-6705.

Reddy, S.M., Cox, E., Iofin, I., Soong, W. and Cohen, J.I., 1998, Varicella-zoster virus (VZV) ORF32 encodes a phosphoprotein that is posttranslationally modified by the VZV ORF47 protein kinase. *J. Virol.* **72**: 8083-8088.

Resnick, L., Herbst, J.S., Ablashi, D., Atherton, S., Frank, B., Rosen, L. and Horwitz, S.N., 1988, Regression of oral hairy leukoplakia after orally administered acyclovir therapy. *JAMA* **259**: 384-388.

Reusser, P., 2001, Oral valganciclovir: a new option for treatment of cytomegalovirus infection and disease in immunocompromised hosts. *Expert. Opin. Investig. Drugs* **10**: 1745-1753.

Roizman, B., 1996, Herpesviridae. In *Herpesviridae* (B.N. Fields, D.M. Knipe, and P.M. Howley, Eds.), pp. 2221-2230. Philadelphia.

Safrin, S., Cherrington, J. and Jaffe, H.S., 1999, Cidofovir. Review of current and potential clinical uses. *Adv. Exp. Med. Biol.* **458**: 111-120.

Selleseth, D.W., Talarico, C.L., Miller, T., Lutz, M.W., Biron, K.K. and Harvey, R.J., 2003, Interactions of 1263W94 with other antiviral agents in inhibition of human cytomegalovirus replication. *Antimicrob. Agents Chemother.* **47**: 1468-1471.

Smith, R.F. and Smith, T.F., 1989, Identification of new protein kinase-related genes in three herpesviruses, herpes simplex virus, varicella-zoster virus, and Epstein-Barr virus. *J. Virol.* **63**: 450-455.

Springer, K.L. and Weinberg, A., 2004, Cytomegalovirus infection in the era of HAART: fewer reactivations and more immunity. *J. Antimicrob. Chemother.* **54**: 582-586.

Stamminger, T., Gstaiger, M., Weinzierl, K., Lorz, K., Winkler, M. and Schaffner, W., 2002, Open reading frame UL26 of human cytomegalovirus encodes a novel tegument protein that contains a strong transcriptional activation domain. *J. Virol.* **76**: 4836-4847.

Sullivan, V., Talarico, C.L., Stanat, S C., Davis, M., Coen, D.M. and Biron, K.K., 1992, A protein kinase homologue controls phosphorylation of ganciclovir in human cytomegalovirus-infected cells. *Nature* **358**: 162-164.

Sundstrom, P., Juto, P., Wadell, G., Hallmans, G., Svenningsson, A., Nystrom, L., Dillner, J. and Forsgren, L., 2004, An altered immune response to Epstein-Barr virus in multiple sclerosis: a prospective study. *Neurology* **62**: 2277-2282.

Taber, D.J., Ashcraft, E., Baillie, G.M., Berkman, S., Rogers, J., Baliga, P. K., Rajagopalan, P. R., Lin, A., Emovon, O., Afzal, F., and Chavin, K. D. 2004. Valganciclovir prophylaxis in patients at high risk for the development of cytomegalovirus disease. *Transpl. Infect. Dis.* **6**, 101-109.

Talarico, C. L., Burnette, T. C., Miller, W. H., Smith, S. L., Davis, M. G., Stanat, S. C., Ng, T.I., He, Z., Coen, D.M., Roizman, B. and Biron, K.K., 1999, Acyclovir is phosphorylated by the human cytomegalovirus UL97 protein. *Antimicrob.Agents Chemother.* **43**: 1941-1946.

Thorley-Lawson, D.A., 2001, Epstein-Barr virus: exploiting the immune system. *Nat. Rev. Immunol.* **1**: 75-82.

Thorley-Lawson, D.A. and Gross, A., 2004, Persistence of the Epstein-Barr virus and the origins of associated lymphomas. *N. Engl. J. Med.* **350**: 1328-1337.

Townsend, L.B., Devivar, R.V., Turk, S.R., Nassiri, M.R. and Drach, J.C., 1995, Design, synthesis, and antiviral activity of certain 2,5,6-trihalo-1-(beta-D-ribofuranosyl) benzimidazoles. *J. Med. Chem.* **38**: 4098-4105.

Townsend, L.B., Gudmundsson, K.S., Daluge, S.M., Chen, J.J., Zhu, Z., Koszalka, G.W., Boyd, L., Chamberlain, S.D., Freeman, G.A., Biron, K.K. and Drach, J.C., 1999, Studies designed to increase the stability and antiviral activity (HCMV) of the active benzimidazole nucleoside, TCRB. *Nucleosides Nucleotides* **18**: 509-519.

Townsend, L.B. and Revankar, G.R., 1970, Benzimidazole nucleosides, nucleotides, and related derivatives. *Chem. Rev.* **70**: 389-438.

Underwood, M.R., Ferris, R.G., Selleseth, D.W., Davis, M.G., Drach, J.C., Townsend, L.B., Biron, K.K. and Boyd, F.L., 2004, Mechanism of action of the ribopyranoside benzimidazole GW275175X against human cytomegalovirus. *Antimicrob. Agents Chemother.* **48**: 1647-1651.

Underwood, M.R., Harvey, R.J., Stanat, S.C., Hemphill, M.L., Miller, T., Drach, J.C., Townsend, L.B. and Biron, K.K., 1998, Inhibition of human cytomegalovirus DNA maturation by a benzimidazole ribonucleoside is mediated through the UL89 gene product. *J. Virol.* **72**: 717-725.

Valantine, H.A., 2004, The role of viruses in cardiac allograft vasculopathy. *Am. J. Transplant.* **4**: 169-177.

van Zeijl, M., Fairhurst, J., Baum, E.Z., Sun, L. and Jones, T.R., 1997, The human cytomegalovirus UL97 protein is phosphorylated and a component of virions. *Virology* **231**: 72-80.

Vancikova, Z. and Dvorak, P., 2001, Cytomegalovirus infection in immunocompetent and immunocompromised individuals—a review *Curr. Drug Targets. Immune. Endocr. Metabol. Disord.* **1**: 179-187.

Walling, D.M., Flaitz, C.M. and Nichols, C.M., 2003, Epstein-Barr virus replication in oral hairy leukoplakia: response, persistence, and resistance to treatment with valacyclovir. *J. Infect. Dis.* **188**: 883-890.

Wang, L.H., Peck, R.W., Yin, Y., Allanson, J., Wiggs, R., and Wire, M.B., 2003, Phase I safety and pharmacokinetic trials of 1263W94, a novel oral anti-human cytomegalovirus agent, in healthy and human immunodeficiency virus-infected subjects. *Antimicrob. Agents Chemother.* **47**: 1334-1342.

Weiland, K.L., Oien, N.L., Homa, F. and Wathen, M.W., 1994, Functional analysis of human cytomegalovirus polymerase accessory protein. *Virus Res* **34**: 191-206.

Whitley, R.J., 2004, Congenital cytomegalovirus infection: epidemiology and treatment. *Adv. Exp. Med. Biol.* **549**: 155-160.

Williams, S.L., Hartline, C.B., Kushner, N.L., Harden, E.A., Bidanset, D.J., Drach, J.C., Townsend, L.B., Underwood, M.R., Biron, K.K. and Kern, E.R., 2003, *In vitro* activities of benzimidazole D- and L-ribonucleosides against herpesviruses. *Antimicrob. Agents Chemother.* **47**: 2186-2192.

Wolf, D.G., Courcelle, C.T., Prichard, M.N. and Mocarski, E.S., 2001, Distinct and separate roles for herpesvirus-conserved UL97 kinase in cytomegalovirus DNA synthesis and encapsidation. *Proc. Natl. Acad. Sci. U.S.A* **98**: 1895-1900.

Yamazaki, M., Kitamura, R., Kusano, S., Eda, H., Sato, S., Okawa-Takatsuji, M., Aotsuka, S. and Yanagi, K., 2005, Elevated immunoglobulin G antibodies to the proline-rich amino-terminal region of Epstein-Barr virus nuclear antigen-2 in sera from patients with systemic connective tissue diseases and from a subgroup of Sjogren's syndrome patients with pulmonary involvements. *Clin. Exp. Immunol.* **139**: 558-568.

Yankulov, K., Yamashita, K., Roy, R., Egly, J.M., and Bentley, D L., 1995, The transcriptional elongation inhibitor 5,6-dichloro-1-beta-D-ribofuranosylbenzimidazole inhibits transcription factor IIH-associated protein kinase. *J. Biol. Chem.* **270**: 23922-23925.

Zacny, V.L., Gershburg, E., Davis, M.G., Biron, K.K. and Pagano, J.S., 1999, Inhibition of Epstein-Barr virus replication by a benzimidazole L-riboside: novel antiviral mechanism of 5, 6-dichloro-2-(isopropylamino)-1-beta-L-ribofuranosyl-1H-benzimidazole. *J. Virol.* **73**: 7271-7277.

Chapter 2.3

BENZIMIDAZOLE-D-RIBONUCLEOSIDES AS ANTIVIRAL AGENTS THAT TARGET HCMV TERMINASE

J. C. DRACH[1,2], L. B. TOWNSEND[2] and E. BOGNER[3]

[1]*Department of Biolologic and Materials Sciences, School of Dentistry, University of Michigan, Ann Arbor, MI, USA;* [2] *Department of Medical Chemistry,College of Pharmacy, University of Michigan, Ann Arbor, MI, USA;* [3]*Institute of Clinical and Molecular Virology, Friedrich-Alexander University Erlangen-Nuernberg, Schlossgarten 4, 91054 Erlangen, Germany*

Abstract: Human cytomegalovirus (HCMV), one of eight human herpesviruses, can cause life-threatening diseases. HCMV is widespread throughout the population world wide; the seroprevalence in adults varies from 50-100%. HCMV infection is rarely of significant consequence in individuals with a competent immune system. However, although immune control is efficient, it cannot achieve clearance of the virus. HCMV persists lifelong in the infected host and reactivates causing reinfection. In neonates as well as in immunocompromised adults, HCMV is a serious pathogen that can cause fatal organ damage. Different antiviral compounds alone or in combination have been used for treatment of CMV diseases. Nearly all currently available drugs are targeted to the viral DNA polymerase. In clinical use, mutations in the viral DNA polymerase confer resistance to ganciclovir, foscarnet and cidofovir. We have previously provided evidence that the terminase subunits pUL56 and pUL89 as well as the portal protein pUL104 fulfill all requirements for a target of new antiviral therapy. Herein we review the development of benzimidazole-D-ribonucleosides as antiviral drugs and describe their effect on the HCMV terminase as well as on pUL104.

1. INTRODUCTION

Human cytomegalovirus (HCMV) is a member of the herpesvirus family and represents a major human pathogen causing severe disease in newborns and immunocompromised patients, *i.e.* organ transplant recipients and

337

E. Bogner and A. Holzenburg (eds.), New Concepts of Antiviral Therapy, 337–349.

patients with AIDS. One characteristic of herpesviruses is their ability to establish lifelong latency in their hosts, thus reactivation during immunosuppression leads to recurrent episodes of diseases. Today antiviral therapy is limited to suppressing reactivation or treating infections caused by actively replicating virus; it can not be expected to eliminate the latent virus.

HCMV is classified as a ß-herpesvirus and characterized by its narrow host range and prolonged replicative cycle in permissive cells (Britt & Alford, 1996; Mocarski, 1996). Other characteristics of this subfamily are strong species specificity, slow replication and cytopathology. HCMV with a genome of approximately 239 kbp has the ability to encode over 200 proteins and so has the highest coding capacity of the herpesvirus family. Maturational events of HCMV DNA replication and capsid assembly require cleavage of concatenated DNA into genome length followed by insertion into preformed capsids. The following seven steps are involved in this process: (i) The recognition of viral DNA by a specific protein, (ii) binding the DNA at specific sequence motifs (packaging signals, *e.g.* pac1 and pac2), (iii) translocation of the DNA-protein complex to the procapsid, (iv) interaction of the DNA-protein complex with the portal protein, (v) import of one unit-length genome into capsid by ATP hydrolysis of one terminase subunit, (vi) completion of the packaging process by cutting off excess DNA (two strand nicking) and finally (vii) budding of the nucleocapsids through nuclear membranes. This early budding event is found throughout all herpesviruses analysed to date (Roizman & Knipe, 2001; Mettenleiter, 2002). Furthermore, it is now a common view that the capsids obtain an initial envelope after budding into the perinuclear space. This envelope enables the particles to fuse with the outer nuclear membrane to release naked nucleocapsids into the cytoplasm. These nucleocapsids traffic to the trans-Golgi network to acquire their final envelope with processed glycoproteins (Radsak et al., 1996).

To date nearly all available drugs are inhibitors of the viral DNA polymerase and do not affect most of the processes just described. Due to multiple problems caused by the current available drugs developing new antiviral targets, which are non-nucleosidic and have a different mechanism of action, are needed. Consequently, to broaden therapy of HCMV infections and to circumvent current mechanisms of drug resistance, an inhibitor of HCMV terminase would be of great value, because it would act subsequent to DNA synthesis and block the first steps in viral maturation.

2. DISCOVERY AND DEVELOPMENT OF BENZIMIDAZOLE NUCLEOSIDES AS ANTIVIRAL AGENTS

2.1 From Vitamin B_{12} to Drugs for HCMV Infections

Interest in benzimidazole nucleosides began in the late 1940's when 5,6-dimethyl-1-(α-D-ribofuranosyl)benzimidazole was found as a constituent of vitamin B_{12} (Beavan et al., 1949). This led chemists to prepare a number of analogs including 5,6-chloro-1-(ß-D-ribofuranosyl)benzimidazole (DRB, Fig. 1) as potential drugs (Tamm et al., 1954). Although initial data were interpreted to mean that DRB and analogs had antiviral activity - especially against influenza viruses, subsequent work established that the activity against viruses was not selective (Tamm I. & Sehgal, 1978).

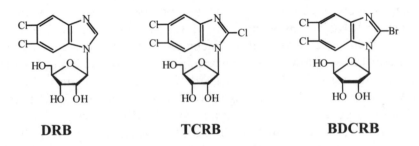

DRB   **TCRB**   **BDCRB**

Figure 1. Structures of benzimidazole-D-ribonucleosides.

Our work with benzimidazole nucleosides originated with the synthesis of the 2-chloro analog of DRB (TCRB, Fig. 1) as a potential anti-cancer agent by Townsend's group in the late 1960's (Townsend and Revankar, 1970). Although inactive as an anti-cancer compound, TCRB was highly selective against HCMV at non-cytotoxic concentrations when evaluated for antiviral activity (Townsend et al., 1995). This was a surprising result because the addition of chlorine at the 2-position of DRB to give TCRB (see Fig. 1) not only made the molecule active against HCMV, it also greatly reduced cytotoxicity compared to DRB. Many analogs then were synthesized and evaluated leading to the conclusion that TCRB, its 2-bromo homolog (BDCRB, Fig. 1), and their 5'-deoxy analogs were among the most active compounds (Krosky et al., 2002).

2.2 Antiviral Activity

The spectrum of antiviral activity of TCRB and BDCRB is very specific and is limited to cytomegaloviruses (CMV's). The compounds were equally active against human and rhesus CMV (North et al., 2004) less active against guinea pig CMV (Nixon and McVoy, 2004), and weakly active or inactive against rodent CMV's. Surprisingly neither compound has significant activity against other herpesviruses including herpes simplex virus types 1 and 2, varicella zoster virus, and human herpes virus-6.

Likewise, the compounds are inactive against respiratory viruses including influenza A, influenza B, respiratory syncytial virus, adenovirus strains 5 and 7. They also are inactive against measles virus; enteroviruses such as coxsackie A9, coxsackie B1, ECHO viruses 7 and 9, polio virus; human immunodeficiency virus; and human papiloma virus.

Activity (IC_{50} or IC_{90}'s) against laboratory strains of HCMV such as Towne and AD169 ranged from 0.1 μM to approximately 3 μM in various plaque and yield reduction assays. This is generally two to five fold more active than ganciclovir (GCV) included in the same assay and is approximately 100-fold lower than concentrations cytotoxic to cultured cells or human progenitor cells (Nassiri et al., 1996). Activity against clinical isolates was similar to that seen against laboratory strains. TCRB and BDCRB also were active against strains of HCMV carrying clinically significant GCV-resistance mutations in the UL97 gene, and against HCMV strains bearing mutations in DNA polymerase associated with GCV, foscarnet or cidofovir resistance.

2.3 From BDCRB to Clinical Candidates

Despite the excellent *in vitro* activity of TCRB, BDCRB and their 5'-deoxy analogs against HCMV, pharmacokinetic and drug metabolism studies in rats and monkeys showed that TCRB and BDCRB were metabolized too rapidly by glycosidic bond cleavage to be good clinical drug candidates (Good et al., 1994). These observations hastened and expanded an analog synthesis program that proceeded concurrently with the biological studies. A number of N-1 modified D- and L-carbohydrate, carbocyclic, and acyclic analogs with various substituents in the 2,4,5 and/or 6-positions were synthesized and evaluated (for examples see Chulay et al., 1999; Gudmundsson et al., 1997; Townsend et al., 1996 and 1999; Zou et al., 1997). From a large number of compounds, two ultimately progressed to clinical evaluation. GW275175X and 1263W94 (maribavir) showed good *in vitro* antiviral activity and selectivity; moreover *in vivo* pharmacokinetic, pharmacological, and toxicological profiles were sufficient for clinical evaluation.

GW275175X is the D-ribopyranosyl homolog of BDCRB; it acts by a mechanism similar to BDCRB (Underwood et al., 2004; see following sections of this chapter). In contrast, maribavir is a L-ribosyl nucleoside (Fig. 2) that acts by a different mechanism (see Chapter 2.2 by K.K. Biron).

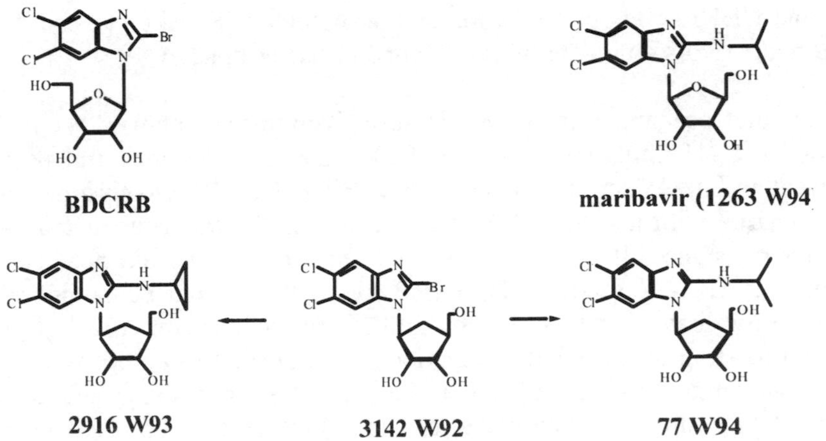

Figure 2. Structures of benzimidazole D- and L-ribosylnucleosides (respectively, top row) and carbocyclic analogs (bottom row).

The discovery route from BDCRB to maribavir was not direct but involved the intermediate discovery of the activity of the carbocyclic analog of BDCRB, 3142W92 (Fig. 2), (Townsend et al., 1996). Resolution of this compound into its component + (or D) and − (or L) enantiomers revealed the surprising result that the − enantiomer was the active compound. This discovery led to the synthesis of other − enantiomers such as 2916W93 and 77W94, and ultimately to the L-riboside, maribavir.

3. INITIAL MODE OF ACTION STUDIES WITH BDCRB

Mechanism of action studies with TCRB and BDCRB in our laboratories established that these compounds did not inhibit DNA, RNA, or protein synthesis. Rather, the compounds acted late in the viral replication cycle on viral assembly by inhibition of the processing of concatemeric DNA into monomeric genome length DNA (Underwood et al., 1998). Selection of drug-resistant virus and marker transfer studies established that resistance mapped to amino acids 344 and 355 of gene UL89 (exon 2) indicating that its gene product was the target for the compounds and that it is involved in HCMV DNA processing (Underwood et al., 1998). Studies with a different HCMV isolate selected for greater resistance to TCRB established that the product of another gene, UL56, was involved in the action of TCRB and BDCRB. These data also first suggested to us that the gene products of UL89 and UL56 act as part of a maturation complex (Krosky et al., 1998) similar to the two-subunit "terminase" found in bacteriophage.

In fact, the mechanism in HCMV is most likely more complex. We have identified a L21F mutation in gene UL104 (that encodes the viral portal protein) of an HCMV strain resistant to TCRB and BDCRB (Komazin et al., 2004). A virus with just the L21F mutation in UL104 was constructed and tested for resistance. It was found that this mutation alone did not confer resistance of HCMV to BDCRB, TCRB, or to the related benzimidazole nucleoside maribavir – which acts by a different mechanism. Although we have data to suggest the UL104 mutation is not needed to compensate for drug-resistant mutation in UL56, it is possible this mutation is needed to compensate for drug-resistant mutations in UL89 or in UL56 and UL89 together.

4. EFFECT ON THE HCMV TERMINASE

Terminases are enzymes that are responsible for the packaging of unit-length genomes into the interior of procapsids. They represent multifunctional heterooligomers whereas one protein catalyzes the ATP-dependent translocation into preformed capsids while the other induces cleavage of DNA concatemers into unit-length genomes. Terminases are highly conserved throughout many double-stranded DNA viruses, including bacteriophages (*e.g.* T3, T4, λ, Φ29), herpesviruses and adenoviruses. The HCMV terminase consists of the large subunit pUL56 and the small pUL89 (Bogner, 2002, Hwang & Bogner, 2002). While pUL56 mediated sequence-specific DNA binding and ATP-hydrolysis, pUL89 is required for duplex nicking (Bogner et al., 1998; Scheffczik et al., 2002; Scholz et al., 2003).

As described above, BDCRB inhibits HCMV DNA processing via the involvement of the terminase subunits pUL89 and pUL56 (Krosky et al., 1998; Underwood et al., 1998; Krosky et al., 2002). Experiments in the presence of BDCRB revealed that this drug targeted to pUL89 has an effect on the nicking activity of pUL89 and therefore leads to the inhibition of viral DNA cleavage (Scheffczik et al., 2002).

Recently, we used BDCRB for analyzing its role in ATP hydrolysis. In order to determine whether the protein is active, the capacity to hydrolyze ATP was analyzed by bioluminometric ATPase activity assays (Savva et al., 2004). This method enables a continuous monitoring of ATP hydrolysis by combining the enzyme activity of proteins with the firefly luciferase system. Another feature is the ability to use this method as a high throughput assay. Interestingly BDCRB reduced the pUL56-associated ATPase activity by approximately 90% (Fig. 3), thus indicating that one reason for the inhibitory effect in viral maturation is the prevention of ATP hydrolysis (Savva et al., 2004).

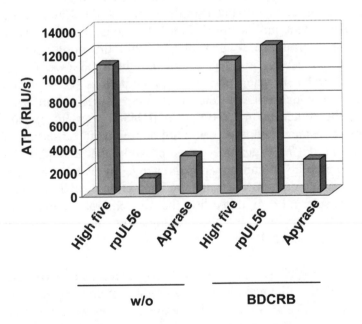

Figure 3. Influence of BDCRB on the pUL56-associated ATPase activity. Extracts from High five cells, *Apyrase* as well as rpUL56 were preincubated without (w/o) or with 100 μM BDCRB in the reaction mixture. The reaction was started by addition of ATP followed by detection of ATP hydrolysis using the luciferase system.

Concerning the short half-life of BDCRB *in vivo* (Good et al., 1994) we have analyzed other benzimidazole analogs with this assay. We identified one compound with an identical inhibitory effect as BDCRB, 2-bromo-4,5,6-trichloro-1-(2,3,5-tri-O-acctyl-ß-D-ribofuranosyl)benzimidazole (BTCARB). However, we do not have data on its oral bioavailability or pharmacokinetics in experimental animals.

Furthermore, DNA packaging also requires the involvement of a so-called portal protein. Portal proteins are oligomers that form a channel at one vertex of the capsid. These proteins are like terminases found in most dsDNA bacteriophages and herpesviruses. The channel enables the insertion of DNA into the capsid. In HCMV the portal protein is encoded by UL104, that assembles into high molecular weight complexes (Dittmer and Bogner, 2005). The protein has the ability to bind DNA, but is not able to hydrolyze ATP. In the case of bacteriophage $\Phi 29$ it was demonstrated that at the end of packaging an internal force is 50 pN (Smith et al., 2001). Since DNA packaging is ATP-dependent and the portal protein has no enzymatic activity the interaction of the terminase subunit pUL56 with the portal protein pUL104 is essential. Recently, a direct interaction between pUL56 and pUL104 was identified by us using co-immunoprecipitations (Dittmer and Bogner, 2005). This is consistent with the observations by Komazin et al. (2004) that found the mutation in UL104 in a BDCRB-resistant mutant. Even though a drug-resistant phenotype could not be found after insertion of the mutation into the HCMV genome, it is possible that the pUL104 mutation could compensate a conformational change in pUL56 or in a UL56-UL89 complex. We demonstrated that the interaction of pUL104 with pUL56 could be specifically inhibited by the tetrachlorobenzimidazole ribonucleoside (Cl_4RB) but not by the virologically inactive, control dimethyl benzimidazole analog (Dittmer et al., 2005).

Both compounds, the ATPase inhibitory BTCRB as well as Cl_4RB, were analyzed concerning their influence on viral maturation. Interestingly, in the presence of BTCARB or Cl_4RB no replication active particles were observed in infected cells (Fig. 4). The activity (IC_{50}) of both compounds against HCMV AD169 are approximately 0.06 in plaque reduction assays. This is several hundreds-fold lower than cytotoxic concentrations.

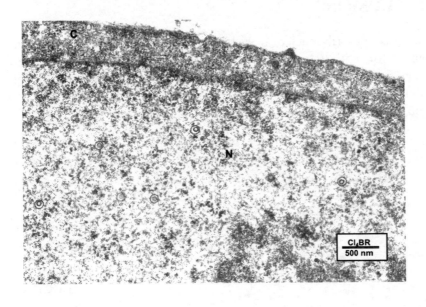

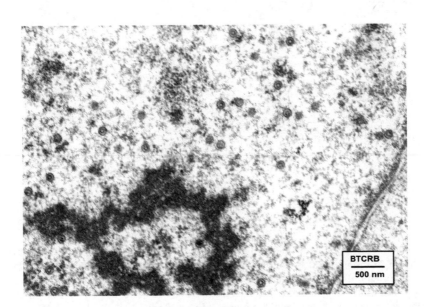

Figure 4: Electron microscopy analysis of ultrathin section of HCMV infected cells in the presence of 50 µM Cl₄RB or 50 µM BTCRB. After 3 days cells were fixed in 4% paraformaldehyde and 2.5% glutaraldehyde with 1% tannin and prepared for electron microscopy. Only empty B-capsids were found in the nucleus.

5. CONCLUSIONS

Benzimidazole ribonucleosides have become the most promising compounds for antiviral treatment of HCMV. The advantage of these compounds is the targeting of unique, virus specific processes absent in host cells, thus leading to higher efficacy and greater safety than currently approved drugs (Fig. 5).

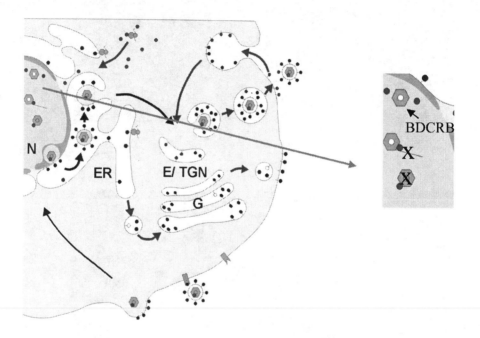

Figure 5: Mode of action of a terminase inhibitor during HCMV maturation. After synthesis in the cytoplasm at the smooth ER pUL56 and pUL89 are translocated into the nucleus. In the late stage of infection capsid assembly is performed. DNA synthesis occurs and a large number of concatamers are found in the nucleus. BDCRB targeted to pUL56 and pUL89 prevents the cleavage of concatenated DNA and packaging in procapsids.

ACKNOWLEDGEMENTS

We thank all present and past members of our and Dr. Karen K. Biron's laboratories for their many contributions to these studies and to Dr. Biron and other colleagues for helpful discussions. The research has been supported by the Wilhelm Sander Foundation (No. 2004.031.1).

REFERENCES

Beavan G.R., Holiday E.R., Johnson E.A., Ellis B., Mamalis P., Petrow V. and Sturgeon B. 1949, The chemistry of anti-pernicious anaemia factors Part III. 5:6-Disubstituted benzimidazoles as products of acid hydrolysis of vitamin B_{12}. *J. Pharmacy Pharmacol.* 1949; **1**: 957-970

Bogner, E., 2002, Human cytomegalovirus terminase as a target for antiviral chemotherapy. *Rev. Med. Virol.* **12**: 115-127

Bogner, E., Radsak, K., and Stinski, M.F,. 1998, The gene product of human cytomegalovirus open reading frame UL56 binds the pac motif and has specific nuclease activity. *J. Virol.* **72**: 2259-2264

Britt, W.J. and Alford, C.A., 1996, Cytomegalovirus. In *Fields Virology* (N. Fields, D.M., Knipe, and P.M. Howley, eds.), 3th ed. Lippincott Williams and Wilkins, Philadelphia,, pp. 2493-2523

Chulay, J., Biron, K. Wang, L. Underwood, M., Chamberlain, S., Frick, L., Good, S., Davis, M., Harvey, R., Townsend, L., Drach, J., and Koszalka, G., 1999 Development of novel benzimidazole riboside compounds for treatment of cytomegalovirus disease. *Adv. Exp. Med. Biol.* **458**: 129-134

Dittmer, A. and Bogner, E., 2005, Analysis of the quaternary structure of the putative HCMV portal protein pUL104. *Biochemistry* **44**: 759-765

Dittmer, A., Drach, J.C., Townsend, L.B., Fischer, A. and Bogner, E., 2005, Interaction of the putative HCMV portal protein pUL104 with the large terminase subunit pUL56 and its inhibition by Benzimidazole-D-Ribonucleosides. *J. Virol.* **79**: in press

Good, S.S., Owens, B.S., Townsend, L.B. and Drach, J.C., 1994, The disposition in rats and monkey of 2-bromo-5,6-dichloro-1-(ß-D-ribofuranosyl)-benzimidazole (BDCRB) and its 2,5,6-trichloro congner (TCRB). *Antiviral Research* **23**: Suppl. I, 103

Gudmundsson, K.S., Drach, J.C., Wotring, L.L., and Townsend, L,B., 1997, Synthesis and antiviral activity of certain 5'-modified analogs of 2,5,6,-trichloro-1-(ß-D-ribofuranosyl)-benzimidazole (TCRB). *J. Med. Chem.* **40**: 785-793

Hwang, J.-S. and Bogner, E,. 2002, ATPase activity of the terminase subunit pUL56 of human cyto megalovirus. *J. Biol. Chem.* **27**: 6943-6948

Komazin, G., Townsend, L.B. and Drach, J.C., 2004, Role of a mutation in human cytomegalovirus gene UL104 in resistance to benzimidazole ribonucleosides. *J. Virol.* **78**: 710-715

Krosky, P.M., Underwood, M.R., Turk, S.R., Feng, K. W-H, Jain, R.K., Ptak, R.G., Westerman, A.C., Biron, K.K., Townsend, L.B. and Drach, J.C., 1998, Resistance of human cytomegalovirus to benzimidazole ribonucleosides maps to two open reading frames: UL89 and UL56. *J. Virol.* **72**: 4721-4728

Krosky P.M., Borysko K.., Nassiri M.R., Devivar R.V., Ptak R.G., Davis M.G., Biron K.K., Townsend L.B., Drach J.C., 2002, Phosphorylation of ß-D-ribosylbenzimidazoles is not required for activity against human cytomegalovirus. *Antimicrob. Agents Chemother.* **46**: 478-486

Mettenleiter, TC., 2002, Herpesvirus assembly and egress. *J. Virol.* **76**: 1537-1547

Mocarski, E. S., Jr,. 1996, Cytomegalovirus and their replication. In *Fields Virology* (N. Fields, D.M., Knipe, and P.M. Howley, eds.), 3th ed. Lippincott Williams and Wilkins, Philadelphia., pp. 2447-2492

Nassiri, M.R., Emerson, S.G., Devivar, R.V., Townsend, L.B., Drach, J.C., and Taichman, R.S., 1996, Comparison of benzimidazole nucleosides and ganciclovir on the in vitro proliferation and colony formation of human bone marrow progenitor cells. *Brit. J. Haematol.* **93**: 273-279

Nixon, D.E. and McVoy, M.A., 2004, Dramatic effects of 2-bromo-5,6-dichloro-1-beta-D-ribofuranosyl benzimidazole on the genome structure, packaging, and egress of guinea pig cytomegalovirus. *J. Virol.* **78**: 1623-1635

North, T.W., Sequar, G., Townsend, L.B., Drach, J.C., and Barry, P,A., 2004, Rhesus cytomegalovirus is similar to human cytomegalovirus in susceptibility to benzimidazole nucleosides. *Antimicrob. Agents Chemother.* **48**: 2760-2765

Radsak, K., Eickmann, M., Mockenhaupt, T., Bogner, E., Kern, H., Eis-Hübinger, A. and Reschke, M., 1996, Retrieval of human cytomegalovirus glycoprotein B from infected cell surface for virus envelopment. *Arch.Virol.* **141**: 557-572

Roizman, B. and Knipe, D., 2001, Herpes simplex viruses and their replication. In *Fields Virology* (D.M., Knipe, and P.M. Howley, eds.), 3th ed. Lippincott Williams and Wilkins, Philadelphia, pp. 2399-2460

Savva, C.G.W., Holzenburg, A. and Bogner, E., 2004, Three-dimensional structure of the terminase subunit pUL56 of human cytomegalovirus. *FEBS Lett.* **563**: 135-140

Scheffczik, H., Savva, C.G.W., Holzenburg, A., Kolesnikova, L. and Bogner, E., 2002, The terminase subunits pUL56 and pUL89 of human cytomegalovirus are DNA-metabolizing proteins with toroidal structure. *Nuc Acids Res.* **30**: 1695-1703

Scholz, B., Rechter, S., Drach, J.C., Townsend, L.B. and Bogner, E., 2003, Identification of the ATP-binding site in the terminase subunit pUL56 of human cytomegalovirus. *Nuc.Acids Res.* **31**: 1426-1433

Smith, D.e., Tans, S:J., Smith, S.B., Frimes, S., Anderson, D.L. and Bustamante, C., 2001, The bacteriophage straight phi29 portal motor can package DNA against an internal force. *Nature* **413**: 748-752

Tamm I., Folkers K., Shunk C.H., Horsfall F.L., 1954, Inhibition of influenza virus multiplication by n-glycosides of benzimidazoles. *J. Exp. Med.* **99**: 227-250

Tamm I. and Sehgal P.B., 1978, Halobenzimidazole ribosides and RNA synthesis of cells and viruses. *Adv. Virus Res.* **22**: 187-258

Townsend L.B. and Revankar G.R., 1970, Benzimidazole nucleosides, nucleotides, and related derivatives. *Chem. Rev.* **70**: 389-438

Townsend L.B., Devivar R.V., Turk S.R., Nassiri M.R., and Drach J.C., 1995, Design, synthesis and antiviral activity of certain 2,5,6-trihalo-1-(ß-D-ribofuranosyl)benzimid-azoles. *J. Med. Chem.*, **38**: 4098-4105

Townsend, L.B., Drach, J.C., Good, S.S., DaLuge, S.M., and Martin, M.C.,1996, Therapeutic Nucleosides. U.S. Patent 5,534,535 issued 7/9/96.

Townsend, L.B., Gudmundsson, K.S., Daluge, S.M., Chen, J.J., Zhu, Z., Koszalka, G.W., Boyd, L., Chamberlain, S.D., Freeman, G.A., Biron, K.K., and Drach., J.C., 1999, Studies designed to increase the stability and antiviral activity (HCMV) of the active benzimidazole nucleoside, TCRB. *Nucleosides and Nucleotides*, **18**: 509-519

Underwood, M.R., Harvey, R.J., Stanat, S.C., Hemphill, M.L., Miller, T., Drach, J.C., Townsend, L.B. and Biron, K.K., 1998, Inhibition of human cytomegalovirus DNA maturation by a benzimidazole ribonucleoside is mediated through the UL89 gene product. *J. Virol.* **72**: 717-72

Underwood, M.R., Ferris, R.G., Selleseth, D.W., Davis, M.G., Drach, J.C., Townsend, L.B., Biron, K.K. and Boyd, F.L., 2004, Mechanism of Action of the Ribopyranoside Benzimidazole GW275175X Against Human Cytomegalovirus. *Antimicrob. Agents Chemother.* **48**: 1647-1651

Zou, R., Drach, J.C. and Townsend, L.B., 1997 Design, Synthesis and Antiviral Evaluation of 2-Substituted 4,5-Dichlro- and 4,6-Dichloro-1-(ß-D-ribofuranosyl)benzimidazoles as Potential Agents for Human Cytomegalovirus Infections. *J. Med. Chem.* **40**: 802-810

Chapter 2.4

RECENT DEVELOPMENTS IN ANTI-HERPESVIRAL THERAPY BASED ON PROTEIN KINASE INHIBITORS

T. HERGET[1] and M. MARSCHALL[2]
[1]Merck KGaA, Darmstadt, Germany; [2]Institute for Clinical and Molecular Virology, University of Erlangen-Nürnberg, Schlossgarten 4,91054 Erlangen, Germany)

Abstract: Although modern approaches of antiviral research have lead to many successful antiviral drugs, these antivirals have certain limitations. Most of them rapidly select for resistance, frequently induce an unsatisfactorily high degree of adverse side effects and tend to be active against only one or a few related viruses. Therefore, the development of novel antiviral strategies is a challenging goal of current investigations. Protein kinases are now considered as potential antiviral targets with a high perspective to improve the efficacy of therapy. With regard to herpesviral infections, both viral and cellular protein kinases are crucial regulatory factors determining the efficiency of virus reproduction. Beyond that, all known viruses are dependent on or at least strongly influenced by protein kinase activities. Thus, pharmacological protein kinase inhibitors, targeting either viral or cellular protein kinases or both, may be active against a variety of viruses, which commonly require similar kinase activities. This approach might even provide novel drugs against viral variants resistant to conventional antiviral therapy. The specific chances and challenges of protein kinase inhibitor-based antiviral strategies, particularly those opening various novel possibilities of combination therapies, are focused in this review.

1.   INTRODUCTION

Viral replication is a complex process relying on a sophisticated network of interacting proteins encoded by the viral and host genome. Protein kinases are important regulators of the replication program of many human pathogenic viruses. During herpesviral infections, both viral and cellular

351

E. Bogner and A. Holzenburg (eds.), New Concepts of Antiviral Therapy, 351–371.
© 2006 *Springer. Printed in the Netherlands.*

protein kinases provide activities crucial for efficient virus reproduction. The specific phosphorylation of substrate proteins generally induces activation, inactivation or other functions of the substrate and thus may directly regulate herpesvirus-host cell interaction. There is a variety of prominent examples of protein kinases determining replication of human herpesviruses, such as MAP kinases, cyclin-dependent kinases and herpesvirus-encoded protein kinases, such as pUL97 of human cytomegalovirus (HCMV). The identification of essential phosphorylation steps and the validation of protein kinases as antiviral targets are challenging first-line goals for biotech and pharma industry. Once an appropriate target is validated, the management of an extensive and successful screening procedure is an often underestimated recipe. The identification and development of small molecules, selectively inhibiting one or only a few distinct protein kinases, will eventually lead to a pre-clinical candidate. Usually from a broad spectrum of drug candidates, i.e. compounds derived from various chemical classes with anti-herpesviral activity in cell culture and animal models, there is only a small number of candidates entering clinical trials.

A particular chance for the development of protein kinase inhibitors for anti-herpesviral therapy is the obvious necessity of novel combination therapies with reduced side-effects. Such drugs, which might be applicable also in long-term regimens and particularly for infection of infants are urgently needed. Novel anti-herpesviral protein kinase inhibitors are unlikely to interfere with conventional drugs such as inhibitors of viral DNA synthesis. Furthermore, there are promising attempts to use combinations of protein kinase inhibitors and inhibitors with twofold specificity, e.g. interfering with both virus-encoded kinases and cellular protein kinases which may be required for efficient viral replication. Addressing crucial host kinases may not only lead to improved therapy of herpesviral diseases but may yield broad-spectrum antiviral drugs interfering with protein kinase activities required by various herpesviruses and potentially also nonrelated other human pathogenic viruses.

Among the large number of widely distributed viruses belonging to the family of Herpesviridae, there are eight members, which selectively infect humans and cause a variety of clinically significant diseases (Table 1). Although most herpesviral diseases are self-limiting in immunocompetent individuals, they can be associated with severe and even life-threatening outcome in immunocompromised patients. Particularly problematic is the infection of immunosuppressed patients, such as transplant recipients, tumor or AIDS patients. In some cases, particularly with human cytomegalovirus (HCMV), the infection of infants and neonates is a nonresolved problem in clinical virology. In addition to acute infection, herpesviruses establish life-long persistent infections (viral latency) that may reactivate and eventually result in recurrent disease and virus shedding. In the absence of vaccination, which is presently available only for varicella zoster virus (VZV), a strong need exists for means of safe and effective antiviral therapy. Currently, anti-herpesviral therapy is primarily applied against infections with herpes simplex viruses types 1 and 2 (HSV-1, HSV-2), VZV and HCMV (Table 1). The clinical application of anti-herpesviral drugs, however, faces severe limitations by the induction of adverse side effects and selection of resistant viruses. Thus, novel antiviral targets and therapy approaches are needed for improved treatment.

Table 1: Herpesvirus infections of humans

Virus	Tropism of infection	Seroprev.*	Associated diseases	Common treatments#
HSV-1	neurotropic, lytic replication in epithelia, latency in ganglia	70-90%	cutaneous herpes, encephalitis, genital h.;	ACV (iv/po/top), VAL (po), FAM (po), PCV (iv);
HSV-2		10-30%	genital herpes, encephalitis;	ACV (iv/po/top), VAL (po), FAM (po);
VZV		85-95%	varicella, zoster	ACV (iv/po)
HCMV	lytic replication in fibroblasts, epithelial, endothelial and smooth muscle cells, latency in granulocyte-macrophage progenitors	40-90%	mononucleosis, hepatitis, congenital CID; immunocompromised: retinitis, pneumonitis, colitis, encephalitis	GCV (iv/po), VALGCV (po), FOS (iv/top), CDV (iv), FOM (invit)
EBV	lymphotropic, lytic replication in epithelial cells, latency in B cells	85-95%	mononucleosis, hepatitis, Burkitt's lymphoma, nasopharyngeal carcinoma	ACV (iv/po)δ
HHV-6	lymphotropic	60-100%	roseola infantum	GCV (iv/po)
HHV-7	lymphotropic	60-100%	roseola infantum	GCV (iv/po)
HHV-8	lymphotropic, replication in B cells, epithelial and spindle cells of KS	~5%	Kaposi's sarcoma	radiation, cytotoxic drugs, IFN-α, GCV

* Seroprevalence in healthy adults in industrialized countries
# ACV, acyclovir; VAL, valacyclovir; FAM, famciclovir; PCV, penciclovir; GCV, ganciclovir; VALGCV, valganciclovir; FOS, foscarnet; FOM, fomivirsen; iv, intravenous; po, oral; top, topical; invit, intravitreal
δ Clinical efficacy controversially discussed; proven inhibitory effect only on lytic replication

2.   CURRENT STATE OF ANTI-HERPESVIRAL THERAPY

2.1   Inhibitors of DNA Replication

In recent years, the demand for new antiviral strategies has increased markedly and there are numerous novel drugs that are available or are being developed (De Clercq E., 2004). Antiviral therapy is primarily available for members of four major groups of clinically important viruses: human immunodeficiency virus type 1 (HIV-1), herpesviruses, hepatitis viruses and influenza viruses. The discovery of the inhibitor of viral DNA replication acyclovir (ACV), more than 25 years ago (Schaeffer et al., 1978), represents a milestone in the management of herpesvirus infections. Fig. 1 provides an overview of developments in antiviral therapy in front of important periods of virus epidemics. While the first anti-herpesviral drugs (nucleoside analogues like idoxuridine) failed to reach clinical importance (with the exception of topical treatment), intravenously administered ACV was shown to possess remarkable efficacy and safety with systemic administration. Since then, ACV has become the therapy of choice for HSV and VZV infections. In a similar fashion, the DNA replication inhibitor ganciclovir (GCV) became the gold standard for management of infections with HCMV, particularly in immunosuppressed transplant recipients. The use of inhibitors of DNA replication in anti-herpesviral therapy is summarized in Table 1. Nucleoside analogues are actually prodrugs which undergo specific modifications in infected cells. GCV needs to be activated intracellular by phosphorylation. In HCMV-infected cells, this important pace-maker reaction is performed by the HCMV UL97-encoded protein kinase (pUL97). The central role of this protein kinase for GCV therapy has been recognized a decade ago (Littler et al., 1992; Sullivan et al., 1992). It is striking to note that pUL97 does not phosphorylate natural nucleosides (Michel et al., 1996). Its physiological role during HCMV replication is the phosphorylation of viral and cellular protein substrates (Kawaguchi et al., 1999; Marschall et al., 2003; Krosky et al., 2003b; Marschall et al., 2005). However, during GCV therapy, pUL97 mediates the phosphorylation of GCV to its monophosphate which subsequently becomes further phosphorylated by cellular enzymes involved in nucleotide metabolism. The resulting GCV triphosphate obstructs viral DNA synthesis by inhibiting the viral DNA polymerase activity by competing with the natural nucleoside triphosphate dGTP. A premature chain termination of the evolving DNA strand is the consequence. The latter aspect is the reason why replication and repair of cellular DNA is partially also affected in HCMV infected cells by phosphorylated GCV thereby causing cytotoxicity. The selectivity of therapy of HCMV disease with GCV, and also penciclovir (Zimmermann et al., 2000), is therefore mainly provided by the special mode of action of the viral kinase pUL97. Thus, pUL97 is essential for GCV therapy and virus resistance to GCV frequently results from UL97 mutation.

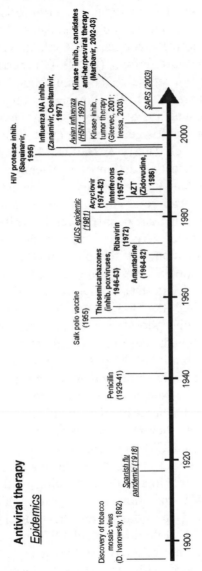

Figure 1: Hallmarks of human virus epidemics and antiviral therapy. Important starting points of virus epidemics and developents of antiviral strategies are listed in the time from 1892 to 2005. The periods indicated in brackets cover the time between compound discovery, development, approval as a medical drug and the start of clinical use.

2.2   Novel approaches including herpesviral protein kinases as antiviral targets

Poor oral bioavailability of GCV and other first-generation anti-herpesviral drugs and the emergence of drug-resistant virus mutants have driven the search for novel drug candidates. Many of these are also nucleoside or nucleotide analogues that inhibit viral DNA synthesis, e.g. famciclovir (FAM) and cidofovir (CDV). Others are targeted to viral gene products, which do not regulate DNA replication but viral gene expression. An interesting example is the antisense oligonucleotide ISIS 2922 (fomivirsen, FOM), which is a sequence-specific inhibitor of transcription of the major immediate early gene region thereby blocking the onset of the HCMV replication cycle (Perry & Balfour, 1999). Other novel approaches are directed to the selective action of viral enzymes, such as proteases or protein kinases. Protein kinases encoded by human herpesviruses possess important physiological roles during viral replication. Some of these functions have been elucidated only recently while others have still to be discovered. Functions of herpesviral protein kinases are focused on several stages of viral replication, i.e. the phosphorylations directed to tegument proteins of the infecting virions, immediate early and early transactivators, processivity factors of viral DNA polymerases, as well as main mediators of cell regulation (RNAP II, EF-1d). Another function is part of viral late replication, i.e. the phosphorylation of proteins involved in viral nuclear capsid export. Combined, these complex functions are potential targets for small molecule inhibitors of herpesviral protein kinases. Thus, inhibitor compounds derived from independent chemical classes, such as indolocarbazoles, benzimidazole ribosides, quinazolines and others, were shown to possess strong anti-herpesviral activity *in vitro*. Particularly, novel strategies of antiviral therapy were exemplified by the development of inhibitors targeting the HCMV protein kinase pUL97 (Marschall et al., 2002; Biron et al., 2002, Herget et al., 2004). In this context, a number of recent publications contributed to the understanding of the physiological role of pUL97 during HCMV infection. pUL97 is a multifunctional viral protein kinase with profound importance for the efficiency of viral replication (Prichard et al., 1999). Inhibition of pUL97 kinase activity or deletion of ORF-UL97 from the viral genome lead to defects in the early and late events of replication associated with viral DNA synthesis and nucleocytoplasmic capsid export (Wolf et al., 2001; Marschall et al., 2003; Krosky et al., 2003a; Krosky et al., 2003b). Thus, specific inhibitors of the pUL97 activity are prospective candidates for the development of novel drugs for the treatment of HCMV-associated disease.

Table 2: Pharmacological inhibitors of protein kinases and phosphatases

(a): approved drugs for human diseases

drug:	substance type:	targeted kinase / phos.:	therapy:	status:
CsA (cyclosporine)	small-mol. inhib.	PP2B phosphatase	immunosuppression	approved 1983 [a]
Fasudil (HA1077)	small-mol. inhib.	Rho-kinase (ROCK)	cerebral vasospasm	approved 1995 [b]
Herceptin (trastuzumab)	antibody	Her2	breast cancer	approved 1998 [c]
Rapamune (rapamycin)	small-mol. inhib.	mTOR	immunosuppression	approved 1999 [d]
Gleevec (imatinib)	small-mol. inhib.	ABL (PDGFR, KITR)	leukemia	approved 2001 [e]
Iressa (gefitinib)	small-mol. inhib.	EGFR	lung cancer (NSCLC)	approved 2003 [f]
Tarceva (erlotinib)	small-mol. inhib.	EGFR	lung cancer (NSCLC)	approved 2004 [g]
Erbitux (cetuximab)	antibody	EGFR	colorectal cancer	approved 2004 [h]

(b): candidates for novel antiviral drugs

candidate compound:	chemical class:	targeted kinase:	sensitive viruses:	status:
Maribavir (1263W94)	benzimidazole riboside	pUL97/pUL27	HCMV	clinical phase I /II [i]
G66976, NGIC-I	indolocarbazoles	pUL97	HCMV	pre-clinical [k]
Ax7376, Ax7396, Ax7543	quinazolines (gefinitib-der.)	pUL97/EGFR	HCMV	pre-clinical [l]
Roscovitine	purine-derived cdk inhib.	cdk 1/2/5/7	HCMV, HSV-1, HIV-1	pre-clin. [m], clin. phase II [n]
Flavopyridol	flavonoid	cdk 1/2/4/9	HCMV, HSV-1, HIV-1	pre-clin. [m], clin. phase II [n]

[a] reviewed by Ponticelli C, 2004, J Nephrol
[b] Schering / Asahi Kasei; Takanashi et al., 2001, Acta Neurochir Suppl
[c] Genentech / Roche; Cobleigh et al., 1999, J Clin Oncol; Vogel et al., 2002, J Clin Oncol
[d] Novartis; reviewed by Cohen P, 2002, Nature Rev Drug Discovery
[e] Novartis; reviewed by Cohen P, 2002, Nature Rev Drug Discovery
[f] Astra Zeneca; reviewed by Cohen et al., 2003, Oncologist
[g] OSI Pharmaceuticals; reviewed by Fabian et al., 2005, Nature Biotechnol
[h] ImClone Systems / Merck / Bristol-Myers Squibb; reviewed by Wilkes GM, 2005, Clin J Oncol Nurs
[i] ViroPharma / GSK; Lalezari et al, 2002, Antimicrob Agents Chemother ; Wang et al, 2003, Antimicrob Agents Chemother
[k] Marschall et al, 2002, J Gen Virol; Zimmermann et al, 2000, Antivir Res; Slater et al, 1999, Bioorg Med Chem Lett
[l] Herget et al, 2004, Antimicrob Agents Chemother
[m] reviewed by Schang LM, 2004, Biochim Biophys Acta; Schang LM, 2002, J Antimicrob Chemother; Schang LM, 2001, Antivir Chem Chemother
[n] presently analyzed in clinical trials for tumor therapy

2.3   Cellular protein kinases as putative antiviral targets

During the replication of herpesviruses, important regulatory functions are not solely fulfilled by herpesvirus-encoded protein kinases but additionally by various cellular protein kinases which are specifically activated or induced in infected cells. Presently, there is an increasing number of publications pointing to the importance of families of cellular protein kinases for replication of human herpesviruses. Main attention has been drawn to the MAPKs (mitogen-activated protein kinases) and CDKs (cyclin-dependent kinases) (Fortunato et al., 2000; Schang LM., 2002). Members of these protein kinase families have been shown to be crucial for replication efficiency of alpha-, beta- and gamma-herpesviruses. Moreover, several examples illustrated that the development of selective inhibitors of these protein kinases (CDK inhibitors, MAPK inhibitors, EGFR inhibitors and others) eventually possess markedly high antiviral activity against HCMV, HSV-1 and/or Epstein-Barr virus (EBV) (Schang LM., 2002; Cohen P., 2002; Schang LM., 2004; Klebl et al., 2005). Thus, a very promising development is the generation of protein kinase inhibitors with selectivity for virus-activated cellular kinases or herpesvirus-encoded kinases and also those with multipotent inhibitory activity. Approved drugs and candidate compounds targeting protein kinases are listed in Table 2 and Fig. 2.

3.   DEVELOPMENT OF DRUGS AGAINST PROTEIN KINASES

The characterization of protein kinases, and also phosphatases, has revealed new targets for the development of drugs for several indications, including diabetes, inflammatory disorders and cancer (Table 2). The identification of cellular and virally encoded kinases as being essential for HCMV pathogenesis makes them attractive targets for antiinfective therapies (Müller G., 2003; Klebl et al., 2005).

The development of a novel antiviral drug initiates with the identification of a target protein, the modulation of which might inhibit or reverse disease progression. Several techniques are used for the identification and validation of novel virally encoded targets – such as allelic-exchange mutagenesis or high-density mutagenesis – that can be used to knock-out a particular gene from the viral genome. Furthermore, analysing the mRNA and protein expression profile of cellular genes in infected and non-infected cells with the help of microarrays allows the identification of de-regulated genes, which are potential targets for antiviral therapy. Then, the use of antisense or small interfering (si) RNA to down-regulate mRNA expression might help to validate essential genes involved in viral replication and pathogenesis. Transfection of host cells with a mutated, inactive version of the protein kinase, which exerts a dominant-negative effect, may further validate the kinase as a drug target.

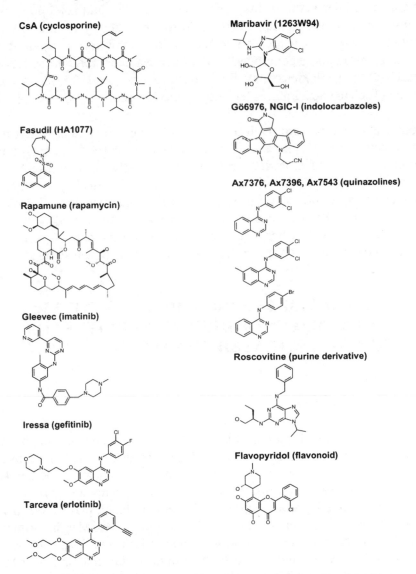

Figure 2: Chemical structures of protein kinase and phosphatase inhibitors.

After validation of a kinase, small compounds that modify its activity can be identified by screening compound libraries using purified enzyme. An example for successful screening and characterization procedures is described in Fig. 3. Kinase inhibitors, so-called "hits", from these bio-chemical screens are further selected by criteria such as physical properties – for example cellular permeability, microsomal stability and solubility. The most promising compounds are then tested for their ability to inhibit viral replication *in vitro* and to determine the minimal inhibitory concentration (MIC). In addition, these compounds are further evaluated for cytotoxicity in cultured cell lines (toxicity profiling) and are also used for identifying inhibition of related host kinases (selectivity profiling).

Compounds that perform well in these enzymatic and cellular assays are further selected on the basis of a favourable eADME- (early drug absorption, distribution, metabolism and excretion) and toxicity profile. Medicinal chemistry helps to optimize these compounds with promising pharmaco-kinetics and pharmacodynamic properties (PPKIs, pharmacological protein kinase inhibitors). Positive candidates, known as "lead compounds", are tested in animal models when available. Successful candidates – so-called "pre-clinical candidates" – are then further evaluated in clinical settings.

4.   QUINAZOLINES AND OTHER COMPOUNDS AS EXAMPLES OF DRUG CANDIDATES FOR HCMV-ASSOCIATED DISEASE

4.1   Quinazolines

The HCMV-encoded protein kinase pUL97 represents an important determinant of viral replication and pUL97-specific kinase inhibitors proved to be powerful tools for the control of HCMV replication. Recently, evidence was provided that three related quinazoline compounds are potent inhibitors of the pUL97 kinase activity being able to block pUL97 substrate phosphorylation (Herget et al., 2004). Replication of HCMV in primary human fibroblasts was suppressed with high efficiency. IC_{50} values of these quinazoline compounds were in the range of ganciclovir (2-4 μM). Importantly, a strong inhibitory effect of quinazolines was demonstrated even for clinical HCMV isolates including GCV- and CDV-resistant viruses.

A

Protein kinase in-cell activity assay

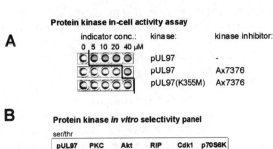

B

Protein kinase *in vitro* selectivity panel

ser/thr

	pUL97	PKC	Akt	RIP	Cdk1	p70S6K
Ax7376	+	-	-	-	-	-

tyr

	Abl	Src	Lck	Kit	EGFR	PDGF	InsR	Met
Ax7376	-	-	-	-	+	-	-	-

tyr and ser/thr

	ERK1	p38	JNK
Ax7376	-	-	-

C

HCMV GFP-based replication assay

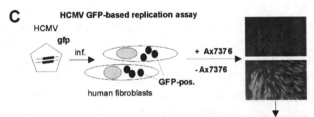

(a) cell lysate: automated GFP fluorometry
(inhibitor screenings)

(b) fixed cells: GFP-based FACS
(test % replication inhibition / cytotoxicity)

(c) infectious supernatant: GFP plaque assay
(test inhibition of virus release)

D

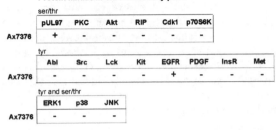

E

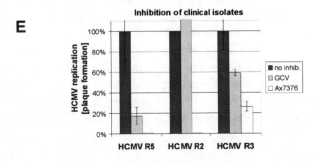

Figure 3: Initial steps of drug development at the preclinical stage: example of quinazoline compounds possessing anti-HCMV activity. (A) Inhibition of the cytomegaloviral protein kinase pUL97 by quinazoline Ax7376 was determined by large screening panels using the protein kinase in cell-activity assay (Marschall et al., 2001; Mett et al., 2005). Kinase activity was determined in UL97-transfected cells by the use of a specific indicator substrate (GCV) which becomes cytotoxic after phosphorylation (conversion to a red colour signal). Mutant K355M was used as a catalytically inactive control. (B) A selectivity panel characterizing the inhibitory potential of Ax7376 was based on *in vitro* kinase assays (Mett et al., 2005). +, inhibition of kinase activity to 10% or less; -, lower or no inhibitory effect (C) A replication assay for HCMV, based on green fluorescent protein (GFP) expression of a recombinant virus, is used for the quantification of the antiviral effect of novel drugs (Marschall et al., 2000). (D) The antiviral effect of Ax7376 towards laboratory strain HCMV AD169-GFP was demonstrated by the GFP-based replication assay. GCV (20 µM) served as a positive control. (E) The anti-HCMV effect of Ax7376 towards clinical HCMV isolates was demonstrated by conventional plaque reduction assays. R5, GCV-sensitive; R2, GCV-high level resistant; R3, GCV-low level resistant (Herget et al., 2004).

Moreover, in contrast to GCV, resistance formation against quinazolines was not observed. Quinazolines inhibit viral early-late protein synthesis but have no obvious effects at other stages of the replication cycle such as viral entry. The mode of action of selected quinazoline compounds in blocking HCMV replication was confined to the inhibition of the pUL97 protein kinase. Importantly, these compounds were also effective in inhibiting GCV and CDV therapy-resistant clinical isolates of HCMV. Further experimental details and results obtained with quinazoline compounds are given in Fig. 3.

Interestingly, quinazolines are not strictly selective in inhibition of pUL97 but also show a strong inhibitory effect towards the cellular tyrosine kinase EGFR. The drug Iressa (gefitinib, ZD1839), a EGFR kinase inhibitor approved for tumor therapy (Cohen et al., 2003), also belongs to the class of quinazolines. It might be rather beneficial for an anti-HCMV drug with respect to efficacy and resistance formation to block both pUL97 and the EGFR, since EGFR represents an HCMV binding and signaling receptor (Wang et al., 2003b). Thus, for the treatment of HCMV infection, a promising approach is using a combination therapy with drugs addressing different target proteins or domains. In this context, one of the limitations of pUL97 inhibitors is given by the fact that they are not suitable for usage in conjunction with UL97-dependent prodrugs like GCV. However, the use of protein kinase inhibitors possessing twofold specificity, e.g. for pUL97 and in addition for a cellular protein kinase which plays an important role in HCMV replication (such as EGFR), might be particularly efficient. Furthermore, it is conceivable that GCV will be continued to be used as first-line treatment, but upon resistance formation, kinase inhibitors will be applied for follow-up treatment. The use of combination therapies with protein kinase inhibitors (e.g. those approved for other clinical applications such as tumor therapy) might also open new avenues for antiviral treatment. In this respect, quinazolines with a favourable eADME / toxicity profile are quite promising and derivatives of this class are in late stage of clinical development or already approved for treatment of cancer patients (Iressa/ gefitinib, Tarceva/erlotinib). Indeed, also gefitinib showed a concentration-dependent inhibition of pUL97 activity in *in vitro* kinase assays and to some degree blocked HCMV replication in HFF cells. However, the potency of gefitinib was lower than that of quinazoline hit compounds (Herget et al., 2004). Thus, quinazolines are promising canditdates which encourage a detailed further investigation of pharmacological, toxicological and antiviral properties in preclinical and clinical settings.

4.2 Indolocarbazoles

Independent detailed investigations led to the validation of pUL97 as an antiviral target (Marschall et al., 2002; Krosky et al., 2003a). A strong inhibitory effect of indolocarbazole compounds (e.g. NGIC-I) on pUL97 kinase activity and on the *in vitro* replication of HCMV was previously reported (Marschall et al., 2001; Slater et al., 1999; Zimmermann et al., 2000). However, the excellent antiviral potency of distinct indolocarbazoles *in vitro* seemed to be accompanied by disadvantageous pharmacological properties *in vivo* such as poor pharmacokinetics and bioavailability (Slater et al., 2001). Thus, further preclinical developments await continuation.

4.3 Benzimidazoles

Another pUL97-inhibiting compound, 1263W94 (maribavir), belonging to the chemical class of benzimidazole L-ribosides, has been characterized by several investigators (Biron et al., 2002; Chulay et al., 1999; McSharry et al., 2001; in more detail described in chapter 2.2). In preclinical and phase I/II clinical studies, maribavir possessed clear antiviral activity (Lalezari et al., 2002) and very promising pharmacokinetic profiles (Koszalka et al., 2002) accompanied by a low degree of severe adverse effects (Wang et al., 2003a). Since maribavir is a potent inhibitor of UL97 kinase activity *in vitro*, its main target of action in HCMV-infected cells was postulated to be pUL97 (Biron et al., 2002). However, selection and analysis of maribavir-resistant HCMV variants surprisingly mapped the resistance-conferring mutation to a gene of unknown function (UL27) while UL97 was unchanged (Komazin et al., 2003). This points to a more complex and controversially discussed mode of action of maribavir. Nevertheless, the promising status of an antiviral strategy targeting pUL97 seems undoubted (Coen & Schaffer, 2003).

5.   FUTURE PERSPECITVES FOR COMBINATION THERAPIES BASED ON PROTEIN KINASE INHIBITORS

5.1   Conventional anti-herpesviral drugs in combination with novel pharmacological protein kinase inhibitors

Combination therapy with antiviral drugs is a field that has been profoundly improved after introduction of highly active antiretroviral therapy (HAART) of HIV-1 positive individuals. Since 1996, striking decreases in the number of progressive AIDS and death have been reported in people with HIV-1 coincident with the introduction and widespread uptake of HAART (Mocroft et al., 1998; Palella et al., 1998; Porter et al., 2003). An important basis of HAART is the individual composition of drug combinations due to the specific requirements of the patient. These may take into account both medical and virological parameters, in particular the resistance profile of the respective HIV-1 isolate. Thus, it is an intriguing advantage of HAART to include combinations of antiviral drugs targeted to different viral enzymes. This therapeutic strategy enables the interference with more than one regulatory step of viral replication and specifically favours the combination of drugs possessing different modes of action, e.g. inhibitors of viral reverse transcriptase, protease, integrase, fusion activity etc. This principle might be similarly helpful for novel regimens of anti-herpesviral therapy using protein kinase inhibitors. Conventional therapy of herpesvirus infections is mainly based on drugs interfering with viral DNA synthesis, such as ACV, GCV, CDV, FOS and others (Table 1). Therefore, the introduction of pharmacologically optimized inhibitors of viral and/or cellular protein kinases might widely broaden the spectrum of therapeutic possibilities. It seems promising to combine the established inhibitors of viral DNA synthesis with novel pharmacological protein kinase inhibitors (PPKIs), particularly in those cases where conventional therapy is ineffective, e.g. after monotherapy-based selection of drug-resistant virus. Combination therapy might allow to reduce the effective dosing of conventional drugs, such as GCV, which might be a major advantage in those cases dealing with severe side-effects, e.g. in long-term antiviral therapy. Here, important experiences can be transferred from therapies or clinical trials with antitumor drugs having the potential to be used in combination (Herbst et al., 2004). In case of antiviral combination therapy, another important advantage of reducing drug concentrations might be given in the therapy of pregnant women and infants. This seems particularly prospective with the use of highly specific inhibitors of virus-encoded protein kinases which might be well-tolerated due to the lack of inhibition of cellular enzymes.

5.2 PPKIs with high selectivity for herpesviral or cellular protein kinases

Five PPKIs have so far been approved for clinical use in tumor therapies (Fasudil, Rapamune, Gleevec, Iressa and Tarceva; Table 2, Fig. 2) and a far greater number is undergoing human clinical trials in the near future. In the field of cancer, in which much of the effort has been directed, PPKIs are proving to be well-tolerated compared with conventional chemotherapy treatments. This is encouraging, because several PPKIs that have entered human clinical trials turned out to lack their initially predicted selectivity (Davies et al., 2000; Bain et al., 2003; Daub et al., 2004). Nevertheless, there is a number of PPKIs recognizing their primary target with rather high specificity (Fabian et al., 2005) and there is a constant attempt of researchers in this field to improve PPKI selectivity as a safety aspect for clinical trials. For inhibitors of herpesviral protein kinases, the monospecific mode of action is also not always provided, although at the first glance it seems easier to develop small molecules with specificity for a virus-encoded enzyme which may possess unique properties distinct from most cellular enzymes. However, in practical research, several herpesviral protein kinases show properties which are very similar to cellular kinases and inhibitor profiles are frequently overlapping. Therefore, it is an attractive aim to develop highly selective PPKIs, some of which may be directed to cellular target kinases, others to single members of the herpesviral protein kinase family, particularly when both, viral and cellular protein kinases are required for an efficient herpesviral replication. Once this aim is achieved, a combination of monospecific PPKIs might enable to reach high efficacy of virus inhibiton even at very limited concentrations of each drug. In some cases, this strategy might include the use of PPKIs already approved for tumor therapy. By combination therapy treatment with these drugs, it seems suggestive to reduce or eliminate adverse side effects usually seen during intensive therapy trials with GCV such as long-term treatment. Thus, combination therapy approaches with highly selective PPKIs may have theoretically several advantages over current clinical HCMV treatments.

5.3   PPKIs with twofold specificity for both, herpesviral and cellular protein kinases

The selectivity of PPKIs in antiviral activity is a crucial point of drug development and in many cases the evidence for monospecificity of action is difficult to provide. It is an inherent property of many protein kinase-directed small molecule inhibitors to interact with groups of kinases possessing similar inhibitor binding sites rather than with a single kinase. However, this situation is not in all cases a draw-back in development, but may rather become an advantage when considering the points of drug efficacy and the probability of inducing drug resistance. As far as efficacy is concerned, it is a major experience from many current PPKI characterization studies that inhibitors with twofold, or even multifold, specificity may possess highest disease-inhibiting capacity. Prominent examples are CDK inhibitors, such as Roscovitine and Flavopyridol, which both target at least four cellular protein kinases and are interesting candidates in clinical trials for tumor and antiviral therapy (Table 2). There might be a number of further examples for twofold/multifold-specificity PPKIs possessing high antiviral or antitumor efficacy. However, relatively little effort has been made to exactly characterize the specificities of potential PPKI drug candidates *in vivo*. Thus, monospecificity of action may not necessarily be advantageous for kinase inhibitors. To the contrary, several lines of evidence argue for the advantages of PPKIs with twofold specificity, e.g. the lower predicted frequency of viral resistance selected by antiviral treatment. The requirement to induce resistance mutations in two or more kinase target genes may significantly reduce the frequency over a monospecific mode of action. In this respect, it seems to be a favourable situation in drug design when dealing with two or more target kinases, possibly one being virally encoded, the second being of cellular origin. Generally, the involvement of a cellular enzyme within the mode of action of an antiviral drug almost always reduces the probability of resistance formation to a minimum (Daub et al., 2004). Thus, the combined inhibitory activity of a drug against a cellular and additionally a viral protein kinases may render the occurrence of drug resistance particularly unlikely.

5.4   Paninhibitors: PPKIs with broad inhibitory potential against herpesviruses and nonrelated other human pathogenic viruses

In the novel field of development of PPKIs with antiviral activity another aspect seems highly attractive, namely the identification of PPKIs which show broad inhibitory activity against several human pathogenic viruses ("paninhibitors"). Some viruses, such as members of the herpesvirus family, encode their own protein kinases and thus, drugs targeting one of these kinases might also be active against the homologous kinases of other herpesviruses. For example, the indolocarbazole NGIC-I possesses *in vitro* inhibitory activity against the protein kinases of human and rat cytomegaloviruses and inhibits replication of both viruses (M. Marschall, unpublished data). Medicinal chemistry might greatly support the success in identifying further broad-spectrum inhibitors useful as antiviral drugs against human herpesviruses.

In addition, almost all viruses are dependent on the activity of cellular protein kinases. Inhibition of these cellular protein kinases by PPKIs may result in a strong reduction of viral replication efficiency. This reduction may not be restricted to one specific virus but may concern several viruses. Hereby, even nonrelated viruses belonging to different families may underly inhibition by a particular drug if this drug targets a cellular protein kinase jointly required for viral replication. For example, MAP kinases have been shown to be involved in diverse regulatory steps during the replication cycle of many human pathogenic viruses, such as HCMV, HSV-1, EBV, HIV-1, HBV, influenza A viruses and others. Therefore, PPKIs targeting MAP kinases, CDKs, PKC or other cellular protein kinases regulating several major intracellular pathways might represent promising candidates for the development of antiviral paninhibitors.

It has been demonstrated that CDKs are important for the replication of herpesviruses (HCMV, HSV-1, HSV-2 and VZV), HIV-1, adeno- and papillomaviruses (Schang LM., 2002). A pronounced feature of the role of CDKs during viral replication is the involvement within various steps of viral transcription and regulation of the replication cycle (Schang LM., 2004). Pharmacological CDK inhibitors target interaction between CDKs and viral replication. Furthermore they may target the pathogenic mechanisms of the diseases induced by CDK-dependent viruses. Since CDKs are required for replication and pathogenesis of many viruses they are especially attractive drug targets and CDK inhibitors are therefore considered as promising drug candidates for the development of paninhibitors.

Apart of a direct need for the activity of a cellular kinase for the replication of a virus there are other possibilities, how inhibition of a cellular kinase can indirectly prevent viral replication. The process of a viral infection generally induces apoptosis of the infected cell. However, pathogens manage to inhibit the activation of the apoptotic program in order to allow replication in host cells (Koul et al, 2004). Inhibiting apoptosis is often accomplished by the activity of kinases, e.g. MAP kinases. Inhibition of these kinases will leverage apoptosis and thus prevent replication and spreading of the virus in a very elegant manner.

6.   CONCLUSIONS

During the past few years, several new agents have been developed to improve control of herpesviral infections. Most of them belong to the classical nucleoside analogues, others are entirely novel chemical structures with unique mechanisms of action. Presently, a number of promising protein kinase inhibitors with highly interesting antiviral and pharmacological characteristics is at the early stages of preclinical development. Within some of these chemical classes, there are protein kinase inhibitors which are already approved drugs for tumor therapy. It seems suggestive that at least a small selection of the novel pharmacological protein kinase inhibitors will proceed into clinical trials for efficacy as antiviral agent in the near future. This should provide new opportunities to face the clinical problems of antiviral drug resistance, the increasing need for anti-herpesviral drugs in transplant or cancer patients and the continuous concerns about anti-herpesviral therapy in children and pregnant women.

ACKNOWLEDGEMENTS

The authors are grateful to Thomas Stamminger for critically reading the manuscript and helpful suggestions.

REFERENCES

Bain J, McLauchlan H, Elliott M & Cohen P, 2003, The specificity of protein kinase inhibitors: an update. *Biochem J* **371**: 199-204.
Biron KK, Harvey RJ, Chamberlain SC, Good SS, Smith III AA, Davis MG, Talarico CL, Miller WH, Ferris R, Dornsife RE, Stanat SC, Drach JC, Townsend LB & Koszalka GW, 2002, Potent and selective inhibition of human cytomegalovirus replication by 1263W94, a benzimidazole L-riboside with a unique mode of action. *Antimicrob Agents Chemother* **46**: 2365-2372.
Chulay J, Biron K, Wang L, Underwood M, Chamberlain S, Frick L, Good S, Davis M, Harvey R, Townsend L, Drach J & Koszalka G, 1999, Development of novel benzimodazole riboside compounds for treatment of cytomegalovirus disease. *Adv Exp Med Biol* **458**: 129-134.

Coen DM & Schaffer PA, 2003, Antiherpesvirus drugs: a promising spectrum of new drugs and drug targets. *Nature Rev Drug Disc* **2**: 278-288.

Cohen P, 2002, Protein kinases - the major drug targets of the twenty-first century? *Nature Rev Drug Disc* **1**: 309-315.

Cohen MII, Williams GA, Sridhara R, Chen G & Pazdur R, 2003, FDA drug approval summary: gefitinib (ZD1839) (Iressa) tablets. *Oncologist* **8**: 303-306.

Daub H, Specht K & Ullrich A, 2004, Strategies to overcome resistance to targeted protein kinase inhibitors. *Nature Rev Drug Disc* **3**: 1001-1010.

Davies SP, Reddy H, Caivano M & Cohen P, 2000, Specificity and mechanism of action of some commonly used protein kinase inhibitors. *Biochem J* **351**: 95-195.

De Clercq E, 2004, Antivirals and antiviral strategies. *Nature Rev Microb* **2**: 704-720.

Fabian, MA, Biggs III, WH, Treiber DK, Atteridge CE, Azimioara MD, Benedetti MG, Carter TA, Ciceri P, Edeen PT, Floyd M, Ford JM, Galvin M, Gerlach JM, Grotzfeld RM, Herrgard S, Insko DE, Insko MA, Lai AG, Lelias JM, Mehta SA, Milanov ZV, Velasco AM, Wodicka LM, Patel HK, Zarrinkar PP & Lockhart DJ, 2005, A small molecule-kinase interaction map for clinical kinase inhibitors. *Nature Biotech* **23**: 329-336.

Fortunato EA, McElroy AK, Sanchez V & Spector DH, 2000, Exploitation of cellular signaling and regulatory pathway by human cytomegalovirus. *Trends Microbiol* **8**: 111-119.

Herbst RS, Fukuoka M & Baselga J, 2004, Gefitinib – a novel targeted approach to treating cancer. *Nature Rev Cancer* **4**: 956-965.

Herget T, Freitag M, Morbitzer M, Stamminger T & Marschall M, 2004, A novel chemical class of pUL97 protein kinase-specific inhibitors with strong anti-cytomegaloviral activity. *Antimicrob Agents Chemother* **48**: 4154-4162.

Kawaguchi Y, Matsumara T, Roizman B & Hirai K, 1999, Cellular elongation factor 1 is modified in cells infected with representative alpha-, beta-, or gammaherpesviruses. *J Virol* **73**: 4456-4460.

Klebl B, Kurtenbach A, Salassidis K, Daub H & Herget T, 2005, Host cell targets in HCV therapy: novel strategies or proven practice? *Antivir Chem Chemother* **16**: 69-90.

Komazin G, Ptak RG, Emmer BT, Townsend LB & Drach JC, 2003, Resistance of human cytomegalovirus to the benzimidazole L-ribonucleoside maribavir maps to UL27. *J Virol* **77**: 11499-11506.

Koszalka GW, Johnson NW, Good SS, Boyd L, Chamberlain SC, Townsend LB, Drach LC & Biron KK, 2002, Preclinical and toxicology studies of 1263W94, a potent and selective inhibitor of human cytomegalovirus replication. *Antimirob Agents Chemother* **46**: 2373-2380.

Koul A, Herget T, Klebl B & Ulrich A, 2004, Interplay between mycobacteria and host signalling pathways. *Nature Rev Microb* **2**: 189-202.

Krosky PM, Baek, MC & Coen DM, 2003a, The human cytomegalovirus UL97 protein kinase, an antiviral drug target, is required at the stage of nuclear egress. *J Virol* **77**: 905-914.

Krosky PM, Baek MC, Jahng WJ, Barrera I, Harvey RJ, Biron KK, Coen DM & Sethna PB (2003b). The human cytomegalovirus UL44 protein is a substrate for the UL97 protein kinase. *J Virol* **77**: 7720-7727.

Lalezari JP, Aberg JA, Wang LH, Wire MB, Miner R, Snowden W, Talarico CL, Shaw S, Jacobsen MA & Drew WL, 2002, Phase I dose escalation trial evaluating the pharmacokinetics, anti-human cytomegalovirus (HCMV) activity, and safety of 1263W94 in human immunodeficiency virus-infected men with asymptomatic HCMV shedding. *Antimicrob Agents Chemother* **46**: 2969-2976.

Marschall M, Freitag M, Weiler S, Sorg G & Stamminger T, 2000, Recombinant GFP-expressing human cytomegalovirus as a tool for screening of antiviral agents. *Antimicrob Agents Chemother* **44**: 1588-1597.

Marschall M, Stein-Gerlach M, Freitag M, Kupfer R, van den Bogaard M & Stamminger T, 2001, Inhibitors of human cytomegalovirus replication drastically reduce the activity of the viral protein kinase pUL97. *J Gen Virol* **82**: 1439-1450.

Marschall M, Stein-Gerlach M, Freitag M, Kupfer R, van den Bogaard M & Stamminger T, 2002, Direct targeting of human cytomegalovirus protein kinase pUL97 by kinase inhibitors is a novel principle of antiviral therapy. *J Gen Virol* **83**: 1013-1023.

Marschall M, Freitag M, Suchy P, Romaker D, Kupfer R, Hanke M & Stamminger T, 2003, The protein kinase pUL97 of human cytomegalovirus interacts with and phospohorylates the DNA polymerase processivity factor pUL44. *Virology* **311**: 60-71.

Marschall M, Marzi A, aus dem Siepen P, Jochmann R, Freitag M, S. Auerochs, Lischka P, Leis M & Stamminger T, 2005, Cellular p32 recruits cytomegalovirus kinase pUL97 to redistribute the nuclear lamina. *J Biol Chem* **in press**.

McSharry JJ, McDonough A, Olson B, Talarico C, Davis M & Biron KK, 2001, Inhibition of ganciclovir-susceptible and -resistant human cytomegalovirus clinical isolates by the benzimidazole L-riboside 1263W94. *Clin Diagnostic Lab Immunol* **8**: 1279-1281.

Mett H, Hölscher K, Degen H, Esdar C, Felden de Neumann B, Flicke B, Freudenreich T, Holzer G, Schinzel S, Stamminger T, Stein-Gerlach M, Marschall M & Herget T, 2005, Identification of inhibitors for a virally encoded protein kinase by two different screening systems: *in vitro* kinase assay and in-cell-activity assay. *J Biomol Screening* **10**: 36-45.

Michel D, Pavic I, Zimmermann A, Haupt E, Wunderlich K, Heuschmid M & Mertens T, 1996, The UL97 gene product of human cytomegalovirus is an early-late protein with nuclear localization but is not a nucleoside kinase. *J Virol* **70**: 6340-6346.

Mocroft A, Vella S, Benfield TL, Chiesi A, Miller V, Gargalianos P, d'Arminio Monforte A, Yust I, Bruun JN, Phillips AN & Lundgren JD, 1998, Changing patterns of mortality across Europe in patients infected with HIV-1. EuroSIDA Study Group. *Lancet* **352**:1725-30.

Müller G, 2003, Medicinal chemistry of target family-directed masterkeys. *Drug Discov Today* **8**: 681-691.

Palella FJ Jr, Delaney KM, Moorman AC, Loveless MO, Fuhrer J, Satten GA, Aschman DJ & Holmberg SD, 1998, Declining morbidity and mortality among patients with advanced human immunodeficiency virus infection. HIV Outpatient Study Investigators. *N Engl J Med* **338**: 853-60.

Perry CM & Balfour JAB, 1999, Fomivirsen. *Drugs* **57**: 375-380.

Porter K, Babiker A, Bhaskaran K, Darbyshire J, Pezzotti P, Porter K & Walker AS; CASCADE Collaboration, 2003, Determinants of survival following HIV-1 sero-conversion after the introduction of HAART. *Lancet* **362**: 1267-74.

Prichard MN, Gao N, Jairath S, Mulamba G, Krosky P, Coen DM, Parker BO & Pari GS, 1999, A recombinant human cytomegalovirus with a large deletion in UL97 has a severe replication deficiency. *J Virol* **73**: 5663-5670.

Schaeffer HJ, Beacuchamp L, de Miranda P, Elion GB, Bauer DJ & Collins P, 1978, 9-(hydroxy-ethoxymethyl)guanine activity against viruses of the herpes group. *Nature* **272**: 583-585.

Schang LM, 2002, Cyclin-dependent kinases as cellular targets for antiviral drugs. *J Antimicrob Chemother* **50**: 779-792.

Schang LM, 2004, Effects of pharmacological cyclin-dependent kinase inhibitors on viral transcription and replication. *Biochim Biophys Acta* **1697**: 197-209.

Slater MJ, Cockerill S, Baxter R, Bonser RW, Gohil K, Gowrie C, Robinson J. E, Littler E, Parry N, Randall R & Snowden W, 1999, Indolocarbazoles: potent, selective inhibitors of human cytomegalovirus replication. *Bioorg Med Chem* **7**: 1067-1074.

Slater MJ, Cockerill S, Baxter R, Bonser RW, Gohil K, Robinson E, Parry N, Randall R & Snowden W, 2001, *14th Int Conf Antiviral Res Seattle* **abstract 69**.

Wang LH, Peck RW, Yin Y, Allanson J, Wiggs R & Wire MB, 2003a, Phase I safety and pharmacokinetic trials of 1263W94, a novel oral anti-human cytomegalovirus agent, in healthy and human immunodeficiency virus-infected subjects. *Antimicrob Agents Chemother* **47**: 1334-1342.

Wang X, Huong SM, Chiu ML, Raab-Traub N & Huang ES, 2003b, Epidermal growth factor receptor is a cellular receptor for human cytomegalovirus. *Nature* **424**: 456-461.

Wolf DG, Courcelle CT, Prichard MN & Mocarski ES, 2001, Distinct and separate roles for herpesvirus-conserved UL97 kinase in cytomegalovirus DNA synthesis and encapsidation. *Proc Natl Acad Sci USA* **98**: 1895-1900.

Zimmermann A, Michel D, Pavic I, Hampl W, Lüske A, Neyts J, De Clerq E, & Mertens T, 1997, Phosphorylation of aciclovir, genciclovir, penciclovir and S2242 by the cytomegalovirus UL97 protein: a quantitative analysis using recombinant vaccinia viruses. *Antiviral Res* **36**: 35-42.

Zimmermann A, Wilts H, Lenhardt M, Hahn M & Mertens T, 2000, Indolocarbazoles exhibit strong antiviral activity against human cytomegalovirus and are potent inhibitors of the pUL97 protein kinase. *Antiviral Res* **48**: 49-60.

Chapter 2.5

IMMUNE THERAPY AGAINST PAPILLOMAVIRUS-RELATED TUMORS IN HUMANS

L. GISSMANN[1]

[1]*Deutsches Krebsforschungszentrum Heidelberg, Schwerpunkt Infektionen und Krebs, Im Neuenheimer Feld 242, 69120 Heidelberg, Germany*

Abstract: Human papillomaviruses (HPV) represent a very heterogeneous group of infectious agents that can infect the skin and mucosae where they induce exophytic warts or atypical flat epithelial lesions. Persistent infection by certain papillomaviruses is linked to cancer of the anogenital or respiratory tract. This association is best documented in case of the so called high-risk HPV types (e.g. HPV 16, 18, 31, 45) and the development of cancer of the uterine cervix which arises through well defined precursors. Based on observations of the natural history of HPV infections and by the aid of animal models, the role of the immune system in prevention of infection as well as in controlling virus persistence and progression of the precursor lesions is generally acknowledged. Therefore there has been growing interest in HPV-specific vaccines among academic researchers and corporations. The development of prophylactic vaccinations is well under way but, despite several clinical trials with a variety of different vaccines (proteins, peptides, recombinant vectors), progress in immune therapies for treatment of HPV-related lesions has so far been disappointing.

1.   INTRODUCTION

There are members of the papillomaviridae family in humans and in several vertebrate species, such as cattle, rabbits, dogs and birds (Sundberg, 1987). In man, papillomaviruses form a large group of individual members ("types"; HPV 1, 2 etc.): as yet more than 100 types are fully characterized by complete sequencing of the approximately 8000 bp circular DNA

373

E. Bogner and A. Holzenburg (eds.), New Concepts of Antiviral Therapy, 373–394.

© *2006 Springer. Printed in the Netherlands.*

genome. Many additional isolates have been identified and partially characterized (de Villiers *et al.*, 2004). Human papillomaviruses can be classified according to the site of their preferred occurrence (at the skin or anogenital or oral mucosa) where they induce epithelial hyperproliferations that manifest themselves as skin warts, laryngeal papillomas, genital warts (condylomata acuminata) or intraepithelial neoplasia (see below). Some of these epithelial lesions can, depending on the HPV type involved and after long time of persistence and under rare circumstances, progress into malignant tumors (IARC, 1995).

Despite the diversity the individual papillomavirus genomes are organized in a similar fashion (Bernard *et al.*, 1994) and there is a high degree of nucleotide and protein sequence homology. The papillomavirus genomes consist of three functionally distinct regions, i.e. the late region containing the two structural genes (L1 and L2), the up to seven open reading frames that code for the early proteins responsible for virus DNA replication and control of viral gene expression, and an approximately 1 kb non-coding region containing cis-acting regulatory sequences such as the origin of DNA replication or promoter elements (for review see Scheffner *et al.*, 1994).

2.   BIOLOGY OF PAPILLOMAVIRUS INFECTIONS

Papillomaviruses replicate exclusively in differentiating epithelial cells (Stubenrauch and Laimins, 1999). It is assumed that intracellular control of viral gene expression is the major determinant for this cell tropism since cellular receptors for the viruses seem to be present on cells of different origin. Papillomavirus infection takes place within the basal cells of the epithelium. A few copies of the viral DNA persist within this compartment for various length of time. Upon differentiation of the cells vegetative virus replication is initiated and production of virus particles following expression of the structural proteins L1 and L2 takes place. A key event appears to be the activity of the viral proteins E6 and E7 which induce S-phase in the differentiated cells thus facilitating the replication of the viral DNA (for review see (Munger *et al.*, 2004). Other viral non-structural proteins are required in addition for the amplification of the genome or for the regulation of viral gene expression (Scheffner *et al.*, 1994).

Papillomaviruses are transmitted through direct contact with infected individuals or contaminated objects. HPV infections of the anogenital tract are mostly sexually transmitted and induce either genital warts (mostly related to HPV types 6 or 11 (Gissmann *et al.*, 1983)) or so called intraepithelial neoplasias. Depending on the HPV type such lesions are putative precursors for cancer of the anogenital tract best documented for cervical intraepithelial neoplasias that may progress to invasive cancer (for review see (IARC, 1995). Papillomaviruses involved in malignant transformation are called high-risk types the prototypes of which are HPV 16 and HPV 18 (Adams *et al.*; Boshart *et al.*, 1984). Additional 12 HPV types can be found in cervical cancer biopsies. Taken together more than 99% of tumors analyzed in a recent study were shown to be positive for HPV DNA (Walboomers *et al.*, 1999).

Cervical cancer arises from well defined precursor lesions (cervical intraepithelial neoplasia: CIN). CIN is often detected by cytology ("Pap test") performed during the screening programs that exist in most industrialized countries (Sawaya *et al.*, 2001). In case of cellular abnormalities suggesting a high grade CIN (also named high-grade squamous intraepithelial lesion: HSIL) colposcopy is indicated and eventually the final diagnosis will be made by histology of a colposcopically guided biopsy. Because of the relatively poor sensitivity of the Pap test (Naryshkin, 1997) and the close association between persistent high-risk HPV and development of high grade CIN (Ho *et al.*, 1998; Nobbenhuis *et al.*, 1999) detection of virus infection is discussed as alternative method for cervical cancer screening. HPV infection is diagnosed on the basis of DNA detection. Cervical swabs are tested for the presence of genomes of high-risk HPV types by nucleic acid hybridization with or without previous amplification by PCR (Wieland and Pfister, 1996). Intraepithelial neoplasias similar to CIN occur also within the vagina, at the vulva, penis or perianal region (called VAIN, VIN, PIN or AIN, respectively) but the natural history is best studied for cervical lesions. There are approximately 1.2 million new cases of low grade and 300,000 cases of high grade CIN each year in the U.S. In contrast the HPV infection rate is at least 30 times higher as can be estimated from the number of abnormal smears (IARC, 1995).

Cervical cancer arises in about 90% of cases from squamous epithelial cells, in some instances it is of glandular origin. Both tumor types are related to HPV infection but the prevalence of particular types varies between the two variants. Altogether cervical cancer represents about 10% of cases of invasive tumors in women worldwide (estimated for 1990) accounting for approximately 370,000 new cases per year (estimated annual death rate is 210,000). There is a more than a threefold higher incidence of this disease in developing countries (Pisani *et al.*, 1999). This difference is mostly due to the missing Pap screening programs in resource-poor settings.

The viral proteins E6 and E7 through their interaction with several cellular proteins (most importantly the cellular tumor suppressor gene products p53 and pRB) are pivotal in cellular transformation. Virus persistence for many years is required for development of cancer (Tommasino, 2001). Very likely the continued expression of the viral oncogenes E6 and E7 leads to accumulation of cellular mutations (e.g. overexpression of cellular oncogenes) finally resulting in invasive cell growth. In addition there is evidence that these proteins interfere with cytokine-induced signal transduction and thus may prevent the virus-specific immune surveillance (Koromilas *et al.*, 2001). Most importantly it was shown that the viral oncoproteins E6 and E7 are required not only in the initial phases of transformation but also for maintaining the proliferative state of a tumor cell (von Knebel Doeberitz *et al.*, 1988). This feature makes them unique targets for the induction of a virus-specific anti-tumor immune response (see below).

HPV infections are mostly transient suggesting the involvement of the immune system in clearing of the infection. Direct evidence is provided by the presence of virus-specific serum antibodies, T helper and cytotoxic T lymphocytes (CTL) in infected individuals (reviewed by (Daemen *et al.*, 2000; Konya and Dillner, 2001; Scott *et al.*, 2001)). Often infiltration of T cells (Aiba *et al.*, 1986), macrophages, natural killer (NK) cells and CD4+ T lymphocytes (Coleman *et al.*, 1994) accompany a "spontaneous regression" of warts in humans and of papillomavirus-induced lesions in rabbits, dogs or cattle a (Jensen *et al.*, 1997; Knowles *et al.*, 1996; Nicholls *et al.*, 2001; Okabayashi *et al.*, 1991). Warts responding to treatment with IFN-α or IFN-γ showed a remarkable increase in infiltrating Th1 inflammatory cells whereas warts of non-responding patients showed a depletion of Langerhans cells and a reduced surface expression of MHC II and diminished levels of the cytokines IL-1a and 1b, GM-CSF (Arany and Tyring, 1996). In genital warts patients treated with imiquimod similar observations have been made. Imiquimod is an immune response modifier that triggers the innate immune system through interaction with the toll-like receptor 7 (TLR 7) (Arany *et al.*, 1999; Hemmi *et al.*, 2002). Progression of lesions proved to be related to IL-6 and TNF-α (Arany *et al.*, 2001). There appears to be an inverse correlation between the expression of the early HPV 16 protein E7 within a lesion and its ability to respond to the immune treatment (Arany and Tyring, 1996).

Several interactions of the HPV E6 and E7 gene products with cellular molecules that are involved at different levels of the IFN-regulating pathways have been reported (reviewed by (Koromilas *et al.*, 2001)). The transcriptional activity of the IRF-3/CBP complex on the IFN-β gene is reduced by binding of the E6 protein to IRF-3 and the transcription factor CBP (Patel *et al.*, 1999; Ronco *et al.*, 1998; Zimmermann *et al.*, 1999). Through binding with the HPV E7 protein the cellular protein IRF-9 is prevented from nuclear translocation and its interaction with the Stat1/Stat2 heterodimers resulting in a decreased transcription of IFN-α (Barnard and McMillan, 1999). These interactions are stronger with the oncoproteins of the high-risk HPV types 16 and 18 than with the corresponding proteins of HPV 6 (Arany *et al.*, 1999; Ronco *et al.*, 1998). These data corroborate *in vivo* results that demonstrate a reduced response of HPV 16 or HPV 18 versus HPV 6 or 11 positive cervical lesions to IFN-α treatment (Schneider *et al.*, 1987).

Persistent HPV infections are the prerequisite for malignant progression of cervical neoplasia (Remmink *et al.*, 1995). As suggested by Stern and colleagues such infections are facilitated by the papillomaviruses through stealth, specific interference with the innate immune system and certain effects on the adaptive immunity (Stern *et al.*, 2001) whose role in keeping an HPV infection in check is largely unknown (for review see (1995; Konya and Dillner, 2001).

About half of newly infected patients develop antibodies against papillomavirus structural proteins that are considered as marker of persisting infection. It is still controversial whether naturally occurring antibodies protect against reinfection with the same virus (Buonamassa *et al.*, 2002). However, the fact that the different HPV types originally classified as genotypes later on proved also to represent individual serotypes strongly suggests that the antibody-based immune recognition of virus particles was indeed a driving force for immune escape. In addition, neutralizing antibodies induced after vaccination in animal models and human studies have been shown to be a surrogate marker for protection against papillomavirus infection and associated diseases although under experimental conditions antibody titers are up to 100 fold higher than during natural exposure to the virus.

Antibodies against the early HPV proteins E2, E6 and E7 occur rarely during the natural virus life cycle and appear late during malignant progression of HPV-induced lesions and may become useful as diagnostic markers (Galloway, 1992; Jochmus-Kudielka *et al.*, 1989; Meschede *et al.*, 1998).

3. MODELS FOR HUMAN PAPILLOMAVIRUS INFECTIONS

Under defined experimental conditions, the cottontail rabbit papillomavirus (CRPV) induces papillomas in domestic rabbits that may persist, regress spontaneously or become malignant (Kreider and Bartlett, 1981). Vaccination with combinations of the early genes E1, E2, E6 and E7 of

CRPV completely protects against challenge by virus or DNA and, when given after virus infection, induces regression of existing papillomas and prevents malignant progression (Alcami and Koszinowski, 2000; Arany *et al.*, 1999). Post challenge immunization with E7 but not with E6 delayed the onset of carcinoma development of CRPV-induced papillomas (de Jong *et al.*, 2002). Protective immunity after vaccination with the CRPV L1 protein is based on neutralizing antibodies (Breitburd *et al.*, 1995) but not on cell-mediated immunity.

The papillomas developing as a consequence of the canine oral papillomavirus (COPV) at the mucosae of dogs resemble the HPV 6 or 11-related genital warts. Naturally occurring or experimentally induced warts usually regress spontaneously after several weeks and the developing humoral immunity protects against reinfection (for review see (Moore, 2004). Protective immunity depends upon neutralizing antibodies specific for the L1 major structural protein (Ghim *et al.*, 2000). Dogs become immune against virus challenge by vaccination with inactivated virus particles, by virus-like particles (VLPs, see below) or sub-particle structures (capsomeres, see below) produced by recombinantly expressed L1 protein or by L1 DNA (Bell *et al.*, 1994; Nicholls *et al.*, 2001; Suzich *et al.*, 1995; Yuan *et al.*, 2001).

In cattle, 6 different papillomaviruses (BPV 1-6) have been described, two of which (BPV 2 and 4) are related to cancer under natural conditions (for review see (Campo, 1995)). The first vaccines against papillomavirus infection were reported in cattle and consisted of inactivated purified virus. Protection was shown to be accompanied by the production of neutralizing antibodies (Jarrett *et al.*, 1990).

Animal models proved to be particularly informative for the analysis of immune responses against HPV particles. Development of attenuated or inactivated papillomaviruses as vaccines is not feasible as they cannot be grown to large titers in cell culture systems. Instead virus-like particles (VLP) can be generated that are based on the observation that viral structural proteins are able to self-assemble into capsids that are indistinguishable in morphology from infectious virus particles (Hagensee *et al.*, 1994). In most studies VLPs were generated by expression of L1-recombinant baculo-viruses in insect cells, but also expression in yeast, in mammalian cells and even in plants proved to be successful (Biemelt *et al.*, 2003; Bosch *et al.*, 1995; Buonamassa *et al.*, 2002; Heino *et al.*, 1995; Warzecha *et al.*, 2003; Zhou *et al.*, 1991), for review see (Da Silva *et al.*, 1999). Production of VLPs of several animal and human papillomaviruses have been reported, i.e. canine oral papillomavirus, bovine papillomavirus (BPV) types and 4, cottobtail rabbit papillomavirus, rabbit oral papillomavirus, and 14 different HPV types (Adams *et al.*; Bouwes Bavinck *et al.*, 2000; Christensen *et al.*,

2000; Combita *et al.*, 2002; Cubie *et al.*, 1998), for review see (Breitburd *et al.*, 1995).

4.   EXPERIMENTAL MODELS FOR HPV-SPECIFIC IMMUNE THERAPY

Immune therapy of papillomavirus-induced lesions was demonstrated in case of BPV 4 induced warts in calves that were immunized post-challenge with recombinant E7 protein (Campo *et al.*, 1993). Therapeutic activity was also observed for the L2 minor structural protein. This result was unexpected since late proteins were assumed to be confined to the superficial layers of differentiating epithelium (Jarrett *et al.*, 1991). Similarly surprising were the reported therapeutic effects of complete virus particles both in cattle and in a human trial (Zhang *et al.*, 2000). There is no animal model system for infection by human papillomaviruses but their oncogenic properties can be studied *in vivo* by tumor growth of HPV-transformed rodent cells (Borysiewicz *et al.*, 1996; Feltkamp *et al.*, 1993) following inoculation into syngeneic laboratory animals (mice or rats). Transformation of cells depends upon the expression of the viral proteins E6 and E7 (for review see (Gissmann, 2003; Munger *et al.*, 2004)). Immunizations with the HPV oncoproteins in various formulations have been successfully performed in a prophylactic or therapeutic experimental setting when (challenge with transformed cells after or before vaccination; for review see (Berry and Palefsky, 2003; Eiben *et al.*, 2003; 1995). Such studies permitted the analysis of the role of T helper and CTL responses and the identification of HPV E6/E7-specific T cell epitopes restricted to the HLA background of the animals used (in most studies C57/Bl6 or Balb/c mice). Since this information is of limited value for human studies, transgenic mice that express a particular human HLA allele have also been used. Such immunization experiments permit the evaluation of a vaccine within a quasi human background and resulted in the identification of HPV-specific epitopes for a number HLA haplotypes (for review see (Eiben *et al.*, 2003)).

5.   THERAPEUTIC VACCINES

Several vaccination trials in young sexually active women using a combination of HPV 16+18 VLPs or a tetravalent vaccine (HPV 6/11/16/18) have demonstrated protection against persistent HPV infection and cervical neoplasias (Harper *et al.*, 2004; Koutsky *et al.*, 2002; Villa *et al.*, 2005). It is expected that prophylactic vaccinations to be given to adolescent girls before their first sexual contact will reduce the worldwide incidence of cervical cancer. Since it will take many years before this goal can be reached (Schiller and Davies, 2004) it is mandatory to develop strategies for immune therapy for those women who are already infected and will not benefit from the prophylactic vaccines.

Development of HPV-specific therapeutic vaccines is based on the fact that in persistently infected cells the early viral genes E6 and E7 are expressed. Numerous studies have demonstrated that these proteins are required to maintain the proliferative state of HPV transformed cells and prevent their apoptosis (Butz *et al.*, 2003; DeFilippis *et al.*, 2003). Therefore the viral oncoproteins E6 and E7 are the prime target for an immune therapeutic strategy of HPV-related neoplasia.

As summarized in the table, several therapeutic vaccination trials (phase I or II) with cervical cancer patients, individuals affected by intraepithelial neoplasias (CIN, VIN, VAIN, AIN) and healthy volunteers using a variety of different antigen formulations have been published (listed in PubMed up to May 2005). In general, immunizations were well tolerated by the recipients and major side effects, if any, were not attributable to the vaccine. In cervical cancer patients immune responses (CTL, Th1, DTH, production of cytokines) were measured in some of the recipients but clinical response was not observed in cancer patients. Due to the usually heavily compromised immune system within these late stage cancer patients this result was not unexpected. In addition, advanced cancer is unlikely to respond to immune therapy as sole treatment since tumor cells have developed means to escape from immune surveillance (e.g. down-regulation of MHC I molecules (Stern *et al.*, 2000). It remains to be investigated whether patients that have received standard treatment may benefit from vaccination in terms of increased recurrent disease-free intervals.

On the other hand, there were some promising results in cases with less advanced intraepithelial lesions, i.e. precursors to cervical, vulval or perianal cancer (CIN, VIN, AIN). However, all studies demonstrated a poor correlation between clinical response and immune response within the individual patients. This implies that we need more information about the immune effector functions in order to improve HPV-specific immunotherapy. As with anti-cancer immune therapies in general, the immune tolerance against the tumor cells must be overcome (for discussion see (Pardoll and Allison, 2004). Since in HPV-transformed cells the tumor antigen is of non-self origin there is, unlike in other tumors with no known infectious etiology, no risk to inducing autoimmune responses against normal cells. Thus successful therapeutic immunization against HPV-positive tumors can serve as proof-of principle for the concept of cancer immunotherapy as a whole. It is quite likely that induction of local inflammation as indication for an ongoing innate immune response may be required to push the adaptive immune system against the persistently infected cells of the epithelial lesions.

6. PROTEINS

Purified proteins are inefficient in inducing CD8+ T cell response as they are mostly processed through the MHC class II pathway which directs the immune response towards the Th2 phenotype. Addition of appropriate adjuvants can prevent this bias (Hariharan *et al.*, 1998; Kim *et al.*, 2002), however, only very few are licensed for use in humans.

Immunization with fusion proteins of HPV 16 E7 linked to a variety of other proteins such as the Haemophilus influenzae lipoprotein D, the Mycobacterium bovis hsp65 or calreticulin (Chu *et al.*, 2000; Hallez *et al.*, 2004; Hsieh *et al.*, 2004) was shown to abolish the need for adjuvants in preclinical mouse models. The hsp65-E7 and lipoprotein D vaccines were further developed for use in humans and were applied in separate studies to patients with AIN or CIN (see table) inducing clinical response and antibody/CTL response in some of them (Adams *et al.*, 2001; Berry and Palefsky, 2003; Goldstone *et al.*, 2002; Hallez *et al.*, 2004). Fusion proteins were also used in chimeric virus-like particles (CVLP) that consist of the early protein E7 linked either to the L1 or to the L2 protein (Greenstone *et al.*, 1998; Müller *et al.*, 1997). In the former case expression leads to generation of chimeric capsids with L1 and E7-specific sequences in a 1:1 ratio. The alternative concept requires co-expression of L1 to generate viral capsids that contain the (L2-fused) E7 sequences in much lower molarity than L1 (estimated 12 molecules L2 against 360 copies of L1 per particle). The advantage of this kind of CVLPs is a higher tolerance towards longer inserts, which in case of L1 CVLPs are restricted to 50-60 amino acids (Greenstone *et al.*, 1998; Müller *et al.*, 1997). HPV 16 CVLPs were designed as combinatory vaccine aiming to stimulate the formation of neutralizing antibodies and E7-specific cytotoxic T cells. The expectation was confirmed in a mouse model and in a clinical trial where both humoral and T cell responses could be induced (Schäfer *et al.*, 1999). The independence of adjuvants can be explained by the efficient activation of dendritic cells by CVLPs as demonstrated in *in vitro* studies (Freyschmidt *et al.*, 2004).

The induction of B-and T cell responses after immunization of healthy human volunteers with the HPV 16 L2/E6/E7 fusion protein in the absence of adjuvant was unexpected. The authors speculate that this effect may be due to formation of aggregates of the vaccine during the purification procedure (de Jong *et al.*, 2002).

7.   PEPTIDES

Application of short peptides derived from tumor antigens is a straightforward strategy to induce a clinically relevant T cell response (for review see (Brinkman *et al.*, 2004). The major advantage of this approach is based on the easy chemical synthesis thus high purity of the manufactured peptides.

The initial peptide-based immunotherapies in patients used HLA-A2 restricted CTL epitopes together with a universal HLA-DR helper epitope (Muderspach *et al.*, 2000; Steller *et al.*, 1998; van Driel *et al.*, 1999) (for review see (Salit *et al.*, 2002)). Despite the induction of specific T cells the authors reported only poor clinical responses. It was argued that this was due to the failure of the vaccine to induce a strong sustained immune response because of the absence of a strong T helper response (Zwaveling *et al.*, 2002). Long overlapping peptides (32-35mer) that cover the whole protein may represent the next generation vaccine. In a mouse study HPV 16 E7 35mer peptide containing a CTL and a T helper epitope was shown to be superior in the induction of E7-specific CTL (compared to the minimal CTL epitope) even in the absence of T helper functions in the experimental animals, suggesting that such long peptides are preferably taken up and presented by dendritic cells (Zwaveling *et al.*, 2002).

Minigenes that code for different T cell epitopes is another concept that has been applied in clinical studies (Klencke *et al.*, 2002). A eukaryotic expression vector was used that codes for 13 C-terminal amino acids of the HPV 16 E7 protein that contain several overlapping CTL HLA-A2-restricted CTL epitopes. The polygene was fused to a leader sequence to facilitate secretion of the peptide. The DNA was encapsulated into microparticles consisting of a biodegradable polymer to allow better delivery to APCs (Hedley *et al.*, 1998).

8.   RECOMBINANT VECTORS

Recombinant vaccinia (rVaccinia) is the prototype for vector-mediated delivery of HPV genes. Vaccinia itself induces a strong immune response and provides adjuvant effects (Borysiewicz *et al.*, 1996; Meneguzzi *et al.*, 1991). HPV 16+18 E6/E7 recombinant vaccinia were already tested in several clinical studies (Adams *et al.*, 2001; Baldwin *et al.*, 2003;

Borysiewicz *et al.*, 1996; Corona Gutierrez *et al.*, 2004; Davidson *et al.*, 2003; Kaufmann *et al.*, 2002; Smyth *et al.*, 2004) (see table 1). More recent approaches to vector-based delivery of HPV genes include adenoviruses and alphaviruses that, however, have not yet reached the state of clinical studies (Cassetti *et al.*, 2004; Daemen *et al.*, 2000; He *et al.*, 2000). The same holds true for HPV-recombinant bacteria, i.e. attenuated Listeria monocytogenes or Salmonella typhimurium and Bacille Calmette-Guerin (BCG). Attenuated salmonellae have the potential for a mucosal immunization which makes them suitable vaccine candidates particularly for developing countries (Revaz *et al.*, 2001). For a comprehensive review of HPV-based vectors the reader is referred to the excellent article by Eiben *et al.* (Eiben *et al.*, 2003).

9.   AUTOLOGOUS DC

Murine DCs loaded with protein were shown to induce strong T cell responses upon inoculation into syngeneic animals (De Bruijn *et al.*, 1998; Freyschmidt *et al.*, 2004; Navabi *et al.*, 1997). Similar effects were observed when DCs of HPV 16 or 18 positive cervical cancer patients were loaded with the corresponding E7 protein leading to the *in vitro* induction of E7-specific Th1 CD8+ CTL and CD4+ responses against autologous tumor cells and cultured MHC I matched cervical cancer cell (Kaufmann *et al.*, 2001; Santin *et al.*, 1999). In a study of 15 stage IV cervical cancer patients inoculation of autologous DCs loaded *ex vivo* with HPV E7 protein induced no objective clinical response (Ferrara *et al.*, 2003) whereas in a case report of a single patient with a metastatic HPV 18 positive adenocarcinoma of the cervix a similar protocol delayed the progression of the disease (Santin *et al.*, 2002).

10.   CONCLUSION

Whereas the development of prophylactic vaccinations against certain human papilloma viruses aiming at the prevention of cervical cancer via neutralizing antibodies is well under way, progress in immune therapies for treatment of HPV-related lesions is lagging behind. Of the almost 30 studies published as of today (see table 1) some yielded promising clinical responses in a portion of patients, yet there is no clear picture about the kind of immune response within the clinical responders (T helper cell, CTL). More information about the immune mechanisms against HPV infections will be needed to design improved vaccines that may include the use of efficient adjuvants. In a post-exposure prophylaxis strategy one should consider tackling persistent infections in young women instead of clinically apparent disease, which reasonably should include the induction of neutralizing.

Table 1. Summary of Monoclonal Antibodies Used in the Present Study

Vaccine	Patients	Results	Reference
HPV 16+18 E6-E7 rVaccinia	5 studies: HPV 16 pos. early or advanced cervical cancer, high grade CIN, VIN, AIN n=109	no environmental contamination by rVaccinia; HPV-specific antibodies and/or T cell response in some pts , reduction of lesion size in 10/12 VIN pts	Adams *et al.,* 2001; Baldwin *et al., 2003;* Borysiewicz *et al.,* 1996; Davidson *et al.,2003;* Kaufmann *et al.,* 2002)
BPV E2 rVaccinia	HPV 16 or 18 pos. CIN I-III, n=36	antibodies and cytotoxic T cells to vaccine in all pts, 87% of pts. free of lesion	(Corona Gutierrez *et al.,* 2004)
HPV 16 GST-E7 protein in Algammulin	recurrent cervical cancer, n=5	antibodies to E7 in 2 pts, E7-specific lymphoproliferation in2/3 pts; all pts had progressive disease	(Frazer *et al.,* 1999)
HPV 16 E7 CTL+Th peptide (lipidated)	refractory cervical or vaginal cancer, n=12	HPV E7-specific CTLs in 2/3 pts, no clinical responses	(Steller *et al.,* 1998)
HPV 16 E7 CTL + Th peptides in Montanide ISA 51	treatment-refractory cervical cancer, n=19	no induction of CTL, response to Th epitope in 4/12 pts, stable disease for one year in 2 pts	(Ressing *et al.,* 2000; van Driel *et al., 1999)*
HPV 16 E7 peptides in IFA	HPV 16 pos. CIN II/III, (VIN III), n=18	reduction in virus load in 12/18 pts; E7-specific CT in 10/16 pts; 3 complete and 6 partial responders	(Muderspach *et al.,* 2000)
HPV 6 L2-E7 in Alhydrogel®	2 studies; healthy males or male genital wart pts (n=69)	T cell response and/or antibodies in most recipients, CR in 5/27 pts	(Lacey *et al.,* 1999)
HPV 16 E6-E7 in ISCOMATRIX™	CIN (n=31)	antibodies and T cell response in most pts	(Frazer *et al ,* 2004)
HPV 16 L2-E6-E7	healthy individuals (30 males, 10 females)	vaccine-specific antibodies in 24/32, T cell responses in 25/32 individuals	(de Jong *et al.,* 2002)

HPV 16 L2-E6-E7 + HPV 16+18 E6-E7 rVaccinia	VIN III (VAIN III), n=29	antibodies and T cell response in most pts, 6 clinical responders	(Smyth *et al.*, 2004)
HPV 16 E7-*H.influenzae* lipoprotein D in AS02B	HPV 16 pos. CIN I (III), n=7	antibodies to vaccine in all pts, CD8+ T cells in 5 pts	(Hallez *et al.*, 2004)
HPV 16 E7-Hsp65	2 studies; genital warts, anal HSIL n=100	CR in some pts, effect not HPV type specific	(Berry and Palefsky, 2003; Goldstone *et al.*, 2002)
HPV 16 E7-specific minigenes, in microparticles	3 studies; anal HSIL, cervical HSIL, n=154	T cell response in most pts tested, clearance of lesions only in young women	(Garcia *et al.*, 2004; Klencke *et al.*,2002;Sheets *et al.*, 2003)
HPV 16 L1-E7 CVLPs	CIN II/III, n=24	CR in 37% (vs. 25% in placebo; n.s.), Th1 response in 54% (0 in placebo)	(Nieland *et al.*, 2004)
autologous DCs pulsed with cervical cancer lysate	advanced cervical cancer, n=9	induction of HPV 16-specific CTL and T-helper responses	(Adams *et al.*, 2003)
autologous DCs pulsed with HPV 16 (18) E7 protein	2 studies; HPV 16 or 18 pos. late stage cervical cancer, n=16	Antibodies and T cell response in some pts., performance status of one patient improved	(Ferrara *et al.*, 2003; Santin *et al.*, 2002)
HPV 6 L1 VLPs	genital warts, n=32	HPV 6-specific DTH reaction in 28/28, antibodies to HPV 6 VLPs in 31/32 pts, complete regression in 25/33pts	(Zhang *et al.*, 2000)

AIN: anal intraepithelial neoplasia

CIN: cervical intraepithelial neoplasia

CR: clinical response

CVLPs: chimeric virus-like particles

DTH: delayed type hypersensitivity

HSIL: high-grade squamous intraepithelial lesion

HSP: heat shock protein

pts: patients

rVaccinia: recombinant vaccinia

VIN: vulval intraepithelial neoplasia

VLPs: virus-like particles

Please note the English spelling of acknowledgements.

REFERENCES

Adams, M., Borysiewicz, L., Fiander, A., Man, S., Jasani, B., Navabi, H., Lipetz, C., Evans, A. S., and Mason, M., 2001, Clinical studies of human papilloma vaccines in pre-invasive and invasive cancer. *Vaccine* **19**: 2549-2556.

Adams, M., Navabi, H., Jasani, B., Man, S., Fiander, A., Evans, A. S., Donninger, C., and Mason, M., 2003, Dendritic cell (DC) based therapy for cervical cancer: use of DC pulsed with tumour lysate and matured with a novel synthetic clinically non-toxic double stranded RNA analogue poly [I]:poly [C(12)U] (Ampligen R). *Vaccine* **21**: 787-790.

Aiba, S., Rokugo, M., and Tagami, H., 1986, Immunohistologic analysis of the phenomenon of spontaneous regression of numerous flat warts. *Cancer* **58**: 1246-1251.

Alcami, A., and Koszinowski, U. H,. 2000, Viral mechanisms of immune evasion. *Trends Microbiol* **8**: 410-418.

Arany, I., Muldrow, M., and Tyring, S. K., 2001, Correlation between mRNA levels of IL-6 and TNF alpha and progression rate in anal squamous epithelial lesions from HIV-positive men. *Anticancer Res* **21**: 425-428.

Arany, I., and Tyring, S. K., 1996, Status of local cellular immunity in interferon-responsive and -nonresponsive human papillomavirus-associated lesions. *Sex. Transm. Dis.* **23**: 475-480.

Arany, I., Tyring, S. K., Stanley, M. A., Tomai, M. A., Miller, R. L., Smith, M. H., McDermott, D. J., and Slade, H. B., 1999, Enhancement of the innate and cellular immune response in patients with genital warts treated with topical imiquimod cream 5%. *Antiviral Res.* **43**: 55-63.

Baldwin, P. J., van der Burg, S. H., Boswell, C. M., Offringa, R., Hickling, J. K., Dobson, J., Roberts, J. S., Latimer, J. A., Moseley, R. P., Coleman, N., et al, 2003, Vaccinia-expressed human papillomavirus 16 and 18 e6 and e7 as a therapeutic vaccination for vulval and vaginal intraepithelial neoplasia. *Clin. Cancer Res.* **9**: 5205-5213.

Barnard, P., and McMillan, N. A., 1999, The human papillomavirus E7 oncoprotein abrogates signaling mediated by interferon-alpha. *Virology* **259**: 305-313.

Bell, J. A., Sundberg, J. P., Ghim, S. J., Newsome, J., Jenson, A. B., and Schlegel, R., 1994, A formalin-inactivated vaccine protects against mucosal papillomavirus infection: a canine model. *Pathobiology* **62**: 194-198.

Bernard, H.-U., Chan, S.-Y., and Delius, H., 1994, Evolution of Papillomaviruses, In Human Pathogenic Papillomaviruses, H. zur Hausen, ed. (Berlin: Springer-Verlag), pp. 33-54.

Berry, J. M., and Palefsky, J. M., 2003, A review of human papillomavirus vaccines: from basic science to clinical trials. *Front. Biosci.* **8**: 333-345.

Biemelt, S., Sonnewald, U., Galmbacher, P., Willmitzer, L., and Muller, M., 2003, Production of human papillomavirus type 16 virus-like particles in transgenic plants. *J. Virol.* **77**: 9211-9220.

Borysiewicz, L. K., Fiander, A., Nimako, M., Man, S., Wilkinson, G. W., Westmoreland, D., Evans, A. S., Adams, M., Stacey, S. N., Boursnell, M. E., et al., 1996, A recombinant vaccinia virus encoding human papillomavirus types 16 and 18, E6 and E7 proteins as immunotherapy for cervical cancer. *Lancet* **347**: 1523-1527.

Bosch, F. X., Manos, M. M., Munoz, N., Sherman, M., Jansen, A. M., Peto, J., Schiffman, M. H., Moreno, V., Kurman, R., and Shah, K. V., 1995, Prevalence of human papillomavirus in cervical cancer: a worldwide perspective. International biological study on cervical cancer (IBSCC) Study Group [see comments]. *J. Natl. Cancer Inst.* **87**: 796-802.

Boshart, M., Gissmann, L., Ikenberg, H., Kleinheinz, A., Scheurlen, W., and zur Hausen, H., 1984, A new type of papillomavirus DNA, its presence in genital cancer biopsies and in cell lines derived from cervical cancer. *Embo J.* **3**: 1151-1157.

Bouwes Bavinck, J. N., Stark, S., Petridis, A. K., Marugg, M. E., Ter Schegget, J., Westendorp, R. G., Fuchs, P. G., Vermeer, B. J., and Pfister, H., 2000, The presence of antibodies against virus-like particles of epidermodysplasia verruciformis-associated humanpapillomavirus type 8 in patients with actinic keratoses. *Br. J. Dermatol.* **142**: 103-109.

Breitburd, F., Kirnbauer, R., Hubbert, N. L., Nonnenmacher, B., Trin-Dinh-Desmarquet, C., Orth, G., Schiller, J. T., and Lowy, D. R., 1995, Immunization with viruslike particles from cottontail rabbit papillomavirus (CRPV) can protect against experimental CRPV infection. *J. Virol.* **69**: 3959-3963.

Brinkman, J. A., Fausch, S. C., Weber, J. S., and Kast, W. M., 2004, Peptide-based vaccines for cancer immunotherapy. *Expert. Opin. Biol. Ther.* **4**: 181-198.

Buonamassa, D. T., Greer, C. E., Capo, S., Yen, T. S., Galeotti, C. L., and Bensi, G., 2002, Yeast coexpression of human papillomavirus types 6 and 16 capsid proteins. *Virology* **293**: 335-344.

Butz, K., Ristriani, T., Hengstermann, A., Denk, C., Scheffner, M., and Hoppe-Seyler, F., 2003, siRNA targeting of the viral E6 oncogene efficiently kills human papillomavirus-positive cancer cells. *Oncogene* **22**: 5938-5945.

Campo, M. S., 1995, Infection by bovine papillomavirus and prospects for vaccination. *Trends Microbiol.* **3**: 92-97.

Campo, M. S., Grindlay, G. J., O'Neil, B. W., Chandrachud, L. M., McGarvie, G. M., and Jarrett, W. F., 1993, Prophylactic and therapeutic vaccination against a mucosal papillomavirus. *J. Gen. Virol.* **74**: 945-953.

Cassetti, M. C., McElhiney, S. P., Shahabi, V., Pullen, J. K., Le Poole, I. C., Eiben, G. L., Smith, L. R., and Kast, W. M., 2004, Antitumor efficacy of Venezuelan equine encephalitis virus replicon particles encoding mutated HPV16 E6 and E7 genes. *Vaccine* **22**: 520-527.

Christensen, N. D., Cladel, N. M., Reed, C. A., and Han, R., 2000, Rabbit oral papillomavirus complete genome sequence and immunity following genital infection. *Virology* **269**: 451-461.

Chu, N. R., Wu, H. B., Wu, T., Boux, L. J., Siegel, M. I., and Mizzen, L. A., 2000, Immunotherapy of a human papillomavirus (HPV) type 16 E7-expressing tumour by administration of fusion protein comprising Mycobacterium bovis bacille Calmette-Guerin (BCG) hsp65 and HPV16 E7. *Clin. Exp. Immunol.* **121**: 216-225.

Coleman, N., Birley, H. D., Renton, A. M., Hanna, N. F., Ryait, B. K., Byrne, M., Taylor-Robinson, D., and Stanley, M. A., 1994, Immunological events in regressing genital warts. *Am. J. Clin. Pathol.* **102**: 768-774.

Combita, A. L., Bravo, M. M., Touze, A., Orozco, O., and Coursaget, P., 2002, Serologic response to human oncogenic papillomavirus types 16, 18, 31, 33, 39, 58 and 59 virus-like particles in colombian women with invasive cervical cancer. *Int. J. Cancer* **97**: 796-803.

Corona Gutierrez, C. M., Tinoco, A., Navarro, T., Contreras, M. L., Cortes, R. R., Calzado, P., Reyes, L., Posternak, R., Morosoli, G., Verde, M. L., and Rosales, R., 2004, Therapeutic vaccination with MVA E2 can eliminate precancerous lesions (CIN 1, CIN 2, and CIN 3) associated with infection by oncogenic human papillomavirus. *Hum. Gene Ther.* **15**: 421-431.

Cubie, H. A., Plumstead, M., Zhang, W., de Jesus, O., Duncan, L. A., and Stanley, M. A., 1998, Presence of antibodies to human papillomavirus virus-like particles (VLPs) in 11-13-year-old schoolgirls. *J. Med. Virol.* **56**: 210-216.

Da Silva, D. M., Velders, M. P., Rudolf, M. P., Schiller, J. T., and Kast, W. M., 1999, Papillomavirus virus-like particles as anticancer vaccines. *Curr. Opin. Mol. Ther.* **1**: 82-88.

Daemen, T., Pries, F., Bungener, L., Kraak, M., Regts, J., and Wilschut, J., 2000, Genetic immunization against cervical carcinoma: induction of cytotoxic T lymphocyte activity with a recombinant alphavirus vector expressing human papillomavirus type 16 E6 and E7. *Gene Ther.* **7**: 1859-1866.

Davidson, E. J., Boswell, C. M., Sehr, P., Pawlita, M., Tomlinson, A. E., McVey, R. J., Dobson, J., Roberts, J. S., Hickling, J., Kitchener, H. C., and Stern, P. L., 2003, Immunological and clinical responses in women with vulval intraepithelial neoplasia vaccinated with a vaccinia virus encoding human papillomavirus 16/18 oncoproteins. *Cancer Res.* **63**: 6032-6041.

De Bruijn, M. L., Schuurhuis, D. H., Vierboom, M. P., Vermeulen, H., de Cock, K. A., Ooms, M. E., Ressing, M. E., Toebes, M., Franken, K. L., Drijfhout, J. W., et al., 1998 Immunization with human papillomavirus type 16 (HPV16) oncoprotein-loaded dendritic cells as well as protein in adjuvant induces MHC class I-restricted protection to HPV16-induced tumor cells. *Cancer Res.* **58**: 724-731.

de Jong, A., O'Neill, T., Khan, A. Y., Kwappenberg, K. M., Chisholm, S. E., Whittle, N. R., Dobson, J. A., Jack, L. C., St Clair Roberts, J. A., Offringa, R., et al., 2002, Enhancement of human papillomavirus (HPV) type 16 E6 and E7-specific T-cell immunity in healthy volunteers through vaccination with TA-CIN, an HPV16 L2E7E6 fusion protein vaccine. *Vaccine* **20**: 3456-3464.

de Villiers, E. M., Fauquet, C., Broker, T. R., Bernard, H. U., and zur Hausen, H., 2004, Classification of papillomaviruses. *Virology* **324**: 17-27.

DeFilippis, R. A., Goodwin, E. C., Wu, L., and DiMaio, D., 2003, Endogenous human papillomavirus E6 and E7 proteins differentially regulate proliferation, senescence, and apoptosis in HeLa cervical carcinoma cells. *J. Virol.* **77**:1551-1563.

Eiben, G. L., da Silva, D. M., Fausch, S. C., Le Poole, I. C., Nishimura, M. I., and Kast, W. M., 2003, Cervical cancer vaccines: recent advances in HPV research. *Viral Immunol.* **16**: 111-121.

Feltkamp, M. C., Smits, H. L., Vierboom, M. P., Minnaar, R. P., de Jongh, B. M., Drijfhout, J. W., ter Schegget, J., Melief, C. J., and Kast, W. M., 1993, Vaccination with cytotoxic T lymphocyte epitope-containing peptide protects against a tumor induced by human papillomavirus type 16-transformed cells. *Eur. J. Immunol.* **23**: 2242-2249.

Ferrara, A., Nonn, M., Sehr, P., Schreckenberger, C., Pawlita, M., Durst, M., Schneider, A., and Kaufmann, A. M., 2003, Dendritic cell-based tumor vaccine for cervical cancer II: results of a clinical pilot study in 15 individual patients. *J. Cancer Res. Clin. Oncol.* **129**: 521-530.

Frazer, I. H., Quinn, M., Nicklin, J. L., Tan, J., Perrin, L. C., Ng, P., O'Connor, V. M., White, O., Wendt, N., Martin, J., et al. (2004). Phase 1 study of HPV16-specific immunotherapy with E6E7 fusion protein and ISCOMATRIX adjuvant in women with cervical intra-epithelial neoplasia. *Vaccine* **23**: 172-181.

Frazer, I. H., Thomas, R., Zhou, J., Leggatt, G. R., Dunn, L., McMillan, N., Tindle, R. W., Filgueira, L., Manders, P., Barnard, P., and Sharkey, M., 1999, Potential strategies utilised by papillomavirus to evade host immunity. *Immunol. Rev.* **168**: 131-142.

Freyschmidt, E. J., Alonso, A., Hartmann, G., and Gissmann, L., 2004, Activation of dendritic cells and induction of T cell responses by HPV 16 L1/E7 chimeric virus-like particles are enhanced by CpG ODN or sorbitol. *Antivir. Ther.* **9**: 479-489.

Galloway, D. A., 1992,. Serological assays for the detection of HPV antibodies, In The epidemiology of human papillomavirus and cervical cancer, N. Munoz, F. X. Bosch, K. V. Shah, and A. Meheus, eds. (Lyon: IARC Scientific Publications), pp. 147-161.

Garcia, F., Petry, K. U., Muderspach, L., Gold, M. A., Braly, P., Crum, C. P., Magill, M., Silverman, M., Urban, R. G., Hedley, M. L., and Beach, K. J., 2004, ZYC101a for treatment of high-grade cervical intraepithelial neoplasia: a randomized controlled trial. *Obstet. Gynecol.* **103**: 317-326.

Ghim, S., Newsome, J., Bell, J., Sundberg, J. P., Schlegel, R., and Jenson, A. B., 2000, Spontaneously regressing oral papillomas induce systemic antibodies that neutralize canine oral papillomavirus. *Exp. Mol. Pathol.* **68**: 147-151.

Gissmann, L., 2003, Human papillomaviruses, In The vaccine book, B. R. Bloom, and P. H. Lambert, eds. (Amsterdam: Academic Press), pp. 311-322.

Gissmann, L., Wolnik, L., Ikenberg, H., Koldovsky, U., Schnurch, H. G., and zur Hausen, H., 1983, Human papillomavirus types 6 and 11 DNA sequences in genital and laryngeal papillomas and in some cervical cancers. *Proc. Natl. Acad. Sci. USA* **80**: 560-563.

Goldstone, S. E., Palefsky, J. M., Winnett, M. T., and Neefe, J. R., 2002, Activity of HspE7, a novel immunotherapy, in patients with anogenital warts. *Dis. Colon Rectum.* **45**: 502-507.

Greenstone, H. L., Nieland, J. D., de Visser, K. E., De Bruijn, M. L., Kirnbauer, R., Roden, R. B., Lowy, D. R., Kast, W. M., and Schiller, J. T., 1998, Chimeric papillomavirus virus-like particles elicit antitumor immunity against the E7 oncoprotein in an HPV16 tumor model. *Proc. Natl. Acad. Sci. USA* **95**: 1800-1805.

Hagensee, M. E., Olson, N. H., Baker, T. S., and Galloway, D. A., 1994, Three-dimensional structure of vaccinia virus-produced human papillomavirus type 1 capsids. *J. Virol.* **68**: 4503-4505.

Hallez, S., Simon, P., Maudoux, F., Doyen, J., Noel, J. C., Beliard, A., Capelle, X., Buxant, F., Fayt, I., Lagrost, A. C., et al., 2004, Phase I/II trial of immunogenicity of a human papillomavirus (HPV) type 16 E7 protein-based vaccine in women with oncogenic HPV-positive cervical intraepithelial neoplasia. *Cancer Immunol. Immunother.* **53**: 642-650.

Hariharan, K., Braslawsky, G., Barnett, R. S., Berquist, L. G., Huynh, T., Hanna, N., and Black, A., 1998, Tumor regression in mice following vaccination with human papillomavirus E7 recombinant protein in PROVAX. *Int. J. Oncol.* **12**: 1229-1235.

Harper, D. M., Franco, E. L., Wheeler, C., Ferris, D. G., Jenkins, D., Schuind, A., Zahaf, T., Innis, B., Naud, P., and De Carvalho, N. S., 2004, Efficacy of a bivalent L1 virus-like particle vaccine in prevention of infection with human papillomavirus types 16 and 18 in young women: a randomised controlled trial. *The Lancet* **364**: 1757-1765.

He, Z., Wlazlo, A. P., Kowalczyk, D. W., Cheng, J., Xiang, Z. Q., Giles-Davis, W., and Ertl, H. C., 2000, Viral recombinant vaccines to the E6 and E7 antigens of HPV-16. *Virology* **270**: 146-161.

Hedley, M. L., Curley, J., and Urban, R., 1998, Microspheres containing plasmid-encoded antigens elicit cytotoxic T-cell responses. *Nat. Med.* **4**: 365-368.

Heino, P., Dillner, J., and Schwartz, S., 1995, Human papillomavirus type 16 capsid proteins produced from recombinant Semliki Forest virus assemble into virus-like particles. *Virology* **214**: 349-359.

Hemmi, H., Kaisho, T., Takeuchi, O., Sato, S., Sanjo, H., Hoshino, K., Horiuchi, T., Tomizawa, H., Takeda, K., and Akira, S., 2002, Small anti-viral compounds activate immune cells via the TLR7 MyD88-dependent signaling pathway. *Nat. Immunol.* **3**: 196-200.

Ho, G. Y., Bierman, R., Beardsley, L., Chang, C. J., and Burk, R. D., 1998, Natural history of cervicovaginal papillomavirus infection in young women. *N. Engl. J. Med.* **338**: 423-428.

Hsieh, C. J., Kim, T. W., Hung, C. F., Juang, J., Moniz, M., Boyd, D. A., He, L., Chen, P. J., Chen, C. H., and Wu, T. C., 2004, Enhancement of vaccinia vaccine potency by linkage of tumor antigen gene to gene encoding calreticulin. *Vaccine* **22**: 3993-4001.

IARC, 1995, Human Papillomaviruses, Vol 64 (Lyon: IARC Lyon France).

Jarrett, W. F., O'Neil, B. W., Gaukroger, J. M., Laird, H. M., Smith, K. T., and Campo, M. S., 1990, Studies on vaccination against papillomaviruses: a comparison of purified virus, tumour extract and transformed cells in prophylactic vaccination. *Vet .Rec.* **126**: 449-452.

Jarrett, W. F., Smith, K. T., O'Neil, B. W., Gaukroger, J. M., Chandrachud, L. M., Grindlay, G. J., McGarvie, G. M., and Campo, M. S., 1991, Studies on vaccination against papillomaviruses: prophylactic and therapeutic vaccination with recombinant structural proteins. *Virology* **184**: 33-42.

Jensen, E. R., Selvakumar, R., Shen, H., Ahmed, R., Wettstein, F. O., and Miller, J. F., 1997, Recombinant Listeria monocytogenes vaccination eliminates papillomavirus-induced tumors and prevents papilloma formation from viral DNA. *J. Virol.* **71**: 8467-8474.

Jochmus-Kudielka, I., Schneider, A., Braun, R., Kimmig, R., Koldovsky, U., Schneweis, K. E., Seedorf, K., and Gissmann, L., 1989, Antibodies against the human papillomavirus type 16 early proteins in human sera: correlation of anti-E7 reactivity with cervical cancer. *J. Natl. Cancer Inst .***81**: 1698-1704.

Kaufmann, A. M., Nieland, J., Schinz, M., Nonn, M., Gabelsberger, J., Meissner, H., Muller, R. T., Jochmus, I., Gissmann, L., Schneider, A., and Durst, M., 2001, HPV16 L1E7 chimeric virus-like particles induce specific HLA-restricted T cells in humans after *in vitro* vaccination. *Int. J. Cancer* **92**: 285-293.

Kaufmann, A. M., Stern, P. L., Rankin, E. M., Sommer, H., Nuessler, V., Schneider, A., Adams, M., Onon, T. S., Bauknecht, T., Wagner, U., et al., 2002, Safety and immunogenicity of TA-HPV, a recombinant vaccinia virus expressing modified human papillomavirus (HPV)-16 and HPV-18 E6 and E7 genes, in women with progressive cervical cancer. Clin *Cancer Res.* **8**: 3676-3685.

Kim, T. Y., Myoung, H. J., Kim, J. H., Moon, I. S., Kim, T. G., Ahn, W. S., and Sin, J. I., 2002, Both E7 and CpG-oligodeoxynucleotide are required for protective immunity against challenge with human papillomavirus 16 (E6/E7) immortalized tumor cells: involvement of CD4+ and CD8+ T cells in protection. *Cancer Res.* **62**: 7234-7240.

Klencke, B., Matijevic, M., Urban, R. G., Lathey, J. L., Hedley, M. L., Berry, M., Thatcher, J., Weinberg, V., Wilson, J., Darragh, T., et al., 2002, Encapsulated plasmid DNA treatment for human papillomavirus 16-associated anal dysplasia: a Phase I study of ZYC101. *Clin. Cancer Res.* **8**: 1028-1037.

Knowles, G., O'Neil, B. W., and Campo, M. S. (1996). Phenotypical characterization of lymphocytes infiltrating regressing papillomas. *J. Virol.* **70**: 8451-8458.

Konya, J., and Dillner, J., 2001, Immunity to oncogenic human papillomaviruses. *Adv. Cancer Res.* **82**: 205-238.

Koromilas, A. E., Li, S., and Matlashewski, G., 2001, Control of interferon signaling in human papillomavirus infection. *Cytokine Growth Factor Rev.* **12**: 157-170.

Koutsky, L. A., Ault, K. A., Wheeler, C. M., Brown, D. R., Barr, E., Alvarez, F. B., Chiacchierini, L. M., and Jansen, K. U., 2002, A controlled trial of a human papillomavirus type 16 vaccine. *N. Engl. J. Med.* **347**: 1645-1651.

Kreider, J. W., and Bartlett, G. L., 1981, The Shope papilloma-carcinoma complex of rabbits: a model system of neoplastic progression and spontaneous regression. *Adv. Cancer Res.* **35**: 81-110.

Lacey, C. J., Thompson, H. S., Monteiro, E. F., O'Neill, T., Davies, M. L., Holding, F. P., Fallon, R. E., and Roberts, J. S., 1999, Phase IIa safety and immunogenicity of a therapeutic vaccine, TA-GW, in persons with genital warts. *J. Infect. Dis.* **179**: 612-618.

Meneguzzi, G., Cerni, C., Kieny, M. P., and Lathe, R., 1991, Immunization against human papillomavirus type 16 tumor cells with recombinant vaccinia viruses expressing E6 and E7. *Virology* **181**: 62-69.

Meschede, W., Zumbach, K., Braspenning, J., Scheffner, M., Benitez-Bribiesca, L., Luande, J., Gissmann, L., and Pawlita, M. M., 1998, Antibodies against early proteins of human papillomaviruses as diagnostic markers for invasive cervical cancer. *J. Clin. Microbiol.* **36**: 475-480.

Moore, R. A., 2004, Canine papillomavirus - a mucosal model of human disease. *Papillomavirus Rep.* **15**: 55-62.

Muderspach, L., Wilczynski, S., Roman, L., Bade, L., Felix, J., Small, L. A., Kast, W. M., Fascio, G., Marty, V., and Weber, J., 2000, A phase I trial of a human papillomavirus (HPV) peptide vaccine for women with high-grade cervical and vulvar intraepithelial neoplasia who are HPV 16 positive. Clin *Cancer Res.* **6**: 3406-3416.

Müller, M., Zhou, J., Reed, T. D., Rittmüller, C., Burger, A., Gabelsberger, J., Braspenning, J., and Gissmann, L., 1997, Chimeric papillomavirus-like particles. *Virology* **234**: 93-111.

Munger, K., Baldwin, A., Edwards, K. M., Hayakawa, H., Nguyen, C. L., Owens, M., Grace, M., and Huh, K., 2004, Mechanisms of human papillomavirus-induced oncogenesis. *J. Virol.* **78**: 11451-11460.

Naryshkin, S., 1997, The false-negative fraction for Papanicolaou smears: how often are "abnormal" smears not detected by a "standard" screening cytologist? *Arch. Pathol. Lab. Med.* **121**: 270-272.

Navabi, H., Jasani, B., Adams, M., Evans, A. S., Mason, M., Crosby, T., and Borysiewicz, L., 1997, Generation of *in vitro* autologous human cytotoxic T-cell response to E7 and HER-2/neu oncogene products using ex-vivo peptide loaded dendritic cells. *Adv. Exp. Med. Biol.* **417**: 583-589.

Nicholls, P. K., Moore, P. F., Anderson, D. M., Moore, R. A., Parry, N. R., Gough, G. W., and Stanley, M. A., 2001, Regression of canine oral papillomas is associated with infiltration of CD4+ and CD8+ lymphocytes. *Virology* **283**: 31-39.

Nieland, J. D., Jochmus, I., Baur, S., Friese, K., Gabelsberger, J., Gissmann, L., Hillemanns, P., Höpfl, R., Ikenberg, H., Jänicke, F., et al., 2004, Medigene's HPV 16 L1/E7 chimeric virus-like particle vaccine, Paper presented at: 21st International Papillomavirus Conference (Mexico City).

Nobbenhuis, M. A., Walboomers, J. M., Helmerhorst, T. J., Rozendaal, L., Remmink, A. J., Risse, E. K., van der Linden, H. C., Voorhorst, F. J., Kenemans, P., and Meijer, C. J., 1999, Relation of human papillomavirus status to cervical lesions and consequences for cervical-cancer screening: a prospective study. *The Lancet* **354**: 20-25.

Okabayashi, M., Angell, M. G., Christensen, N. D., and Kreider, J. W., 1991, Morphometric analysis and identification of infiltrating leucocytes in regressing and progressing Shope rabbit papillomas. *Int. J. Cancer* **49**: 919-923.

Pardoll, D., and Allison, J., 2004, Cancer immunotherapy: breaking the barriers to harvest the crop. *Nat. Med.* **10**: 887-892.

Patel, D., Huang, S. M., Baglia, L. A., and McCance, D. J., 1999, The E6 protein of human papillomavirus type 16 binds to and inhibits co-activation by CBP and p300. *Embo J.* **18**: 5061-5072.

Pisani, P., Parkin, D. M., Bray, F., and Ferlay, J., 1999, Estimates of the worldwide mortality from 25 cancers in 1990. *Int. J. Cancer* **83**: 18-29.

Remmink, A. J., Walboomers, J. M., Helmerhorst, T. J., Voorhorst, F. J., Rozendaal, L., Risse, E. K., Meijer, C. J., and Kenemans, P., 1995, The presence of persistent high-risk HPV genotypes in dysplastic cervical lesions is associated with progressive disease: natural history up to 36 months. *Int. J. Cancer* **61**: 306-311.

Ressing, M. E., van Driel, W. J., Brandt, R. M., Kenter, G. G., de Jong, J. H., Bauknecht, T., Fleuren, G. J., Hoogerhout, P., Offringa, R., Sette, A., et al., 2000, Detection of T helper responses, but not of human papillomavirus-specific cytotoxic T lymphocyte responses, after peptide vaccination of patients with cervical carcinoma. *J. Immunother* **23**: 255-266.

Revaz, V., Benyacoub, J., Kast, W. M., Schiller, J. T., De Grandi, P., and Nardelli-Haefliger, D., 2001, Mucosal vaccination with a recombinant Salmonella typhimurium expressing human papillomavirus type 16 (HPV16) L1 virus-like particles (VLPs) or HPV16 VLPs purified from insect cells inhibits the growth of HPV16-expressing tumor cells in mice. *Virology* **279**: 354-360.

Ronco, L. V., Karpova, A. Y., Vidal, M., and Howley, P. M., 1998, Human papillomavirus 16 E6 oncoprotein binds to interferon regulatory factor-3 and inhibits its transcriptional activity. *Genes Dev.* **12**: 2061-2072.

Salit, R. B., Kast, W. M., and Velders, M. P., 2002, Ins and outs of clinical trials with peptide-based vaccines. *Front. Biosci.* **7**: 204-213.

Santin, A. D., Bellone, S., Gokden, M., Cannon, M. J., and Parham, G. P., 2002, Vaccination with HPV-18 E7-pulsed dendritic cells in a patient with metastatic cervical cancer. *N. Engl. J. Med.* **346**: 1752-1753.

Santin, A. D., Hermonat, P. L., Ravaggi, A., Chiriva-Internati, M., Zhan, D., Pecorelli, S., Parham, G. P., and Cannon, M. J., 1999, Induction of human papillomavirus-specific CD4(+) and CD8(+) lymphocytes by E7-pulsed autologous dendritic cells in patients with human papillomavirus type 16- and 18-positive cervical cancer. *J. Virol.* **73**: 5402-5410.

Sawaya, G. F., Brown, A. D., Washington, A. E., and Garber, A. M., 2001, Current Approaches to Cervical-Cancer Screening. *N. Engl. J. Med.* **344**: 1603-1607.

Schäfer, K., Müller, M., Faath, S., Henn, A., Osen, W., Zentgraf, H., Benner, A., Gissmann, L., and Jochmus, I., 1999, Immune response to human papillomavirus 16 L1E7 chimeric virus-like particles: induction of cytotoxic T cells and specific tumor protection. *Int. J. Cancer* **81**: 881-888.

Scheffner, M., Romanczuk, H., Münger, K., Huibregtse, J. M., Mietz, J. A., and Howley, P. M., 1994, Functions of Human Papillomavirus Proteins, In Human Pathogenic Papillomaviruses, H. zur Hausen, ed. (Berlin: Springer-Verlag), pp. 83-99.

Schiller, J. T., and Davies, P., 2004, Delivering on the promise: HPV vaccines and cervical cancer. *Nat .Rev. Microbiol.* **2**: 343-347.

Schneider, A., Papendick, U., Gissmann, L., and De Villiers, E. M., 1987, Interferon treatment of human genital papillomavirus infection: importance of viral type. *Int. J. Cancer* **40**: 610-614.

Scott, M., Nakagawa, M., and Moscicki, A. B., 2001, Cell-mediated immune response to human papillomavirus infection. *Clin. Diagn. Lab. Immunol.* **8**: 209-220.

Sheets, E. E., Urban, R. G., Crum, C. P., Hedley, M. L., Politch, J. A., Gold, M. A., Muderspach, L. I., Cole, G. A., and Crowley-Nowick, P. A., 2003 Immunotherapy of human cervical high-grade cervical intraepithelial neoplasia with microparticle-delivered human papillomavirus 16 E7 plasmid DNA. *Am. J. Obstet. Gynecol.* **188**: 916-926.

Smyth, L. J., Van Poelgeest, M. I., Davidson, E. J., Kwappenberg, K. M., Burt, D., Sehr, P., Pawlita, M., Man, S., Hickling, J. K., Fiander, A. N., et al., 2004, Immunological responses in women with human papillomavirus type 16 (HPV-16)-associated anogenital intraepithelial neoplasia induced by heterologous prime-boost HPV-16 oncogene vaccination. *Clin Cancer Res.* **10**: 2954-2961.

Steller, M. A., Gurski, K. J., Murakami, M., Daniel, R. W., Shah, K. V., Celis, E., Sette, A., Trimble, E. L., Park, R. C., and Marincola, F. M., 1998, Cell-mediated immunological responses in cervical and vaginal cancer patients immunized with a lipidated epitope of human papillomavirus type 16 E7. *Clin. Cancer Res.* **4**: 2103-2109.

Stern, P. L., Brown, M., Stacey, S. N., Kitchener, H. C., Hampson, I., Abdel-Hady, E. S., and Moore, J. V., 2000, Natural HPV immunity and vaccination strategies. *J. Clin. Virol.* **19**: 57-66.

Stern, P. L., Faulkner, R., Veranes, E. C., and Davidson, E. J., 2001, The role of human papillomavirus vaccines in cervical neoplasia. Best Pract Res Clin Obstet Gynaecol **15**: 783-799.

Stubenrauch, F., and Laimins, L. A., 1999, Human papillomavirus life cycle: active and latent phases. *Semin .Cancer Biol.* **9**: 379-386.

Sundberg, J. P., 1987, Papillomavirus Infections in Animals, In Papillomaviruses and Human Disease, K. Syrjänen, L. Gissmann, and L. G. Koss, eds. (Berlin: Springer Verlag), pp. 40-103.

Suzich, J. A., Ghim, S. J., Palmer-Hill, F. J., White, W. I., Tamura, J. K., Bell, J. A., Newsome, J. A., Jenson, A. B., and Schlegel, R., 1995, Systemic immunization with papillomavirus L1 protein completely prevents the development of viral mucosal papillomas. *Proc. Natl. Acad. Sci. USA* **92**: 11553-11557.

Tommasino, M., 2001, Early Genes of Human papillomaviruses, In Encyclopedic References of Cancer, M. Schwab, ed. (Berlin: Springer), pp. 266-272.

van Driel, W. J., Ressing, M. E., Kenter, G. G., Brandt, R. M., Krul, E. J., van Rossum, A. B., Schuuring, E., Offringa, R., Bauknecht, T., Tamm-Hermelink, A., et al., 1999, Vaccination with HPV16 peptides of patients with advanced cervical carcinoma: clinical evaluation of a phase I-II trial. *Eur. J. Cancer* **35**: 946-952.

Villa, L. L., Costa, R. L., Petta, C. A., Andrade, R. P., Ault, K. A., Giuliano, A. R., Wheeler, C. M., Koutsky, L. A., Malm, C., Lehtinen, M., et al., 2005, Prophylactic quadrivalent human papillomavirus (types 6, 11, 16, and 18) L1 virus-like particle vaccine in young women: a randomised double-blind placebo-controlled multicentre phase II efficacy trial. *Lancet Oncol.* **6**: 271-278.

von Knebel Doeberitz, M., Oltersdorf, T., Schwarz, E., and Gissmann, L., 1988, Correlation of modified human papilloma virus early gene expression with altered growth properties in C4-1 cervical carcinoma cells. *Cancer Res.* **48**: 3780-3786.

Walboomers, J. M., Jacobs, M. V., Manos, M. M., Bosch, F. X., Kummer, J. A., Shah, K. V., Snijders, P. J., Peto, J., Meijer, C. J., and Munoz, N., 1999, Human papillomavirus is a necessary cause of invasive cervical cancer worldwide. *J. Pathol.* **189**: 12-19.

Warzecha, H., Mason, H. S., Lane, C., Tryggvesson, A., Rybicki, E., Williamson, A. L., Clements, J. D., and Rose, R. C., 2003, Oral immunogenicity of human papillomavirus-like particles expressed in potato. *J. Virol.* **77**: 8702-8711.

Wieland, U., and Pfister, H., 1996, Molecular diagnosis of persistent human papilloma virus infections. *Intervirology* **39**: 145-157.

Yuan, H., Estes, P. A., Chen, Y., Newsome, J., Olcese, V. A., Garcea, R. L., and Schlegel, R, 2001, Immunization with a pentameric L1 fusion protein protects against papillomavirus infection. *J. Virol.* **75**: 7848-7853.

Zhang, L. F., Zhou, J., Chen, S., Cai, L. L., Bao, Q. Y., Zheng, F. Y., Lu, J. Q., Padmanabha, J., Hengst, K., Malcolm, K., and Frazer, I. H., 2000, HPV6b virus like particles are potent immunogens without adjuvant in man. *Vaccine* **18**: 1051-1058.

Zhou, J., Sun, X. Y., Stenzel, D. J., and Frazer, I. H., 1991, Expression of vaccinia recombinant HPV 16 L1 and L2 ORF proteins in epithelial cells is sufficient for assembly of HPV virion-like particles. *Virology* **185**: 251-257.

Zimmermann, H., Degenkolbe, R., Bernard, H. U., and O'Connor, M. J., 1999, The human papillomavirus type 16 E6 oncoprotein can down-regulate p53 activity by targeting the transcriptional coactivator CBP/p300. *J. Virol.* **73**: 6209-6219.

Zwaveling, S., Ferreira Mota, S. C., Nouta, J., Johnson, M., Lipford, G. B., Offringa, R., van der Burg, S. H., and Melief, C. J. 2002, Established human papillomavirus type 16-expressing tumors are effectively eradicated following vaccination with long peptides. *J. Immunol.* **169**: 350-358.

3. Concepts of therapy for emerging viruses

Chapter 3.1

THE SARS CORONAVIRUS RECEPTOR ACE2 A POTENTIAL TARGET FOR ANTIVIRAL THERAPY

J. H. KUHN[1,2], S. R. RADOSHITZKY[1], W. LI[1], S. KEE WONG[1], H. CHOE[3] and M. FARZAN[1]

[1]*Department of Microbiology and Molecular Genetics, Harvard Medical School, New England Primate Research Center, Southborough, MA, USA;* [2]*Department of Biology, Chemistry, Pharmacy, Freie Universität Berlin, Berlin, Germany;* [3]*Department of Pediatrics, Children's Hospital, Harvard Medical School, Boston, MA, USA*

Abstract: In recent years, coronaviruses have received increasing attention from clinical infectious disease experts. The discovery of a coronavirus as the etiological agent of severe acute respiratory syndrome (SARS-CoV) and the identification of another coronavirus that causes upper respiratory tract diseases in children (HCoV-NL63) emphasized the importance of improving our understanding of these agents. The identification of angiotensin-converting enzyme 2 (ACE2), the cell-surface receptor for both SARS-CoV and HCoV-NL63, has opened the field for the development of new treatments and prophylactics. In this review, we describe the discovery of this coronavirus receptor and its interaction with coronaviral surface proteins in light of viral entry processes and host tropism.

1. INTRODUCTION

1.1 Coronaviruses

The viral order *Nidovirales* groups numerous viruses of vertebrates and invertebrates that share common characteristics such as genomic organization, transcription and replication strategies, and disease manifestations. Many of these viruses cause acute and chronic upper

E. Bogner and A. Holzenburg (eds.), New Concepts of Antiviral Therapy, 397–418.

respiratory gastrointestinal, hepatic, and neurological diseases in both humans and animals (Enjuanes *et al.*, 2000). The nidoviruses are compartmentalized into the three viral families *Arteriviridae*, *Coronaviridae*, and *Roniviridae*. Nidoviral pathogens that infect humans belong exclusively to the family *Coronaviridae*, which contains the two genera *Coronavirus* and *Torovirus* (Enjuanes *et al.*, 2000).

Coronaviruses *sensu stricto* (viruses belonging to the genus *Coronavirus* rather than all members of the family *Coronaviridae*) are spherical to slightly pleomorphic viruses with a diameter of 100-160nm. The virions contain one copy of a 20-33kb-long capped and polyadenylated single-stranded RNA of positive polarity, which is helically encapsidated by nucleocapsid (N) proteins (Enjuanes *et al.*, 2000). The filamentous ribonucleocapsids are surrounded by matrix (M) proteins, which form icosahedral cores. These cores are wrapped in envelopes formed during coronaviral budding from a host cell. The membranes contain distinct, club- or petal-shaped protrusions, identified as spike (S) proteins. It is those proteins that give the virions a crown-like appearance (Latin: *coronae*) in electron microscopic images (Enjuanes *et al.*, 2000). S proteins are the major antigenic determinants of coronaviruses (Daniel *et al.*, 1994; Gallagher *et al.*, 2001; Koo *et al.*, 1999; Moore *et al.*, 1997; Song *et al.*, 1998).

Coronaviruses attach to cell-surface receptors via their S proteins, which then mediate fusion with the host-cell membrane. Subsequently, a nested set of subgenomic RNAs is produced. Genome replication occurs in the cytoplasm and results in high-frequency recombination. After replication and maturation, coronaviruses bud from the endoplasmic reticulum-Golgi intermediate compartment (Enjuanes *et al.*, 2000).

Three distinct genetic and serological groups of coronaviruses are acknowledged, but the grouping is undergoing revision (Gonzalez *et al.*, 2003). Several coronaviruses are known to cause disease in humans. Human coronavirus 229E (HCoV-229E), a group 1 virus, and human coronavirus OC43 (HCoV-OC43), a group 2 virus, cause mild upper respiratory infections that result in self-resolving common colds in otherwise healthy individuals, or severe pneumonia in immunocompromised people (Denison, 1999; Pene *et al.*, 2003). Human coronavirus NL63 (HCoV-NL63; also referred to as HCoV-NH and HCoV-NL) has recently been identified as a group 1 virus causing conjunctivitis and sometimes serious respiratory infections in children (Esper *et al.*, 2005b; Fouchier *et al.*, 2004; van der Hoek *et al.*, 2004). Epidemiological analyses suggest this virus to be associated with Kawasaki disease (Esper *et al.*, 2005a). Another group 2 coronavirus (CoV-HKU1) was recently isolated from a 71-year old man with pneumonia (Woo *et al.*, 2005). Several more pathogenic human coronaviruses (B814, HCoV-OC16, HCoV-OC37, and HCoV-OC48) await characterization (McIntosh, 2005).

1.2 Severe acute respiratory syndrome virus

Severe acute respiratory syndrome (SARS) was first described in November of 2002, when inhabitants of Guangdong Province, People's Republic of China, presented with an influenza-like illness that began with dyspnea, myalgia, and pyrexia, and was often followed by acute pneumonia, respiratory failure, and death (Peiris *et al.*, 2003). The novel disease was transmitted via droplets and fomites and through direct contact of patients with uninfected individuals (Olsen *et al.*, 2003; Yu *et al.*, 2004). The outbreak spread all over Asia, Europe, and North America. At the end of the mini-pandemic in July 2003, a total of 8,096 cases had been recorded, of which 774 (9.6%) had died (Cherry, 2004). A novel coronavirus, severe acute respiratory syndrome virus (SARS-CoV; species *Severe acute respiratory syndrome coronavirus*) was identified as the etiologic agent (Drosten *et al.*, 2003; Fouchier *et al.*, 2003; Ksiazek *et al.*, 2003; Kuiken *et al.*, 2003; Mayo, 2002). Molecular analyses demonstrated that SARS-CoV clusters with group 2 coronaviruses (Gibbs *et al.*, 2004; Gorbalenya *et al.*, 2004).

SARS reemerged in Guangdong Province in 2003-2004, when it infected four individuals, all of which recovered (Fleck, 2004; Song *et al.*, 2005). All other cases recorded to date trace back to accidental laboratory infections. One and two nonfatal cases, respectively, were recorded in laboratories in the Republic of Singapore (Lim *et al.*, 2004) and Taiwan (Normile, 2004) in 2003. The last cases occurred in 2004 in laboratories in Anhui Province (nine cases including one fatality) and in Beijing (two cases), People's Republic of China.

The natural reservoir of SARS-CoV remains elusive. Chinese ferret-badgers (*Melogale moschata*), Himalayan palm civets (*Paguma larvata*), and racoon dogs (*Nyctereutes procyonoides*) have been suggested as host candidates, because they carry SARS-CoV-like viruses (Guan *et al.*, 2003). Domestic cats (*Felis domesticus*), ferrets (*Mustela putorius furo*), and Himalayan palm civets have been proposed as potential carriers because they are susceptible to SARS-CoV infection and because they can transmit the agent (Martina *et al.*, 2003; Wu *et al.*, 2005). BALB/c mice (Subbarao *et al.*, 2004; Wentworth *et al.*, 2004), hamsters (Roberts *et al.*, 2005), and cynomolgus and rhesus macaques (Rowe *et al.*, 2004) are currently being used as animal models for SARS-CoV infection.

As of today, vaccines have not been approved for the prevention of human SARS-CoV infections, although several candidates are clinically evaluated.

Sera obtained from SARS survivors contain high titers of SARS-CoV-neutralizing antibodies. SARS patients were reported to clinically improve after the administration of such reconvalescent sera (Li *et al.*, 2003a; Pearson *et al.*, 2003). Neutralizing antibodies, administered to BALB/c mice intraperitoneally one day before intranasal challenge with 10^4 50% tissue culture infective doses ($TCID_{50}$) of the SARS-CoV Urbani strain, reduced viral load by four orders of magnitude (Sui *et al.*, 2005). These data suggest that antibodies could be used to prevent and treat SARS, and also that the development of a SARS vaccine is plausible. Promising results in vaccine development have been obtained with inactivated whole virus (He *et al.*, 2004b; Takasuka *et al.*, 2004), recombinant N or S protein administered either as a DNA vaccine (Yang *et al.*, 2004; Zhao *et al.*, 2005), via replication-deficient adenoviruses (Zakhartchouk *et al.*, 2005) or *Lactococcus lactis* (Pei *et al.*, 2005), as well as via S protein-expressing vaccinia viruses (Bisht *et al.*, 2004; Chen *et al.*, 2005). In all these cases, development of neutralizing antibodies was reported after immunization, and in the majority, these vaccines protected mice from challenge with infectious SARS-CoV. However, mice do not develop a SARS-like disease after viral infection. SARS-CoV infection only leads to high levels of replication in their respiratory tracts, and virus is cleared within seven days post infection (Subbarao *et al.*, 2004). Hence, it remains unclear whether the vaccine candidates tested would also be efficacious in primates or, later, in humans. Studies in nonhuman primates have been sparse, but at least one group achieved protection of African green monkeys (*Chlorocebus aethiops*) against SARS-CoV challenge after vaccination with an attenuated parainfluenza virus expressing SARS-CoV S protein (Bukreyev *et al.*, 2004).

Obvious targets for antiviral agents include the SARS-CoV protease, mRNA cap-1 methyl-transferase, NTPase/helicase, and the transcriptase-replicase complex. However, with the exception of interferons, which have been shown effective in inhibiting SARS-CoV replication in tissue cultures, no clinically approved antiviral drug could be identified for the treatment of SARS so far (Tan *et al.*, 2004).

1.3   Coronaviral spike proteins and host-cell receptors

The host spectrum and virulence of a specific coronavirus is largely determined by its S protein (Kuo *et al.*, 2000; Phillips *et al.*, 1999; Sanchez *et al.*, 1999). Coronaviral S proteins measure 12-24 nm in length. They are spaced widely apart, and are evenly distributed all over the surface of the coronavirion (Enjuanes *et al.*, 2000). They are type I transmembrane and class I fusion proteins (Bosch *et al.*, 2003; Gallagher, 2001), and consist of distinct N-terminal (S1) and C-terminal (S2) domains, which mediate receptor binding and virus-cell fusion, respectively (Gallagher, 2001). Upon binding of the cell-surface receptor, a conformational change of the S1 domain induces the exposure of a fusion peptide embedded in the S2 domain, which then induces the reorganization of S2's unusually large heptad repeats into coiled-coils. This extreme conformation switch forces the virion membrane into close apposition to the cell membrane and allows fusion and hence penetration of the viral core into the cell (Colman *et al.*, 2003; Dimitrov, 2004; Lai *et al.*, 1997).

Only few coronavirus cell-surface receptors have been identified. Aminopeptidase N (APN, CD13) was shown to be the receptor for canine coronavirus, feline infectious peritonitis virus, HCoV-229E, porcine epidemic diarrhea virus, and porcine transmissible gastroenteritis virus, all of which are group 1 coronaviruses (Delmas *et al.*, 1992; Oh *et al.*, 2003; Yeager *et al.*, 1992). Members of the pleiotropic family of carcinoembryonic antigen-cell adhesion molecules were identified as receptors for the group 2 pathogen murine hepatitis virus (Williams *et al.*, 1991), whereas bovine group 2 coronaviruses bind to 9-*O*-acetylated sialic acids (Holmes *et al.*, 1994).

2. IDENTIFICATION OF ACE2 AS THE RECEPTOR OF SARS CORONAVIRUS AND HUMAN CORONAVIRUS NL63

Full-length and partial S proteins of different SARS-CoV isolates were cloned, expressed, and characterized by several research groups only months after the discovery of the virus (Babcock *et al.*, 2004; Bisht *et al.*, 2004; Li *et al.*, 2003b; Simmons *et al.*, 2004; Wong *et al.*, 2004; Xiao *et al.*, 2003). SARS-CoV S protein was found to have a molecular weight of about 170-200 kD in PAGE analyses. The protein was suggested to undergo posttranslational modifications since the determined weight was much higher than the molecular weight predicted from the amino acid sequence (Bisht *et al.*, 2004; Xiao *et al.*, 2003). SARS-CoV S protein is a typical class I fusion protein. However, there is little (20-27%) similarity between the amino acid sequence of S protein of SARS-CoV and those of other coronaviral S proteins (Rota *et al.*, 2003). Furthermore, in contrast to most other S proteins, SARS-CoV S protein is not cleaved by a host cell protease into S1 and S2 subunits, although domains corresponding to S1 and S2 have been identified (Marra *et al.*, 2003; Rota *et al.*, 2003; Xiao *et al.*, 2003).

SARS-CoV S protein alone is sufficient to mediate virus-cell fusion. The SARS-CoV S1 domain was identified as the receptor-binding region of the attachment protein, whereas the S2 domain was found to mediate fusion with the host cell membrane (Li *et al.*, 2003b; Xiao *et al.*, 2003). However, it was impossible to predict which cell-surface receptor SARS-CoV might utilize to fuse with host cells, in part because of the limited sequence homology between the SARS-CoV S1 domain and the S1 proteins of other coronaviruses. In fact, the observed severity and lethality of SARS-CoV infections and the elucidated limited S1 sequence homology with other S proteins suggested a receptor different from all other known coronavirus receptors.

Our laboratory employed a straight-forward approach, coimmuno-precipitation, to identify the SARS-CoV cell surface receptor (Li *et al.*, 2003b). We first created a codon-optimized gene encoding the SARS-CoV Urbani strain S1 domain (amino acid residues 12-672) fused to the Fc domain of human IgG1 (S1-Ig). Codon optimization resulted in markedly enhanced expression levels compared to wild-type S protein (Moore *et al.*, 2004). The construct bound to the surface of African green monkey kidney (Vero E6) cells, which are highly susceptible to SARS-CoV infection, with high affinity, but not to human embryonic kidney (HEK 293T) cells, in which SARS-CoV replicates poorly (Li *et al.*, 2003b). Encouraged by this observation, we lysed metabolically labelled Vero E6 cells with a detergent (0.3% n-decyl-β-D-maltopyranoside in PBS), and incubated the lysate with SARS-CoV S1-Ig and Sepharose-Protein A beads. Bound proteins were separated by PAGE, and individual bands, which were not present in control reactions, were analyzed by trypsin digestion and mass spectrometry. We identified a promising SARS-CoV receptor candidate, the 110kD human angiotensin-converting enzyme 2 (ACE2) (Li *et al.*, 2003b). A soluble form of ACE2, but not of the closely related ACE1, blocked binding of S1-Ig to the surface of Vero E6 cells. Antibodies to ACE2, but not those to ACE1, blocked SARS-CoV (strain Urbani) replication in Vero E6 cells. SARS-CoV replicated more efficiently in HEK 293T cells after ACE2 transfection. Last, we demonstrated syncytia formation between S protein-expressing HEK 293T cells and HEK 293T cells expressing ACE2 (Li *et al.*, 2003b), suggesting that the syncytia observed in SARS-CoV-infected Vero E6 cells, nonhuman primates, and patients (Drosten *et al.*, 2003; Fouchier *et al.*, 2003; Kuiken *et al.*, 2003) develop through S protein-ACE2 interactions. Commercially available antibodies targeting ACE2 blocked syncytia formation (Li *et al.*, 2003b). Stably ACE2-expressing HEK 293T cells were more transducible with S protein-pseudotyped simian immunodeficiency virus than VeroE6 cells (Moore *et al.*, 2004), which is consistent with the idea that the abundance of ACE2 on the cell surface determines the efficiency of SARS-CoV infection. Others have identified a recombinant human single-chain variable region fragment (scFV) against the SARS-CoV (strain TOR2) S1 domain from nonimmune human antibody libraries. This scFV (named 80R scFV) not only inhibited the formation of syncytia between HEK 293T cells expressing ACE2 and HEK 293T cells expressing the S protein, but also efficiently neutralized infection of Vero E6 cells with the SARS-CoV Urbani strain. 80R scFV competed with a soluble form of ACE2 for association with the S1 domain, and was found to bind S1 with high affinity (K_d=32.3 nM) in plasmon resonance studies. A human IgG1 form of 80R scFV (80R) bound S1 with higher affinity (K_d=1.59 nM) and neutralized the Urbani strain of SARS-CoV with a 20-fold higher efficiency than 80R scFV (Sui *et al.*, 2004). ACE2 was also independently identified as a SARS-CoV receptor by another group, which transduced HeLa cells with a retrovirus cDNA library from Vero E6 cells, followed by flow cytometry to select transduced cells that bound to purified S protein (residues 14-502) (Wang *et al.*, 2004).

ACE2, a carboxy-metalloprotease not related to any other coronavirus receptor but distantly related to ACE, was discovered in 2000 (Donoghue *et al.*, 2000; Tipnis *et al.*, 2000). The enzyme is a type I transmembrane protein with a single metalloprotease active site with a HEXXH zinc binding motif. (Donoghue *et al.*, 2000; Tipnis *et al.*, 2000). The physiological function of ACE2 remains enigmatic. The enzyme has been shown to cleave a variety of regulatory peptides *in vitro*, among them angiotensin I and II, des-Arg-bradykinin, kinetensin, and neurotensin (Donoghue *et al.*, 2000; Vickers *et al.*, 2002). Some cleavage products have been shown to be potent vasodilators with antidiuretic effects. This finding gave rise to the theory that ACE2 counterbalances the actions of ACE1, which mediates vasoconstriction (Yagil *et al.*, 2003). Furthermore, targeted disruption of ACE2 in mice resulted in severe cardiac contractility defects (Crackower *et al.*, 2002).

Despite high expression levels of ACE2, only low levels of SARS-CoV replication were found in cardiac tissue (Farcas *et al.*, 2005). This finding suggested that another receptor or a coreceptor is necessary for successful cellular infection with SARS-CoV. Indeed, CD209L (DC-SIGNR, L-SIGN) has been identified as a molecule that can facilitate SARS-CoV infection (Jeffers *et al.*, 2004; Marzi *et al.*, 2004). However, the tissue distribution of ACE2 matches the tissue tropism of SARS-CoV. For example, high levels of ACE2 expression were detected in the gastrointestinal tract, kidneys, and lungs (Ding *et al.*, 2004; Donoghue *et al.*, 2000; Hamming *et al.*, 2004; Harmer *et al.*, 2002), all of which are targets of SARS-CoV (Hamming *et al.*, 2004). Others have shown that the level of ACE2 expression in different cell lines correlates with their susceptibility for SARS-CoV infection

(Hattermann *et al.*, 2005; Hofmann *et al.*, 2004) or their susceptibility for transduction with lentiviral pseudotypes carrying SARS-CoV S proteins (Nie *et al.*, 2004). Transfection of ACE2 into SARS-CoV-refractory cell lines conferred susceptibility to infection (Mossel *et al.*, 2005).

Taken together, these results imply that ACE2 is the major cell-surface receptor for SARS-CoV. They also suggest that tissues, which express ACE2 but are rather refractory to SARS-CoV infection, limit virus replication at a step following the interaction of the S protein with ACE2.

We and others also determined the receptor-binding domain (RBD) of SARS-CoV S1 (Babcock *et al.*, 2004; Wong *et al.*, 2004), which consists of only 193 amino acids (S1 residues 318-510; see Figure 1).

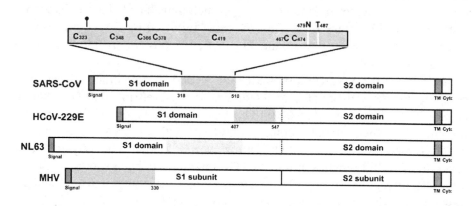

Figure 1: Receptor-binding domains of coronaviruses. Representation of the S proteins of SARS-CoV, HCoV-229E, HCoV-NL63 (NL63), and mouse hepatitis virus (MHV), aligned by their S2 domains. Dark gray indicates leader and transmembrane sequences. Light gray indicates receptor-binding domain. Future experiments may define the receptor-binding domain of NL63 more precisely than what is indicated. The receptor-binding domain of SARS-CoV is shown with N-glycosylation sites (small circles) and cysteines indicated. Residues that are critical to the high affinity interaction of SARS-CoV with human ACE2 (asparagine 479 and threonine 487) are shown as white bars.

An IgG1 Fc fusion protein (RBD-Ig) bound ACE2 more efficiently than full-length S1-Ig (Wong *et al.*, 2004). At least two amino acid residues (glutamic acid 452 and aspartic acid 454) are crucial for S1-ACE2 interaction (Wong *et al.*, 2004). In HEK 293T cells stably expressing ACE2, RBD-Ig blocked transduction of cells with S protein-pseudotyped simian immunodeficiency virus at a lower concentration ($IC_{50}=10$ nM) than full-length S1-Ig ($IC_{50}=50$nM) (Wong *et al.*, 2004). RBD-Ig in combination with Freund's complete adjuvant induced strong antibody responses in intradermally immunized NZW rabbits. These antibodies completely inhibited infection of Vero E6 cells with 100 $TCID_{50}$ SARS-CoV at a serum dilution of 1:10,240, and blocked the interaction of S1 with commercially available ACE2 (He *et al.*, 2004a). Similarly, the immunization of mice and rabbits with inactivated SARS-CoV induced antibodies that specifically recognize the RBD, block S1-RBD interaction, and prevent cell transduction with S protein-pseudotyped HIV-1 (He *et al.*, 2004b). Attenuated modified vaccinia virus Ankara expressing the SARS-CoV HKU38849 strain S protein induced neutralizing antibodies in immunized BALB/c mice, rabbits, and Chinese rhesus macaques. The antibodies were demonstrated to recognize a neutralizing epitope consisting of S protein's amino acid residues 400-600. This strongly suggested that the RBD was targeted by the antibodies. In fact, RBD-Ig could absorb the majority of neutralizing antibodies after incubation with antibody-containing sera (Chen *et al.*, 2005). Finally, the conformationally sensitive epitope recognized by the potent SARS-CoV-neutralizing antibody 80R (Sui *et al.*, 2004) was identified in S protein's N-terminal amino acid residues 324-503, which overlap the RBD, and 80R scFV precipitated RBD-Ig as efficiently as protein A (Sui *et al.*, 2005). Combined, these results strongly support the idea that SARS-CoV neutralizing antibodies target the receptor-binding domain of S protein and that virus neutralization occurs mainly through the interruption of the S1-ACE2 interaction.

Deletion studies demonstrated that the cytoplasmic domain of ACE2 is not important for ACE2-S1 interaction. Additionally, soluble ACE2 consisting only of its ectodomain blocked transduction with S protein-pseudotyped HIV-1 (Hofmann *et al.*, 2004). We demonstrated that an enzymatically inactive ACE2 variant, which contains two asparagine residues in place of the active-site histidine residues (ACE2-NN-Ig), still binds S protein and facilitates entry of S protein-pseudotyped lentiviruses. This finding demonstrated that the proteolytic activity of ACE2 plays no role in SARS-CoV cell penetration (Moore *et al.*, 2004). Also, binding of the S protein to ACE2 did not alter ACE2's enzymatic activity (Li *et al.*, 2005; Moore *et al.*, 2004). Finally, we showed that purified ACE2-NN-Ig prevented HEK 293T cell transduction with S-protein-pseudotyped simian immunodeficiency virus (IC_{50} = 2 nM), and that it blocked infection of VeroE6 cells with infectious SARS-CoV (Moore *et al.*, 2004).

The SARS-CoV-neutralizing antibody 80R blocked HEK 293T cell transduction with HIV-1 pseudotypes carrying the SARS-CoV strain TOR2 or strain SZ3 S proteins, but did not prevent transduction with pseudotypes carrying the strain GD03T0013 S protein (Sui *et al.*, 2005). This observation implied that the RBDs differ in key residues and suggested the possibility that different RBDs have different affinities to ACE2 of different species. A comparison of the RBD amino acid sequences of reported distinct SARS-

Jens H. Kuhn et al.

SARS-CoV isolates and SARS-CoV-like viruses from Himalayan palm civets revealed only few amino acid changes (see Table 1) (Sui *et al.*, 2005).

Table 1. Amino acid changes in the receptor binding domains (RBDs) of the spike (S) proteins of various SARS-CoV or SARS-CoV-like isolates according to (Sui et al., 2005)

SARS-CoV isolate	Origin	GenBank accession No.	Amino acid at position 344	Amino acid at position 360	Amino acid at position 472	Amino acid at position 479	Amino acid at position 480	Amino acid at position 487
GD01	Early phase 2002-2003	AY278489	R/K	F	L	N	D	T
TOR2	Middle/ Late phase 2002-2003	AY274119	K	F	L	N	D	T
SZ3 (SARS-CoV-like)	Himalayan palm civet	AY304486, AY881174	R	S	L	K	D	S
GD 03T00 13	2003-2004	AY525636	R	S	P	N	G	S

The tissue distribution of murine and rat ACE2 was found to be comparable to that of human ACE2 (Gembardt *et al.*, 2005; Komatsu *et al.*, 2002). However, we found that murine ACE2 bound less efficiently to the SARS-CoV S1 domain and supported S protein-mediated infection less efficiently than human ACE2. Likewise, HEK 293T cell syncytia formation occurred at much higher efficiency when human ACE2, rather than murine ACE2, was expressed. Murine NIH 3T3 cells expressing human ACE2 supported SARS-CoV replication at a level of one magnitude higher than cells expressing murine ACE2. Rat ACE2, on the other hand, did not support S protein-mediated infection nor HEK 293T cell syncytia formation. (Li *et al.*, 2004). Interestingly, the exchange of only four amino acid residues of rat ACE2 for the equivalent residues found in human ACE2 (residues 82-84 and 353) transformed rat ACE2 into an efficient receptor for the SARS-CoV S1 domain (Li *et al.*, 2005).

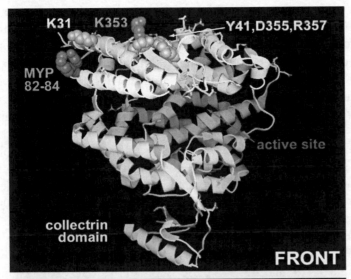

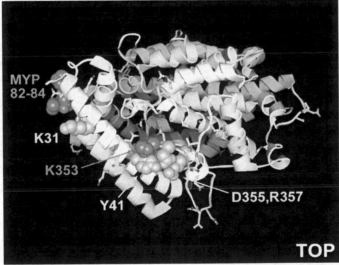

Figure 2: The S protein-binding site on human ACE2. (Upper panel) The crystal structure of ACE2 is shown, oriented with the C-terminal "collectrin" domain downward and viewed from the front of the cleft bearing the enzymatic active site. Residues of rat ACE2 whose alteration to the corresponding human residues converted rat ACE2 to an efficient SARS-CoV receptor are shown in dark gray spheres. Human ACE2 residues whose alteration substantially decreased S1-Ig association are shown in light gray spheres. Residues whose alteration did not affect S1-Ig association are shown in stick figure format. Low-resolution electron density associated with the collectrin domain is represented by a small β-sheet and α-helix at the base of the figure. (Lower panel) A view identical to that in in Upper panel except that the molecule has been rotated 90 degrees about the horizontal axis to present a "virus-eye" view of ACE2.

The findings led us to hypothesize that the host tropism of SARS-CoV and SARS-CoV-like viruses, as well as the severity of induced disease in different species might be dependent on RBD and ACE2 mutations. Experiments from our laboratory support this hypothesis. As shown in Figure 2, we localized the S protein-binding domain of ACE2 to α-helix 1 and to a loop of β-sheet 5 (Li *et al.*, 2005), both of which are located afar from the enzyme's active site but within the catalytic domain (Dales *et al.*, 2002; Towler *et al.*, 2004). S proteins from the different SARS-CoV strains isolated during the 2002-2003 and the 2003-2004 outbreaks as well as S proteins from the SARS-CoV-like viruses isolated from Himalayan palm civets bound to Himalayan palm civet ACE2 efficiently. However, as shown in Figure 3, the latter two proteins bound to human ACE2 much less efficiently than the S protein of the virus causing the initial 2002-2003 outbreak (Li *et al.*, 2005). Exchanging specific residues of human ACE2 for those found in Himalayan palm civet ACE2 markedly enhanced binding of the S proteins of the 2003-2004 and Himalayan palm civet viruses. Likewise, exchanging residues 479 and 487 of these S protein's RBDs for residues found in the S protein of virus isolates from the 2002-2003 outbreak (see Table 1) increased the binding efficiency to human ACE2 (Li *et al.*, 2005). Collectively, these data suggest that the apparent lack of severity and human-to-human transmission associated with the small SARS outbreak in 2003-2004 is in part due to incomplete adaptation of the associated virus isolate to human ACE2, and that this adaptation is mediated mainly by two amino acid residues. These results also provide biochemical support for the hypothesis that the palm civet was a source of SARS-CoV.

The tissue tropism of human coronavirus NL63 (HCoV-NL63) and SARS-CoV seems to be identical. Recently, using HIV-1 pseudotypes, it was shown that HCoV-NL63 also uses ACE2 as a cell-entry receptor, albeit with lower affinity (Hofmann *et al.*, 2005). This finding is surprising because the spike proteins of the two viruses share no amino acid identity. Additionally, this finding suggests that SARS is caused by an unidentified virulence factor of SARS-CoV, which is not encoded by the relatively harmless HCoV-NL63 genome. Clearly, further studies of the HCoV-NL63 entry mechanism are necessary. (Hofmann *et al.*, 2005)

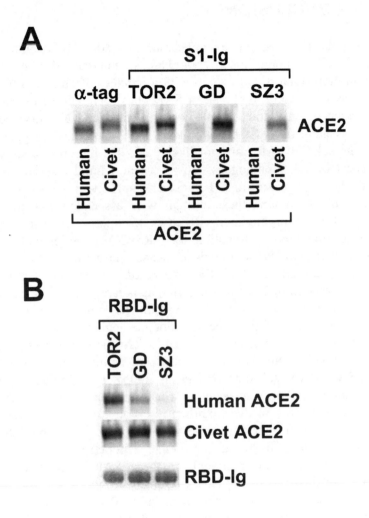

Figure 3: Differential association of three S proteins with human and palm-civet ACE2. (A) 293T cells were transfected with plasmids encoding human or palm-civet ACE2, radiolabeled, and lysed. ACE2 proteins were immunoprecipitated with an antibody, which recognizes a tag present at the carboxy-terminus of each receptor, or with S1-Ig variants containing the S1 domains of TOR2, GD, or SZ3 S proteins, and analyzed by SDS-PAGE. TOR2 SARS-CoV was isolated from humans infected during the 2002-2003 outbreak; GD, from a small outbreak in 2003-2004 which resulted in mild symptoms and no human-to-human transmission; and SZ3, from palm civets. (B) An experiment similar to that in (A) except that S-protein residues 318-510, comprising the receptor-binding domains (RBDs) of the indicated S proteins fused to the Fc domain of human IgG1 (RBD-Ig), were used to immunoprecipitate human or palm-civet ACE2. The bottom panel shows a Coomassie-stained SDS-PAGE gel of the individual RBDs used in this experiment.

3.   CONCLUSIONS

During its first recognized appearance in 2002, SARS-CoV quickly spread globally among human populations, and caused a novel severe disease in more than 8,000 people with a lethality of almost 10% (Cherry, 2004). Only one natural outbreak of SARS has been recorded since. During this emergence, only four people were infected and all survived (Fleck, 2004). This gave rise to the notion that SARS-CoV has disappeared; that future outbreaks of SARS are not likely; and that research funds would be better spent on other pathogens. However, the natural reservoir of SARS-CoV still has to be identified. Himalayan palm civets are suspected to be carriers of the virus, and these animals were associated epidemiologically with many, but not all, human SARS cases. In experiments, Himalayan palm civets were shown to be susceptible to SARS-CoV infection and replication. However, other animals like domestic cats and ferrets were also shown to be susceptible (Martina *et al.*, 2003; Wu *et al.*, 2005); and Chinese ferret badgers and racoon dogs were suggested as natural carriers of SARS-CoV-like viruses (Guan *et al.*, 2003). Our experiments demonstrated that only two amino acid residues of the receptor binding domain of SARS-CoV spike protein determine its efficiency to bind to its cell-surface receptor, ACE2. These two residues are variable in different animals and hence allow or prohibit SARS-CoV cell entry (Li *et al.*, 2005). This finding suggests that only few mutations need to occur in a naturally circulating SARS-CoV or SARS-CoV-like virus in the unknown reservoir to allow for cross-species transmission and further human SARS outbreaks. On the other hand, experiments performed in our and other laboratories also suggest that specific prevention and treatment of these SARS outbreaks could be achievable. In this review, we have presented strong support for the role of ACE2 as the sole cell-entry determinant of SARS-CoV, and we described the ability of neutralizing antibodies to disrupt the S-ACE2 interaction (Hattermann *et al.*, 2005; Li *et al.*, 2003b; Mossel *et al.*, 2005; Nie *et al.*, 2004; Sui *et al.*, 2004; Wang *et al.*, 2004). These results suggest that future vaccine candidates should elicit antibody responses to the S protein receptor binding domain. In terms of treatment, several promising research directions could be taken. First, based on our findings that the SARS-CoV S protein receptor binding domain can prevent cell transduction with S protein-pseudotyped lentiviruses (Wong *et al.*, 2004), modified polypeptides resembling S protein fragments could be developed to bind to ACE2 and prevent binding of SARS-CoV to its host cells. Second, the development of specific inhibitors of ACE2, which do not contain S protein fragments, is feasible. For example, the identification of an inhibitor, which prevents binding of SARS-CoV to ACE2 and ACE2 proteolytic activity, was recently reported (Huentelman *et al.*, 2004). However, this inhibitor is most likely associated with severe side effects since it interrupts ACE2's physiological, as of yet unknown, function. We demonstrated that SARS-CoV binds to ACE2 at a site remote of the enzyme's active site (Li *et al.*, 2005), and we showed further that SARS-CoV binding does not interfere with ACE2's enzymatic activity (Li *et al.*, 2005; Moore *et al.*, 1997). Hence, better inhibitors could be identified, which only block S-ACE2 interaction but do not alter the physiological function of ACE2. Last, it has been shown that soluble ACE2 can prevent transduction of ACE2-expressing cells with SARS-CoV S protein-pseudotyped lentiviruses (Hofmann *et al.*, 2004; Li *et al.*, 2005), suggesting that a polypeptide consisting of the SARS-CoV-binding domain of ACE2 could be used as a therapeutic. We have shown that enzymatically inactive ACE2 is still able to interrupt S protein-mediated entry of lentiviral pseudotypes (Moore *et al.*, 2004). This should allow for the synthesis of ACE2 variants or fragments thereof, which would prevent SARS-CoV infection but would not interfere with the predicted role of ACE2 as a regulator of blood pressure (Yagil *et al.*, 2003).

The surprising discovery of ACE2 as a receptor of another coronavirus, HCoV-NL63, raises the question whether antivirals or vaccines could be developed, which target both this virus and SARS-CoV at the same time. Evidently, it is necessary to define the receptor-binding domain of HCoV-NL63's spike protein and the interacting region of ACE2 to find an answer.

REFERENCES

Babcock, G. J., Esshaki, D. J., Thomas, W. D., Jr. and Ambrosino, D. M., 2004, Amino acids 270-510 of the severe acute respiratory syndrome coronavirus spike protein are required for interaction with the receptor. *J. Virol.* **78**:4552-4560

Bisht, H., Roberts, A., Vogel, L., Bukreyev, A., Collins, P. L., Murphy, B. R., Subbarao, K. and Moss, B., 2004, Severe acute respiratory syndrome coronavirus spike protein expressed by attenuated vaccinia virus protectively immunizes mice. *Proc. Natl. Acad. Sci. USA* **101**:6641-6646

Bosch, B. J., van der Zee, R., de Haan, C. A. and Rottier, P. J., 2003, The coronavirus spike protein is a class I virus fusion protein: structural and functional characterization of the fusion core complex. *J. Virol.* **77**:8801-8811

Bukreyev, A., Lamirande, E. W., Buchholz, U. J., Vogel, L. N., Elkins, W. R., St Claire, M., Murphy, B. R., Subbarao, K. and Collins, P. L., 2004, Mucosal immunisation of African green monkeys (Cercopithecus aethiops) with an attenuated parainfluenza virus expressing the SARS coronavirus spike protein for the prevention of SARS. *Lancet* **363**:2122-2127

Chen, Z., Zhang, L., Qin, C., Ba, L., Yi, C. E., Zhang, F., Wei, Q., He, T., Yu, W., Yu, J., Gao, H., Tu, X., Gettie, A., Farzan, M., Yuen, K. Y. and Ho, D. D., 2005, Recombinant modified vaccinia virus Ankara expressing the spike glycoprotein of severe acute respiratory syndrome coronavirus induces protective neutralizing antibodies primarily targeting the receptor binding region. *J. Virol.* **79**:2678-2688

Cherry, J. D., 2004, The chronology of the 2002-2003 SARS mini pandemic. *Paediatr. Respir. Rev.* **5**:262-269

Colman, P. M. and Lawrence, M. C., 2003, The structural biology of type I viral membrane fusion. *Nat. Rev. Mol. Cell Biol.* **4**:309-319

Crackower, M. A., Sarao, R., Oudit, G. Y., Yagil, C., Kozieradzki, I., Scanga, S. E., Oliveira-dos-Santos, A. J., da Costa, J., Zhang, L., Pei, Y., Scholey, J., Ferrario, C. M., Manoukian, A. S., Chappell, M. C., Backx, P. H., Yagil, Y. and Penninger, J. M., 2002, Angiotensin-converting enzyme 2 is an essential regulator of heart function. *Nature* **417**:822-828

Dales, N. A., Gould, A. E., Brown, J. A., Calderwood, E. F., Guan, B., Minor, C. A., Gavin, J. M., Hales, P., Kaushik, V. K., Steward, M., Tummino, P. J., Vickers, C. S., Ocain, T. D. and Patane, M. A., 2002, Substrate-based design of the first class of angiotensin-converting enzyme-related carboxypeptidase (ACE2) inhibitors. *J. Am. Chem. Soc.* **124**:11852-11853

Daniel, C., Lacroix, M. and Talbot, P. J., 1994, Mapping of linear antigenic sites on the S glycoprotein of a neurotropic murine coronavirus with synthetic peptides: a combination of nine prediction algorithms fails to identify relevant epitopes and peptide immunogenicity is drastically influenced by the nature of the protein carrier. *Virology* **202**:540-549

Delmas, B., Gelfi, J., L'Haridon, R., Vogel, L. K., Sjostrom, H., Noren, O. and Laude, H., 1992, Aminopeptidase N is a major receptor for the entero-pathogenic coronavirus TGEV. *Nature* **357**:417-420

Denison, M. R., 1999, The common cold. Rhinoviruses and coronaviruses. In *Viral Infections of the Respiratory Tract* (R. Dolin and F. P. Wright, ed.). New York, Marcel Dekker, pp. Dimitrov, D. S., 2004, Virus entry: molecular mechanisms and biomedical applications. *Nat. Rev. Microbiol* **2**:109-122

Ding, Y., He, L., Zhang, Q., Huang, Z., Che, X., Hou, J., Wang, H., Shen, H., Qiu, L., Li, Z., Geng, J., Cai, J., Han, H., Li, X., Kang, W., Weng, D., Liang, P. and Jiang, S., 2004, Organ distribution of severe acute respiratory syndrome (SARS) associated coronavirus (SARS-CoV) in SARS patients: implications for pathogenesis and virus transmission pathways. *J. Pathol.* **203**:622-630

Donoghue, M., Hsieh, F., Baronas, E., Godbout, K., Gosselin, M., Stagliano, N., Donovan, M., Woolf, B., Robison, K., Jeyaseelan, R., Breitbart, R. E. and Acton, S., 2000, A novel angiotensin-converting enzyme-related carboxypeptidase (ACE2) converts angiotensin I to angiotensin 1-9. *Circ. Res.* **87**:E1-9

Drosten, C., Gunther, S., Preiser, W., van der Werf, S., Brodt, H. R., Becker, S., Rabenau, H., Panning, M., Kolesnikova, L., Fouchier, R. A., Berger, A., Burguiere, A. M., Cinatl, J., Eickmann, M., Escriou, N., Grywna, K., Kramme, S., Manuguerra, J. C., Muller, S., Rickerts, V., Sturmer, M., Vieth, S., Klenk, H. D., Osterhaus, A. D., Schmitz, H. and Doerr, H. W., 2003, Identification of a novel coronavirus in patients with severe acute respiratory syndrome. *N. Engl. J. Med.* **348**:1967-1976

Enjuanes, L., Spaan, W., Snijder, E. and Cavanagh, D., 2000, *Nidovirales*. In *Virus Taxonomy - Seventh Report of the International Committee for the Taxonomy of Viruses* (M. H. V. Regenmortel, C. M. Fauquet, D. H. L. Bishop, E. B. Carsten, M. K. Estes, S. M. Lemon, J. Maniloff, M. A. Mayo, D. J. McGeoch, C. R. Pringle and R. B. Wickner, ed.). New York, Academic Press, pp. 835-849

Esper, F., Shapiro, E. D., Weibel, C., Ferguson, D., Landry, M. L. and Kahn, J. S., 2005a, Association between a Novel Human Coronavirus and Kawasaki Disease. *J. Infect. Dis.* **191**:499-502

Esper, F., Weibel, C., Ferguson, D., Landry, M. L. and Kahn, J. S., 2005b, Evidence of a novel coronavirus that is associated with respiratory tract disease in infants and young children. *J. Infect. Dis.* **191**:492-498

Farcas, G. A., Poutanen, S. M., Mazzulli, T., Willey, B. M., Butany, J., Asa, S. L., Faure, P., Akhavan, P., Low, D. E. and Kain, K. C., 2005, Fatal severe acute respiratory syndrome is associated with multiorgan involvement by coronavirus. *J. Infect. Dis.* **191**:193-197

Fleck, F., 2004, SARS virus returns to China as scientists race to find effective vaccine. *Bull. World Health Organ.* **82**:152-153

Fouchier, R. A., G., H. N., Bestebroer, T. M., Niemeyer, B., de Jong, J. C., Simon, J. H. and Osterhaus, A. D., 2004, A previously undescribed coronavirus associated with respiratory disease in humans. *Proc. Natl. Acad. Sci. USA* **101**:6212-6216

Fouchier, R. A., Kuiken, T., Schutten, M., van Amerongen, G., van Doornum, G. J., van den Hoogen, B. G., Peiris, M., Lim, W., Stohr, K. and Osterhaus, A. D., 2003, Aetiology: Koch's postulates fulfilled for SARS virus. *Nature* **423**:240

Gallagher, T. M., 2001, Murine coronavirus spike glycoprotein. Receptor binding and membrane fusion activities. *Adv. Exp. Med. Biol.* **494**:183-192

Gallagher, T. M. and Buchmeier, M. J., 2001, Coronavirus spike proteins in viral entry and pathogenesis. *Virology* **279**:371-374

Gembardt, F., Sterner-Kock, A., Imboden, H., Spalteholz, M., Reibitz, F., Schultheiss, H.-P., Siems, W.-E. and Walther, T., 2005, Organ-specific distribution of ACE2 mRNA and correlating peptidase activity in rodents. *Peptides,* **in press**

Gibbs, A. J., Gibbs, M. J. and Armstrong, J. S., 2004, The phylogeny of SARS coronavirus. *Arch. Virol.*

Gonzalez, J. M., Gomez-Puertas, P., Cavanagh, D., Gorbalenya, A. E. and Enjuanes, L., 2003, A comparative sequence analysis to revise the current taxonomy of the family *Coronaviridae. Arch. Virol.* **148**:2207-2235

Gorbalenya, A. E., Snijder, E. and Spaan, W. J. M., 2004, Severe acute respiratory syndrome coronavirus phylogeny: toward consensus. *J. Virol.* **78**:7863-7866

Guan, Y., Zheng, B. J., He, Y. Q., Liu, X. L., Zhuang, Z. X., Cheung, C. L., Luo, S. W., Li, P. H., Zhang, L. J., Guan, Y. J., Butt, K. M., Wong, K. L., Chan, K. W., Lim, W., Shortridge, K. F., Yuen, K. Y., Peiris, J. S. and Poon, L. L., 2003, Isolation and characterization of viruses related to the SARS coronavirus from animals in southern China. *Science* **302**: 276-278

Hamming, I., Timens, W., Bulthuis, M. L., Lely, A. T., Navis, G. J. and van Goor, H., 2004, Tissue distribution of ACE2 protein, the functional receptor for SARS coronavirus. A first step in understanding SARS pathogenesis. *J. Pathol.* **203**:631-637

Harmer, D., Gilbert, M., Borman, R. and Clark, K. L., 2002, Quantitative mRNA expression profiling of ACE2, a novel homologue of angiotensin converting enzyme. *FEBS Lett.* **532**:107-110

Hattermann, K., Muller, M. A., Nitsche, A., Wendt, S., Donoso Mantke, O. and Niedrig, M., 2005, Susceptibility of different eukaryotic cell lines to SARS-coronavirus. *Arch. Virol.* **150**:1023-1031

He, Y., Zhou, Y., Liu, S., Kou, Z., Li, W., Farzan, M. and Jiang, S., 2004a, Receptor-binding domain of SARS-CoV spike protein induces highly potent neutralizing antibodies: implication for developing subunit vaccine. *Biochem. Biophys. Res. Commun.* **324**: 773-781

He, Y., Zhou, Y., Siddiqui, P. and Jiang, S., 2004b, Inactivated SARS-CoV vaccine elicits high titers of spike protein-specific antibodies that block receptor binding and virus entry. *Biochem. Biophys. Res. Commun.* **325**:445-452

Hofmann, H., Geier, M., Marzi, A., Krumbiegel, M., Peipp, M., H., F. G., Gramberg, T. and Pöhlmann, S., 2004, Susceptibility of SARS coronavirus S protein-driven infection correlates with expression of angiotensin-converting enzyme 2 and infection can be blocked by soluble receptor. *Biochem. Biophys. Res. Commun.* **319**:1216-1221

Hofmann, H., Pyrc, K., van der Hoek, L., Geier, M., Berkhout, B. and Pöhlmann, S., 2005, Human coronavirus NL63 employs the severe acute respiratory syndrome coronavirus receptor for cellular entry. *Proc. Natl. Acad. Sci. USA*, **in press**

Holmes, K. V. and Dveksler, G. S., 1994, Specificity of coronavirus/receptor interaction. In *Cell receptors for animal viruses* (E. Wimmer, ed.). Cold Spring Harbor, Cold Spring Harbor Press, pp. 403-443

Huentelman, M. J., Zubcevic, J., Hernandez Prada, J. A., Xiao, X., Dimitrov, D. S., Raizada, M. K. and Ostrov, D. A., 2004, Structure-based discovery of a novel angiotensin-converting enzyme 2 inhibitor. *Hypertension* **44**:903-906

Jeffers, S. A., Tusell, S. M., Gillim-Ross, L., Hemmila, E. M., Achenbach, J. E., Babcock, G. J., Thomas, W. D., Jr., Thackray, L. B., Young, M. D., Mason, R. J., Ambrosino, D. M., Wentworth, D. E., Demartini, J. C. and Holmes, K. V., 2004, CD209L (L-SIGN) is a receptor for severe acute respiratory syndrome coronavirus. *Proc. Natl. Acad. Sci. USA* **101**:15748-15753

Komatsu, T., Suzuki, Y., Imai, J., Sugano, S., Hida, M., Tanigami, A., Muroi, S., Yamada, Y. and Hanaoka, K., 2002, Molecular cloning, mRNA expression and chromosomal localization of mouse angiotensin-converting enzyme-related carboxypeptidase (mACE2). *DNA Seq.* **13**:217-220

Koo, M., Bendahmane, M., Lettieri, G. A., Paoletti, A. D., Lane, T. E., Fitchen, J. H., Buchmeier, M. J. and Beachy, R. N., 1999, Protective immunity against murine hepatitis virus (MHV) induced by intranasal or subcutaneous administration of hybrids of tobacco mosaic virus that carries an MHV epitope. *Proc. Natl. Acad. Sci. USA* **96**:7774-7779

Ksiazek, T. G., Erdman, D., Goldsmith, C. S., Zaki, S. R., Peret, T., Emery, S., Tong, S., Urbani, C., Comer, J. A., Lim, W., Rollin, P. E., Dowell, S. F., Ling, A. E., Humphrey, C. D., Shieh, W. J., Guarner, J., Paddock, C. D., Rota, P., Fields, B., DeRisi, J., Yang, J. Y., Cox, N., Hughes, J. M., LeDuc, J. W., Bellini, W. J. and Anderson, L. J., 2003, A novel coronavirus associated with severe acute respiratory syndrome. *N. Engl. J. Med.* **348**:1953-1966

Kuiken, T., Fouchier, R. A., Schutten, M., Rimmelzwaan, G. F., van Amerongen, G., van Riel, D., Laman, J. D., de Jong, T., van Doornum, G., Lim, W., Ling, A. E., Chan, P. K., Tam, J. S., Zambon, M. C., Gopal, R., Drosten, C., van der Werf, S., Escriou, N., Manuguerra, J. C., Stohr, K., Peiris, J. S. and Osterhaus, A. D., 2003, Newly discovered coronavirus as the primary cause of severe acute respiratory syndrome. *Lancet* **362**: 263-270

Kuo, L., Godeke, G. J., Raamsman, M. J., Masters, P. S. and Rottier, P. J., 2000, Retargeting of coronavirus by substitution of the spike glycoprotein ectodomain: crossing the host cell species barrier. *J. Virol.* **74**:1393-1406

Lai, M. M. and Cavanagh, D., 1997, The molecular biology of coronaviruses. *Adv. Virus Res.* **48**:1-100

Li, G., Chen, X. and Xu, A., 2003a, Profile of specific antibodies to the SARS-associated coronavirus. *N. Engl. J. Med.* **349**:508-509

Li, W., Greenough, T. C., Moore, M. J., Vasilieva, N., Somasundaran, M., Sullivan, J. L., Farzan, M. and Choe, H., 2004, Efficient replication of severe acute respiratory syndrome coronavirus in mouse cells is limited by murine angiotensin-converting enzyme 2. *J. Virol.* **78:**11429-11433

Li, W., Moore, M. J., Vasilieva, N., Sui, J., Wong, S. K., Berne, M. A., Somasundaran, M., Sullivan, J. L., Luzuriaga, K., Greenough, T. C., Choe, H. and Farzan, M., 2003b, Angiotensin-converting enzyme 2 is a functional receptor for the SARS coronavirus. *Nature* **426:**450-454

Li, W., Zhang, C., Sui, J., Kuhn, J. H., Moore, M. J., Luo, S., Wong, S. K., Huang, I. C., Xu, K., Vasilieva, N., Murakami, A., He, Y., Marasco, W. A., Guan, Y., Choe, H. and Farzan, M., 2005, Receptor and viral determinants of SARS-coronavirus adaptation to human ACE2. *Embo J.* **24:**1634-1643

Lim, P. L., Kurup, A., Gopalakrishna, G., Chan, K. P., Wong, C. W., Ng, L. C., Se-Thoe, S. Y., Oon, L., Bai, X., Stanton, L. W., Ruan, Y., Miller, L. D., Vega, V. B., James, L., Ooi, P. L., Kai, C. S., Olsen, S. J., Ang, B. and Leo, Y. S., 2004, Laboratory-Acquired Severe Acute Respiratory Syndrome. *N. Engl. J. Med.* **350:**1740-1745

Marra, M. A., Jones, S. J., Astell, C. R., Holt, R. A., Brooks-Wilson, A., Butterfield, Y. S., Khattra, J., Asano, J. K., Barber, S. A., Chan, S. Y., Cloutier, A., Coughlin, S. M., Freeman, D., Girn, N., Griffith, O. L., Leach, S. R., Mayo, M., McDonald, H., Montgomery, S. B., Pandoh, P. K., Petrescu, A. S., Robertson, A. G., Schein, J. E., Siddiqui, A., Smailus, D. E., Stott, J. M., Yang, G. S., Plummer, F., Andonov, A., Artsob, H., Bastien, N., Bernard, K., Booth, T. F., Bowness, D., Czub, M., Drebot, M., Fernando, L., Flick, R., Garbutt, M., Gray, M., Grolla, A., Jones, S., Feldmann, H., Meyers, A., Kabani, A., Li, Y., Normand, S., Stroher, U., Tipples, G. A., Tyler, S., Vogrig, R., Ward, D., Watson, B., Brunham, R. C., Krajden, M., Petric, M., Skowronski, D. M., Upton, C. and Roper, R. L., 2003, The Genome sequence of the SARS-associated coronavirus. *Science* **300:**1399-1404

Martina, B. E., Haagmans, B. L., Kuiken, T., Fouchier, R. A., Rimelzwaan, G. F., van Amerongen, G., Peiris, J. S., Lim, W. and Osterhaus, A. D., 2003, SARS virus infection of cats and ferrets. *Nature* **425:**915

Marzi, A., Gramberg, T., Simmons, G., Moller, P., Rennekamp, A. J., Krumbiegel, M., Geier, M., Eisemann, J., Turza, N., Saunier, B., Steinkasserer, A., Becker, S., Bates, P., Hofmann, H. and Pöhlmann, S., 2004, DC-SIGN and DC-SIGNR interact with the glycoprotein of Marburg virus and the S protein of severe acute respiratory syndrome coronavirus. *J. Virol.* **78:**12090-12095

Mayo, M. A., 2002, A summary of taxonomic changes recently approved by ICTV. *Arch. Virol.* **147:**1655-1663

McIntosh, K., 2005, Coronaviruses in the Limelight. *J. Infect. Dis.* **191:**489-491

Moore, K. M., Jackwood, M. W. and Hilt, D. A., 1997, Identification of amino acids involved in a serotype and neutralization specific epitope within the s1 subunit of avian infectious bronchitis virus. *Arch. Virol.* **142:**2249-2256

Moore, M. J., Dorfman, T., Li, W., Wong, S. K., Li, Y., Kuhn, J. H., Coderre, J., Vasilieva, N., Han, Z., Greenough, T. C., Farzan, M. and Choe, H., 2004, Retroviruses pseudotyped with the severe acute respiratory syndrome coronavirus spike protein efficiently infect cells expressing angiotensin-converting enzyme 2. *J. Virol.* **78:**10628-10635

Mossel, E. C., Huang, C., Narayanan, K., Makino, S., Tesh, R. B. and Peters, C. J., 2005, Exogenous ACE2 expression allows refractory cell lines to support severe acute respiratory syndrome coronavirus replication. *J. Virol.* **79:**3846-3850

Nie, Y., Wang, P., Shi, X., Wang, G., Chen, J., Zheng, A., Wang, W., Wang, Z., Qu, X., Luo, M., Tan, L., Song, X., Yin, X., Chen, J., Ding, M. and Deng, H., 2004, Highly infectious SARS-CoV pseudotyped virus reveals the cell tropism and its correlation with receptor expression. *Biochem. Biophys. Res. Commun.* **321**:994-1000

Normile, D., 2004, Second Lab Accident Fuels Fears About SARS. *Science* **303**:26

Oh, J. S., Song, D. S. and Park, B. K., 2003, Identification of a putative cellular receptor 150kD polypeptide for porcine epidemic diarrhea virus in porcine enterocytes. *J. Vet. Sci.* **4**:269-275

Olsen, S. J., Chang, H. L., Cheung, T. Y., Tang, A. F., Fisk, T. L., Ooi, S. P., Kuo, H. W., Jiang, D. D., Chen, K. T., Lando, J., Hsu, K. H., Chen, T. J. and Dowell, S. F., 2003, Transmission of the severe acute respiratory syndrome on aircraft. *N. Engl. J. Med.* **349**:2416-2422

Pearson, H., Clarke, T., Abbott, A., Knight, J. and Cyranoski, D., 2003, SARS: what have we learned? *Nature* **424**:121-126

Pei, H., Liu, J., Cheng, Y., Sun, C., Wang, C., Lu, Y., Ding, J., Zhou, J. and Xiang, H., 2005, Expression of SARS-coronavirus nucleocapsid protein in Escherichia coli and Lactococcus lactis for serodiagnosis and mucosal vaccination. *Appl. Microbiol. Biotechnol.*

Peiris, J. S., Yuen, K. Y., Osterhaus, A. D. and Stohr, K., 2003, The severe acute respiratory syndrome. *N. Engl. J. Med.* **349**:2431-2441

Pene, F., Merkat, A., Vabret, A., Rozenberg, F., Buzyn, A., Dreyfus, F., Cariou, A., Freymuth, F. and Lebon, P., 2003, Coronavirus 229E-related pneumonia in immunocompromised patients. *Clin. Infect. Dis.* **37**:929-932

Phillips, J. J., Chua, M. M., Lavi, E. and Weiss, S. R., 1999, Pathogenesis of chimeric MHV4/MHV-A59 recombinant viruses: the murine coronavirus spike protein is a major determinant of neurovirulence. *J. Virol.* **73**:7752-7760

Roberts, A., Vogel, L., Guarner, J., Hayes, N., Murphy, B., Zaki, S. and Subbarao, K., 2005, Severe acute respiratory syndrome coronavirus infection of golden Syrian hamsters. *J. Virol.* **79**:503-511

Rota, P. A., Oberste, M. S., Monroe, S. S., Nix, W. A., Campagnoli, R., Icenogle, J. P., Penaranda, S., Bankamp, B., Maher, K., Chen, M. H., Tong, S., Tamin, A., Lowe, L., Frace, M., DeRisi, J. L., Chen, Q., Wang, D., Erdman, D. D., Peret, T. C., Burns, C., Ksiazek, T. G., Rollin, P. E., Sanchez, A., Liffick, S., Holloway, B., Limor, J., McCaustland, K., Olsen-Rasmussen, M., Fouchier, R., Gunther, S., Osterhaus, A. D., Drosten, C., Pallansch, M. A., Anderson, L. J. and Bellini, W. J., 2003, Characterization of a novel coronavirus associated with severe acute respiratory syndrome. *Science* **300**: 1394-1399

Rowe, T., Gao, G., Hogan, R. J., Crystal, R. G., Voss, T. G., Grant, R. L., Bell, P., Kobinger, G. P., Wivel, N. A. and Wilson, J. M., 2004, Macaque model for severe acute respiratory syndrome. *J. Virol.* **78**:11401-11404

Sanchez, C. M., Izeta, A., Sabchez-Morgado, J. M., Alonso, S., Sola, I., Balasch, M., Plana-Duran, J. and Enjuanes, L., 1999, Targeted recombination demonstrates that the spike gene of transmissible gastroenteritis coronavirus is a determinant of its enteric tropism and virulence. *J. Virol.* **73**:7607-7618

Simmons, G., Reeves, J. D., Rennekamp, A. J., Amberg, S. M., Piefer, A. J. and Bates, P., 2004, Characterization of severe acute respiratory syndrome-associated coronavirus (SARS-CoV) spike glycoprotein-mediated viral entry. *Proc. Natl. Acad. Sci. USA* **101**:4240-4245

Song, C., Lee, Y., Lee, C., Sung, H., Kim, J., Mo, I., Izumiya, Y., Jang, H. and Mikami, T., 1998, Induction of protective immunity in chickens vaccinated with infectious bronchitis virus S1 glycoprotein expressed by a recombinant baculovirus. *J. Gen. Virol.* **79**:719-723

Song, H. D., Tu, C. C., Zhang, G. W., Wang, S. Y., Zheng, K., Lei, L. C., Chen, Q. X., Gao, Y. W., Zhou, H. Q., Xiang, H., Zheng, H. J., Chern, S. W., Cheng, F., Pan, C. M., Xuan, H., Chen, S. J., Luo, H. M., Zhou, D. H., Liu, Y. F., He, J. F., Qin, P. Z., Li, L. H., Ren, Y. Q., Liang, W. J., Yu, Y. D., Anderson, L., Wang, M., Xu, R. H., Wu, X. W., Zheng, H. Y., Chen, J. D., Liang, G., Gao, Y., Liao, M., Fang, L., Jiang, L. Y., Li, H., Chen, F., Di, B., He, L. J., Lin, J. Y., Tong, S., Kong, X., Du, L., Hao, P., Tang, H., Bernini, A., Yu, X. J., Spiga, O., Guo, Z. M., Pan, H. Y., He, W. Z., Manuguerra, J. C., Fontanet, A., Danchin, A., Niccolai, N., Li, Y. X., Wu, C. I. and Zhao, G. P., 2005, Cross-host evolution of severe acute respiratory syndrome coronavirus in palm civet and human. *Proc. Natl. Acad. Sci. USA* **102**:2430-2435

Subbarao, K., McAuliffe, J., Vogel, L., Fahle, G., Fischer, S., Tatti, K., Packard, M., Shieh, W. J., Zaki, S. R. and Murphy, B., 2004, Prior infection and passive transfer of neutralizing antibody prevent replication of severe acute respiratory syndrome coronavirus in the respiratory tract of mice. *J. Virol.* **78**:3572-3577

Sui, J., Li, W., Murakami, A., Tamin, A., Matthews, L. J., Wong, S. K., Moore, M. J., Tallarico, A. S., Olurinde, M., Choe, H., Anderson, L. J., Bellini, W. J., Farzan, M. and Marasco, W. A., 2004, Potent neutralization of severe acute respiratory syndrome (SARS) coronavirus by a human mAb to S1 protein that blocks receptor association. *Proc. Natl. Acad. Sci. USA* **101**:2536-2541

Sui, J., Li, W., Roberts, A., Matthews, L. J., Murakami, A., Vogel, L., Wong, S. K., Subbarao, K., Farzan, M. and Marasco, W. A., 2005, Evaluation of Human Monoclonal Antibody 80R for Immunoprophylaxis of Severe Acute Respiratory Syndrome by an Animal Study, Epitope Mapping, and Analysis of Spike Variants. *J. Virol.* **79**:5900-5906

Takasuka, N., Fujii, H., Takahashi, Y., Kasai, M., Morikawa, S., Itamura, S., Ishii, K., Sakaguchi, M., Ohnishi, K., Ohshima, M., Hashimoto, S., Odagiri, T., Tashiro, M., Yoshikura, H., Takemori, T. and Tsunetsugu-Yokota, Y., 2004, A subcutaneously injected UV-inactivated SARS coronavirus vaccine elicits systemic humoral immunity in mice. *Int. Immunol.* **16**:1423-1430

Tan, E. L., Ooi, E. E., Lin, C. Y., Tan, H. C., Ling, A. E., Lim, B. and Stanton, L. W., 2004, Inhibition of SARS coronavirus infection in vitro with clinically approved antiviral drugs. *Emerg. Infect. Dis.* **10**:581-586

Tipnis, S. R., Hooper, N. M., Hyde, R., Karran, E., Christie, G. and Turner, A. J., 2000, A human homolog of angiotensin-converting enzyme. Cloning and functional expression as a captopril-insensitive carboxypeptidase. *J. Biol. Chem.* **275**:33328-33243

Towler, P., Staker, B., Prasad, S. G., Menon, S., Tang, J., Parsons, T., Ryan, D., Fisher, M., Williams, D., Dales, N. A., Patane, M. A. and Pantoliano, M. W., 2004, ACE2 X-ray structures reveal a large hinge-bending motion important for inhibitor binding and catalysis. *J. Biol. Chem.* **279**:17996-18007

van der Hoek, L., Pyrc, K., Jebbink, M. F., Vermeulen-Oost, W., Berkhout, R. J., Wolthers, K. C., Wertheim-van Dillen, P. M., Kaandorp, J., Spaargaren, J. and Berkhout, B., 2004, Identification of a new human coronavirus. *Nat. Med.* **10**:368-373

Vickers, C., Hales, P., Kaushik, V., Dick, L., Gavin, J., Tang, J., Godbout, K., Parsons, T., Baronas, E., Hsieh, F., Acton, S., Patane, M., Nichols, A. and Tummino, P., 2002, Hydrolysis of biological peptides by human angiotensin-converting enzyme-related carboxypeptidase. *J. Biol. Chem.* **277**:14838-14843

Wang, P., Chen, J., Zheng, A., Nie, Y., Shi, X., Wang, W., Wang, G., Luo, M., Liu, H., Tan, L., Song, X., Wang, Z., Yin, X., Qu, X., Wang, X., Qing, T., Ding, M. and Deng, H., 2004, Expression cloning of functional receptor used by SARS coronavirus. *Biochem. Biophys. Res. Commun.* **315**:439-444

Wentworth, D. E., Gillim-Ross, L., Espina, N. and Bernard, K. A., 2004, Mice susceptible to SARS coronavirus. *Emerg. Infect. Dis.* **10**:1293-1296

Williams, R. K., Jiang, G. S. and Holmes, K. V., 1991, Receptor for mouse hepatitis virus is a member of the carcinoembryonic antigen family of glycoproteins. *Proc. Natl. Acad. Sci. USA* **88**:5533-5536

Wong, S. K., Li, W., Moore, M. J., Choe, H. and Farzan, M., 2004, A 193-amino acid fragment of the SARS coronavirus S protein efficiently binds angiotensin-converting enzyme 2. *J. Biol. Chem.* **279**:3197-3201

Woo, P. C. Y., Lau, S. K. P., Chu, C.-M., Chan, K.-H., Tsoi, H.-W., Huang, Y., Wong, B. H. L., Poon, R. W. S., Cai, J. J., Luk, W.-K., Poon, L. L. M., Wong, S. S. Y., Guan, Y., Peiris, J. S. M. and Yuen, K.-Y., 2005, Characterization and Complete Genome Sequence of a Novel Coronavirus, Coronavirus HKU1, from Patients with Pneumonia. *J. Virol.* **79**: 884-895

Wu, D., Tu, C., Xin, C., Xuan, H., Meng, Q., Liu, Y., Yu, Y., Guan, Y., Jiang, Y., Yin, X., Crameri, G., Wang, M., Li, C., Liu, S., Liao, M., Feng, L., Xiang, H., Sun, J., Chen, J., Sun, Y., Gu, S., Liu, N., Fu, D., Eaton, B. T., Wang, L. F. and Kong, X., 2005, Civets are equally susceptible to experimental infection by two different severe acute respiratory syndrome coronavirus isolates. *J. Virol.* **79**:2620-2625

Xiao, X., Chakraborti, S., Dimitrov, A. S., Gramatikoff, K. and Dimitrov, D. S., 2003, The SARS-CoV S glycoprotein: expression and functional characterization. *Biochem. Biophys. Res. Commun.* **3212**:1159-1164

Yagil, Y. and Yagil, C., 2003, Hypothesis: ACE2 modulates blood pressure in the mammalian organism. *Hypertension* **41**: 871-873

Yang, Z. Y., Kong, W. P., Huang, Y., Roberts, A., Murphy, B. R., Subbarao, K. and Nabel, G. J., 2004, A DNA vaccine induces SARS coronavirus neutralization and protective immunity in mice. *Nature* **428**:561-564

Yeager, C. L., Ashmun, R. A., Williams, R. K., Cardellichio, C. B., Shapiro, L. H., Look, A. T. and Holmes, K. V., 1992, Human aminopeptidase N is a receptor for human coronavirus 229E. *Nature* **357**:420-422

Yu, I. T., Li, Y., Wong, T. W., Tam, W., Chan, A. T., Lee, J. H., Leung, D. Y. and Ho, T., 2004, Evidence of airborne transmission of the severe acute respiratory syndrome virus. *N. Engl. J. Med.* **350**:1731-1739

Zakhartchouk, A. N., Viswanathan, S., Mahony, J. B., Gauldie, J. and Babiuk, L. A., 2005, Severe acute respiratory syndrome coronavirus nucleocapsid protein expressed by an adenovirus vector is phosphorylated and immunogenic in mice. *J. Gen. Virol.* **86**:211-215

Zhao, P., Cao, J., Zhao, L. J., Qin, Z. L., Ke, J. S., Pan, W., Ren, H., Yu, J. G. and Qi, Z. T., 2005, Immune responses against SARS-coronavirus nucleocapsid proteins induced by DNA vaccine. *Virology* **331**:128-135

Chapter 3.2

THERAPY OF EBOLA AND MARBURG VIRUS INFECTIONS

M. BRAY[1]

[1]*Biodefense Clinical Research Branch, Office of Clinical Research, National Institute of Allergy and Infectious Diseases, National Institutes of Health, Bethesda, Maryland 20892., USA*

Abstract: The filoviruses, Marburg and Ebola, cause fulminant hemorrhagic fever in humans and nonhuman primates. The agents replicate to high titer in macrophages and dendritic cells, suppressing type I interferon responses, and disseminate to these and other cell types in tissues throughout the body, causing extensive necrosis and inducing an intense systemic inflammatory response resembling septic shock. Effective therapies are urgently needed to deal with laboratory accidents, natural epidemics and the threat of bioterrorism. An immediate goal should be the development of postexposure prophylaxis for persons who have been exposed to Marburg or Ebola virus, but have not yet become ill. This article first reviews current research aimed at blocking individual steps in the filoviral replication pathway, then describes efforts to characterize and modify damaging host responses. No direct inhibitors of filoviral replication have yet been identified, but some interventions that counteract suppression of type I interferon or modify inflammatory responses have shown benefit in laboratory animal models of lethal infection.

1.   INTRODUCTION

The filoviruses, Marburg and Ebola, are nonsegmented, negative-strand RNA viruses that cause fulminant hemorrhagic fever in humans and nonhuman primates. In epidemics caused by these agents in central Africa, case fatality rates have ranged from 50 to 90%. Marburg virus was first recognized by the medical community in 1967, when it was introduced into

E. Bogner and A. Holzenburg (eds.), New Concepts of Antiviral Therapy, 419–452.

Europe in a shipment of monkeys from Uganda, infecting 25 vaccine plant workers and 7 of their medical attendants (Martini *et al.*, 1968). Other than a handful of laboratory accidents, all filoviral infections of humans since that time have occurred in Africa. Only a few additional cases of Marburg hemorrhagic fever were detected until 1999, when a large outbreak took place in the Democratic Republic of Congo; a new epidemic in Angola has claimed more than 300 victims at the time of writing.

Ebola virus was first recognized in 1976, when the Zaire and Sudan species caused separate hospital-based epidemics in those two countries (Sanchez, 2001; Bray, 2002). Both agents largely disappeared from view until the mid-1990s, but they have since caused epidemics with increasing frequency (Pourrut *et al.*, 2005). A third Ebola species, Ivory Coast virus, has so far caused only a single known human infection (Formenty *et al.*, 1999). The fourth, the enigmatic Reston agent, caused outbreaks of lethal illness among quarantined macaques imported from the Phillipines from 1989 through 1995 (Jahrling *et al.*, 1990; Miranda *et al.*, 1999). No disease was observed in workers exposed to sick animals, but the agent's virulence for humans remains undetermined.

The animal reservoir of the filoviruses has not been identified. As for the other viruses that cause severe hemorrhagic fever, humans are only accidental hosts (Bray, 2005; Pourrut *et al.*, 2005). The lack of adaptation of these pathogens to humans helps to explain both the rapidly overwhelming nature of infection and the inefficiency of person-to-person transmission, which requires direct contact with virus-containing material (Dowell *et al.*, 1999; Bray and Geisbert, 2005). Nonhuman primates are also highly susceptible to lethal infection, but the rapid progression of illness and high mortality in these animals indicates that they cannot act as natural reservoirs. For unknown reasons, the Zaire species of Ebola virus has been spreading among wild primates in central Africa, and a number of recent outbreaks in humans have been initiated by direct contact with a sick or dead gorilla, chimpanzee or other animal (Peterson *et al.*, 2004; Pourrut *et al.*, 2005; Rouquet *et al.*, 2005).

Even though less than 3000 cases of Marburg and Ebola virus infection have been recognized since their discovery, these agents are a cause of global concern because of the highly lethal nature of infection, the risk that infected individuals will transport these pathogens out of Africa, and the possibility that they could be used as bioterror weapons (Bray, 2003). No specific treatment is currently available for filoviral hemorrhagic fever, so the development of effective therapies is a high priority for biodefense research. Efforts have followed two general approaches (Bray and Paragas, 2002). The first aims to identify pharmacologic agents such as nucleoside analogs, peptides, antisense oligonucleotides or other substances that directly inhibit viral replication. These efforts have had only limited success, but recent advances in understanding the viral replication pathway may lead to more rapid progress. The other therapeutic strategy attempts to prevent or mitigate illness by modifying host responses. This approach recognizes that the virulence of filoviruses for humans stems in large part from their ability to suppress innate antiviral defenses, particularly the type I IFN response, and their induction of a systemic inflammatory syndrome (Mahanty and Bray, 2004; Bray and Geisbert, 2005). Interventions therefore aim to stimulate innate defenses while blocking damaging inflammatory responses. In contrast to the first approach, this strategy has had some success in protecting filovirus-infected animals, but so far efficacy has been limited to pre- or postexposure prophylaxis.

This chapter begins with a summary of the basic characteristics and replication cycle of the filoviruses, followed by discussions of the clinical features and pathogenesis of filoviral hemorrhagic fever, the nature of host responses to infection, experience with treatment of human disease, and available methods for *in vitro* and *in vivo* drug testing. The remainder of the chapter reviews principal research findings from the two approaches to therapy just described, and concludes by suggesting promising areas for further investigation.

2. FILOVIRAL CLASSIFICATION, VIRION STRUCTURE AND REPLICATION PATHWAY

2.1 Classification

The two filoviral genera, Marburgvirus and Ebolavirus, make up the family *Filoviridae* in the order *Mononegavirales*. The family name is derived from the Latin word *filum* (thread), reflecting the agents' unique filamentous morphology (Figure 1). Four different species of Ebola virus are recognized, as listed above, but all isolates of Marburg virus are currently grouped as one species.

All filoviruses share the same genomic organization, but Marburg and Ebola differ significantly in nucleotide sequence and lack antigenic cross-reactivity (Bray, 2002; Feldmann *et al.*, 2003). Filoviral genomes are approximately 19 kb in length. The linear arrangement of the seven viral genes and the mechanisms of transcription and genome replication resemble those of the better-known rhabdo- and paramyxoviruses. Transcription of each viral gene is controlled by conserved initiation and termination sequences in noncoding regions at the 3' and 5' ends (Sanchez, 2001). The ends of the genome contain conserved, complementary sequences, which apparently function as cis-acting regulators of genomic replication and transcription and may also play a role in genome packaging.

2.2   Virion structure

Filovirus virions appear as long filamentous threads under the electron microscope (Figure 1). Viral particles vary considerably in length, but have a constant diameter of 80 nm. Because host-cell enzymes cannot transcribe or copy the negative-sense RNA genome, all components of the viral replication complex must be carried within the virion. The central ribonucleoprotein (RNP) core of the virion is comprised of the genomic RNA molecule and its encapsidating nucleoproteins, NP and VP30. Two other proteins, VP35 and the RNA-dependent RNA polymerase, or L protein, are also present within the virion, but in lower copy number. The viral envelope, derived from the host cell membrane, is linked to the RNP by two matrix proteins, VP24 and VP40, which apparently interact with the C-termini of trimeric spikes of the embedded virion surface glycoprotein (GP) (Figure 1).

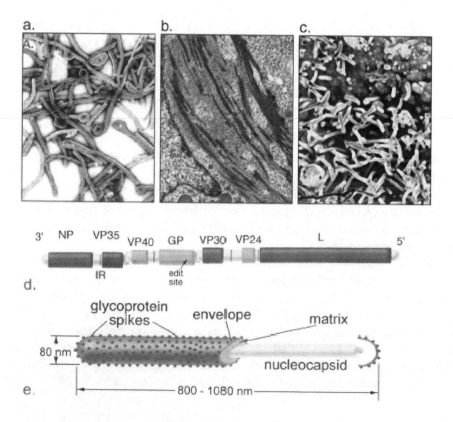

Figure 1: A-C : Ebola Zaire virus. A. Transmission electron micrograph of free virions. B. Cytoplasmic inclusion body within an infected hepatocyte, showing viral nucleocapsids. C. Scanning electron micrograph showing virions budding from the surface of an infected primary human umbilical vein endothelial cell. D. Sequence of genes along the single-stranded negative-sense RNA genome of Ebola virus (see text for abbreviations). Intergenic regions (IR) contain initiation and termination signals. The edit site in the GP gene of Ebola virus consists of a series of 6 adenosine residues. E. Structure of an Ebola or Marburg virion. The RNP core consists of the genomic RNA molecule and its encapsidating nucleoproteins, NP and VP30, which are linked by matrix proteins, VP24 and VP40, to the virion envelope, a lipid bilayer derived from the host cell. The virion surface contains embedded trimeric spikes of the viral GP. (Images A-C courtesy of Tom Geisbert, USAMRIID.).

2.3   Replication Cycle

Infection begins with the attachment of the virion to the cell surface (Fig. 2). A specific receptor molecule has not been identified. Studies employing pseudotyped retroviruses bearing Marburg or Ebola GP on their surface indicated that the human folate receptor-α was a filovirus binding site, but subsequent investigations failed to support this hypothesis (Chan *et al.,* 2001; Simmons *et al.*, 2003).

The asialoglycoprotein was also identified as a receptor for Marburg virus, but because this molecule is not expressed by a number of cell lines that are susceptible to filovirus infection, its role in pathogenesis remains uncertain (Becker, Spiess, and Klenk, 1995). In fact, it now appears that the initiation of filoviral infection may not require a specific receptor, since the heavily glycosylated filoviral GP can bind to cell-surface lectins, including the dendritic cell surface protein DC-SIGN and macrophage C-type lectin (Marzi *et al.*, 2004; Takada *et al.*, 2004).

Following attachment, filovirus virions exploit the cellular endocytosis machinery, in which engulfment in an endosome is followed by a drop in pH, a conformational change in GP, fusion of viral and cell membranes and release of the genome and internal virion proteins into the cytoplasm. Transcription is then performed by a replication complex, which in the case of Ebola virus is made up of four proteins (VP35, VP30, NP and L); VP30 is not required for Marburg virus replication (Boehmann *et al.*, 2005; Theriault *et al.*, 2004). Transcription initiates at the 3' end of the genome, resulting in synthesis of a leader RNA and seven polyadenylated mRNAs (Sanchez, 2001). Accumulation of the first two gene products (NP and VP35) in some way triggers a switch to the production of full-length, positive-sense "antigenomes", which are then used as templates for genome synthesis. The production of new genomes and their encapsidation leads to the accumulation of nucleocapsids in large cytoplasmic inclusion bodies visible in stained tissue sections (Figures 1 and 2).

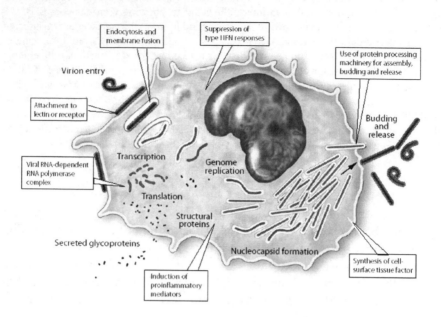

Figure 2: The filovirus replication cycle. Boxed labels denote processes that may represent targets for therapeutic intervention. Virions attach to cell-surface lectins and/or a specific receptor, and are then taken up by endocytosis. Fusion with the cell membrane within endosomes releases the RNP and other components of the replication complex into the cytoplasm. Viral mRNA is translated by the host cell machinery to produce virion structural proteins, plus, in the case of Ebola virus, a C-terminally truncated, secreted form of the virion surface glycoprotein (GP). Full-length GP is glycosylated in the Golgi apparatus, transported to the cell membrane and cleaved to form two units that remain linked by a disulfide bond as GP1,2. Virion assembly begins with the encapsidation of new genomic RNA molecules by NP and VP30. Production of filamentous virions takes place at the cell membrane, in association with lipid rafts. The process is apparently driven by VP40, which interacts with cellular proteins, bringing about a conformational change that results in its self-association into a progressively enlarging structure that forms the internal scaffolding of the virion envelope. VP24 may help link this structure to the central RNP core and/or to the C-terminal cytoplasmic tails of GP1,2 .

The assembly of new virions begins on the inner surface of the plasma membrane, where nucleocapsids apparently become linked through two matrix proteins, VP24 and VP40, to the cytoplasmic tail of GP2 molecules embedded in the cell surface. This process localizes to cholesterol- and glycosphingolipid-rich areas termed lipid rafts, which may provide a stabilizing structure for the assembly process (Bavari *et al.*, 2002; Aman *et al.*, 2003; Panchal *et al.*, 2003). Like a number of other pathogens, Ebola and Marburg virus apparently bring about their own release from the cell by "hijacking" systems normally employed for degradation or excretion of proteins. Budding of new virions appears to be driven primarily by VP40, which is able to self-oligomerize on contact with lipid bilayers. The expression of that gene alone in transfected cells results in the extrusion of filamentous virus-like particles (VLP) that closely resemble infectious virus; their production is markedly enhanced by co-expression of GP in the same cells (Jasenosky and Kawaoka, 2004; Licata *et al.*, 2004; Irie, Licata, and Harty, 2005). Two overlapping "late-domain" motifs, PTAP and PPEY, in the N terminus of VP40 interact with proteins involved in the endosomal pathway: the ubiquitin ligase Nedd4, which marks proteins for degradation or excretion, and Tsg101, a component of the ESCRT-1 complex that directs ubiquitinated proteins to endosomes (Jasenosky and Kawaoka, 2004). Optimum VLP formation requires the presence of these sequences, suggesting that specific interactions between virus and host proteins are required for virion release and may represent targets for antiviral therapy (Irie, Licata, and Harty, 2005). The primary product encoded by the Ebola virus GP gene is a truncated form of the protein that is released from infected cells ("soluble" glycoprotein, sGP) (Feldmann *et al.*, 1999). Synthesis of the full-length virion surface GP requires the insertion of an additional adenosine at an "editing site" during transcription (Volchkov *et al.*, 1995). The release of large amounts of sGP has been proposed to play a role in the remarkable virulence of Ebola virus for primates, but since Marburg virus is just as virulent, but does not encode a similar protein, the actual contribution of sGP to pathogenesis remains unknown. The GP of both Ebola and Marburg virus becomes abundantly glycosylated with N- and O-linked carbohydrates during transit through the Golgi apparatus, and is then cleaved by a cellular furin-like enzyme to produce two units, extracellular GP1 and transmembrane GP2, that remain linked by a disulfide bond as GP1,2 (Volchkov *et al.*, 1998).

3.   FILOVIRAL HEMORRHAGIC FEVER

3.1   Clinical Syndrome

Marburg and Ebola viruses cause similar diseases in humans (Martini, 1973; Bwaka *et al.*, 1999; Mahanty and Bray, 2004). After an incubation period of some 5-10 days, illness begins abruptly, with headache, fever, muscle pain, vomiting, diarrhea and other nonspecific signs and symptoms. Over the next week, persistent fever and worsening prostration and stupor are accompanied by a fall in blood pressure. Hemorrhagic phenomena generally take the form of petechiae, ecchymoses, conjunctival hemorrhages and oozing from venipuncture sites. Massive bleeding is rare. The onset of clinical deterioration is marked by tachypnea, anuria, a fall in body temperature and the development of coma as progressive hypotension leads to intractable shock. Death usually occurs 6-9 days after the onset of symptoms.

Data from African patients infected with Ebola Zaire virus indicate that fatal infection is characterized by a steadily rising titer of circulating virus and the absence of a detectable antibody response, while in survivors, peak viral titers are significantly lower and IgM anti-Ebola antibodies are detectable during the second week of illness (Ksiazek *et al.*, 1999; Sanchez *et al.*, 2004; Towner *et al.*, 2004). Limited data from Africa indicate that persons who survive infection also show evidence of early, strong proinflammatory cytokine responses, while in fatal cases these are seen later in infection, often accompanied by anti-inflammatory mediators such as interleukin(IL)-10 (Baize *et al.*, 1999, 2002). Studies performed during outbreaks in Gabon also revealed the presence of proinflammatory cytokines and Ebola-specific IgG antibodies in blood samples from some persons who were in direct contact with Ebola Zaire patients, but did not become ill, suggesting that these individuals were able to eliminate infection through an early, brisk inflammatory response (Leroy *et al.*, 2000, 2001).

Because of the extreme difficulty of performing clinical studies under outbreak conditions in central Africa, most data on the clinicopathologic features of filoviral hemorrhagic fever come from studies in laboratory animals, especially from cynomolgus macaques infected with Ebola Zaire virus from the 1995 outbreak (Bray *et al.*, 2001; Geisbert *et al.*, 2003b, c). Inoculation of a standard 1000 pfu challenge is followed in 3-4 days by the onset of fever and diminished activity, the appearance of a hemorrhagic macular rash on day 4-5, obtundation by day 6 and death on day 7-8. As in humans, mild hemorrhagic phenomena are common, but profuse bleeding is rare. Virus is detectable in the serum by day 3, and titers exceed 10^7 pfu/mL by day 5. Infection is also characterized by lymphopenia, neutrophilia, DIC and massive "bystander" apoptosis of lymphocytes and NK cells. The model thus appears to differ from Ebola hemorrhagic fever in humans principally in that death occurs in all cases, usually 3-4 days after the onset of visible symptoms, while humans are ill for 1-2 weeks and a small percentage survive infection.

Pathologic changes in the macaque model include the infection and necrosis of macrophages and dendritic cells in the spleen, liver, lymph nodes and other lymphoid tissues; multifocal necrosis of the liver, adrenal cortices and other organs; and extensive apoptosis of lymphocytes. Abnormalities in clinical laboratory tests include neutrophilia, lymphopenia, marked thrombocytopenia and prolongation of coagulation times, with circulating fibrin degradation products and other features of disseminated intravascular coagulation (DIC). Serum levels of liver-associated enzymes, particularly aspartate aminotransferase, are elevated.

3.2   Target cells and viral dissemination

Like all microbial pathogens, filoviruses are confronted by macrophages and dendritic cells as soon as they enter the body. Both types of cells employ a variety of innate immune mechanisms for antiviral defense, but dendritic cells also specialize in initiating adaptive immune responses. Far from being destroyed by these cells, filoviruses replicate within them and impair their function, so that they are able to initiate inflammation and coagulation, but cannot prevent viral dissemination (Figure 3) (Mahanty and Bray, 2004; Bray and Geisbert, 2005). Viral replication and spread are aided by suppression of type I interferon (IFN) responses (see below). In consequence, infection spreads rapidly to similar cells in tissues throughout the body, causing the release of massive quantities of proinflammatory cytokines, chemokines and other mediators that produce a syndrome of refractory hypotension and DIC resembling septic shock (Hensley *et al.*, 2002; Bray and Mahanty, 2003). Infected dendritic cells are unable to undergo maturation or initiate adaptive immune responses (Mahanty *et al.*, 2003b; Bosio *et al.*, 2004).

In addition to these indirect effects, filoviruses cause direct tissue injury, since replication in macrophages, dendritic cells and parenchymal cells of the liver, adrenal cortices and other organs results in extensive necrosis, contributing to fatal disease (Figure 3). Careful studies in nonhuman primates have shown that endothelial cells remain uninfected by virus until the later stages of disease, perhaps only becoming susceptible to infection after they have become activated by circulating inflammatory mediators (Geisbert *et al.*, 2003c). Although lymphocytes also remain uninfected, they undergo massive apoptosis through a "bystander" mechanism, further impairing immune function (Geisbert *et al.*, 2000; Geisbert *et al.*, 2003b; Reed *et al.*, 2004).

3.3   Host response to infection

The principal early host defense against viral infection is the production and release of type I IFN, which binds to cell-surface receptors and sets off a train of reactions, including the expression of a large number of new proteins, that result in an "antiviral state" (Katze *et al.*, 2002; Weber *et al.*, 2004). Not unexpectedly, viruses have evolved a variety of mechanisms to block or evade these responses. Considerable evidence now indicates that suppression of IFN responses plays a major role in the ability of the filoviruses to cause rapidly overwhelming infection in primates.

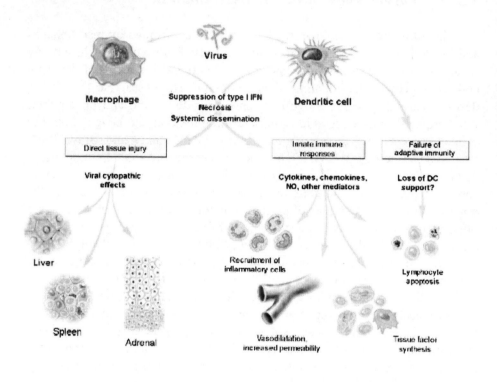

Figure 3: Pathogenic mechanisms of filoviral hemorrhagic fever, based on studies of lethal Ebola Zaire virus infection of macaques. Virus initially replicates in macrophages and dendritic cells (DC), causing their necrosis; suppression of type I IFN production aids viral dissemination. Proinflammatory cytokines, chemokines, tissue factor and other mediators produced by these cells cause increased vascular permeability and coagulopathy. Filoviruses also spread to parenchymal cells in the liver, adrenal cortex and other organs, causing multifocal necrosis, while destruction of macrophages and DC causes extensive injury to the spleen and other lymphoid tissues. Massive 'bystander' apoptosis of lymphocytes is apparently brought about through mediator effects and loss of DC support. NO: nitric oxide.

Most data on viral interactions with the type I IFN system come from studies in cultured cells. Primary human macrophages infected with Ebola Zaire virus produced only small amounts of IFN-α late in infection, and failed to release IFN when exposed to double-stranded RNA (dsRNA)

(Harcourt *et al.*, 1998, 1999). Basler et al. have identified such suppression with the Ebola VP35 protein, which can functionally replace the influenza NS1 to inhibit IFN production by infected cells (Basler *et al.*, 2000). VP35 appears to act by blocking the recognition of dsRNA, preventing the phosphorylation and nuclear translocation of interferon response factor-3 and the transcription of a series of IFN-related genes (Basler *et al.*, 2003; Hartman, Towner, and Nichol, 2004; Basler, 2005). Further confirmation of the anti-IFN activity of VP35 was provided by a study in which a recombinant alphavirus encoding the protein induced significantly less IFN-α production by infected cells than the control virus (Bosio *et al.*, 2003).

The mouse model of Ebola Zaire virus infection has been particularly useful in exploring interactions between filoviruses and type I IFN responses (Bray, 2001; Bray *et al.*, 2002; Gupta *et al.*, 2004). In marked contrast to primates, normal adult mice are solidly resistant to all wild-type filoviruses. However, they can be rendered sensitive to lethal infection by Ebola Zaire and Ebola Sudan virus through treatment with anti-IFN-α antibodies, and knockout mice lacking the STAT-1 protein or the cell-surface IFN-α/β receptor are rapidly killed by these and other filoviruses (Bray, 2001). Further evidence of the critical role of type I IFN responses has been obtained using a variant of Ebola Zaire virus that is lethal for adult mice, which was obtained through sequential passage in suckling mice of virus from the 1976 outbreak (Bray *et al.*, 1998; Gibb *et al.*, 2001). The LD_{50} of this "mouse-adapted virus" by the intraperitoneal route is approximately one virion; infection induces only a low level of IFN-α production, that first becomes detectable in serum when the animals are already ill. Remarkably, mice remain healthy when inoculated subcutaneously with the same virus; in this case, infection results in the rapid production of large amounts of IFN-α (Mahanty *et al.*, 2003a).

3.4 Experience with human therapy

Almost all patients with Marburg or Ebola hemorrhagic fever in African outbreaks have received general supportive treatment, and only a few descriptions of attempts to provide specific therapy can be found in the medical literature. One researcher accidentally exposed to Ebola Sudan virus in the laboratory became severely ill, but survived infection after treatment with human convalescent serum and IFN-α (Emond *et al.*, 1977). It is not clear that either therapy contributed to the fortunate outcome. A Russian laboratory worker accidentally infected with Marburg virus also survived after prolonged treatment (Nikiforov *et al.*, 1994). Because the nature of his infection was not recognized until after the onset of illness, he did not receive immune serum or IFN, but underwent several rounds of extracorporeal blood treatment with hemosorbents and dialysis. These measures produced transient improvements in his condition, but it is again difficult to conclude from this single case that therapy was responsible for survival. Several Russian laboratory workers who were possibly exposed to Ebola Zaire virus in the laboratory have been treated with a preparation of equine anti-Ebola immunoglobulin and recombinant IFN (see below) (Kudoyarova-Zubavichene *et al.*, 1999).

The only published report of an attempt to provide specific therapy during the course of a natural outbreak describes the use of whole blood transfusions from convalescent survivors toward the end of the 1995 Ebola Zaire epidemic (Mupapa *et al.*, 1999). The initial impression was highly favorable, since seven of the eight laboratory-confirmed patients who received blood survived infection. However, subsequent analysis showed that all of these patients would have been expected to survive, even without therapy, since they had already lived for at least 11 days since the outset of symptoms at the time they were treated (Sadek *et al.*, 1999).

4. EVALUATING NEW DRUGS FOR FILOVIRAL INFECTIONS

4.1 *In vitro* testing

The evaluation of a candidate antiviral drug for filoviral infections is commonly performed by assessing its ability to prevent viral cytopathic effect, as measured by the uptake of neutral red dye or another indicator molecule by virus-infected cells in culture. Such work must necessarily be performed under BSL-4 containment. Recently, this approach has been supplemented by the construction of a recombinant Ebola Zaire virus encoding green fluorescent protein (GFP), that can potentially be used to obtain a direct read-out of the extent of viral replication (Towner *et al.*, 2005).

Because few laboratories possess the level of biocontainment required to work with filoviruses, other methods have been developed to study portions of the filoviral replication pathway. One approach is the construction of "pseudotyped" viruses that encode the replicative machinery of vesicular stomatitis virus (VSV), murine leukemia virus or human immunodeficiency virus (HIV)-1, but express the Marburg or Ebola GP on the virion surface (Takada *et al.*, 1997; Kobinger *et al.*, 2001; Manicassamy *et al.*, 2005;

Yonezawa *et al.*, 2005). These chimeric agents provide valuable information about such GP-related functions as virion binding, fusion and entry into target cells, and can therefore be used for the initial evaluation of inhibitors of these activities.

A method that can be used to study transcription and genome replication is the "minigenome" system, in which cultured cells are transfected with plasmids encoding individual components of the filoviral replication complex. A substrate for transcription is provided by generating an RNA transcript containing a reporter gene flanked by filoviral initiation and termination sequences (Muhlberger *et al.*, 1999; Weik *et al.*, 2002; Groseth *et al.*, 2005). Although somewhat cumbersome, this system could potentially be used without the constraints of BSL-4 containment to identify compounds that inhibit the viral RNA-dependent RNA polymerase, induce a high error rate or cause chain termination.

Transfection of cultured cells with plasmids encoding filoviral proteins can also be employed to study later steps in replication, including the assembly and budding of new virions. As described above, this method has been used to demonstrate that expression of Ebola virus VP40 is sufficient to cause production and release of filamentous virus-like particles from the cell surface, and that specific late-domain sequences in the N-terminus of VP40 are essential for this process (Jasenosky and Kawaoka, 2004). This system could be used to screen for inhibitors of protein-protein interactions involved in the budding process.

Another approach to finding effective therapies for filovirus infections is based on the study of less virulent viruses with similar replication mechanisms. This strategy recognizes that it would be more financially advantageous for a pharmaceutical company to develop a drug that is active against both filoviruses and another, more common pathogen than to identify a compound that only inhibits Ebola and Marburg virus. One such "surrogate" agent is respiratory syncytial virus (RSV), a nonsegmented, negative-strand RNA virus that causes severe illness in infants and immunocompromised adults (Huggins *et al.*, 1999). It should be noted, however, that although the use of surrogates may identify lead compounds for further study, such predictions are not always accurate. For example, even though ribavirin is active against RSV, it does not inhibit filovirus replication.

4.2   Animal models

Three types of laboratory animals – mice, guinea pigs and nonhuman primates – are in use for testing antiviral drugs and vaccines against Marburg and Ebola infection and to study filovirus pathogenesis. Although the principal target cells of infection are similar in these diverse species, these animals differ significantly in their inherent resistance to filoviruses. Nonhuman primates are exquisitely sensitive to all wild-type Marburg and Ebola isolates, including Ebola Reston virus, but guinea pigs and mice can only be used as experimental models after a virus has been "adapted" to them through serial passage. This difference in innate resistance to infection apparently reflects the ability of filoviruses to suppress type I IFN responses in primates, but not in rodents; it frequently causes divergent outcomes in tests of drug and vaccine efficacy (Geisbert *et al.*, 2002).

Filovirus replication in mice can be studied by a number of approaches. Marburg virus and the Zaire and Sudan species of Ebola virus cause lethal infection in newborn mice, but do not cause visible illness in adult animals. Severe combined immunodeficient (SCID) mice also differ from primates in their sensitivity to filoviruses, in that they develop slowly progressive disease after inoculation of most filoviral species, dying 3-6 weeks after challenge. By contrast, gene-knockout mice lacking the STAT-1 protein or the cell-surface IFN-α/β receptor develop rapidly lethal disease after injection of the same viruses (Bray, 2001). Most recent studies in mice have employed the mouse-adapted variant of Ebola Zaire '76 virus described above, that causes rapidly lethal disease in normal adult mice when inoculated by the intraperitoneal route (Bray *et al.*, 1998). The pathologic features of infection resemble those in primates, except that signs of coagulopathy are much less prominent (Bray *et al.*, 2001; Gibb *et al.*, 2001). As noted, even though normal mice can be killed by the intraperitoneal injection of a single particle of this "mouse-adapted virus", they are solidly resistant to massive doses of the same virus injected subcutaneously – a property that makes this model especially useful for studying mechanisms of susceptibility and resistance to filovirus infection.

Guinea pigs are inherently less resistant to filoviral infections than mice, since they develop a mild febrile illness after inoculation with wild-type Marburg or Ebola Zaire or Sudan virus, and these agents can readily be adapted to lethal virulence in guinea pigs through serial passage. The major pathologic features of fatal infection resemble those in mice and primates, except that signs of coagulopathy are again lacking (Connolly *et al.*, 1999). Guinea pigs have frequently been employed for vaccine testing, but the lack of specific reagents or detailed knowledge of immune function makes them less useful for this purpose than mice. Their comparatively large size is also a distinct disadvantage for antiviral drug testing, since even a young animal may require 10-20 times as much of an experimental compound as a weanling mouse.

In marked contrast to rodents, nonhuman primates are highly susceptible to severe or lethal hemorrhagic fever after challenge with wild-type filoviruses, including Ebola Reston. Cynomolgus and rhesus macaques are used most often for laboratory studies. Ebola Zaire virus infection is uniformly lethal for these animals, while the Sudan and Reston species and Marburg virus cause a similar disease, but with a longer clinical course and generally less than 100% mortality. Interestingly, the mouse-adapted variant of Ebola Zaire virus appears to be somewhat attenuated for primates, since two of three macaques inoculated with the agent developed only low-level viremia and survived infection (Bray *et al.,* 2001).

5. APPROACHES TO THERAPY OF FILOVIRAL INFECTIONS

Effective therapies are needed to deal with two situations. The more urgent, but perhaps more easily achievable goal is to develop forms of postexposure prophylaxis that can be used to treat persons who have been infected with Marburg or Ebola virus, but have not yet become ill. Such treatment could be given to individuals who have been in direct contact with patients in an African outbreak or to a researcher accidentally infected in the laboratory, and might also be used to protect people exposed to a filovirus in a bioterror attack. Postexposure prophylaxis could potentially be achieved in two different ways: treatment with a compound that directly inhibits filovirus replication, or administration of a substance that bolsters innate immunity and blocks damaging host responses. Possible targets for both approaches are discussed below.

The second, far more difficult problem is to develop specific interventions that can rescue persons who are already ill with filoviral hemorrhagic fever. In contrast to treatment during the incubation period, halting the progression of disseminated infection probably cannot be achieved solely by modifying host responses, but will require therapy that blocks viral replication. Given the extreme severity of these conditions, a combination of pharmacologic agents with different mechanisms of action may be required to cure patients with full-blown disease.

5.1 Direct inhibition of viral replication

The following section proceeds through the stages in the filoviral replication cycle, from attachment to the cell surface through the release of new progeny virions, and describes current efforts to develop inhibitors of each step. The discussion includes both "small molecules," such as nucleoside analogues, that are produced through chemical synthesis, and substances generated in biological systems, such as antibodies and lectins.

5.1.1   Prevention of virion binding

The fact that high viremia in the absence of an antibody response is predictive of death in patients with Ebola hemorrhagic fever suggests that neutralization of circulating virus would improve survival. To date, efforts to develop such therapies have been based principally on the use of polyvalent or monoclonal antibodies. The most experience with polyvalent antiserum has been obtained with immunoglobulin produced in Russia by hyperimmunizing horses, sheep or goats with Ebola Zaire virus (Kudoyarova-Zubavichene *et al.*, 1999). These studies suggest that the efficacy of antibody therapy depends both on the donor species and the animal model in which it is tested. Thus, pre- or immediate postexposure treatment with equine anti-Ebola immunoglobulin prevented the death of most baboons infected with Ebola Zaire virus and protected guinea pigs infected with an adapted variant of the same agent, but even large doses of the same product administered soon after challenge did no more than delay the onset of illness in mice and cynomolgus macaques by 1-2 days (Jahrling *et al.*, 1996, 1999). Despite these mixed reviews, this immunoglobulin is now approved in Russia for the treatment of laboratory-acquired Ebola virus infections.   It has been administered to several workers thought to have been exposed to Ebola virus, who remained well, but without proof that they were actually infected (Kudoyarova-Zubavichene *et al.,* 1999).

These and other results suggest that antibody therapy is more likely to be successful in species in which infection is characterized by a longer incubation period and lower peak serum viral titer (guinea pigs and baboons) than in animals in which the same virus causes rapidly progressive disease with a high peak viral load (mice and monkeys). The efficacy of heterologous antiserum may also be limited by immune recognition and clearance of the foreign protein and by incompatibility between the donor IgG and the recipient's Fc receptors (Wilson *et al.*, 2001). The latter consideration suggests that efficacy can be improved by using homologous immune serum, and in contrast to the results cited above, passive transfer of serum from immune to naïve mice protected the recipients against mouse-adapted Ebola Zaire virus (Gupta *et al.*, 2001). Treatment up to 2 days after infection was remarkably effective, preventing the death of both normal and SCID mice.

These experimental successes with polyvalent antisera have led to efforts to develop monoclonal antibodies (mabs) for use in humans. Initial experiments in mice showed that mabs specific for any of five different epitopes on the Ebola Zaire GP were protective when inoculated up to 2 days after virus challenge (Wilson *et al.*, 2000). Some antibodies were protective even though they failed to neutralize virus in plaque-reduction assays, suggesting that the *in vitro* test does not always reflect mechanisms of protection *in vivo*. Human mabs that react strongly with the Ebola Zaire GP, sGP or NP have been produced from phage-display libraries constructed using mRNA from survivors of the 1995 Ebola Zaire outbreak (Maruyama *et al.*, 1999a, b). One antibody that neutralized Ebola Zaire virus *in vitro* at an inhibitory concentration of 0.3 µg/ml also protected guinea pigs when given before or within an hour after virus challenge (Parren *et al.*, 2002). Some treated animals survived despite developing a high viremia, providing further evidence that antibodies can act through mechanisms other than direct neutralization. However, successful protection of nonhuman primates with this or other antibodies has yet to be reported.

Another approach to preventing the binding of a filovirus to target cells involves the synthesis of bispecific IgG, in which one arm of the antibody molecule binds a viral antigen, while the other has specificity for a host protein. One such "heteropolymer" has been described that recognizes both the Marburg virus GP and the complement receptor-1 on human erythrocytes (Nardin *et al.*, 1998). Ideally, such a molecule could clear virions from plasma by attaching them to the surface of red blood cells, which would then be taken up and eliminated by phagocytes in the spleen and other tissues. *In vitro* tests showed that such antibodies could bind inactivated Marburg virus to erythrocytes, but their *in vivo* efficacy has not been reported.

Since filoviruses may make use of DC-SIGN, macrophage surface C-type lectin and other lectins to attach to target cells, another therapeutic strategy involves administering naturally occurring lectins that bind carbohydrate moieties on the filoviral surface GP and can thus serve as "decoys" for the virus, in a manner similar to naturally occurring mannose-binding lectin, which serves as a natural antimicrobial defense in the bloodstream (Ji *et al.*, 2005). Proof-of-concept studies have employed cyanovirin-N, an 11K protein produced by cyanobacteria, which has attracted interest because it binds high-mannose oligosaccharides on the human immunodeficiency virus (HIV) gp120 with high affinity and neutralizes some isolates of influenza virus (Boyd *et al.*, 1997; O'Keefe *et al.*, 2003). Even though cyanovirin has only limited activity against Ebola Zaire virus in cell culture, it was still able to able to slow the progression of disease in mice (Barrientos *et al.*, 2003; Barrientos and Gronenborn, 2005). Efforts are now under way to characterize the oligosaccharides on the Marburg and Ebola GP that are recognized by CV-N and to isolate other naturally occurring lectins with higher affinity for filoviral GPs.

Another approach to blocking infection of target cells by Marburg or Ebola virus is to administer a substance that can compete for binding to DC-SIGN or other C-type lectins. This has recently been attempted using dendrimer technology, in which a hyperbranched polymer molecule is conjugated to multiple units of a high-mannose oligosaccharide (Lasala *et al.*, 2003; Rojo and Delgado, 2004). In proof-of-concept experiments using pseudotyped retroviral particles expressing the Ebola virus GP, a dendrimer molecule blocked virus entry at an IC_{50} of less than 1 µM. Evaluation against live virus or in an animal model has not yet been reported.

5.1.2   Prevention of membrane fusion and entry

Once a filovirus virion has bound to a target cell, its surface GP must undergo a conformational change to permit the viral envelope to fuse with the cell membrane, releasing the RNP and viral proteins into the cytoplasm. Some pathogens, such as human immunodeficiency virus (HIV)-1, carry out fusion at the cell surface, but filoviruses apparently perform this process within endosomes, in a step requiring acidification (Wool-Lewis and Bates, 1998; Chandran *et al.*, 2005).

Efforts to develop filoviral fusion inhibitors have been based in part on the successful "T20" HIV-1 inhibitor (Cooley and Lewin, 2003). In situ mutagenesis of the Ebola Zaire GP expressed by pseudotyped viruses has identified regions of the protein that interact with each other to bring about a conformational change and membrane fusion, and are required for viral entry (Watanabe *et al.*, 2000). An oligopeptide mimicking one of these fusion sequences blocked infection by a pseudotyped retrovirus, but the 50% inhibitory concentration exceeded 1 mg/ml, a level that probably could not be achieved *in vivo*. Successful therapy with a fusion inhibitor may therefore require the development of methods of directing such peptides into endosomes, so as to increase their concentration at sites of fusion.

As noted, filovirus entry into target cells has also recently been shown to involve cleavage of GP1 by cellular cathepsins (Chandran *et al.*, 2005). It is not yet known if this process is essential for infection, or whether alternative entry pathways may be available to these pathogens. Attempts to block infection *in vivo* through treatment with cathepsin inhibitors, or the characterization of infection in knockout mice lacking these enzymes, has not yet been reported.

5.1.3 Interference with transcription and genome replication

No compounds have yet been identified that directly inhibit the filoviral RNA-polymerase complex. Although the nucleoside analog ribavirin blocks transcription of a wide range of DNA and RNA viruses, including a number of nonsegmented negative-sense RNA agents, it does not inhibit filoviruses. Thus, treatment with ribavirin failed to alter the course of illness of Ebola- or Marburg-infected guinea pigs or monkeys infected with Ebola Zaire virus (Huggins *et al.*, 1999; Ignatiev *et al.*, 2000). The reason for its lack of activity against filoviruses is not clear.

Another approach that has been successfully employed to inhibit transcription and genome replication of some negative-strand RNA viruses makes use of antisense oligonucleotides complementary to sequences in viral genomic or mRNA. In the case of RSV, for example, inhibition was achieved by covalently linking the antisense molecule to a 2'-5'oligoadenylate sequence, so that binding activated cellular 2'-5' oligoadenylate-dependent RNase L, resulting in cleavage of the RNA target (Torrence *et al.*, 1997; Adah *et al.*, 2001). However, the same approach failed to demonstrate sequence-specific inhibition of Ebola Zaire virus replication (M Bray, P Torrence, unpublished data). Effective antisense therapy may require improved identification of critical regions of the viral genome that are accessible to oligonucleotide binding, stabilization of antisense molecules and development of methods to deliver them efficiently into infected cells.

5.1.4 Prevention of viral mRNA cap methylation

Production of mRNA for efficient translation in eukaryotes requires the attachment of a 5'-5'-linked guanosine cap to the 5' end of the growing mRNA strand, followed by methylation of the cap and adjacent nucleotides at one or more sites by (guanine-7-)methyltransferase. For the latter process, S-adenosylmethionine (SAM) is the methyl donor, and the reaction is driven forward by removal of the product, S-adenosylhomocysteine (SAH), through its hydrolysis by SAH hydrolase. Since the mRNA of many viruses is also capped, inhibitors of the latter enzyme can therefore indirectly block viral replication by causing the cytoplasmic SAH/SAM ratio to rise, impairing methylation (Cools and De Clercq, 1990; Oxenrider *et al.*, 1993).

A number of adenosine analogs have been identified that act through this mechanism to suppress the replication of a broad range of DNA and RNA viruses, including the filoviruses (De Clercq, 1987, 1998; Huggins *et al.*,

1999). Two of them, carbocyclic 3-deaza-adenosine (C-c3Ado) and 3-deazaneplanocin A (c3-Npc A), have *in vitro* 50% inhibitory concentrations (IC50) against Ebola Zaire virus of 30 and 2 μM, respectively. Both were highly effective in treating Ebola-infected mice when administered thrice daily, beginning on day −1, 0 or +1, but treatment begun on day 2 was less protective (Huggins *et al.*, 1999). Unexpectedly, equivalent or better results were obtained by treating mice once, with a somewhat larger, but nontoxic dose of either drug, on any day from 0 to +2 (Bray *et al.*, 2000). A single dose given as early as 2 days before virus challenge was partially protective.

At present, c3-Npc A is the most potent form of postexposure prophylaxis for Ebola Zaire infection in mice, since a single dose of 1 mg/kg prevents illness and death. However, it is clear that the mechanism of action involves more than just impaired cap methylation, since the beneficial effect can be eliminated by administering antibodies to murine IFN-α/β. As discussed below, further experiments showed that treatment causes massively increased production of IFN-α in infected, but not in uninfected animals (Bray *et al.*, 2002). Unfortunately, trials in nonhuman primates infected with Ebola Zaire virus did not result in either enhanced IFN production or mitigation of illness; the cause of this species-specific difference in drug effect is not known.

5.1.5   Interference with protein processing, viral assembly and exit

Pharmacologic blockade of a viral protease has proven to be a successful strategy for HIV therapy, but it does not appear to be a viable option for the filoviruses, since they do not encode a protease. On the other hand, host-cell enzymes that process viral proteins may constitute a therapeutic target. As noted above, cleavage of GP by host-cell cathepsins occurs during viral entry, and another proteolysis event occurs during maturation of the nascent protein, when GP is cleaved into GP1,2 by a cellular furin-like protease (Volchkov *et al.*, 1998). However, even if these steps are essential for filoviral maturation, it appears doubtful that effective treatment could be based entirely on inhibition of these ubiquitous cellular enzymes.

Other types of interactions between viral and host proteins may be more rewarding targets for antiviral attack. Current interest focuses on VP40, which appears to be the "driver" behind virion assembly. As described above, the expression of VP40 in plasmid-transfected cells is sufficient to cause the extrusion of filamentous VLPs from the cell surface (Harty *et al.*, 2000; Jasenosky *et al.*, 2001; Jasenosky and Kawaoka, 2004). Since VP40 interacts with specific components of the cellular protein-processing machinery, including Nedd4 and Tsg101, it is possible that a compound that interferes with these protein-protein interactions could block viral replication. Harty et al. used this approach to show that budding of vesicular stomatitis virus (VSV) is blocked by treatment with proteosome inhibitors (Harty *et al.*, 2001), but this strategy has not yet been tested for the filoviruses.

5.1.6   Destruction of virus infected cells

The expression of viral antigens on the surface of infected cells makes them vulnerable to destruction by the immune system. Antibodies to viral antigens can contribute to this process by enhancing complement activation and facilitating killing by natural killer cells. Although the virion GP is assumed to be the most important target for such antibodies, other virion components may also become accessible to immune recognition in the course of virion assembly and budding. Limited evidence of this was provided by the observation that mabs specific for the Marburg virus VP40 protein caused complement-mediated lysis of infected cells, even though they did not neutralize the virus *in vitro* (Razumov *et al.*, 1998). In a single reported experiment, these antibodies protected guinea pigs against lethal Marburg virus challenge.

Two other strategies that could be used to attack filovirus-infected cells also involve the use of antibodies that recognize antigens on the cell surface. The first is based on studies in the HIV field, which have shown that a variety of toxins coupled to antibodies against viral antigens ("immunotoxins") are capable of destroying virus-infected cells (Berger *et al.*, 1998; Pincus *et al.*, 2001). Recently this approach has been extended to include the labelling of monoclonal antibodies with an alpha-particle-emitting radionuclide, which has shown efficacy in eliminating infected cells in murine models of bacterial and fungal infection (Dadachova *et al.*, 2004).

The other approach to attacking infected cells takes advantage of the fact that the assembly and budding of new virions may result in the display of cellular molecules that are normally concealed from the immune system. One of these, phosphatidylserine, is the target of a monoclonal antibody that has shown promise in experimental cancer therapy (Ran *et al.*, 2005), and is now being tested for antiviral activity. One caveat regarding these approaches to filovirus therapy is that their potential ability to cause the simultaneous destruction of a vast number of infected host cells may result in severe tissue damage, particularly hepatotoxicity.

5.2   Modulation of host response

Detailed studies in mice and nonhuman primates are helping to define the mechanisms by which filoviruses cause rapidly lethal disease. Mouse experiments have shown that suppression of type I IFN responses plays a critical role in permitting viral dissemination. Studies in nonhuman primates, by contrast, have probably been most useful in revealing that inflammatory mediators released by virus-infected macrophages and dendritic cells are responsible for the vasodilatation, increased vascular permeability and DIC of filoviral hemorrhagic fever. As described below, this work paid off with the recent demonstration that Ebola-infected macaques could be protected from death through blockade of the extrinsic coagulation pathway. By revealing an interlocking relationship among viral replication, inflammation and coagulation, this work gives reason to hope that a number of approaches could be used to prevent the development of hemorrhagic fever in humans who are infected with Marburg or Ebola virus, but have not yet become ill.

5.2.1   Viral suppression of type I IFN responses

Since most, if not all viruses have acquired mechanisms of blocking or evading IFN in the course of their evolution, it is reasonable to assume that the filoviruses inhibit these responses in their unidentified reservoir species, as a means of prolonging infection and increasing the likelihood of transmission to new hosts.   One might therefore explain the extraordinary virulence of Marburg and Ebola virus for humans and nonhuman primates by proposing that their ability to suppress type I IFN responses proves "too successful" when the agents encounter the primate immune system, resulting in rapidly overwhelming infection (Bray, 2005).

Efforts to develop therapies that compensate for filovirus-induced suppression of type I IFN responses have so far been more successful in rodent models than in nonhuman primates. For example, pre- or postexposure prophylaxis of mice with the cross-species-active chimeric B/D form of human IFN-α prevents the development of illness after challenge with mouse-adapted Ebola Zaire virus (Bray, 2001). The IFN inducer polyICLC and synthetic oligodeoxynucleotides containing unmethylated CpG motifs, which stimulate macrophages to release IFN-α and other proinflammatory cytokines, are also effective for pre- or early postexposure prophylaxis in mice (Klinman *et al.*, 1999; Bray, 2001).

As noted, the induction of a strong type I IFN response is the mechanism by which the adenosine analogs c3-Npc A and C-c3Ado protect Ebola-infected mice. The quantity of circulating IFN induced by treatment increases progressively when the dose is delayed from day 0 to day 1 and day 2. Since they do not induce IFN production in normal mice, these data indicate that these compounds exert their effect only in virus-infected cells (Bray *et al.*, 2000, 2002). A likely mechanism of action is that inhibition of cellular SAH hydrolase results in impaired mRNA cap methylation, causing the accumulation of untranslated viral mRNA transcripts, some of which may associate with negative-sense genomic strands to form dsRNA that acts as a strong stimulus for IFN production.

This previously unrecognized effect of SAH hydrolase inhibitors may help to explain their broad-spectrum antiviral activity. Efforts are now under way to determine whether single-dose therapy with c3-Npc A is also effective against other RNA virus infections in rodents. Understanding the mechanism by which adenosine analogs induce IFN production in virus-infected, but not in uninfected cells, and the identification of compounds that can achieve such an effect in primates, would represent a genuine breakthrough in antiviral therapy, since treatment could selectively eliminate foci of viral replication without causing the systemic signs and symptoms associated with IFN therapy.

Despite considerable evidence that filoviruses suppress type I IFN responses in primate cells, little effort has been made to evaluate IFN treatment or other countermeasures in nonhuman primates. In the only reported attempt at IFN therapy, four rhesus macaques were given large daily doses of recombinant human IFN-α 2b, beginning immediately after challenge with Ebola Zaire virus, resulting in a 1-2-day delay in the onset of illness, viremia and death (Jahrling *et al.*, 1999). Since natural IFN responses include IFN-β and multiple subtypes of IFN-α, it is evident that a variety of strategies, including the administration of a mixture of IFNs or the use of IFN inducers such as polyICLC or CpG, might improve on the results of this pilot study.

5.2.2 Modulation of inflammation/coagulation

The fever, shock and hemorrhage that characterize lethal filovirus infection are caused by inflammatory mediators released from virus-infected macrophages and other cells (Bray and Mahanty, 2003; Bray and Geisbert, 2005). In Ebola-infected cynomolgus macaques, these processes develop so quickly that the animals become ill within 3 days after virus challenge and die on day 6-8, a course too rapid to permit development of antigen-specific immune responses. In humans, by contrast, the process is slower, such that death usually occurs during the second week of illness, providing additional time for the development of adaptive responses required for survival. Limited data from Africa suggest that the outcome of illness may in fact be determined soon after the entry of virus into the body, by the nature and timing of innate immune responses (Baize *et al.*, 1999, 2002). Taken together, these findings suggest that modulation of innate responses could slow viral dissemination and aid development of antigen-specific immunity.

A recent experiment demonstrated that such intervention can be surprisingly effective. Using the uniformly lethal model of Ebola Zaire virus infection in macaques, Geisbert *et al.*, treated 9 animals with daily doses of recombinant nematode anticoagulant factor C2 (rNAPc2), an inhibitor of TF-factor VIIa interaction, beginning either immediately after or the day following virus challenge (Geisbert *et al.*, 2003a). Even though rNAPc2 had no direct antiviral activity in cell culture, therapy was quite beneficial: three of the treated animals survived infection, while the average time to death of those that did become ill was significantly greater than that of placebo-treated animals. Treatment markedly reduced physical signs of coagulopathy and the levels of D-dimers and other markers of DIC in the blood, and also caused significant decreases in blood levels of the proinflammatory cytokines IL-6 and MCP-1. Mean peak circulating viral titers in surviving macaques were 100-fold lower than in the control group.

These wide-ranging effects of treatment with a substance that was only expected to block the extrinsic coagulation pathway have important implications for the development of effective immunomodulatory therapies. First, the results clearly demonstrate that host responses to filoviral replication are damaging, since triggering of the extrinsic coagulation pathway by infected macrophages in untreated macaques resulted in more severe disease than in animals in which that pathway was blocked. These findings also serve as a reminder that viral replication, inflammation and coagulation should not be thought of as independent processes. Instead, they form such a tightly interlinked matrix that the inhibition of coagulation can "feed back" to suppress the first two components. The mechanism of communication between the induction of coagulation and other macrophage activities may be signalling from the cytoplasmic tails of cell-surface TF molecules (Ruf, 2004). Since these responses begin with the first infected macrophage, it is obviously essential to characterize filovirus infections at that level.

Efforts to modify inflammatory responses in filovirus-infected animals have otherwise been limited to a series of small experiments performed in Russia, the results of which have yet to be confirmed by other investigators. Ignat'ev *et al.*, first reported that when Marburg-infected guinea pigs were treated with the compound desferoxamin (desferal), which allegedly blocks the induction of endothelial cell adhesion molecules by TNF-α, three of six animals survived infection (Ignat'ev *et al.*, 1996). Similarly, when five Marburg-infected guinea pigs were treated daily, beginning on day 3 postinfection, with anti-TNF-α serum, three of them survived, while all control animals died'(Ignat'ev, 2000). The same group recently reported that treatment of Marburg-infected guinea pigs with the licensed anti-TNF-α mab, Remicade®, worsened the outcome if begun soon after virus challenge, by shortening the mean time to death, while therapy begun on day 3 led to 50% survival (Ignat'ev, Bukin, and Otrashevskaia, 2004). If confirmed in larger, placebo-controlled experiments, these results would be in agreement with observations from clinical trials in septic shock, which suggest that proinflammatory cytokine responses play a protective role early in infection, but are harmful if they continue for too long or are expressed in an unregulated manner (Hotchkiss and Karl, 2003).

5.2.3 Strategies employing genetically-engineered viruses

In contrast to Marburg and Ebola virus, the vast majority of micro-organisms encountered by humans evoke innate responses that prevent the development of disease. The ability of the primate immune system to block most viral infections suggests that it might be possible to suppress a filovirus infection by "sending in" another microbe that induces protective responses in the same target cell population (Margolis, 2003). Current laboratory methods now make it possible to design an agent for this purpose.

An example of a genetically engineered virus that might be used in this manner is VSV, which is nonpathogenic for humans. A chimeric VSV, in which the GP gene is replaced by that of Ebola or Marburg virus, has proven highly effective as a conventional vaccine against these agents in laboratory animals (Garbutt *et al.*, 2004; Jones *et al.*, 2005). Since the vaccine virus displays a filoviral surface GP, it might also be effective if administered soon after filovirus infection, since the virions will attach to the same cell types targeted by the authentic pathogens, but their replication will evoke strong innate responses, including the production of large amounts of type I IFN, that could suppress filoviral replication in the same or neighbouring cells. Therapeutic efficacy could potentially be improved by engineering the gene encoding the VSV M protein to induce an enhanced IFN response (Stojdl *et al.*, 2003). Such a chimeric agent could thus act as both a drug and a vaccine, by inducing both immediate protection and long-term immunity in those who have been infected with a filovirus, but have not yet become ill.

6. TARGETS FOR FURTHER RESEARCH

The extreme virulence of Marburg and Ebola virus for humans poses a major challenge for the development of antiviral therapy. At present, it appears that the most easily achievable goal is the development of an effective form of postexposure prophylaxis. Because early, rapid viral dissemination and many features of the subsequent illness result from interactions between filoviruses and macrophages and dendritic cells, it will be essential to obtain a detailed understanding of these events at the molecular level.

As regards the development of direct inhibitors of filovirus replication, it is not possible at present to identify the approach that is most likely to lead to a licensed medication. Since the level of viremia in Ebola hemorrhagic fever patients is a critical determinant of survival or death, it is clearly reasonable to pursue the development of antibodies and lectins that bind to virus in the bloodstream and prevent its spread to new target cells. Because it would also be highly desirable to inhibit the filoviral RNA-dependent RNA polymerase complex, determination of the structures of its components and their mode of interaction is also an important scientific objective. Such efforts should follow leads provided by studies of related negative-strand RNA viruses that are easier to manipulate in the laboratory. Those attempting to devise means of preventing filoviruses from exploiting host protein-processing pathways for entry into cells and the assembly and release of new virions should also monitor the success of efforts to employ such strategies against HIV and other pathogens.

As regards the indirect suppression of Marburg and Ebola virus replication through modulation of host responses, several approaches have shown promise. It will be essential to learn in detail how filoviruses suppress type I IFN responses and the role that such inhibition plays in disease. In particular, the identification of compounds that stimulate the production of IFN by virus-infected primate cells, in the manner of adenosine analogues in mice, could lead to major progress in the treatment of severe viral infections. The recent discovery that inhibition of the extrinsic coagulation pathway has widespread effects, including suppression of viral replication, should also be exploited for the development of new therapeutic approaches. By "correcting" damaging host responses, such interventions may prove beneficial even in the absence of specific antiviral therapy.

REFERENCES

Adah, S. A., Bayly, S. F., Cramer, H., Silverman, R. H. and Torrence, P. F., 2001, Chemistry and biochemistry of 2',5'-oligoadenylate-based antisense strategy. *Curr Med Chem* **8**: 1189-212.

Aman, M. J., Bosio, C. M., Panchal, R. G., Burnett, J. C., Schmaljohn, A. and Bavari, S., 2003, Molecular mechanisms of filovirus cellular trafficking. *Microbes Infect* **5**: 639-49.

Baize, S., Leroy, E. M., Georges, A. J., Georges-Courbot, M. C., Capron, M., Bedjabaga, I., Lansoud-Soukate, J. and Mavoungou, E., 2002, Inflammatory responses in Ebola virus-infected patients. *Clin Exp Immunol* **128**:163-8.

Baize, S., Leroy, E. M., Georges-Courbot, M. C., Capron, M., Lansoud-Soukate, J., Debre, P., Fisher-Hoch, S. P., McCormick, J. B. and Georges, A. J., 1999, Defective humoral responses and extensive intravascular apoptosis are associated with fatal outcome in Ebola virus-infected patients. *Nat Med* **5**: 423-6.

Barrientos, L. G. and Gronenborn, A. M., 2005, The highly specific carbohydrate-binding protein cyanovirin-N: structure, anti-HIV/Ebola activity and possibilities for therapy. *Mini Rev Med Chem* **5**: 21-31.

Barrientos, L. G., O'Keefe, B. R., Bray, M., Sanchez, A., Gronenborn, A. M., and Boyd, M. R., 2003, Cyanovirin-N binds to the viral surface glycoprotein, GP1, 2 and inhibits infectivity of Ebola virus. *Antiviral Res* **58**: 47-56.

Basler, C. F. (2005). Interferon antagonists encoded by emerging RNA viruses. In "Modulation of Host Gene Expression and Innate Immunity by Viruses" (P. Palese, Ed.), pp. 197-220. Springer, Dordrecht, The Netherlands.

Basler, C. F., Mikulasova, A., Martinez-Sobrido, L., Paragas, J., Muhlberger, E., Bray, M., Klenk, H. D., Palese, P., and Garcia-Sastre, A., 2003, The Ebola virus VP35 protein inhibits activation of interferon regulatory factor 3. *J. Virol.* **77**: 7945-5796.

Basler, C. F., Wang, X., Muhlberger, E., Volchkov, V., Paragas, J., Klenk, H. D., Garcia-Sastre, A., and Palese, P., 2000, The Ebola virus VP35 protein functions as a type I IFN antagonist. *Proc. Natl. Acad. Sci. USA* **97**: 12289-12294.

Bavari, S., Bosio, C. M., Wiegand, E., Ruthel, G., Will, A. B., Geisbert, T. W., Hevey, M., Schmaljohn, C., Schmaljohn, A., and Aman, M. J., 2002, Lipid raft microdomains: a gateway for compartmentalized trafficking of Ebola and Marburg viruses. *J. Exp. Med.* **195**: 593-602.

Becker, S., Spiess, M., and Klenk, H. D., 1995, The asialoglycoprotein receptor is a potential liver-specific receptor for Marburg virus. *J. Gen. Virol.* **76**: 393-399.

Berger, E. A., Moss, B., and Pastan, I., 1998,. Reconsidering targeted toxins to eliminate HIV infection: you gotta have HAART. *Proc. Natl. Acad. Sci. USA* **95**: 11511-11513.

Boehmann, Y., Enterlein, S., Randolf, A., and Muhlberger, E., 2005, A reconstituted replication and transcription system for Ebola virus Reston and comparison with Ebola virus Zaire. *Virology* **332**: 406-417.

Bosio, C. M., Aman, M. J., Grogan, C., Hogan, R., Ruthel, G., Negley, D., Mohamadzadeh, M., Bavari, S., and Schmaljohn, A., 2003, Ebola and Marburg viruses replicate in monocyte-derived dendritic cells without inducing the production of cytokines and full maturation. *J. Infect. Dis.* **188**: 1630-1638.

Bosio, C. M., Moore, B. D., Warfield, K. L., Ruthel, G., Mohamadzadeh, M., Aman, M. J., and Bavari, S., 2004, Ebola and Marburg virus-like particles activate human myeloid dendritic cells. *Virology* **326**: 280-287.

Boyd, M. R., Gustafson, K. R., McMahon, J. B., Shoemaker, R. H., O'Keefe, B. R., Mori, T., Gulakowski, R. J., Wu, L., Rivera, M. I., Laurencot, C. M., Currens, M. J., Cardellina, J. H., 2nd, Buckheit, R. W., Jr., Nara, P. L., Pannell, L. K., Sowder, R. C., 2nd, and Henderson, L. E., 1997, Discovery of cyanovirin-N, a novel human immunodeficiency virus-inactivating protein that binds viral surface envelope glycoprotein gp120: potential applications to microbicide development. *Antimicrob. Agents Chemother.* **41**: 1521-1530.

Bray, M., 2001, The role of the Type I interferon response in the resistance of mice to filovirus infection. *J. Gen. Virol.* **82**: 1365-1373.

Bray, M., 2002, Filoviridae. In "Clinical *Virology*" (D. D. Richman, Whitley, R. J., Hayden, F. G., Ed.), pp. 875-890. ASM Press, Washington, D.C.

Bray, M., 2003, Defense against filoviruses used as biological weapons. *Antiviral Res* **57**: 53-60.

Bray, M., 2005, Pathogenesis of viral hemorrhagic fever. Curr Opin Immunol.

Bray, M., Davis, K., Geisbert, T., Schmaljohn, C., and Huggins, J., 1998, A mouse model for evaluation of prophylaxis and therapy of Ebola hemorrhagic fever. *J. Infect. Dis.* **178**: 651-661.

Bray, M., Driscoll, J., and Huggins, J. W., 2000, Treatment of lethal Ebola virus infection in mice with a single dose of an S-adenosyl-L-homocysteine hydrolase inhibitor. *Antiviral Res* **45**: 135-147.

Bray, M., and Geisbert, T. W., 2005, Ebola virus: The role of macrophages and dendritic cells in the pathogenesis of Ebola hemorrhagic fever. *Int. J. Biochem. Cel. Biol.* **37**: 1560-1566.

Bray, M., Hatfill, S., Hensley, L., and Huggins, J. W., 2001, Haematological, biochemical and coagulation changes in mice, guinea-pigs and monkeys infected with a mouse-adapted variant of Ebola Zaire virus. *J. Com.p Pathol.* 125: 243-253.

Bray, M., and Mahanty, S., 2003, Ebola hemorrhagic fever and septic shock. *J. Infect. Dis.* **188**:1613-1617.

Bray, M., and Paragas, J., 2002, Experimental therapy of filovirus infections. *Antiviral Res* **54**: 1-17.

Bray, M., Raymond, J. L., Geisbert, T., and Baker, R. O., 2002, 3-deazaneplanocin A induces massively increased interferon-alpha production in Ebola virus-infected mice. *Antiviral Res* **55**: 151-9.

Bwaka, M. A., Bonnet, M. J., Calain, P., Colebunders, R., De Roo, A., Guimard, Y., Katwiki, K. R., Kibadi, K., Kipasa, M. A., Kuvula, K. J., Mapanda, B. B., Massamba, M., Mupapa, K. D., Muyembe-Tamfum, J. J., Ndaberey, E., Peters, C. J., Rollin, P. E., and Van den Enden, E., 1999, Ebola hemorrhagic fever in Kikwit, Democratic Republic of the Congo: clinical observations in 103 patients. *J. Infect. Dis.* **179**: S1-7.

Chan, S. Y., Empig, C. J., Welte, F. J., Speck, R. F., Schmaljohn, A., Kreisberg, J. F., and Goldsmith, M. A., 2001, Folate receptor-alpha is a cofactor for cellular entry by Marburg and Ebola viruses. *Cell* **106**: 117-126.

Chandran, K., Sullivan, N. J., Felbor, U., Whelan, S. P., and Cunningham, J. M., 2005, Endosomal Proteolysis of the Ebola Virus Glycoprotein Is Necessary for Infection. *Science* **308**: 1643-1645.

Connolly, B. M., Steele, K. E., Davis, K. J., Geisbert, T. W., Kell, W. M., Jaax, N. K., and Jahrling, P. B., 1999, Pathogenesis of experimental Ebola virus infection in guinea pigs. *J. Infect. Dis.* **179** (Suppl 1): 203-217.

Cooley, L. A., and Lewin, S. R., 2003, HIV-1 cell entry and advances in viral entry inhibitor therapy. *J Clin Virol* **26**: 121-132.

Cools, M., and De Clercq, E., 1990, Influence of S-adenosylhomocysteine hydrolase inhibitors on S-adenosylhomocysteine and S-adenosylmethionine pool levels in L929 cells. *Biochem Pharmacol* **40**: 2259-2264.

Dadachova, E., Burns, T., Bryan, R. A., Apostolidis, C., Brechbiel, M. W., Nosanchuk, J. D., Casadevall, A., and Pirofski, L., 2004, Feasibility of radioimmunotherapy of experimental pneumococcal infection. *Antimicrob. Agents Chemother.* **48**: 1624-1629.

De Clercq, E., 1987, S-adenosylhomocysteine hydrolase inhibitors as broad-spectrum antiviral agents. *Biochem Pharmacol* **36**: 2567-2575.

De Clercq, E., 1998, Carbocyclic adenosine analogues as S-adenosylhomocysteine hydrolase inhibitors and antiviral agents: recent advances. *Nucleosides Nucleotides* **17**: 625-634.

Dowell, S. F., Mukunu, R., Ksiazek, T. G., Khan, A. S., Rollin, P. E., and Peters, C. J., 1999, Transmission of Ebola hemorrhagic fever: a study of risk factors in family members, Kikwit, Democratic Republic of the Congo, 1995. Commission de Lutte contre les Epidemies a Kikwit. *J. Infect. Dis.* **179** (Suppl 1):87-91.

Emond, R. T., Evans, B., Bowen, E. T., and Lloyd, G., 1977, A case of Ebola virus infection. *Br. Med. J.* **2**: 541-544.

Feldmann, H., Jones, S., Klenk, H. D., and Schnittler, H. J., 2003, Ebola virus: from discovery to vaccine. *Nat. Rev. Immunol.* **3**:677-685.

Feldmann, H., Volchkov, V. E., Volchkova, V. A., and Klenk, H. D., 1999, The glycoproteins of Marburg and Ebola virus and their potential roles in pathogenesis. *Arch. Virol.* (Suppl 15): 159-169.

Formenty, P., Hatz, C., Le Guenno, B., Stoll, A., Rogenmoser, P., and Widmer, A., 1999, Human infection due to Ebola virus, subtype Cote d'Ivoire: clinical and biologic presentation. *J. Infect. Dis.* **179** (Suppl 1): 48-53.

Garbutt, M., Liebscher, R., Wahl-Jensen, V., Jones, S., Moller, P., Wagner, R., Volchkov, V., Klenk, H. D., Feldmann, H., and Stroher, U., 2004, Properties of replication-competent vesicular stomatitis virus vectors expressing glycoproteins of filoviruses and arenaviruses. *J. Virol.* **78**: 5458-5465.

Geisbert, T. W., Hensley, L. E., Gibb, T. R., Steele, K. E., Jaax, N. K., and Jahrling, P. B., 2000, Apoptosis induced *in vitro* and *in vivo* during infection by Ebola and Marburg viruses. *Lab. Invest.* **80**: 171-186.

Geisbert, T. W., Hensley, L. E., Jahrling, P. B., Larsen, T., Geisbert, J. B., Paragas, J., Young, H. A., Fredeking, T. M., Rote, W. E., and Vlasuk, G. P., 2003a, Treatment of Ebola virus infection with a recombinant inhibitor of factor VIIa/tissue factor: a study in rhesus monkeys. *Lancet* **362**: 1953-1958.

Geisbert, T. W., Hensley, L. E., Larsen, T., Young, H. A., Reed, D. S., Geisbert, J. B., Scott, D. P., Kagan, E., Jahrling, P. B., and Davis, K. J., 2003b, Pathogenesis of Ebola hemorrhagic fever in cynomolgus macaques: evidence that dendritic cells are early and sustained targets of infection. *Am. J. Pathol.* **163**: 2347-2370.

Geisbert, T. W., Pushko, P., Anderson, K., Smith, J., Davis, K. J., and Jahrling, P. B., 2002, Evaluation in nonhuman primates of vaccines against Ebola virus. *Emerg. Infect. Dis.* **8**: 503-507.

Geisbert, T. W., Young, H. A., Jahrling, P. B., Davis, K. J., Larsen, T., Kagan, E., and Hensley, L. E., 2003c, Pathogenesis of Ebola hemorrhagic fever in primate models: evidence that hemorrhage is not a direct effect of virus-induced cytolysis of endothelial cells. *Am. J. Pathol.* **163**: 2371-2382.

Gibb, T. R., Bray, M., Geisbert, T. W., Steele, K. E., Kell, W. M., Davis, K. J., and Jaax, N. K., 2001, Pathogenesis of experimental Ebola Zaire virus infection in BALB/c mice. *J. Comp. Pathol.* **125**: 233-242.

Groseth, A., Feldmann, H., Theriault, S., Mehmetoglu, G., and Flick, R., 2005, RNA polymerase I-driven minigenome system for Ebola viruses. *J. Virol.* **79**: 4425-4433.

Gupta, M., Mahanty, S., Bray, M., Ahmed, R., and Rollin, P. E., 2001, Passive transfer of antibodies protects immunocompetent and imunodeficient mice against lethal Ebola virus infection without complete inhibition of viral replication. *J. Virol.* **75**: 4649-4654.

Gupta, M., Mahanty, S., Greer, P., Towner, J. S., Shieh, W. J., Zaki, S. R., Ahmed, R., and Rollin, P. E., 2004, Persistent infection with ebola virus under conditions of partial immunity. *J. Virol.* **78**: 958-967.

Harcourt, B. H., Sanchez, A., and Offermann, M. K., 1998, Ebola virus inhibits induction of genes by double-stranded RNA in endothelial cells. *Virology* **252**: 179-188.

Harcourt, B. H., Sanchez, A., and Offermann, M. K., 1999, Ebola virus selectively inhibits responses to interferons, but not to interleukin-1beta, in endothelial cells. *J. Virol.* **73**: 3491-346.

Hartman, A. L., Towner, J. S., and Nichol, S. T., 2004, A C-terminal basic amino acid motif of Zaire ebolavirus VP35 is essential for type I interferon antagonism and displays high identity with the RNA-binding domain of another interferon antagonist, the NS1 protein of influenza A virus. *Virology* **328**: 177-184.

Harty, R. N., Brown, M. E., McGettigan, J. P., Wang, G., Jayakar, H. R., Huibregtse, J. M., Whitt, M. A., and Schnell, M. J., 2001, Rhabdoviruses and the cellular ubiquitin-proteasome system: a budding interaction. *J. Virol.* **75**: 10623-1069.

Harty, R. N., Brown, M. E., Wang, G., Huibregtse, J., and Hayes, F. P., 2000, A PPxY motif within the VP40 protein of Ebola virus interacts physically and functionally with a ubiquitin ligase: implications for filovirus budding. *Proc. Natl. Acad. Sci. USA* **97**: 13871-13876.

Hensley, L. E., Young, H. A., Jahrling, P. B., and Geisbert, T. W., 2002, Proinflammatory response during Ebola virus infection of primate models: possible involvement of the tumor necrosis factor receptor superfamily. *Immunol. Lett.* **80**: 169-179.

Hotchkiss, R. S., and Karl, I. E., 2003, The pathophysiology and treatment of sepsis. *N.Engl. J. Med.* **348**: 138-150.

Huggins, J., Zhang, Z. X., and Bray, M., 1999, Antiviral drug therapy of filovirus infections: S-adenosylhomocysteine hydrolase inhibitors inhibit Ebola virus *in vitro* and in a lethal mouse model. *J. Infect. Dis.* **179** (Suppl 1): 240-247.

Ignat'ev, G., Steinkassrer, A., Strelsova, M., Atrasheuskaya, A., Agafonov, A., Lubitz, W., 2000, Experimental study on the possibility of treatment of some hemorrhagic fevers. *J. Biotechnol.* **83**: 67-76.

Ignat'ev, G. M., Bukin, E. K., and Otrashevskaia, E. V., 2004, [An experimental study of possibility of treatment of hemorrhagic fever Marburg by Remicade]. *Vestn. Ross. Akad. Med. Nauk.* **11**: 22-24.

Ignat'ev, G. M., Strel'tsova, M. A., Agafonov, A. P., Kashentseva, E. A., and Prozorovskii, N. S., 1996, [Experimental study of possible treatment of Marburg hemorrhagic fever with desferal, ribavirin, and homologous interferon]. *Vopr. Virusol.* **41**: 206-209.

Ignatiev, G. M., Dadaeva, A. A., Luchko, S. V., and Chepurnov, A. A., 2000, Immune and pathophysiological processes in baboons experimentally infected with Ebola virus adapted to guinea pigs. *Immunol. Lett.* **71**: 131-140.

Irie, T., Licata, J. M., and Harty, R. N., 2005, Functional characterization of Ebola virus L-domains using VSV recombinants. *Virology* **336**: 291-298.

Jahrling, P. B., Geisbert, J., Swearengen, J. R., Jaax, G. P., Lewis, T., Huggins, J. W., Schmidt, J. J., LeDuc, J. W., and Peters, C. J., 1996, Passive immunization of Ebola virus-infected cynomolgus monkeys with immunoglobulin from hyperimmune horses. *Arch. Virol.* Suppl 11: 135-140.

Jahrling, P. B., Geisbert, T. W., Dalgard, D. W., Johnson, E. D., Ksiazek, T. G., Hall, W. C., and Peters, C. J., 1990, Preliminary report: isolation of Ebola virus from monkeys imported to USA. *Lancet* **335**: 502-505.

Jahrling, P. B., Geisbert, T. W., Geisbert, J. B., Swearengen, J. R., Bray, M., Jaax, N. K., Huggins, J. W., LeDuc, J. W., and Peters, C. J., 1999, Evaluation of immune globulin and recombinant interferon-alpha2b for treatment of experimental Ebola virus infections. *J. Infect. Dis.* **179** (Suppl 1): 224-234.

Jasenosky, L. D., and Kawaoka, Y., 2004, Filovirus budding. *Virus Res.* **106**: 181-188.

Jasenosky, L. D., Neumann, G., Lukashevich, I., and Kawaoka, Y., 2001, Ebola virus VP40-induced particle formation and association with the lipid bilayer. *J. Virol.* **75**: 5205-5214.

Ji, X., Olinger, G.G., Aris, S., Chen, Y., Gewurz, H., and Spear, G.T., 2005, Mannose-binding lectin binds to Ebola and Marburg envelope glycoproteins, resulting in blocking of virus interaction with DC-SIGN and complement-mediated virus neutralization. *J Gen Virol* **86**: 2535-42.

Jones, S. M., Feldmann, H., Stroher, U., Geisbert, J. B., Fernando, L., Grolla, A., Klenk, H. D., Sullivan, N. J., Volchkov, V. E., Fritz, E. A., Daddario, K. M., Hensley, L. E., Jahrling, P. B., and Geisbert, T. W., 2005, Live attenuated recombinant vaccine protects nonhuman primates against Ebola and Marburg viruses. *Nat. Med.*.**11**: 786-790.

Katze, M. G., He, Y., and Gale, M., Jr., 2002, Viruses and interferon: a fight for supremacy. *Nat.Rev. Immunol.* **2**: 675-687.

Kobinger, G. P., Weiner, D. J., Yu, Q. C., and Wilson, J. M., 2001, Filovirus-pseudotyped lentiviral vector can efficiently and stably transduce airway epithelia *in vivo. Nat. Biotechnol.* **19**: 225-230.

Ksiazek, T. G., Rollin, P. E., Williams, A. J., Bressler, D. S., Martin, M. L., Swanepoel, R., Burt, F. J., Leman, P. A., Khan, A. S., Rowe, A. K., Mukunu, R., Sanchez, A., and Peters, C. J., 1999, Clinical *Virology* of Ebola hemorrhagic fever (EHF): virus, virus antigen, and IgG and IgM antibody findings among EHF patients in Kikwit, Democratic Republic of the Congo, 1995. *J. Infect. Dis.* **179** (Suppl 1): 177-187.

Kudoyarova-Zubavichene, N. M., Sergeyev, N. N., Chepurnov, A. A., and Netesov, S. V., 1999, Preparation and use of hyperimmune serum for prophylaxis and therapy of Ebola virus infections. *J. Infect. Dis.* **179** (Suppl 1): 218-223.

Lasala, F., Arce, E., Otero, J. R., Rojo, J., and Delgado, R., 2003, Mannosyl glycodendritic structure inhibits DC-SIGN-mediated Ebola virus infection in cis and in trans. *Antimicrob. Agents Chemother.* **47**: 3970-3972.

Leroy, E. M., Baize, S., Debre, P., Lansoud-Soukate, J., and Mavoungou, E., 2001, Early immune responses accompanying human asymptomatic Ebola infections. *Clin. Exp. Immunol.* **124**: 453-460.

Leroy, E. M., Baize, S., Volchkov, V. E., Fisher-Hoch, S. P., Georges-Courbot, M. C., Lansoud-Soukate, J., Capron, M., Debre, P., McCormick, J. B., and Georges, A. J., 2000, Human asymptomatic Ebola infection and strong inflammatory response. *Lancet* **355**: 2210-2215.

Licata, J. M., Johnson, R. F., Han, Z., and Harty, R. N., 2004, Contribution of ebola virus glycoprotein, nucleoprotein, and VP24 to budding of VP40 virus-like particles. *J. Virol.* **78**: 7344-7351.

Mahanty, S., and Bray, M., 2004, Pathogenesis of filoviral haemorrhagic fevers. *Lancet Infect. Dis.* **4**: 487-498.

Mahanty, S., Gupta, M., Paragas, J., Bray, M., Ahmed, R., and Rollin, P. E., 2003a, Protection from lethal infection is determined by innate immune responses in a mouse model of Ebola virus infection. *Virology* **312**: 415-424.

Mahanty, S., Hutchinson, K., Agarwal, S., McRae, M., Rollin, P. E., and Pulendran, B., 2003b, Cutting edge: impairment of dendritic cells and adaptive immunity by Ebola and Lassa viruses. *J. Immunol.* **170**: 2797-2801.

Manicassamy, B., Wang, J., Jiang, H., and Rong, L., 2005, Comprehensive analysis of ebola virus GP1 in viral entry. *J. Virol.* **79**: 4793-4805.

Margolis, L. (2003). Cytokines--strategic weapons in germ warfare? Nat Biotechnol 21(1), 15-6.

Martini, G. A., 1973, Marburg virus disease. *Postgrad Med. J.* **49**:, 542-546.

Martini, G. A., Knauff, H. G., Schmidt, H. A., Mayer, G., and Baltzer, G., 1968, A hitherto unknown infectious disease contracted from monkeys. "Marburg-virus" disease. *Ger. Med. Mon.* **13**: 457-470.

Maruyama, T., Parren, P. W., Sanchez, A., Rensink, I., Rodriguez, L. L., Khan, A. S., Peters, C. J., and Burton, D. R., 1999a, Recombinant human monoclonal antibodies to Ebola virus. *J. Infect. Dis.* **179** (Suppl 1): 235-239.

Maruyama, T., Rodriguez, L. L., Jahrling, P. B., Sanchez, A., Khan, A. S., Nichol, S. T., Peters, C. J., Parren, P. W., and Burton, D. R., 1999b, Ebola virus can be effectively neutralized by antibody produced in natural human infection. *J. Virol.* **73**: 6024-6030.

Marzi, A., Gramberg, T., Simmons, G., Moller, P., Rennekamp, A. J., Krumbiegel, M., Geier, M., Eisemann, J., Turza, N., Saunier, B., Steinkasserer, A., Becker, S., Bates, P., Hofmann, H., and Pohlmann, S., 2004, DC-SIGN and DC-SIGNR interact with the glycoprotein of Marburg virus and the S protein of severe acute respiratory syndrome coronavirus. *J. Virol.* **78**: 12090-12095.

Miranda, M. E., Ksiazek, T. G., Retuya, T. J., Khan, A. S., Sanchez, A., Fulhorst, C. F., Rollin, P. E., Calaor, A. B., Manalo, D. L., Roces, M. C., Dayrit, M. M., and Peters, C. J., 1999, Epidemiology of Ebola (subtype Reston) virus in the Philippines, 1996. *J. Infect. Dis.* **179** (Suppl 1): 115-119.

Muhlberger, E., Lotfering, B., Klenk, H. D., and Becker, S., 1998, Three of the four nucleocapsid proteins of Marburg virus, NP, VP35, and L, are sufficient to mediate replication and transcription of Marburg virus-specific monocistronic minigenomes. *J. Virol.* **72**: 8756-8764.

Muhlberger, E., Weik, M., Volchkov, V. E., Klenk, H. D., and Becker, S., 1999, Comparison of the transcription and replication strategies of marburg virus and Ebola virus by using artificial replication systems. *J. Virol.* **73**: 2333-2342.

Mupapa, K., Mukundu, W., Bwaka, M. A., Kipasa, M., De Roo, A., Kuvula, K., Kibadi, K., Massamba, M., Ndaberey, D., Colebunders, R., and Muyembe-Tamfum, J. J., 1999, Ebola hemorrhagic fever and pregnancy. *J. Infect. Dis.* **179** (Suppl 1): 11-12.

Nardin, A., Sutherland, W. M., Hevey, M., Schmaljohn, A., and Taylor, R. P., 1998, Quantitative studies of heteropolymer-mediated binding of inactivated Marburg virus to the complement receptor on primate erythrocytes. *J. Immunol.Methods* **211**: 21-31.

Nikiforov, V. V., Turovskii Iu, I., Kalinin, P. P., Akinfeeva, L. A., Katkova, L. R., Barmin, V. S., Riabchikova, E. I., Popkova, N. I., Shestopalov, A. M., Nazarov, V. P., and *et al.*, 1994, [A case of a laboratory infection with Marburg fever]. *Zh. Mikrobiol. Epidemiol. Immunobiol.* **3**: 104-106.

O'Keefe, B. R., Smee, D. F., Turpin, J. A., Saucedo, C. J., Gustafson, K. R., Mori, T., Blakeslee, D., Buckheit, R., and Boyd, M. R., 2003, Potent anti-influenza activity of cyanovirin-N and interactions with viral hemagglutinin. *Antimicrob. Agents Chemother.* **47**: 2518-2525.

Oxenrider, K. A., Bu, G., and Sitz, T. O., 1993, Adenosine analogs inhibit the guanine-7-methylation of mRNA cap structures. *FEBS Lett.* **316**: 273-277.

Panchal, R. G., Ruthel, G., Kenny, T. A., Kallstrom, G. H., Lane, D., Badie, S. S., Li, L., Bavari, S., and Aman, M. J., 2003, *In vivo* oligomerization and raft localization of Ebola virus protein VP40 during vesicular budding. *Proc. Natl. Acad. Sci. USA* **100**: 15936-15941.

Parren, P. W., Geisbert, T. W., Maruyama, T., Jahrling, P. B., and Burton, D. R., 2002, Pre- and postexposure prophylaxis of Ebola virus infection in an animal model by passive transfer of a neutralizing human antibody. *J. Virol.* **76**: 6408-6412.

Peterson, A. T., Carroll, D. S., Mills, J. N., and Johnson, K. M., 2004, Potential mammalian filovirus reservoirs. *Emerg- Infect .Dis.* **10**: 2073-2081.

Pincus, S. H., Marcotte, T. K., Forsyth, B. M., and Fang, H. (2001). *In vivo* testing of anti-HIV immunotoxins. Methods Mol Biol 166, 277-94.

Pourrot, X., Kumulungui, B., Wittmann, T., Moussavou, C., Delicat, A., Yaba, P., Nkoghe, D., Gonzalez, J.P., and Leroy, E.M., 2005, The natural history of Ebola virus in Africa. *Microbes Infect* 7. 1005-14.

Ran, S., He, J., Huang, X., Soares, M., Scothorn, D., and Thorpe, P. E., 2005, Antitumor effects of a monoclonal antibody that binds anionic phospholipids on the surface of tumor blood vessels in mice. *Clin. Cancer Res.* 11: 1551-1562.

Razumov, I. A., Belanov, E. F., Bukreev, A. A., and Kazachinkskaia, E. I., 1998 [Monoclonal antibodies to Marburg virus proteins and their immunochemical characteristics]. *Vopr. Virusol.* 43: 274-279.

Reed, D. S., Hensley, L. E., Geisbert, J. B., Jahrling, P. B., and Geisbert, T. W., 2004, Depletion of peripheral blood T lymphocytes and NK cells during the course of ebola hemorrhagic Fever in cynomolgus macaques. *Viral Immunol.* 17: 390-400.

Rojo, J., and Delgado, R., 2004, Glycodendritic structures: promising new antiviral drugs. *J. Antimicrob. Chemother.* 54: 579-581.

Rouquet, P., Froment, J. M., Bermejo, M., Yaba, P., Delicat, A., Rollin, P. E., and Leroy, E. M., 2005, Wild animal mortality monitoring and human Ebola outbreaks, Gabon and Republic of Congo, 2001-2003. *Emerg. Infect. Dis.* 11: 283-290.

Ruf, W., 2004, Emerging roles of tissue factor in viral hemorrhagic fever. *Trends Immunol.* 25: 461-4.

Sadek, R. F., Khan, A. S., Stevens, G., Peters, C. J., and Ksiazek, T. G., 1999, Ebola hemorrhagic fever, Democratic Republic of the Congo, 1995: determinants of survival. *J. Infect. Dis.* 179 (Suppl 1): 24-47.

Sanchez, A., Khan, A., Zaki, S., Nabel, G., Ksiazek, T., Peters, C.J., 2001, Filoviridae: Marburg and Ebola viruses. In "Fields *Virology*" (D. Knipe, Howley, P., Ed.), pp. 1279-1304. Lippincott Williams and Wilkins, Philadelphia.

Sanchez, A., Lukwiya, M., Bausch, D., Mahanty, S., Sanchez, A. J., Wagoner, K. D., and Rollin, P. E., 2004, Analysis of human peripheral blood samples from fatal and nonfatal cases of Ebola (Sudan) hemorrhagic fever: cellular responses, virus load, and nitric oxide levels. *J. Virol.* 78: 10370-10377.

Simmons, G., Rennekamp, A. J., Chai, N., Vandenberghe, L. H., Riley, J. L., and Bates, P., 2003, Folate receptor alpha and caveolae are not required for Ebola virus glycoprotein-mediated viral infection. *J. Virol.* 77: 13433-13438.

Stojdl, D. F., Lichty, B. D., tenOever, B. R., Paterson, J. M., Power, A. T., Knowles, S., Marius, R., Reynard, J., Poliquin, L., Atkins, H., Brown, E. G., Durbin, R. K., Durbin, J. E., Hiscott, J., and Bell, J. C., 2003, VSV strains with defects in their ability to shutdown innate immunity are potent systemic anti-cancer agents. *Cancer Cell* 4: 263-275.

Takada, A., Fujioka, K., Tsuiji, M., Morikawa, A., Higashi, N., Ebihara, H., Kobasa, D., Feldmann, H., Irimura, T., and Kawaoka, Y., 2004, Human macrophage C-type lectin specific for galactose and N-acetylgalactosamine promotes filovirus entry. *J. Virol.* 78: 2943-2947.

Takada, A., Robison, C., Goto, H., Sanchez, A., Murti, K. G., Whitt, M. A., and Kawaoka, Y., 1997, A system for functional analysis of Ebola virus glycoprotein. *Proc. Natl. Acad. Sci. USA* 94: 14764-14769.

Theriault, S., Groseth, A., Neumann, G., Kawaoka, Y., and Feldmann, H., 2004, Rescue of Ebola virus from cDNA using heterologous support proteins. *Virus Res.* 106: 43-50.

Torrence, P. F., Xiao, W., Li, G., Cramer, H., Player, M. R., and Silverman, R. H., 1997, Recruiting the 2-5A system for antisense therapeutics. *Antisense Nucleic Acid Drug Dev.* 7: 203-206.

Towner, J. S., Paragas, J., Dover, J. E., Gupta, M., Goldsmith, C. S., Huggins, J. W., and Nichol, S. T., 2005, Generation of eGFP expressing recombinant Zaire ebolavirus for analysis of early pathogenesis events and high-throughput antiviral drug screening. *Virology* **332**: 20-27.

Towner, J. S., Rollin, P. E., Bausch, D. G., Sanchez, A., Crary, S. M., Vincent, M., Lee, W. F., Spiropoulou, C. F., Ksiazek, T. G., Lukwiya, M., Kaducu, F., Downing, R., and Nichol, S. T., 2004, Rapid diagnosis of Ebola hemorrhagic fever by reverse transcription-PCR in an outbreak setting and assessment of patient viral load as a predictor of outcome. *J. Virol.* **78**: 4330-4341.

Volchkov, V. E., Becker, S., Volchkova, V. A., Ternovoj, V. A., Kotov, A. N., Netesov, S. V., and Klenk, H. D., 1995, GP mRNA of Ebola virus is edited by the Ebola virus polymerase and by T7 and vaccinia virus polymerases. *Virology* **214**: 421-430.

Volchkov, V. E., Feldmann, H., Volchkova, V. A., and Klenk, H. D., 1998, Processing of the Ebola virus glycoprotein by the proprotein convertase furin. *Proc. Natl. Acad. Sci. USA* **95**: 5762-5767.

Watanabe, S., Takada, A., Watanabe, T., Ito, H., Kida, H., and Kawaoka, Y., 2000, Functional importance of the coiled-coil of the Ebola virus glycoprotein. *J. Virol.* **74**: 10194-10201.

Weber, F., Kochs, G., and Haller, O., 2004, Inverse interference: how viruses fight the interferon system. *Viral Immunol.* **17**: 498-515.

Weik, M., Modrof, J., Klenk, H. D., Becker, S., and Muhlberger, E., 2002, Ebola virus VP30-mediated transcription is regulated by RNA secondary structure formation. *J. Virol.* **76**: 8532-8539.

Wilson, J. A., Bosio, C. M., and Hart, M. K., 2001, Ebola virus: the search for vaccines and treatments. *Cell Mol .Life Sci.* **58**: 1826-1841.

Wilson, J. A., Hevey, M., Bakken, R., Guest, S., Bray, M., Schmaljohn, A. L., and Hart, M. K., 2000, Epitopes involved in antibody-mediated protection from Ebola virus. *Science* **287**: 1664-1666.

Wool-Lewis, R. J., and Bates, P., 1998, Characterization of Ebola virus entry by using pseudotyped viruses: identification of receptor-deficient cell lines. *J. Virol.* **72**: 3155-3160.

Yonezawa, A., Cavrois, M., and Greene, W. C., 2005, Studies of ebola virus glycoprotein-mediated entry and fusion by using pseudotyped human immunodeficiency virus type 1 virions: involvement of cytoskeletal proteins and enhancement by tumor necrosis factor alpha. *J. Virol.* **79**: 918-926.

4. General concepts of therapy

Chapter 4.1

PROTEASOME INHIBITORS AS COMPLEMENTARY OR ALTERNATIVE ANTIVIRAL THERAPEUTICS

S. PRÖSCH[1] and M. KASPARI[1]

[1] *Institute of Virology, Charité – University Medicine Berlin, Schumannstr. 20/21, 10117 Berlin, Germany*

Abstract: The ubiquitine-proteasome system (UPS) catalyses the orderly degradation of abnormal or short-lived regulatory proteins and thus is critical for cell cycle regulation, signal transduction, transcription regulation, protein sorting, apoptosis as well as for the generation of antigenic peptides for the T-cell mediated immune response. Numerous studies suggest that efficient replication of many, if not all, viruses essentially depends on a functional UPS. Consequently proteasome inhibitors (PI) have been found to interfere with efficient replication of many viruses. The mechanisms by which the UPS is involved in regulation of replication of individual viruses are different. Depending on the virus the UPS has been shown to be important for entry, transport, gene expression, assembly or egress of the virus. The introduction of proteasome inhibitors in the clinical praxis as anti-cancer or anti-inflammatory drugs now rise the question whether these compounds present a new class of supplementary or even alternative antiviral drugs. The following article will give an short overview on the structure and function of the UPS, the currently available PI and the various aspects of virus UPS interaction. Finally the potential of different PI as antiviral approaches for the treatment of virus infections is discussed.

1.   INTRODUCTION

Viruses are obligate intracellular parasites and thus widely depend on the cellular machinery for replication and transcription as well as protein and energy metabolism. Options for antiviral therapy are limited to drugs

455

E. Bogner and A. Holzenburg (eds.), New Concepts of Antiviral Therapy, 455–478.

targeting virus-encoded enzymes like DNA- or RNA-polymerases which control replication of the viral genome (e.g. inhibitors of herpesvirus polymerase, of HIV reverse transcriptase or RNA-dependent RNA polymerases), proteases which catalyse processing of viral (poly)proteins (e.g inhibitors of HIV protease) or viral proteins which are involved in uncoating and viral egress (e.g. inhibitors of Influenza virus neuraminidase or M2/hemagglutinin fusion). These drugs are highly specific for certain viruses or virus families and show a satisfying overall in vivo tolerance. However, prophylaxis or even broad therapeutic treatment over a long period of time favours evolvement of drug resistant mutants, particularly when the drug dose or the drug itself is suboptimal and reduces virus replication only marginally. The molecular basis for this phenomenon is a high mutation rate of viral genomes (especially of RNA viruses) combined with a high reproduction rate leading to millions of progeny viruses in one generation. Consequently, drug-mediated modulation of specific cellular pathways essentially involved in virus replication represents an attractive complementary or even alternative strategy to block virus multiplication. Cellular genes have much lower mutation rates so that the problem of drug-resistance should be limited. However, drugs modulating cellular pathways may have a higher toxicity and thus undesirable side effects could limit their in vivo utility.

The ubiquitin-proteasome system (UPS) is the central cellular mediator of regulated protein turnover in living cells. Apart from degradation of mis-folded proteins, the proteasome also functions in defined processing of inactive precursor proteins (Ciechanover and Schwartz, 1998). Furthermore, by catalysing generation of immunogenic peptides for MHC-I presentation, the UPS plays an important role for the MHC-I-restricted cellular immune response against viruses (Kloetzel, 2004). Moreover, the UPS is essentially involved in regulation of transcription, signalling, cell-cycle, apoptosis and protein sorting which represent crucial processes for viral replication. Therefore, intervention into these elementary cellular processes may have more or less severe consequences for replication of certain viruses. Therefore, it seemed legitimate to investigate whether proteasome inhibitors (PI) are useful complementary or alternative therapeutics for treatment of viral infections for which no or only suboptimal drugs are currently available. A new impetus for these studies was the development of pharmacologically potent PI like PS-341 (Bortezomib) which recently has been introduced in clinical use as an anti-cancer drug in relapsed myeloma patients that are refractory to conventional chemotherapy (Adams, 2002). The clinical data support that partial and time-limited inhibition of the UPS is a calculable option for therapeutic intervention, even though certain adverse and toxic effects depending on the dose and the individual patient situation might occur (Richardson *et al.*, 2003).

The following article will give a short overview concerning the structure of the UPS, its function in cellular metabolism and in replication of certain viruses. Furthermore, currently available PI for research and clinical practice as well as their influence on viral replication will be addressed.

2. THE UBIQUITIN-PROTEASOME SYSTEM AND ITS FUNCTIONS IN CELLULAR METABOLISM AND IMMUNE RESPONSE

2.1 Structure of the ubiquitin-proteasome system

The ubiquitin/adenosine triphosphate (ATP)-dependent proteolytic pathway comprises a cascade of enzymes that first catalyse poly-ubiquitinylation of substrates – ubiquitin-activating enzyme (E1), ubiquitin-conjugating enzyme (E2), ubiquitin ligases (E3) and ubiquitin chain-assembly factor (E4). The poly-ubiquitinylated proteins are finally recognised by the proteasome for degradation or processing (for details see Fig. 1, Ciechanova and Schwartz, 1998). Additionally, the UPS catalyses mono-ubiquitinylation of certain proteins. Mono-ubiquitinylation has been shown to regulate protein function and serves as a signal in protein sorting and membrane receptor recycling but does not target the protein for proteasomal degradation (Fig.1). Interestingly, processes depending on mono-ubiquitinylation of proteins can be suppressed by PI indicating that they directly or indirectly require proteasome activity (Dupre *et al.*, 2001). present in the cytoplasm and the nucleus of all eukaryotic cells. There are two functionally distinct proteasome types – the 20S proteasome and the 26S proteasome. The barrel-shaped 20S proteasome comprises four stacked rings each composed of 7 α or β homodimeric subunits that enclose a central chamber. The two central β-rings harbour three main proteolytic activities, a chymotrypsin-like (CT-L) activity that cleaves after large hydrophobic residues, a trypsin-like (T-L) activity that cleaves after acidic residues and a peptidylglutamyl-peptide hydrolysing (PGPH) or caspase-like activity that also cleaves after acidic residues. There are two further activities that cleave after branched-chain residues or small neutral amino acids. The primary sequence of these proteolytically active β-subunits and their mechanisms of action differ from those of other known cellular proteases.

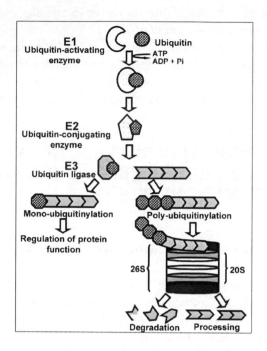

Figure 1: The ubiquitin-proteasome system (UPS). The ubiquitin-proteasome system involves a cascade of enzymes which catalyse mono- or poly-ubiquitinylation of substrates starting with ubiquitin-activating enzyme (E1), followed by ubiquitin-conjugating enzymes (E2) and ubiquitin ligases (E3) which finally mediate covalent conjugation of ubiquitin to the target protein. Repeated cycles of ubiquitin conjugation result in generation of insoluble poly-ubiquitinylated proteins which are then recognised by the proteasome as mis-folded and damaged proteins and are degraded or processed. Eukaryotic cells contain one E1, several E2 and more than 100 different E3 enzymes. The E3 enzymes are suggested to determine the specificity of ubiquitinylation. Mono-ubiquitinylation of proteins is associated with changes in protein function and plays a role in protein transport and sorting, as well as transcription regulation.

Proteasomes are highly conserved multimeric peptidase complexes that are The β -subunits are flanked by two α-rings that guide the entrance of unfolded polypeptide chains into the 20S core (Arendt and Hochstrasser, 1997; Baumeister *et al.*, 1998; Bochtler *et al.*, 1999).

The 26S proteasome is composed of the 20S core particle and two additional 19S regulatory complexes that associate with the outer α-rings and contain the substrate recognition and binding sites. The 19S complex harbours six ATPases and catalyses de-polymerisation of poly-ubiquitin chains as well as unfolding of the protein substrate thus facilitating its entry into the 20S proteasome (Braun *et al.*, 1999; Ferrell *et al.*, 2000). A minimum of four ubiquitin molecules is required for recognition by the

26S proteasome. The ubiquitin chains are then removed from the substrate by de-ubiquitinylating enzymes and re-enter the pool of free ubiquitin.

The 20S proteasome can also associate with the proteasome activator 28/11S regulator (PA28) in an ATP-independent manner thus allowing degradation of non-ubiquitinylated proteins (Baumeister *et al.*, 1998; Rechsteiner *et al.*, 2000; Hoffman *et al.*, 1992). PA28 activates the 20S proteasome by induction of conformational changes which presumably open the core maximally and thus facilitate substrate entry and cleavage product exit without affecting the active sites of the proteases (Stohwasser *et al.*, 2000; Hill *et al.*, 2002). In vivo, the 20S proteasome also forms so-called hybrid proteasomes with one 19S and one PA28 regulatory subunit. However, these hybrid proteasomes are as active as the 19S-20S-19S proteasome (Kopp *et al.*, 2001; Tanahashi *et al.*, 2000).

In the presence of interferon (IFN)-γ (or other cytokines like TNFα and IFN-β) the catalytic β-subunits of the proteasome are replaced by the subunits LMP-2, LMP-7 and MECL-1 thus generating the immuno-proteasome. The immuno-proteasome plays a central role for the proteolytic generation of antigenic peptides for presentation via MHC class I molecules (Kloetzel, 2004). These IFN-inducible subunits modify the cleavage specificity profile of the proteasome by altered cleavage site preference and enhanced generation of specific peptides.

2.2 Function of the proteasome in cellular metabolism

Originally, the proteasome was discovered due to its function in degradation of aged or misfolded proteins into small peptides of 3-20 amino acid residues. However, the proteasome also controls specific degradation of regulatory proteins and processing of inactive precursor proteins.

The UPS is involved in regulation of key cellular processes like cell-cycle progression, transcription, translation, signal transduction, stress response, apoptosis, receptor function and protein sorting by catalysing degradation, processing or translocation of several proteins involved in regulation of these processes, e.g. cyclins and cyclin-dependent kinases, p53, c myc, bcl-2, Bax, Nuclear factor (NF)-κB precursor p100 and p105, I-κB (Inhibitor of NF-κB), JNK and c-jun (Rolfe *et al.*, 1997; Salghetti *et al.*, 1999; Breitschopf *et al.*, 1999; Strous and Govers, 1999; Longva *et al.*, 2002; Baldwin, 1996; Karin and Ben-Neriah, 2000; Dupre *et al.*, 2001; Adams, 2002; Stahl and Barbieri, 2002; Beinke and Ley, 2004). Additionally, the proteasome plays an important role in the cellular immune response by generating peptides for presentation via MHC class I molecules (Kloetzel, 2004).

The most intensively studied proteasome substrate is the transcription factor NF-κB which is crucial for transcription regulation of genes involved in pro-inflammatory processes (e.g. TNF-α, IL-1 and -6, MIP-1α, RANTES, E-selectin and VCAM-1), immune response (e.g. T-cell activation by up-regulation of MHC molecules and CD80/86 on antigen presenting cells), anti-apoptosis (e.g. c.IAP-1/2, AI, Bcl-2 and BCl-XL) and cell cycle progression (cell-cycle regulator cyclin D1) (Zhang and Ghosh, 2001, Li and Verma, 2002; Karin and Li, 2002; Karin et al., 2002). Inhibition of NF-κB activation therefore represents the main mechanism by which PI induce apoptosis and block cell cycle progression or inhibit inflammatory processes.

The most dominant form of NF-κB is a heterodimer consisting of p50 and p65. The p50/p65 heterodimer remains in an inactive state by association with its inhibitors IκBα or IkBβ in the cytoplasm. Various stimuli are able to trigger activation of NF-κB by induction of proteasome-dependent IκB degradation following its phosphorylation by the IKK complex and ubiquitinylation by the SCF (Skp1/Cul1/F-box) E3 ubiquitin ligase complex (Baldwin, 1996; Karin and Ben-Neriah, 2000). The active p50/p65 heterodimer is then translocated into the nucleus where it binds to the promoters of responsive genes. Apart from IκB degradation, the 26S proteasome also controls processing of the NF-κB p105 and p100 precursors to generate the NF-κB p52 and p50 subunits. Furthermore, p50 can be retained in the cytoplasm by binding to its precursor p105. Following stimulation of the cells p105 can be completely degraded by the UPS allowing translocation of p50 into the nucleus (Beinke and Ley, 2004).

3. PROTEASOME INHIBITORS

In the last few years, various low-molecular-weight compounds have been identified that more or less selectively inhibit the UPS through interaction with the active site subunits of the 20S proteasome. These compounds are either synthetic products or naturally occurring substances; the most relevant compounds are listed in Tab. 1. Synthetic PI consist of linear di- or tripeptides linked to different pharmacophores such as benzamides, α-ketoamides, aldehydes, α-ketoaldehydes, vinyl sulfones or boronic acids (Fig. 2). Natural product PI display a broad structural heterogeneity containing both linear as well as non-linear scaffold structures (Fig. 3).

In the early 1980s, serine/cysteine protease and calpain inhibitors belonging to the family of peptide aldehydes were initially discovered to interfere with T-L and CT-L activity of the 20S proteasome, respectively (Wilk et al., 1983; Figueiredo-Pereira et al., 1994). Today, a variety of peptide aldehydes with higher potency and increased selectivity towards the 20S proteasome like MG132, MG115 and PSI have been developed. Binding of peptide aldehydes to the active site β-subunits results in formation of a reversible hemiacetal adduct between the aldehyde group and the hydroxyl group of the N-terminal threonine residue. Due to the highly reactive functional aldehyde group, cross-reactivity with cellular proteases remains a major drawback of this class of inhibitors that limits their in vivo utility.

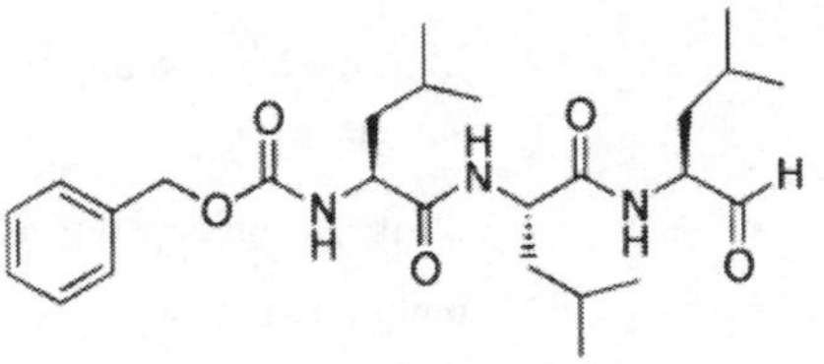

MG132

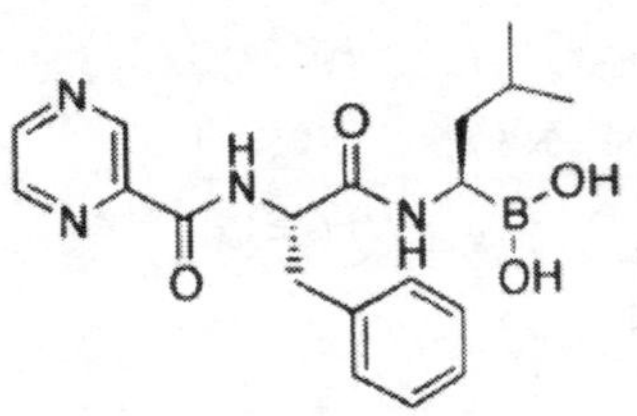

PS-341

Figure 2: Synthetic proteasome inhibitors

Lactacystin

Epoxomicin

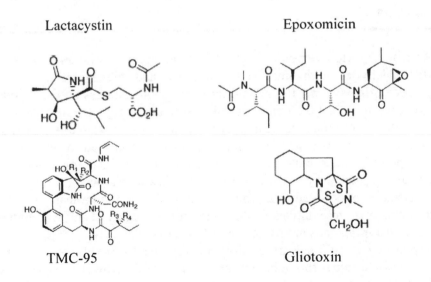

TMC-95

Gliotoxin

Figure 3: Naturally occurring proteasome inhibitors

Table 1: Proteasome inhibitors

CLASS	COMPOUND	NATURE OF INHIBITION	OTHER TARGETS	APPLICATION
Peptide	MG132	reversible	Calpains,	*In vitro* studies
aldehydes	MG115		Cathepsins	*In vitro* studies
Boronic acid	MG262	reversible	-	*In vitro* studies
peptides	PS-341		-	Multiple Myeloma
Lactacystins	β-Lacton	irreversible	Cathepsin A	*In vitro* studies
	PS-519		-	Stroke, asthma, psoriasis, multiple sclerosis (rodents)
Epoxyketones	Epoxomicin	irreversible	-	Cutaneous inflammation
	Eponemycin		-	B16 melanoma (mouse)
Macrocyclic PI	TMC-95	reversible	-	*In vitro* studies
Vinyl sulfones	NLVS	irreversible	Cathepsins	*In vitro* studies
Epipolythiopiperazin toxins	Gliotoxin	reversible	(-)	Graft rejection (mouse)

As a strategy to overcome these limitations, boronic acid derivatives of peptide aldehydes with improved selectivity and higher affinity were developed (Adams *et al.*, 1998; Gardner *et al.*, 2000). Studies on their mode of action revealed that the empty p-orbital of boron accepts the oxygen lone pair of the N-terminal threonine residue to form a stable tetrahedral intermediate. Boronic acid-based inhibitors can be truncated to dipeptides, thus improving important therapeutic characteristics like solubility and membrane permeability. Apart from that, slower dissociation rates of boronic acid peptides lead to a prolonged inhibition of the proteasome. The most common representatives of this class in the context of in vivo studies and clinical applications are MG262 and PS-341 (see below).

Peptide vinyl sulfones irreversibly modify the hydroxyl group of the active site N-terminal threonine (Bogyo *et al.*, 1997). However, since they inhibit both the 20S proteasome and cysteine proteases, these compounds are not regarded as suitable for in vivo applications.

The first naturally occurring proteasome inhibitor identified was the Streptomyces lactacystinaeus metabolite lactacystin (Omura *et al.,* 1991). In aqueous solutions, lactacystin is spontaneously converted into clasto-lactacystin β-lactone that reacts with the hydroxyl group of the β-subunit

N-terminal threonine to form a covalent ester adduct (Fenteany *et al.*, 1995). Thus, lactacystin irreversibly inhibits the CT-L and T-L activity of the proteasome. Although it was originally believed to exclusively target the 20S proteasome, there is increasing experimental evidence that lactacystin also inhibits cathepsin A (Ostrowska *et al.*, 2000). With the aim to reduce non-specific interactions, a synthetic analogue of lactacystin termed PS-519 was designed that has already demonstrated effectiveness against inflammatory diseases in clinical studies (Elliott *et al.*, 1999; Kondagunta *et al.*, 2004).

Screening for anti-tumour agents in mice led to the discovery of peptide α', β'-epoxyketones, another class of natural product PI (Hanada *et al.*, 1992). Epoxomicin and eponemycin, the most common representatives of this class, were isolated from Actinomycetes strain No. Q996-17 and Streptomyces hygroscopicus, respectively. They inhibit the 20S proteasome by forming a morpholino ring complex with the active site threonine residue. Cross-reactions with other cellular targets have not been reported for this class of inhibitors. However, due to the inhibition of all catalytic activities of the 20S proteasome, α', β' epoxyketones might be highly cell toxic and thus of limited therapeutic potential.

A family of non-linear PI was isolated from the fermentation broth of Apiospora montagnei Sacc. TC 1093 (Koguchi *et al.*, 2000). Compounds of this class like the TMC-95 stereoisomers are characterised by a macrocyclic scaffold structure and bind non covalently to the proteasome.

Another non-linear PI class termed epipolythiopiperazin toxins was discovered in Aspergillus fumigatus. Representatives of this class such as gliotoxin consist of a heterobicyclic core containing bisulfide bridges. Proteasome inhibition by gliotoxin can be reversed by dithiothreitol (Kroll *et al.*, 1999).

Generally, PI target the catalytic activities of the proteasome, thus preventing degradation and cleavage of proteins. Consequently, poly-ubiquitinylated proteins accumulate in the cell, precursor proteins remain in their inactive configuration and several proteins are mis-sorted. Signalling pathways are blocked (e.g. the NF-κB pathway, see above) or constitutively activated (e.g. the JNK/c-jun pathway, Nakayama *et al.*, 2001) causing either a decreased or increased expression of responsive genes. The most extensively studied inhibitors are PS-341 (Bortezomib, Velcade) and PS-519. Both compounds have been tested as potential therapeutics for oncological and inflammatory conditions. Bortezomib has demonstrated activity in phase II trials as a treatment option for renal cell cancer, lung cancer, sarcoma and lymphoma (Richardson *et al.*, 2003; Davis *et al.*, 2004; Kondagunta *et al.*, 2004; Maki *et al.*, 2005). Since 2003, bortezomib has been approved for clinical use in patients with multiple myeloma. PS-519 is the most advanced inhibitor in the context of inflammatory diseases like psoriasis, rheumatoid arthritis, asthma, multiple sclerosis, sepsis and vascular restenosis and has entered phase II clinical studies for the indications of acute stroke and myocardial infarction (Phillips *et al.*, 2000; Palombella *et al.*, 1998; Elliott *et al.*, 1999; Kondagunta *et al.*, 2004).

The anti-cancer effect of bortezomib and other PI is mainly based on their ability to stabilise cell cycle inhibitory proteins. Inhibition of NF-κB transcriptional activity furthermore down-regulates expression of various growth, survival and angiogenic factors (Sunwoo *et al.*, 2001). The ability of PI to increase tumour suppressor p53 levels and to activate c-Jun N-terminal kinase (JNK) leading to Fas up-regulation and caspase activation also contributes to the anti-neoplastic, pro-apoptotic effect (Hideshima *et al.*, 2001; Mitsiades *et al.*, 2002; Drexler, 1997, Shinohara *et al.*, 1996). Additionally, it has also been shown that proteasome inhibition enhances the sensitivity of cancer cells for traditional anti-cancer agents (Bold *et al.*, 2001). The anti-inflammatory effect of PS-519 and other PIs is probably based on inhibition of NF-κB activation that plays an important role in the pathogenesis of many inflammatory diseases by controlling expression of cytokines, adhesion molecules and pro-inflammatory molecules.

Given the fundamental metabolic role of the proteasome for all eukaryotic cells, undesirable adverse effects of proteasome inhibition might occur. However, most somatic cells are in a quiescent state and are thus less susceptible for proteasome inhibitors than actively dividing cells like cancer or immune cells. In accordance with this, bortezomib displayed a satisfying overall tolerance in clinical testing. The reported side-effects include fatigue and/or low fever, thrombocytopenia, gastrointestinal symptoms, electrolyte disturbances, headache, anemia, arthralgia, low diarrhea, frequent skin rashes and peripheral neuropathy (Richardson *et al.*, 2003; Davis *et al.*, 2004). Autoradiography experiments in rats showed that bortezomib is present in all organs except CNS, spinal cord, testes and eyes thus sparing these organs adverse effects of proteasome inhibition. Apart from that, proteasome inhibition by bortezomib is reversible and it could be shown that the inhibitor is rapidly metabolised (Adams, 2002). Similarly, PS-519 showed a good bioavailability in all organs except brain tissue. Side-effects have not been reported yet.

4.   INTERACTION BETWEEN VIRUSES AND THE UPS

Interactions between viruses and the UPS are highly complex and variable. Certain viruses essentially require the proteasome for different steps of their replication cycle. For other viruses, however, the proteasome represents a barrier and a degradative factor. In vitro studies also revealed that the interaction between the UPS and the virus presumably depends on the cell type. Most interestingly, the UPS may play different and even opposing roles during different stages of virus replication. Consequently, virus replication is either PI-sensitive or -insensitive and proteasome inhibition either suppresses or even supports replication of the virus depending on the specific situation. So far, antiviral effects of PI have been demonstrated in vitro for members of seven virus families - retroviridae, herpesviridae, parvoviridae, picornaviridae, orthomyxoviridae, paramyxoviridae and coronaviridae. For other viruses it has been shown that certain steps in virus replication are controlled by the UPS, however, an antiviral effect of PI has not been demonstrated yet. Involvement of the UPS in the interaction between the virus and cellular metabolism does not necessarily imply that proteasome inhibition blocks virus replication as shown for Adenovirus (Corbin-Lickfett and Bridge, 2003). The so far described antiviral effects of PI target entry/egress of the virus or interfere with viral transcription/translation which depend on the cellular gene expression machinery. Data arise exclusively from in vitro experiments using virus-infected cell cultures. However, preliminary data from animal experiments performed in our laboratory raise the hope that PI do not only suppress virus replication in vitro but also in vivo (Gille, Prösch., unpublished observation, see below).

4.1 Involvement of the UPS in virus entry

Several studies indicate that proteasome inhibition may block entry of viruses which essentially involve the endocytotic pathway and in particular the late endosome, as has been demonstrated for Influenza virus and a strain of Coronaviruses (Khor *et al.*, 2003; Sieczkarski and Whittaker, 2003; Yu and Lai, 2005). This observation is in accordance with the finding that sorting of proteins into the late endosome via "inwards" budding of membranes is coupled with the protein trafficking and ubiquitin-vacuolar sorting system and requires the functional UPS (Stahl and Barbieri, 2002; Strous and Govers, 1999; Longva *et al.*, 2002).

Productive infection by influenza viruses essentially depends on late endosomes and is thus highly sensitive to proteasome inhibition. Treatment of cells with PI caused a mis-targeting of the virus into so-called sorting and recycling endosomes which are different from early and late endosomes. Accordingly, the virus did not enter into the late endosomes and into the nucleus where replication occurs (Khor *et al.*, 2003; Sieczkarski and

Whittaker, 2003; Yu and Lai, 2005). Since PI do not influence the entry of Vesicular Stomatitis Virus (VSV) and Semliki Forest Virus (SFV) which only require the early and not the late endosome, the target for PI might be localised beyond the early endosome. The exact mechanism of PI-dependent inhibition of influenza virus entry, however, remains unclear up to date. A cellular target for proteasomal degradation seems more likely because viral proteins have not been found to be ubiquitinylated.

Similar to influenza virus, PI also suppress entry and nuclear translocation of Mouse Hepatitis Virus (MHV) strain JHM, a murine Coronavirus, into mouse astrocytoma cells (DBT cell line) (Ros and Kempf, 2004). Treatment of MHV strain JHM infected cells with PI caused a strong accumulation of incoming virus in the early endosome and led to a missorting into the lysosome. MHV may enter its target cells either by plasma membrane fusion or by endocytosis depending on the virus strain and the cell type (Nash and Buchmeier, 1997). Correspondingly, replication of MHV strain JHM which enters DBT cells via the endocytotic pathway was highly sensitive to proteasome inhibition, while MHV strain A59, which enters the cells by fusion, was less efficiently inhibited by PI.

Interestingly, the UPS has also been found to be involved in the pH dependent but endosome-independent entry and nuclear translocation of certain parvoviridae. Replication of at least two members of this virus family, the Minute Virus of Mice (MVM) and the Canine Parvovirus (CPV), but not the distantly related Bovine Parvovirus (BPV), were inhibited by PI. In the presence of PI virus particles aggregated in ring-like structures around the nuclei but were unable to enter into the nucleus (Ros *et al.*, 2002, Ros and Kempf, 2004). The underlying mechanism remains unclear because the proteasome does not seem to be involved in cleavage/processing of capsid protein VP and the virus did not co-localise with the early or late endosomal or the lysosomal compartment. Furthermore, uncoating of parvoviruses occurs in the nuclear core or in the nucleus but not in the cytosol.

4.2   Involvement of UPS in virus maturation and budding

Generally, virus budding at the outer plasma membrane follows the same strategy of "inwards" budding of membranes into so-called multivesicular bodies and association with the protein trafficking and ubiquitin-vacuolar sorting system. Accordingly, budding of certain retroviruses including Human Immunodeficiency Virus (HIV) type 1 and 2, Rous Sarcoma Virus (RSV), Murine Leukemia Virus (MuLV) and Mazon Pfizer Monkey Virus (MPMV) as well as the rhabdovirus VSV and the paramyxovirus SV5 was strongly inhibited by PI (Patniak *et al.*, 2000; Schubert, *et al.*, 2000; Strack *et al.*, 2000; Harty *et al.*, 2001; Ott *et al.*, 2003, recently reviewed in Klinger and Schubert, 2005; Schmitt *et al.*, 2005). Intensive studies on HIV-1 revealed in the absence of a functional UPS retarded budding structures, which accumulated and the released HIV-1 particles were not infectious. It was also shown that processing of the Gag-polyprotein but not the env-precursor, was strongly impaired. As a consequence Gag processing intermediates accumulated in the cell and the intracellular and virus-associated levels of the capsid protein p24 were strongly reduced. Interestingly, the viral protease which normally cleaves the Gag-polyprotein was not targeted by PI (Schubert *et al.*, 2000; Klinger and Schubert, 2005). Despite intensive investigations, the mechanism of PI-dependent inhibition of HIV budding is not completely understood. Several studies demonstrated that the C-terminal portion of Gag proteins of different retroviruses (e.g. p6 of HIV-1 and SIV Gag, p12 of Mo-MLV Gag, p9 of EIAV Gag) is mono-ubiquitinylated. The ubiquitin residue was suggested to play a role in function of the L-domains that are present in these proteins and are required for budding (Ott *et al.*, 1998, 2000, 2002a and b and 2003; Pornillos *et al.*, 2002). Since treatment with PI causes an accumulation of poly-ubiquitinylated proteins, limititation of the endogenous pool of free ubiquitin may prevent Gag ubiquitinylation. However, HIV-1 p6 Gag mutants deficient for ubiquitinylation showed an unchanged phenotype, indicating that ubiquitinylation alone is not sufficient for virus budding (Ott *et al.*, 2000 and 2003). Moreover, mono-ubiquitinylation of Gag alone does not provide an explanation why budding of some retroviruses like Equine Infectious Anemia Virus (EIAV) and Mouse Mamma Tumour Virus (MMTV) are insensitive to PI (Patniak *et al.*, 2002; Ott *et al.*, 2003). The structure of the Gag L-domains and their interaction with different components of the UPS-dependent vacuolar protein-sorting system seems to be more crucial for virus budding. Only viruses with proteins containing an L-domain of the PPXY-type (e.g. MuLV, RSV, Ebola virus) or the PTAP type (e.g. HIV-1 and-2) but not of the YXDL-type (e.g. EIAV) are sensitive to PI (Shehu-Xhilaga *et al.*, 2004; Klinger and Schubert, 2005). By binding to different "bridging molecules", L-domains mediate interaction between the viruses and the vacuolar protein-sorting system required for budding. L-domains of the PPXY-type are able to bind different HECT ubiquitin ligases like WWP1, WWP2 or Itch, while L-domains of the PTAP-motif bind Tsg 101 (e.g. HIV, Ebola virus VP40) which then recruits other components of the endosomal sorting system like ESCRT-I,-II,- III and thus connects the Gag protein to the vacuolar protein sorting system (Garrus *et al.*, 2001; Amit *et al.*, 2004; Martin-Serrano *et al.*, 2001, 2003 and 2005).

Since VP40 of Ebola Virus and Human T-cell Lymphotropic virus (HTLV-1) Gag also contain L-domain motifs of the PPXY type it has been speculated that Ebola virus and HTLV-1 are also sensitive to proteasome inhibition. However, this still has to be proved in an infection model (Harty *et al.*, 2000; Bouamr *et al.*, 2003; Heidecker *et al.*, 2004; Martin-Serrano *et al.*, 2005).

In the paramyxovirus SV5 matrix protein, very recently, a new type of L-domains (FPIV) has been idenitfied which may functionally substitute the PTAP motif in HIV-1 Gag and cause the PI-sensitivity of this virus (Schmitt *et al.*, 2005). The precise role of ubiquitinylation of either viral or cellular targets or of the ubiquitin ligases in virus budding remains unclear, however. Most interestingly, HIV protease inhibitors like nelfinavir and saquinavir display proteasome inhibitory activity (Piccinini *et al.*, 2002 and 2005). This unexpectected activity may be responsible for some side effects of these drugs, whether they contribute to the anti-HIV activity, e.g. by inhibiting NF-κB activation which is involved in positive regulation of HIV gene expression or by inhibiting the budding – is unclear.

An alternative model to explain the PI-sensitivity of HIV suggests that PI facilitate the production of Gag-related defective ribosomal products (DriPs) which are encapsidated together with the normal Gag polyprotein and which may act as competitive non-cleavable substrate inhibitors of the viral protease during Gag-polyprotein processing (Schubert *et al.*, 2000). This hypothesis, however, contradicts the finding that PI are selective only for certain retroviruses.

4.3   Involvement of the UPS in viral gene transcription and translation

The UPS is not only involved in entry and budding of viruses but also in regulation of viral gene expression. Very recently, this effect was demonstrated for Coxsackievirus B3, the most prevalent virus associated with the pathogenesis of myocarditis and myocardiopathy (McManus *et al.*, 1991; Martino *et al.*, 1994). In vitro treatment of virus-infected murine cardiomyocytes with different PIs caused a strong reduction in virus replication. Inhibition occurred at the level of viral RNA transcription and protein synthesis but did not affect virus entry or proteolytic activity of the virus-encoded protease (Luo *et al.*, 2003). However, the mechanism by which PI inhibit coxsackievirus transcription/translation remains unclear so far.

As a further example, Human Cytomegalovirus (HCMV) as well as Rat Cytomegalovirus (RCMV) in vitro-replication in fibroblasts is efficiently suppressed by different reversible as well as irreversible PI (Prösch *et al.*,

2003; Kaspari, Prösch unpublished data). Inhibition of HCMV replication occurs at the immediate early (IE) stage of infection by suppression of the IE gene expression. Virus adsorption as well as entry and trafficking of the capsid/tegument complex into the nucleus are not influenced. Several lines of evidence support the hypothesis that PI specifically block generation of the IE2 messenger while generation of the IE1 messenger is not affected. Both IE1 and IE2 messengers are generated by different splicing of the IE locus, suggesting that the proteasome may be involved in regulation of IE2 mRNA splicing. In accordance with this it was found that expression and localisation of several splice factors are altered in the absence of a functional UPS (Rockel and Mikecz, 2002). Since the IE2 protein is essential for induction of early and late viral proteins and for virus-dependent cellular gene expression, PI are detrimental to HCMV replication. The structure of the IE gene locus of RCMV is very similar to that of HCMV regarding its modular composition and splicing processes, thus, it might be assumed that PI inhibit replication of RCMV and HCMV by the same mechanism.

Remarkably, in a first experiment in which we treated RCMV-infected immunosuppressed rats with MG262, the viral load in all tested organs was clearly reduced (Gille, Prösch, unpublished data). These results raise the hope that the in vitro observed anti-CMV effect of PI is also present in vivo and that PI represent an alternative treatment option against GCV-resistant strains or a complementary drug to ganciclovir/valganciclovir which block virus replication only at the late stage (Prösch *et al.*, 2003).

A third example for involvement of the proteasome in regulation of viral gene expression are Adenoviruses (Corbin-Lickfett and Bridge, 2003). Adenoviruses require the proteasome for initiation of late gene expression by the E4-34kDa protein, subsequently, PI block expression of late genes. However, in vitro replication of Adenovirus was only marginally reduced in the presence of PI, indicating that the UPS function is not essentially required for efficient replication.

Furthermore, transfection experiments suggest that the UPS influences translation of hepatitis C virus, however, it has not been proved whether PI inhibit HCV replication (Kruger *et al.*, 2001).

Table 2: Viruses for which in vitro antiviral activity of proteasome inhibitors has been demonstrated

Virus	Target of antiviral activity	References
Coronaviridae Mouse Hepatitis Virus	Virus entry	Yu and Lai, 2005
Enteroviridae Coxsackievirus B3	Transcription	Luo *et al.*, 2003
Filoviridae * Ebola Virus	Virus budding	Harty *et al.*, 2000 Martin-Serrano *et al.*, 2005
Herpesviridae Human Cytomegalovirus Rat Cytomegalovirus	Transcription IE2 protein expression undefined	Prösch *et al.*, 2003 Kaspari, Prösch, unpubl. Gille, Prösch, unpubl.
Orthomyxoviridae Influenza Virus A	Virus entry	Khor *et al.*, 2003
Paramyxoviridae SV5	Virus budding	Schmitt *et al.*, 2005
Parvoviridae Minute Virus of Mice Canine Parvovirus	Entry	Ros *et al.*, 2002
Rhabdoviridae Vesicular Stomatitis Virus	Virus budding	Harty *et al.*, 2000
Retroviridae Human Immunodeficiency Virus Rous Sarcoma Virus Murine Leukemia Virus Mazon Pfizer Monkey Virus HTLV*	Gag-protein processing, assembly, budding, maturation	Patniak *et al.*, 2000 Strack *et al.*, 2000 Schubert *et al.*, 2000 Ott *et al.*, 2003 Heidecker *et al.*, 2004

* The antiviral effect has only been shown for the budding of virus-like particles, not for infectious virus.

4.4 The UPS in first line defence against viruses - proteasome inhibition and virus (re)activation

The use of PI as antiviral drugs may be contraindicated by the fact that under certain circumstances some viruses may benefit from proteasome inhibition. One example is the Herpes Simplex Virus (HSV) type 1. Establishment and maintenance of latency in neuronal cells has been found to depend on the proteasomal degradation of the viral UL9 protein which may facilitate reactivation of the virus by binding to and unwinding of the origin 3. UL9 interacts with the neurone specific NFB24 protein and is thus recruited to an E3 ubiquitin ligase complex facilitating its ubiquitinylation and proteasomal degradation. Proteasome inhibition allows accumulation of UL9 and thus induce viral replication/reactivation (Eom and Lehmann, 2003). Similarly, reactivation of latent integrated HIV may benefit from proteasome inhibition (Krishnan and Zeichner, 2004; Schwartz *et al.*, 1998). As a reason it has been speculated that key factors required to effectively activate the HIV-1 LTR are degraded or repressed in latently infected cells and that their stabilisation by PI then promotes transcription. Accordingly, latently infected cells show increased expression of various proteasome subunits and inhibition of the UPS may restore this process.

Additionally, for some HIV strains and the Adeno-associated Virus (AAV), a parvovirus, it has been shown that the proteasome may limit their infectivity. Consequently, PI increased their infectivity by preventing viral degradation during entry. This effect is especially relevant for HIV strains entering the cell by fusion, while strains using the endocytotic pathway are less successful (Wei *et al.*, 2005; Schwartz *et al.*, 2001). One explanation might be that the proteasome degrades Gag proteins within the cytoplasm. Similarly, PI support transduction of the recombinant AAV (rAAV). Normally, in less efficiently transduced cells incoming virus is effectively degraded by the proteasome (Douar *et al.*, 2001; Hansen *et al.*, 2000; Duan *et al.*, 2000). Inhibition of the proteasome enhances nuclear accumulation of the virus and viral gene expression (Duan *et al.*, 2000; Jennings *et al.*, 2004). Since rAAV is of great interest as a vector for gene therapy, proteasome inhibition has demonstrated to promote transgene expression (Yan *et al.*, 2002).

5.   CONCLUSIONS

During the last years it became more clearly that the UPS is essentially involved in different steps of virus replication. For some viruses, inhibition of the UPS by PI has been shown fatal at least under certain conditions. This makes PI highly interesting candidates for alternative or complementary treatment of infections with these viruses. All studies so far have been carried out with PI developed as research tools only which are not suitable for clinical use. To evaluate PI as antiviral therapeutics, these experiments have to be confirmed with clinically approved PI like PS-341 and PS-519 and have to be verified in animal settings. Our first in vivo data raise the hope that PI display their antiviral activity not only in vitro but also in vivo.

The therapeutic index of PI as observed in our lab for HCMV and RCMV is low compared to classical antiviral drugs which target virus-specific genes. This emphasises the need to develop PI with increased specificity and reduced toxicity. Alternatively, drugs which target the more specific E3 ubiquitin ligases involved in virus UPS interaction should be more specific and less toxic. Development of such compounds, however, requires the knowledge of viral and cellular factors involved in the interaction between virus and UPS. A further question which has to be addressed is the influence of PI on (re)activation of highly prevalent latent or persistent viruses to exclude that treatment of an acute infection with one virus does not trigger reactivation of another virus. This question is also of substantial interest for the use of PI in anti-cancer therapy because persistent viruses like EBV or HCMV are co-factors for tumour genesis.

Reduced resistance should be one advantage of PI compared to classical antiviral drugs, however, it can not be guaranteed as already shown for other drugs targeting cellular factors. Apart from that, PI are an ideal alternative option for the treatment of viruses resistant to classical antiviral drugs (e.g. GCV-resistant HCMV, Prösch et al., 2003) since they target a cellular factor thus excluding or minimising cross-resistance.

In summary, the present data raise the hope that inhibitors of the UPS support successful treatment or even prevention of severe viral infections as caused e.g. by HIV-1, HCMV or Coxsackievirus because PI are detrimental to replication of these viruses.

ACKNOWLEDGEMENTS

The authors would like to thank Min-Hi Lee for her support in preparing the Figures and M. Raftery for critical reading the paper.

REFERENCES

Adams, J., 2002, Development of the proteasome inhibitor PS-341. *Oncologist* **7**: 9-16

Adams, J., Behnke, M., Chen, S., Cruickshank, A.A., Dick, L.R., Grenier, L., Klunder, J.M., Ma, Y.T., Plamondon, L., and Stein, R.L., 1998, Potent and selective inhibitors of the proteasome: dipeptidyl boronic acids. *Bioorg. Med. Chem. Lett.* **8**: 333-338

Amit, I., Yakir, L., Katz, M., Zwang, Y., Marmor, M.D., Citri, A., Shtiegman, K., Alroy, I., Tuvia, S., Reiss, Y., Roubini, E., Cohen, M., Wides, R., Bacharach, E., Schubert, U., and Yarden, Y., 2004, Tal, a Tsg101-specific E3 ubiquitin ligase regulates receptor endocytosis and retrovirus budding. *Genes Dev.* **18**: 1737-1752

Arendt, C.S., and Hochstrasser, M., 1997, Identification of the yeast 20S proteasome catalytic centers and subunit interactions required for active-site formation. *Proc. Natl. Acad. Sci. USA* **94**: 7156-7161

Baldwin, A.S. Jr., 1996, The NF-κB and IκB proteins: new discoveries and insights. *Ann. Rev. Immunol.* **14**: 649-681

Baumeister, W., Walz, J., Zuhl, F., and Seemüller, E., 1998, The proteasome: paradigm of a self-compartmentalizing protease. *Cell* **92**: 367-380

Beinke, S., and Ley, S.C., 2004, Function of NF-κB1 and NF-κB2 in immune cell biology. *Biochem. J.* **382**: 93-409

Bochtler, M., Ditzel, L., Groll, M., Hartmann, C., and Huber, R., 1999, The proteasome. *Ann. Rev. Biophys. Biomol. Struct.* **28**: 295-317

Bogyo, M., McMaster, J.S., Gaczynska, M., Tortorella, D., Goldberg, A.L., and Ploegh, H., 1997, Covalent modification of the active site threonine of proteasomal beta subunits and the Escherichia coli homolog HslV by a new class of inhibitors. *Proc. Natl. Acad. Sci. USA* **94**: 6629-6634

Bold, R.J., Virudachalam, S., and McConkey, D.J., 2001, Chemosensitization of pancreatic cancer by inhibition of the 26S proteasome. *J. Surg. Res.* **100**: 11-17

Bouamr, F., Melillo, J.A., Wang, M.Q., Nagashima, K., de Los Santos, M., Rein, A., and Goff, S.P., 2003, PPPYEPTAP motif is the late domain of human T-cell leukemia virus type 1 Gag and mediates its functional interaction with cellular proteins Nedd4 and Tsg101. *J. Virol* **77**: 11882-11895

Braun, B.C., Glickman, M., Kraft, R., Dahlmann, B., Kloetzel, P.M., Finley, D., and Schmidt, M., 1999, The base of the proteasome regulatory particle exhibits chaperone-like activity. *Nat. Cell. Biol.* **1**: 221-226

Breitschopf, K., Zeiher, A.M., and Dimmeler, S., 1999, Dephosphorylation targets Bcl-2 for ubiquitin-dependent degradation: a link between the apoptosome and the proteasome pathway. *J. Exp. Med.* **189**: 1815-1822

Ciechanover, A., and Schwartz, A.L., 1998, The ubiquitin-proteasome pathway: the complexity and myriad functions of protein death. *Proc. Natl. Acad. Sci. USA* **95**: 2727-2730

Corbin-Lickfett, K.A., and Bridge, E., 2003, Adenovirus E4-34kDA requires active proteasomes to promote late gene expression. *Virology* **315**: 234-244

Davis, N.B., Taber, D.A., Ansari, R.H., Ryan, C.W., George, C., Vokes, E.E., Vogelzang, N.J., and Stadler, W.M., 2004, Phase II trial of PS-341 in patients with renal cell cancer: a University of Chicago Phase II consortium study. *J. Clin. Oncol.* **22**: 115-119

Douar, A.M., Poulard, K., Stockholm, D., and Danos, O., 2001, Intracellular trafficking of adeno-associated virus vectors. Routing to the late endosomal compartment and proteasome degradation. *J. Virol* **75**: 1824-1833

Drexler, H.C., 1997, Activation of the cell death program by inhibition of proteasome function. *Proc. Natl. Acad. Sci. USA* **94**: 855-860

Duan, D., Yue, Y., Yan, Z., Yang, J., and Engelhardt, J.F., 2000, Endosomal processing limits gene transfer to polarized airway epithelia by adeno-associated virus. *J. Clin. Invest.* **105**: 1573-1587

Dupre, S., Volland, C., and Haguenauer-Tsapis, R., 2001, Membrane transport: ubiquitinylation in endosomal sorting. *Curr. Biol.* **11**: 932-934

Elliott, P.J., Pien, C.S., McCormack, T.A., Chapman, I.D., and Adams, J., 1999, Proteasome inhibition: a novel mechanism to combat asthma. J. Allergy Clin. Immunol. 104:294-300

Eom, C.Y., and Lehmann, I.R., 2003, Replication-initiator protein (UL9) of the herpes simplex virus 1 binds NFB42 and is degraded via the ubiquitin-proteasome pathway. *Proc. Natl. Acad. Sci. USA* **100**: 9803-9807

Fenteany, G., Standaert, R.F., Lane, W.S., Choi, S., Corey, E.J., and Schreiber, S.L., 1995, Inhibition of proteasome activities and subunit-specific amino-terminal threonine modification by lactacystin. *Science.* **268**: 726-731

Ferrell, K., Wilkinson, C.R., Dubiel, W., and Gordon, C., 2000, Regulatory subunit interactions of the 26S proteasome, a complex problem. Trends Biochem. Sci. 25:83-88

Figueiredo-Pereira, M.E., Banik, N., and Wilk, S., 1994, Comparison of the effect of calpain inhibitors on two extralysosomal proteinases: the multicatalytic proteinase complex and m-calpain. *J. Neurochem.* **62**: 1989-1994

Gardner, R.C., Assinder, S.J., Christie, G., Mason, G. G, Markwell, R., Wadsworth, H., McLaughlin, M., King, R., Chabot-Fletcher, M. C., Breton, J. J., Allsop, D., and Rivett, A. J., 2000, Characterisation of peptidyl boronic acid inhibitors of mammalian 20S and 26S proteasomes and their inhibition of proteasomes in cultured cells. *Biochem. J.* **346**: 447-454

Garrus, J.E., von Schwedler, U.K., Pornillos, O.W., 2001, Tsg101 and the vacuolar protein sorting pathway are essential for HIV-1 budding. *Cell* **107**: 55-65.

Hanada, M., Sugawara, K., Kaneta, K., Toda S., Nishiyama, Y., Tomita, K., Yamamoto, H., Konishi, M., and Oki, T., 1992, Epoxomicin, a new antitumor agent of microbial origin. *J. Antibiot.*(Tokyo). **45**: 746-775

Hansen, J., Qing, K., Kwon, H.J., Mah, C., and Srivastava, A., 2000, Impaired intracellular trafficking of adeno-associated virus vectors limits efficient transduction of murine fibroblasts. *J. Virol* **74**: 992-996

Harty, R.N., Brown, M.E., Wang, G., Huiregtse, J., and Hayes, F.P., 2000, A PPxY motif within the VP40 protein of Ebola virus interacts physically and functionally with a ubiquitin ligase: implication for filovirus budding. *Proc. Natl. Acad. Sci. USA* **97**: 13871-13876

Harty, R.N., Brown, M.E., McGettigan, J.P., Wang, G., Jayakar, H.R., Huiregtse, J.M.,Whitt, M.A., and Schnell, M.J., 2001, Rhabdoviruses and the cellular ubiquitin-proteasome system: a budding interaction. *J. Virol* **75**: 10623-10629

Heidecker, G., Lloyd, P.A., Fox, K., Nagashima, K., and Derse, D., 2004, Late assembly motifs of human T-cell leukemia virus Type 1 and their relative roles in particle release. *J. Virol* **78**: 6636-6648

Hideshima, T., Richardson, P., Chauhan, D., Palombella, V.J., Elliott, P.J., Adams, J., and Anderson, K.C., 2001, The proteasome inhibitor PS-341 inhibits growth, induces apoptosis and overcomes drug resistance in human multiple myeloma cells. *Cancer Res.* **61**: 3071-3076

Hill, C.P., Masters, E.I., and Whitby, F.G., 2002, The 11S regulators of 20S proteasome activity. *Curr. Tp. Microbiol. Immunol.* **268**: 73-89

Hoffman, L., Pratt, G., and Rechsteiner, M., 1992, Multiple forms of the 20 S multicatalytic and the 26 S ubiquitin/ATP-dependent proteases from rabbit reticulocyte lysate. *J. Biol. Chem.* **267**: 22362-22368

Jennings, K., Miyamae, T., Fraisler, R., Marinov, A., Katekura, S., Sowders, D., Trappell, B., Wilson, J.M., Gao, G., and Hirsch, R., 2005, Proteasome inhibition enhances AAV-mediated transgene expression in human synoviocytes *in vitro*o and *in vivo*. *Mol. Ther.* **11**: 600-607

Karin, M., and Ben-Neriah, Y., 2000, Phosphorylation meets ubiquitinylation: the control of NF-κkappa B activity. *Ann. Rev. Immunol.* **18**: 621-663

Karin, M., Cao, Y., Grelen, F.R., and Zhi-Wei, L., 2002, NF- B in cancer: from innocent bystander to major aelprill. *Nat. Rev. Cancer* **2**: 301-310

Karin, M., and Lin, A., 2002, NF-κB at the crossroads of life and death. *Nat. Immunol.* **3**: 221-227

Khor, R., McElroy, L.J, and Whittaker, G.R., 2003, The ubiquitin-vacuolar protein sorting system is selectively required during entry of influenza virus into host cells. *Traffic* **4**: 857-868

Klinger, P.P., and Schubert, U., 2005, The ubiquitin-proteasome system in HIV replication: potential targets for antiretroviral therapy. *Expert Rev. Anti Infect. Ther.* **3**: 61-79

Kloetzel, P.M., 2004, The proteasome and MHC class I antigen processing. *Biochim. Biophys. Acta* **1695**: 217-225

Koguchi, Y., Kohno, J., Nishio, M., Takahashi, K., Okuda, T., Ohnuki, T., and Komatsubara, S., 2000, TMC-95A, B, C, and D, novel proteasome inhibitors produced by Apiospora montagnei Sacc. TC 1093. Taxonomy, production, isolation and biological activities. *J. Antibiot.* (Tokyo) **53**: 105-109

Kondagunta, G.V., Drucker, B.J., Schwartz, L., Bacik, J., Marion, S., Russo, P., Mazumdar, M., and Motzer, R.J., 2004, Phase II trial of bortezomib for patients with advanced renal cell carcinoma. *J. Clin. Oncol.* **22**: 3720-3725

Kopp, F., Dahlmann, B., and Kuehn, L., 2001, Reconstitution of hybrid proteasomes from purified PA700-20S complexes and PA28 alpha beta activator: ultrastructure and peptidase activities. *J. Mol. Biol.* **313**: 465-471

Krishnan, V., and Zeichner, S.L., 2004, Host cell expression during human immunodeficiency virus type 1 latency and reactivation and effects of targeting genes that are differentially expressed in viral latency. *J. Virol* **78**: 9458-9473

Kruger, M., Beger, C., Welch, P.J., Barber, J.R., Manns, M.P., and Wong-Staal, F., 2001, Involvement of proteasome α-subunit PSMA7 in hepatitis C virus internal ribosome entry site-mediated translation. *Mol. Cell. Biol.* **21**: 8357-8364

Kroll, M., Arenzana-Seisdedos, F., Bachelerie, F., Thomas, D., Friguet, B., and Conconi, M., 1999, The secondary fungal metabolite gliotoxin targets proteolytic activities of the proteasome. *Chem. Biol.* **6**: 89-698

Li, Q., and Verma, I.M., 2002, NF- B regulation in the immune system. *Nat. Rev. Immunol.* **2** :725-734

Longva, K.E., Blystad, F.D., Stang, E., Larsen, A.M., Johannessen, L.E., and Madshus, I.H., 2002, Ubiquitinylation and proteasomal activity is required for transport of the EGF receptor to inner membranes of multivesicular bodies. *J. Cell Biol.* **156**: 843-854

Luo, H., Zhang, J., Cheung, C., Suarez, A., McManus, B.M., and Yang, D., 2003, Proteasome inhibition reduces coxsackievirus B3 replication in murine cardiomyocytes. *Am. J. Pathol.* **163**: 381-385

McManus, B.M., Chow, L.H., Radio, S.J., Tracy, S.M., Beck, M.A., Chapman, N.M., Klingel, K., and Kandolf, R., 1991, Progress and challenges in the pathological diagnosis of myocarditis. *Eur. Heart J.* **12**: 18-21

Maki, R.G., Kraft, A.S., Scheu, K., Yamada, J., Wadler, S., Antonescu, C.R., Wright, J.J., and Schwartz, G.K., 2005, A multicenter Phase II study of bortezomib in recurrent or metastatic sarcomas. *Cancer* **103**: 1431-1438

Martino, T.A., Liu, P., and Sole, M.J., 1994, Viral infection and the pathogenesis of dilated cardiomyopathy. *Circ. Res.* **74**: 182-188

Martin-Serrano, J., Eastman, S.W., Chung, W., and Bieniasz, P.D., 2005, HECT ubiquitin ligases link viral and cellular PPXY motifs to the protein-sorting pathway. *J. Cell Biol.* **168**: 89-101

Martin-Serrano, J., Zang, T., and Bieniasz, P.D., 2003, Role of ESCRT-I in retroviral budding. *J. Virol* **77**: 4794-4804

Martin-Serrano, J., Zang, T., and Bieniasz, P.D., 2001, HIV-1 and Ebola virus encode small peptide motifs that recruit Tsg101 to sites of particle assembly to facilitate egress. *Nature Med.* **7**: 313-1319

Mitsiades, N., Mitsiades, C.S., Poulaki, V., Chauhan, D., Fanourakis, G., Gu, X., Bailey, C., Joseph, M., Liberman, T.A., Treon, S.P., Munshi, N.C., Richardson, P.G., Hideshima, T., and Anderson, K.C., 2002, Molecular sequelae of proteasome inhibition in human multiple myeloma cells. *Proc. Natl. Acad. Sci. USA* **99**: 14374-14379

Nakayama, K., Furusu, A., Xu, Q., Konta, T., and Kitamura, M., 2001, Unexpected transcriptional induction of monocyte chemoattractant protein 1 by proteasome inhibition: involvement of the c-Jun N-terminal kinase-activator protein 1 pathway. *J. Immunol.* **167**: 145-50

Nash, T.C., and Buchmeier, M.J., 1997, Entry of mouse hepatitis virus into cells by endosomal and nonendosomal pathways. *Virology.* **233**: 1-8.

Omura, S., Fujimoto, T., Matsuzaki, K., Moriguchi, R., Tanaka, H., and Sasaki, Y., 1991, Lactacystin, a novel microbial metabolite, induces neuritogenesis of neuroblastoma cells. *J. Antibiot.* (Tokyo) **44**: 113-116

Ostrowska, H., Wojcik, C., Omura, S., and Worowski, K., 2000, Separation of cathepsin A-like enzyme and the proteasome: evidence that lactacystin / beta-lactone is not a specific inhibitor of the proteasome. *Int. J. Biochem. Cell. Biol.* **32**: 747-757

Ott, D.E., 2002, Potential roles of cellular proteins in HIV-1. *Rev. Med. Virol.* **12**: 3 59-374

Ott, D.E., Coren, L.V., Sowder, R.C. 2nd, Adams. J., Nagashima, K., and Schubert, U., 2002, Equine infectious anemia virus and the ubiquitin-proteasome system. *J. Virol* **76**:3 038-3044

Ott, D.E., Coren, L.V., Chertova, E.N., Gagliardi, T.D., and Schubert, U., 2000, Ubiquitination of HIV-1 and MuLV Gag. *Virology* **278**: 1 11-121

Ott, D.E., Coren, L.V., Copeland, T.D., Kane, B.P., Johnson, D.G., Sowder, R.C. 2nd, Yoshinaka, Y., Orzlan, S., Arthur, L.Q., and henderson, L.E., 1998, Ubiquitin is covalently attached to the p6 Gag proteins of human immunodeficiency virus type 1 and simian immunodeficiency virus and the p12Gag protein of Moloney murine leukemia virus. *J. Virol* **72**: 2962-2968

Ott, D.E., Coren, L.V., Sowder, R.C. 2nd, Adams, J., and Schubert, U., 2003, Retroviruses have differing requirements for proteasome function in the budding process. *J. Virol* **77**: 3384-3393

Palombella, V.J., Conner, E.M., Fuseler, J.W., Destree, A., Davis, J.M., Laroux, F.S., Wolf, R.E., Huang, J., Brand, S., Elliott, P.J., Lazarus, D., McCormack, T., Parent, L., Stein, R., Adams, J., and Grisham, M.B., 1998, Role of the proteasome in and NF-kappaB in streptococcal cell wall-induced polyarthritis. *Proc. Natl. Acad. Sci. USA* **95**: 15671-15676

Patnaik, A., Chau, V., Li, F., Montelaro, R.C., and Wills, J.W., 2002, Budding of equine infectious anemia virus is insensitive to proteasome inhibitors. *J. Virol* **76**: 2641-2647

Phillips, J.B., Williams, A.J., Adams, J., Elliot, P.J., and Tortella, F.C., 2000, Proteasome inhibitor PS-519 reduces infarction and attenuates leukocyte infiltration in a rat model of focal cerebral ischemia. *Stroke* **31**: 1686-1693

Piccinini, M., Rinaudo, M.T., Anselmino, A., Buccinna, B., Ramondetti, C., Dematteis, A., Ricotti, E., Palmisano, L., Mostert, M., and Tovo, P.A., 2005, The HIV protease inhibitors nelfinavir and saquinavir, but not a variety of HIV reverse transcriptase inhibitors, adversely affect human proteasome function. *Antiviral Ther.* **10**: 215-223

Piccinini, M., Rinaudo, M.T., Chiapello, N., Ricotti, E., Baldovino, S., Mostert, M., and Tovo, P.A., 2002, The human 26S proteasome is a target of antiretroviral agents. *AIDS* **16**: 693-700

Pornillos, O., Alam, S.L., Davis, D.R., and Sundquist, W.I., 2002, Structure of the Tsg101 UEV domain in complex with the PTAP motif of the HIV-1 p6 protein. Nat. Struct. Biol. **9**: 812-817

Prösch, S., Priemer, C., Höflich, C., Liebenthal, C., Babel, N., Krüger, D.H., and Volk, H.D., 2003, Proteasome inhibition: a novel tool to suppress human cytomegalovirus replication and virus-induced immune modulation. *Antiviral Ther.* **8**: 555-567

Rechsteiner, M., Realini, C., and Ustrell, V., 2000, The proteasome activator 11S REG (PA28) and class I antigen presentation. *Biochem. J.* **345**: 1-15.

Richardson, P.G., Barlogie, B., Berenson, J., Singhal, S., Jagannath, S., Irwin, D., Rajkumar, S.V., Srkalovic, G., Alsina, M., Alexanian, R., Siegel, D., Orlowski, R. Z., Kuter, D., Limentani, S.A., Lee, S., Hideshima, T., Esseltine, D.L., Kauffman, M., Adams, J., Schenkein, D.P., and Anderson K.C., 2003, A Phase 2 study of bortezomib in relapsed, refractory myeloma. *N. Engl. J. Med.* **348**: 2609-2617

Rockel, T.D., and van Mikecz, A., 2002, Proteasome-dependent processing of nuclear proteins is correlated with their subnuclear localization. *J. Struct. Biol.* **140**: 189-199

Rolfe, M., Chiu, M.I., and Pagano, M., 1997, The ubiquitin-mediated proteolytic pathway as a therapeutic area. *J. Mol. Biol.* **75**: 5-17

Ros, C., and Kempf, C., 2004, The ubiquitin-proteasome machinery is essential for nuclear translocation of incoming minute virus of mice. *Virology* **324**: 50-360

Ros, C., Burckhardt, C.J., and Kempf, C., 2002, Cytoplasmic trafficking of minute virus of mice: low pH requirement, routing to late endosomes, and proteasome interaction. *J. Viol.* **76**: 12634-12645

Salghetti, S.E., Kim, S.Y., and Tansey, W.P., 1999, Destruction of Myc by ubiquitin-mediated proteolysis: cancer-associated and transforming mutations stabilize Myc. *EMBO J.* **18**: 717-726

Schmitt, A.P., Leser, G.P., Morita, E., Sundquist, W.L., and Lamb, R.A., 2005, Evidence for a new viral late-domain core sequence, FPIV, necessary for budding of a paramyxovirus. *J. Virol* **79**: 2988-2997

Schubert, U., Ott, D.E., Chertova, E.N., Welker, R., Tessmer, U., Princiotta, M.F., Bennink, J.R., Kräusslich, H.G., and Yewdell, J.W., 2000, Proteasome inhibition interferes with Gag polyprotein processing, release, and maturation of HIV-1 and HIV-2. *Proc. Natl. Acad. Sci. USA.*, **97**: 13057-13062

Schwartz, O., Marachel, V., Friguet, B., Arenzana-Seisdedos, F., and Heard, J.M., 1998, Antiviral activity of the proteasome on incoming human immunodeficiency virus type 1. *J. Virol* **72**: 3845-3850

Shehu-Xhilaga, M., Ablan, S., Demirov, D.G., Chen, C., Montelaro, R.C., and Freed, E.O., 2004, Late domain-dependent inhibition of equine infectious anemia virus budding. *J. Virol* **78**: 724-732

Shinohara, H., Tomioka, M., Nakano, H., Tone, S., Ito, H., and Kawashima, S., 1996, Apoptosis induction resulting from proteasome inhibition. *Biochem. J.* **317**: 385-388

Sieczkarski, S.B., and Whittaker, G.R., 2003, Differential requirement of Rab5 and Rab7 for endocytosis of influenza and other enveloped viruses. *Traffic* **4**: 333-343

Stahl, P.D., and Barbieri, M.A., 2002, Multivesicular bodies and multivesicular endosomes: the "in and outs" of endosomal traffic. Sci. STKE 2002:PE32

Stohwasser, R, Salzmann U., Giesebrecht, J., Kloetzel, P.M., and Hozhütter, H.G., 2000, Kinetic evidences for facilitation of peptide channelling by the proteasome activator PA28. Eur. J. Biochem. **276**: 6221-6229

Strack, B., Calistri, A., Accola, M.A., Palu, G., and Gottlinger, H.G., 2000, A role for ubiquitin ligase recruitment in retrovirus release. *Proc. Natl. Acad. Sci. USA* **97**: 13063-13068.

Strous, G.J. and Govers, R., 1999, The ubiquitin-proteasome system and endocytosis. *J. Cell Sci.* **112**: 1417-1423

Sunwoo, J.B., Chen, Z., Dong, G., Yeh, N., Crowl Bancroft, C., Sausville, E., Adams, J., Elliott, P., and van Waes, C., 2001, Novel proteasome inhibitor PS-341 inhibits activation of nuclear factor-kappaB, cell survival, tumor growth and angiogenesis in squamous cell carcinoma. *Clin. Cancer Res.* **7**: 1419-1428

Tanahashi, N., Murakami, Y., Minami, Y., Shimbara, N., Hendil, K.B., and Tanaka, K., 2000, Hybrid proteasomes: induction by interferon-γ and contribution to ATP-dependent proteolysis. *J. Biol. Chem.* **275**: 14336-14345

Wei, B.L., Denton, P.W., O'Neill, E., Luo, T., Foster, J.L., and Garcia, J.V., 2005, Inhibition of lysosomal and proteasome function enhances human immunodeficiency virus type 1 infection. *J. Virol* **79**: 5705-5712

Wilk, S., and Orlowski, M., 1983, Evidence that pituitary cation-sensitive neutral endopeptidase is a multicatalytic protease complex. *J. Neurochem.* **40**: 842-849

Yan, Z., Zak, R., Luxton, G.W., Ritchie, T.C., Bantel-Schaal, U., and Engelhardt, J.F., 2002, Ubiquitination of both adeno-associated virus type 2 and 5 capsid proteins affects the transduction efficiency of recombinant vectors. *J. Virol* **76**: 2043-2053

Yu, G.Y., and Lai, M.M.C., 2005, The ubiquitin-proteasome system facilitates the transfer of murine coronavirus from endosome to cytoplasm during virus entry. *J. Virol* **79**: 644-648

Zang, G., and Ghosh, S., 2001, Toll-like receptor-mediated NF-κB activation: a phylogenetically conserved paradigm inEdwards, L., 1990, PhD Thesis *Cancer in Rats*. University of London, England

Chapter 4.2

HUMAN MONOCLONAL ANTIBODIES FOR PROPHYLAXIS AND THERAPY OF VIRAL INFECTIONS

J. TER MEULEN and J. GOUDSMIT
Crucell Holland B.V., Archimedesweg 4, 2301 CA Leiden, The Netherlands

Abstract: Monoclonal antibody (mAb) technology has reached a state of maturity making it possible to rapidly discover and produce fully human mAbs which neutralize any given virus *in vitro*. Several antibodies have also shown good prophylactic and therapeutic efficacy in relevant animal models of viral disease. Surprisingly, only one antiviral mAb has been licensed to date; it is used to prevent infection with respiratory syncytial virus in at-risk infants. Certain paradigms of mAbs prevailing in industry and academia are obstacles to the development of further antiviral antibodies for clinical use. There is a perception that mAbs are much more effective in prophylactic than therapeutic use, offer limited breadth of protection against different viral strains, are prone to select neutralization escape variants and need to be given in high doses, which are still comparatively costly to manufacture. Based on a review of the literature and our own data, we propose that by combining two or more non-competing neutralizing antibodies, viral escape can effectively be controlled and in case of synergistic action of the mAbs it may be possible to significantly reduce the total antibody concentration required for protection. The combination of mAbs with specific antiviral drugs has the potential to expand their use from pre- and post-exposure prophylaxis to therapy of acute and chronic viral infections.

1. INTRODUCTION

Before the advent of specific vaccines and antiviral drugs, passive prophylaxis with pooled normal donor serum ("immune globulin") or pooled pathogen-specific ("immune") serum from recovered patients or immunized volunteers was used very successfully to prevent infections with a variety of

E. Bogner and A. Holzenburg (eds.), New Concepts of Antiviral Therapy, 479–505.

viruses, such as cytomegalovirus, enteroviruses, hepatitis A, hepatitis B, measles, parvovirus B19, rabies, respiratory syncytial virus, varicella and variola (reviewed by Casadevall *et al.* 2004 and Sawyer 2000). Despite high production costs and the inherent risks of transmitting known and unknown blood borne pathogens, many of these sera are still in clinical use. Important indications are the protection of individuals at high risk of infection or at high risk of developing complications after infection. The immune globulin or pathogen-specific immune serum is given either as prophylaxis before or immediately after exposure (post-exposure prophylaxis, PEP) to immuno-compromised or normal persons. It is used less frequently for therapy of acute or chronic infections in immunocompromised patients.

Typical prophylactic indications include RSV-specific antibodies for at-risk infants, hepatitis B or CMV immune serum for transplant patients, varicella zoster and measles virus antiserum for exposed immuno-compromised patients and immunoglobulins for patients with X-linked agammaglobulinemia. However, in the case of highly lethal diseases, immunocompetent persons also benefit from passive immunization. Rabies virus kills 30,000 to 40,000 people annually in Asia and the shortage of affordable high quality rabies immunoglobulin for PEP given in combination with the vaccine is considered a serious global health problem.

CMV-immunoglobulin plus ganciclovir is the treatment of choice for established pneumonia in bone marrow and stem cell transplant patients. Other acute viral infections showing some, mostly only anecdotally documented response to immunotherapy include the complications of smallpox vaccination (eccema vaccinatum, generalized vaccina), Lassa fever, West Nile virus infection, SARS and Ebola virus infection.

Of the chronic viral infections, enteroviral encephalitis and parvovirus B19 anemia in immunosuppressed patients are amenable to immunoglobulin therapy. In contrast, HIV and hepatitis B and C, which affect hundreds of millions of people worldwide, have shown little responsiveness to serum therapy. Possible reasons include low-titered or virus strain-specific neutralizing antibodies in donor sera, generation of neutralization escape variants and requirement of a strong T-cell response for clearance of infected cells.

From the above it can be concluded and has been shown experimentally in animal models, that for pre- and post-exposure prophylaxis of viral infections human immune sera can effectively be replaced by monoclonal antibodies (mAb), provided the mAb offers sufficient breadth of protection against all clinically relevant strains of a virus. It also follows that late PEP or early therapy of an acute infection will be possible in certain circumstances, provided that the main mechanism of viral clearance is antibody mediated and that the mAb reaches sufficient concentrations in the target organs of viral replication (e.g. the brain). Chronic infections, however, are very unlikely to respond to therapy with mAbs unless immune escape is prevented by reducing viral replication independently of antibody action (e.g. by an antiviral drug) and the immune response is modulated such that additional effector mechanisms for viral clearance are stimulated.

Murine mAbs against transplant rejection and blood clotting were the first monoclonal antibodies to enter the clinic in the late 1980s, but the induction of immunological side effects was an obstacle to the development of further products. The advent of recombinant antibody technology in the 1990s then allowed the generation of less immunogenic chimeric mouse-human and later "humanized" mouse mAbs, which has set off an annual 20% growth of the mAb market to currently 30 billion US$/a (Baker 2005). At present 152 novel mAbs are undergoing clinical trials, mostly for cancer or immunological disorders and approximately 10% for infectious diseases. Despite these successes, only a handful of antiviral antibodies are currently in clinical trials, mainly against HBV and HCV. Only one mAb is licensed and marketed for prevention of a viral disease (Palivizumab), which is given to at-risk infants for the prevention of infection with respiratory syncytial virus. While economic considerations with respect to the smaller market size of anti-infectives and comparatively high production costs of mAbs may partly explain this discrepancy, we think that certain scientific paradigms prevailing in industry and academia are important obstacles to the development of further antiviral antibodies for clinical use. The following chapter will therefore try to shed light on the generation and preclinical evaluation of antiviral mAbs and discuss possibilities to improve their properties based on our own data and a review of the literature.

2. GENERATION OF HUMAN MONOCLONAL ANTIBODIES AND EVALUATION OF THEIR ANTIVIRAL PROPERTIES

2.1 Technical approaches

Immunoglobulins are composed of heavy and light chains, the light chain being either κ or λ. The antigen-binding site of an antibody is composed of six hypervariable or complementarity determining regions (CDRs), of which three are located within the variable domains of the light-chain (V_L) and heavy chain (V_H), respectively. In the immune system a large repertoire of different variable domains determining the antibody specificity is generated through combinatorial assembly of germline gene segments (V, D and J), with each B-cell expressing a different antibody specificity. Antigen exposure leads to clonal expansion of those B-cells producing antigen-binding antibodies and induces somatic mutations in the V genes, which results in further selection of clones with improved affinities of the antibodies (affinity maturation).

Of the 18 therapeutic mAbs in clinical use today, three are murine, five are chimeric (murine V region, human F_C region) and the remainder are humanized (mouse CDRs grafted onto a human antibody). Because fully human mAbs will theoretically not induce any immune response in the patient, several *in vivo* and *in vitro* technologies were developed for their generation (reviewed by Hoogenboom 2005). Historically, *in vitro* immortalization of peripheral human B-cells isolated from infected or immunized donors was performed with Epstein Barr virus to subsequently isolate and identify antibody secreting clones (Steinitz *et al.* 1977). This technique has recently been modified to increase the number of transformed IgG+ B-memory cells by *in vitro* stimulation with CpG-oligonucleotides and sorting of memory cells using the surface marker CD22, which has resulted in recovery of a high number of specific antibody clones, which have undergone in-vivo affinity maturation (Traggiai *et al.* 2004). Transgenic mice with human immunoglobulin genes have been generated to study the maturation of antibody genes. They are immunized with the antigen of interest and monoclonal antibodies are rescued by fusing mouse B-cells with mouse myeloma cells (Lonberg 2005). The resulting antibodies are in-vivo affinity-matured.

Ribosome display of antibodies is a PCR-based *in vitro* display technology, in which the individual nascent proteins are coupled to their corresponding mRNA through formation of stable protein-ribosome-mRNA complexes. Thus it is possible to isolate the functional protein (antibody) together with its encoding mRNA, which is reverse transcribed and PCR-amplified for further rounds of selection (Lipovsek *et al.* 2004). Due to the error-prone process of reverse transcription and amplification, the system's most successful application is the affinity maturation of antibodies.

Surface display on yeast (*S. cerevisiae*) allows for selection of antibody repertoires by cell flow cytometry. The V_H and V_L genes are diversified with random mutagenesis to yield high affinity antibodies (Boder *et al.* 2000). Other formats including surface display on E. coli and Bacillus strains, retroviral display, display based on protein-DNA linkage and others have either specific technical disadvantages or are not yet mature enough for commercial use.

Phage display is currently the most popular molecular technology used to tap into the human antibody repertoire (Hoogenboom 2005). As table 1 shows, *in vitro* and *in vivo* neutralizing human monoclonal antibodies have been generated against a variety of virus from diverse families by phage display.

All variable, antigen binding regions of the heavy and light chains, termed V_H and V_L, are cloned from the lymphocytes of a donor and expressed as single-chain antibodies (V_H-V_L) fused to the pIII surface protein of a bacteriophage. The resulting vast repertoire ("library") of different antibody carrying phages ideally mirrors the antibody repertoire of the donor. Cloning of all different antigen binding regions is possible because the hypervariable regions which are responsible for antigen specificity (CDRs) are flanked by constant regions (framework regions, FR) which correspond to families of germline antibody genes for which the genetic sequences are available. Because the heavy and light chains are amplified and cloned by two separate PCR reactions, the library also contains pairings not present in the donor lymphocytes, which further increases its size. The antibody phages are incubated with the antigen of interest and specifically binding phages are rescued together with the genetic information coding for the specific antibody gene. Both the cloning and the selection procedure are very powerful, allowing essentially to generate a repertoire of $10^8 - 10^9$ different molecules and extracting the desired antibody specificity from a single test tube.

For further characterization, antibody genes are recovered from the phages, recloned in appropriate vectors and transiently expressed in eukaryotic cell lines as fully functional monoclonal antibodies of a desired class and subclass. Starting from high-quality donor material, a typical antiviral mAb discovery program may yield 1000-2000 virus binding antibody phages, of which 100-200 are converted to IgG1-molecules for further *in vitro* and *in vivo* screening. Finally, three to four antibodies are selected as product leads and stable production clones of a suitable cell line (e.g. PER.C6®) are generated (figure 1). Excellent laboratory manuals are available describing the technical procedures in detail (Kontermann and Dübel 2001).

Table 1. In-vitro neutralizing human monoclonal antibodies generated by phage display or other methods

Virus	Target	In-vivo neutral.	Reference
CMV	gB, gH	nd	Nejatollahi 2002
Dengue1,2	E	nd	Goncalvez 2004
Ebola	GP	Guineapig	Parren 2002
Hantaan	G1,G2	nd	Koch 2003, Liang 2003
HAV	14S, 70S, subviral particles	nd	Kim 2004
HBV	Pre-S, S	Chimpanzee	Hong 2004[c], Eren 2000[b]
HCV	E1, E2	nd	Keck 2004[a], Habersetzer 1998[a]
HEV	ORF 2	Rhesus	Schofield 2003
HIV-1	gp120 (V2, V3), gp41 CD4/co-receptor complex	Chimpanzee, rhesus SCID- mice	He 2002[b], Moulard 2002, Wang 1999, Purtscher 1994[a]
HIV-2	gp120	nd	Bjorling 1999
Measles	H	nd	De Carvalho Nicacio 2002
Parvo	VP2	nd	Arakelov 1993[a]
Puumala	G2	nd	De Carvalho Nicacio 2000
Rabies	G	Hamsters	Bakker 2005, Hanlon 2001[a], Enssle 1991[a]
Rotavirus	VP4, VP7	nd	Higo-Moriguchi 2004
RSV	F	nd	Johnson 1997[c]
SARS	S1, S2	Mice, ferrets	Greenough 2005[b], ter Meulen 2004, Sui 2004, Traggiai 2004[a]
Vaccinia	p95, p34	nd	Schmaljohn 1999
VZV	gE	nd	Kausmally 2004
Yellow fever	E	nd	Daffis 2005

a Immortalized human B-cells b Transgenic mice c Humanized murine mAb

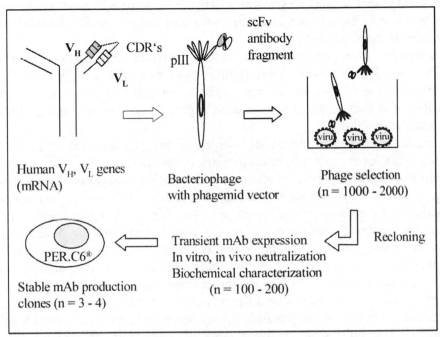

Figure 1. Generation of human monoclonal IgG antibodies from phage display libraries. mRNA is isolated from immunized or infected lymphocytes donors, reverse transcribed into cDNA and all V_H and V_L genes are amplified using a set of approx. 40 consensus primers directed against the conserved framework regions of the variable antibody genes. The antibody genes are cloned in a phagemid vector and infectious phages each expressing a different single-chain antibody molecule (the V_H and V_L chains being linked by an artificial polypeptide spacer) are rescued from bacteria using a defective helper phage (Kramer *et al.* 2003). The resulting "antibody phage library" is screened with viral antigen in different formats for specific binders during 2-3 rounds of selection. The rescued antibody genes are then recloned in appropriate vectors and transiently expressed as fully human IgG1 molecules. These are used to determine the potency, affinity and epitopes of the antibodies. A stably transfected production cell line is generated for each lead antibody. The whole process can be performed in approx. 9 months.

2.2. Libraries and antigens

The quality of the library and the quality of the antigens used for screening are the two major critical factors for a successful discovery of highly potent antibodies. Libraries generated from non-immune donors have been enlarged by introducing mutations in the CDRs of the V_H and V_L chains resulting in a size of up to 10^{10} different specificities, and are readily available for screening (de Kruif *et al.* 1995). Interestingly, these libraries can yield potent neutralizing mAbs reactive with viruses against which the donors have no measurable serum antibody titer, such as SARS coronavirus (van den Brink *et al.* 2005).

Libraries from immunized or infected donors are smaller in size (typically 10^8), but with good timing concerning the sampling of the lymphocytes in relation to the time point of infection they contain a large number of affinity matured and therefore highly potent antibody specificities (Kramer *et al.* 2005).

Neutralizing viral epitopes are located on the glycoproteins in case of enveloped viruses or on the capsid proteins in case of non-enveloped viruses. Often these epitopes are discontinuous and sensitive to denaturation, so that they may not be preserved if purified virions are chemically or physically inactivated for safety reasons before selection. Similarly, recombinant viral envelope proteins may not adopt the native conformation under the expression and purification conditions used. Furthermore, surface proteins of viruses tend to adopt oligomeric configurations in which aminoacids from two molecules may contribute to one epitope (Daffis *et al.* 2005). It is therefore very useful to employ different antigen formats for screening of the libraries, including whole purified virions, virus like particles, recombinant proteins and eukaryotic cells expressing viral surface proteins. By performing the second round of selections on a different antigen format than the first, unspecific reactivity can be reduced. Certain combinations of antigen formats, e.g. selection on cells expressing viral surface proteins followed by selection on whole virions, may yield the majority of neutralizing antibodies recovered from a library (Marissen *et al.* 2005).

2.3.   In vitro and in vivo characterization of antiviral mAbs

Potent neutralizing antibodies need to be selected as quickly and efficiently as possible from the large number of antigen binding phage antibodies, which can be in the order of many thousand. In principle, this is possible by characterizing the molecules biochemically by measuring e.g. their affinities to the virus, or functionally by determining their neutralizing potency. While the term "neutralization" in its broadest sense denotes the inhibitory effect of an antibody on viral infectivity or replication, it is more often specifically used when referring to the inhibition of viral attachment and early entry functions (endocytosis, fusion, uncoating). There has been some debate as to whether neutralization of viruses is due to binding of an antibody to one critical site on the virion (leading to a single-hit kinetic of neutralization) or to epitopes present in multiple copies on the virus surface (multiple-hit kinetic). In the latter case neutralization can be regarded as a function of antibody affinity and concentration, which determine whether a critical number of antigens have reacted with the antibody (so called relative occupancy model of neutralization). Many experimental data point to the multiple-hit kinetic and occupancy model, with 4-7 IgG molecules required to neutralize picornaviruses, 70 to neutralize influenza and 225 to neutralize rabies virus. Interestingly, for several viruses a linear relationship between the number of antibodies required for neutralization and the particle surface area was found, supporting the idea that coating of the virion surface by any antibody is sufficient for neutralization (high occupancy or coating theory). Exhaustive reviews of the subject have been published by Parren & Burton 2001 and Klasse & Sattentau 2002. However, other molecular mechanisms of neutralization have also been proposed which may or may not follow the occupancy model, including induction of conformational changes in the capsid or envelope proteins, aggregation of viral particles, complement fixation and signalling to the interior of the virion or the infected cell.

Neutralization of HIV has been studied in great detail and revealed some unique features of the rare mAbs which are broadly cross-clade neutralizing (reviewed by Ferrantelli and Ruprecht 2002). Long, finger-like CDR3 loops have been found inserting into the narrow CD4 binding pocket of gp120 or displaying a hydrophobic patch which reacts with the membrane-proximal region of gp41 and the viral membrane (mAbs b12 and 2F5/4E10, respectively. Zwick *et al.* 2003, Zwick *et al.* 2005). The latter mAbs, which were recovered by phage display from HIV infected patients, were found to be autoreactive with cardiolipin, which possibly explains that their specificities are rarely detected in humans (Haynes *et al.* 2005). The broadly neutralizing mAb 2G12 was found to bind to the heavily glycosylated, functionally "silent" face of gp120 with an affinity similar to that of a protein-protein interaction. Crystal structure determination revealed that by interdigitation of the Fab-domains (V_H-domain swapping) an increased antigen-binding surface is generated allowing for multivalent interaction with conserved mannose structures (Calarese *et al.* 2005).

Affinity measurement using plasmon surface resonance is widely used to characterize the binding ability of antibodies to their targets and in many cases a correlation can be observed with their biological activity. For virus neutralization, however, it has repeatedly been reported that no such correlation was found (Wu *et al.* 2005, Bakker *et al.* 2005). A possible explanation could be that a low affinity to a functionally important neutralizing epitope results in a higher potency of the mAb compared with a higher affinity to a less important epitope. This may be illustrated with the SARS coronavirus neutralizing mAbs CR3014 and CR3022, which we isolated from a naïve and an immune library, respectively (van den Brink *et al.* 2005, ter Meulen, unpublished). Both block binding of SARS-CoV to its cellular receptor ACE2 on Vero cells, as shown in FACS-experiments using the recombinant expressed S1 subunit of the spike glycoprotein, which contains the receptor binding domain (RBD). Both antibodies bind to the RBD in a non-competing fashion, as shown in competitive ELISA. However, CR3014 has a rather low affinity to recombinant S1 (K_D = 27 nM) but a reasonable potency (7.6 µg/ml for 100% neutralization), whereas CR3022 has a high affinity (300 pM) but lower potency (50 µg/ml). Obviously the mAbs recognize either two different neutralizing epitopes on the RBD, or one binds to another epitope in the S1 subunit thereby inducing a conformational change in the RBD (allosteric effect) or possibly blocks interaction with a putative co-receptor of SARS-CoV.

Therefore we find it preferable to characterize all antibodies first in a rapid screening assay for *in vitro* neutralization, and thereafter define the biochemical characteristics of the most potent ones. However, the neutralizing activity of an antibody may involve factors not present in standard neutralization, i.e. cell culture protection assay, such as complement fixation. Furthermore, if the *in vivo* neutralizing activity involves antibody dependent cellular cytotoxicity (ADCC) or perhaps signalling to infected cells, mAbs with these properties will also be missed. It has been observed that *in vitro* potency may not be predictive of *in vivo* potency, and even non-neutralizing antibodies may confer *in vivo* protection (Iacono-Conors *et al.* 1996, Griffin *et al.* 1997, Mozdzanowska *et al.* 1997). Since high-throughput screening of antibodies in animals is not feasible, there is a need for the development of additional *in vitro* assays for measuring antiviral antibody functions other than blocking of attachment and entry.

Potent neutralizing mAbs will then be further characterized with respect to their target, normally a viral attachment protein on the surface of the virion, which can be identified by Western blot or immunoprecipitation. Binding to known antigenic domains can conveniently be tested in a competition ELISA by blocking binding sites with antibodies whose binding sites are known. If the epitopes recognized are linear, they can be mapped by pepscan analysis with overlapping peptides corresponding to the amino acid (aa) sequence of the protein and by subsequent alanine scan to identify the aa critical for binding. Conformational domains are mapped by sequencing escape variants of the virus generated with subneutralizing antibody concentrations. If the crystal structure of the target protein is known, the 3D localisation of the amino acids critical for neutralization can be determined and the molecular mechanism of neutralisation may be speculated upon (Daffis *et al.* 2005). These neutralization variants are also important to understand the likelihood and mechanism of immune escape induced by a mAb, and how to prevent it by combining two neutralizing mAbs targeting different neutralizing epitopes.

Preclinical development of a mAb requires demonstration of efficacy and safety in animal models. While infected animals replicate many viruses to some titer, they often do not develop a clinical state comparable to the human disease and show a different pathology. In order to limit the number of animals used and unequivocally demonstrate efficacy of an antibody or vaccine, 100% lethal animal test systems have been developed, often based on suckling mice. These systems are convenient to test pre- or post-exposure prophylaxis with a mAb, especially if there is a clear dose-response of virus challenge and mortality as well as mAb concentration and protection. Unfortunately, for many viral infections no small animal models are available which would generate a clinical and pathological state more closely resembling human disease. In case of chronic viral infections such as HIV, hepatitis B and hepatitis C, non-human primates (NHP) or chimpanzees are the most meaningful animal models, which makes evaluation of antivirals extremely costly and time consuming.

The most extensive studies in NHP were performed with anti-HIV mAbs, reviewed by Ferrantelli and Ruprecht 2002. Passive immunization with neutralizing human mAbs – used alone or in combination with other monoclonal or polyclonal neutralizing antibodies – completely prevented infection in some adult animals challenged intravenously or intravaginally, and in neonatal monkeys challenged orally. Most animals that did become infected had low viral RNA loads and were protected from challenge virus-induced acute disease. In a recent study, a cocktail of three human mAbs with potent cross-clade and cross-group HIV-neutralizing activity was given to neonatal rhesus macaques one hour and again eight days after oral HIV challenge; all animals remained virus negative for > 1 year (Ferrantelli *et al.* 2004).

3.   CLINICAL TRIALS WITH ANTIVIRAL MONOCLONAL ANTIBODIES

Current clinical trials with mAbs are targeting well defined groups of patients, in which the medical and economic consequences of unchecked viral replication justify the comparatively high costs of a mAb prophylaxis.

Endogenous reinfection of an orthotopic liver transplant with hepatitis B or C presents a major medical problem, since the virus remains in several other body compartments following removal of the infected liver. Without treatment, reinfection of the transplanted liver occurs rapidly resulting in progressive disease, graft failure, and death. Life-long prophylactic treatment is therefore necessary. A mixture of two fully human monoclonal antibodies, directed against different epitopes of hepatitis B surface antigen (HBsAg) and binding to all major HBV subtypes, is presently being studied in a Phase IIb clinical trial in liver transplant patients (XTL Biopharmaceuticals, Israel). In an earlier phase I clinical study, a total of 27 chronic HBV patients were enrolled. In part A of the study 15 patients in 5 cohorts received a single intravenous infusion of antibodies with doses ranging from 0.26 mg (260 IU) to 40 mg (40,000 IU). All patients completed 16 weeks of follow-up. In the second part of the study (part B), 12 patients in 4 cohorts received 4 weekly infusions of 10, 20, 40, or 80 mg each of the mAb cocktail and were followed for 4 additional weeks. Patients administered doses at an Ab:Ag molar ratio of 1:2 to 1:20 showed a rapid and significant decrease in HBsAg to undetectable levels, with a corresponding reduction of HBV-DNA levels. In part B, the mAbs induced a significant reduction in both HBsAg and HBV-DNA levels repeatedly after administration (Galun *et al.* 2002).

The same company is conducting a phase Ia/b clinical trial with a combination of two anti-HCV-E2 mAbs to prevent hepatitis C re-infection following a liver transplant and for the treatment of chronic HCV. mAb68 and mAb65 immunoprecipitate viral particles from patients' sera infected with different HCV genotypes and incubation of an infectious human serum with Ab68 or Ab65 prevented the serum's ability to infect human liver cells and human liver tissue. A Phase IIa Clinical Trial was performed with Ab68 following liver transplant, which demonstrated the safety and tolerability of the mAb up to 240mg dosed for 12 weeks. The 120mg and 240mg dose groups had a significantly greater reduction in viral load than the placebo group during the first week when dosed daily. This effect was less evident when dosed less frequently than daily. It is thought that a combination of two antibodies that bind to different epitopes is essential to provide broad coverage of virus quasispecies, and to minimize the probability of escape from therapy.

To prevent CMV infection after allogeneic hematopoietic stem cell transplantation, a mAb (MSL-109) specific to the cytomegalovirus (CMV) glycoprotein H with high neutralizing capacity was evaluated in a prospective, randomized, double-blind study (Boeckh *et al.* 2001). Allogeneic hematopoietic stem cell transplantation (HSCT) recipients with positive donor and/or recipient serology for CMV before transplantation received either 60 or 15 mg/kg MSL-109, or placebo intravenously every 2 weeks from day -1 until day 84 after transplantation. CMV pp65 antigenemia, CMV-DNA load in plasma, and viremia by culture were tested weekly. Primary end points were development of pp65 antigenemia at any level and/or viremia for which ganciclovir was given. There was no statistically significant difference in CMV pp65 antigenemia or viremia among patients in the mAb and the placebo group.

In two small phase I studies, two human mAbs and one humanized anti-HIV mAb directed against gp120 and gp41 were reported to transiently lower the viral load in five of seven and three of four patients, respectively (Stiegler *et al.* 2002, Dezube *et al.* 2004). These initially encouraging results led to a recent proof-of-concept study designed to mimick therapeutic vaccination by administering a cocktail of three neutralizing human mAbs (2G12, 2F5, 4E10), interrupting antiretroviral therapy (ART) and measuring the level of viral rebound (Trkola *et al.* 2005). Eight chronically and six acutely HIV-1 infected patients, all with undetectable levels of viremia while on ART, received multiple infusions of the mAbs over an 11-week period. ART was discontinued one day after commencing the antibody treatment. The HIV strains harboured by all patients were shown to be highly sensitive to neutralization with at least two mAbs. No side effects of the mAb therapy were observed, despite 2F5 and 4E10 having been shown to be cross-reactive with cardiolipin (Haynes *et al.* 2005). Viral rebound to pre-ART plasma levels occurred without apparent delay in six chronic and two acute patients. Rebound was delayed in the remaining six patients, but then rose to pre-ART levels in one acute and one chronic subject in the presence of high plasma concentrations of the mAbs. In two acute patients, rebound peaked 2-4 logs below pre-ART levels and remained low after the mAbs were cleared from the circulation. One subject had no detectable viral rebound. Surprisingly, most of the effect of the treatment could be attributed to mAb 2G12 only. Patients with a delay in rebound had very high plasma levels of this mAb with on average 400 times its IC_{90} and more than double the highest effective dose for 2F5 and 4E10 in any patient, probably due to the longer plasma half-life of 2G12. Furthermore, 2G12 was the only antibody that rebound viruses escaped, while remaining sensitive to 2F5 and 4E10. It was therefore speculated that the plasma concentrations achieved for 2F5 and 4E10 were below a crucial threshold needed to control the virus. In conclusion, this study showed that control of HIV by neutralizing mAbs may be possible but will require high concentrations of combinations of antibodies capable of preventing immune escape or combinations with other antiviral drugs reducing replication of the virus.

The combination of mAbs with antiviral drugs in clinical trials is discussed below.

4.   STRATEGIES TO IMPROVE ANTIVIRAL MONOCLONAL ANTIBODIES

Based on experimental data and results from clinical trials, prophylaxis and therapy of viral infections with mAbs is feasible if the breadth of protection against viral strains is sufficient, generation of immune escape variants is prevented and the mAbs are of such high potency that they can be administered at low, economically feasible doses.

4.1   Increasing potency through affinity maturation or altering the avidity

It has been shown that increased binding ability to a critical viral epitope results in greater neutralizing potency of antiviral monoclonal antibodies (Johnson *et al.* 1999). Increased binding may result from increased affinity, as determined by the association (k_{on}) and dissociation (k_{off}) rates in plasmon surface resonance (BIAcore) measurements, or from increased avidity, due to multivalent binding (Alfthan 1998).

The affinities of antibody fragments derived from naïve repertoires of ~10^8 immunoglobulin genes appear to be in the micromolar range, characteristic for primary immune responses (Griffiths *et al.* 1993). *In vivo*, the primary immune response is further improved via a process of mutations and selections called affinity maturation. This process can be mimicked *in vitro* to optimize the antibodies selected by phage display (Chowdhury *et al.* 1999). To improve the affinity, the six CDRs of the heavy and light chains fragments can be subjected to mutagenesis or chain shuffling (recombining a V_H or V_L chain with a library of partner chains), followed by phage based selection (Boder *et al.* 2000, Yang *et al.* 1995, Wu *et al.* 2005, Thompson *et al.* 1996). The affinities achieved *in vitro* have been in the nanomolar or picomolar range, which is comparable to affinities of antibodies isolated from immune libraries.

Although antibody affinity is mediated by both k_{on} and k_{off}, the greatest affinity improvements through *in vitro* maturation have been achieved by reducing k_{off}. Selected or combinatorial point mutations in the CDRs have been reported to reduce k_{off} up to 10-fold and 100-fold respectively, and were reported to improve their *in vitro* HIV neutralization potency (Yang *et al.* 1995, Thompson *et al.* 1996). However, as was recently shown for antibody variants derived from the anti-RSV antibody Palivizumab, k_{on} and k_{off} may have differential effects on the neutralizing potency of F_{ab}-fragments and full IgGs, respectively. A modest four to five-fold increased association rate resulted in a dramatic 20-fold increase in potency, whereas even a 100-fold decrease in dissociation rate did not significantly increase potency (Wu 2005).

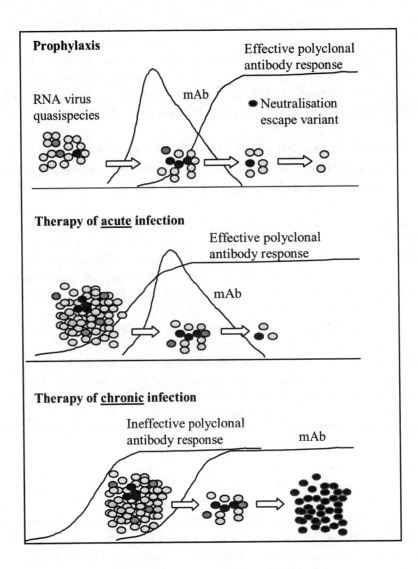

Figure 2. Immune escape induced by antiviral mAbs is less likely to occur in pre- and postexposure prophylaxis and therapy of acute viral infections, because of the effective polyclonal immune and T-cell response. In contrast, in chronic infections the virus reservoir is not eradicated by the mAb and the ineffective polyclonal antibody response will not control escape variants.

The authors suggest that a fast association rate may enable the mAb to neutralize the virus before it infects susceptible cells.

Expressing mAbs as IgM molecules has also been shown to increase the *in vitro* neutralizing potency of anti-HIV mAbs, possibly mediated both by increased avidity and by increased complement fixation (Wolbank *et al.* 2003). However, the IgM-format of recombinant antibodies is currently more difficult to produce and pharmacologically less well evaluated than the standard IgG1 format. To this end, novel antibody formats based on expressing antigen-binding fragments as multivalent molecules (triabodies, tetrabodies) are very interesting recent developments (Hollinger 2005).

4.2   Increasing potency, breadth of protection and preventing immune escape through mixing of non-competing mAbs

Due to their exquisite epitope specificity mAbs may not cover all clinically relevant strains of a virus or rapidly select neutralization escape variants from the quasispecies which RNAs viruses tend to form. However, both in prophylaxis and early therapy of acute infections immune escape is unlikey to occur because the mAb is the first line of defence after which an effective polyclonal immune response should eliminate any escape variants (figure 2). In chronic infections, immune escape variants will be selected by the mAb unless the virus reservoir eradicated.

For acute infections, the immune escape problem can be solved by combining two neutralizing, non-competing mAbs as we have shown for rabies virus. Rabies virus infection in man is 100% fatal, unless the vaccine and human rabies immune globulin (HRIG) are administered immediately after exposure. The disease is most frequently caused by transmission of one of several serotypes of Lyssa virus, genotype 1 through the bite of an infected, rabid canine. Several additional animal species, notably bats, harbour other of the 12 currently known genotypes of Lyssa virus (rabies and rabies-like viruses), some of which have caused disease in man. Therefore, any mAb used for replacement of HRIG must provide the same breadth of protection, most importantly against rabies viruses transmitted by rabid animals ("street viruses"). In addition, there is a hypothetical risk of selecting a viral neutralization escape variant under selective pressure of the mAb, despite only limited replication of rabies virus at the site of inoculation.

To address these problems, we identified and characterized two non-competing human mAbs, binding to antigenic sites I and III of the rabies virus glycoprotein. Their epitopes were mapped by generating neutralization escape variants, which could be neutralized by the respective other mAb.

Furthermore, the mAb combo was shown to have the same breadth of protection as HRIG and high *in vitro* and *in vivo* potency (additive effect). Analysis of the linear epitopes of these antibodies revealed that the majority of natural rabies virus isolates contain both intact epitopes and the few remaining strains contain at least one of the two (Bakker *et al.* 2005). These mAb offer for the first time the possibility to replace HRIG with a safe, potent and cheap recombinant product available in unlimited quantity.

In chronic infections, escape from neutralizing antibodies has readily been observed in humans. Neutralizing antibodies can protect against challenge with HIV-1 *in vivo* if present at appropriate concentrations at the time of virus challenge, but their role in the control of established infection is unclear. Autologous neutralizing antibodies (Nab) can be detected as early as 52 days after detection of HIV-specific antibodies. The viral inhibitory activity of Nab results in complete replacement of neutralization-sensitive virus by successive populations of resistant virus, involving primarily changes in N-linked glycosylation (Wei *et al.* 2003). In a phase I study with a cocktail of two neutralizing mAbs (2F5, 4E10, 2G12), immune escape was observed with one of the mAbs (Trkola *et al.* 2005). Experimentally, even high serum concentrations of neutralizing monoclonal antibodies, either singly or as a cocktail, have little sustained effect on viral load in established HIV-1 infection in hu-PBL-SCID mice. In some instances, virus replication of neutralization-sensitive virus continues even in the presence of high levels of neutralizing antibody. In most instances, neutralization escape occurs in a few days, even from a cocktail of three antibodies that recognize distinct epitopes (Poignard 1999). The authors concluded that humoral immunity is unlikely to play a significant role in the control of established HIV-1 infection in humans. Similar observations have been made in hepatitis B and C (Rehermann and Nascimbeni 2005). It is therefore questionable whether the action of neutralizing mAbs can be improved such that the immune escape problem in chronic infection can be controlled by antibody alone.

In certain instances, combining mAbs may not only increase the breadth of protection, but also lead to an increase of the *in vitro* and *in vivo* potency of the mixture disproportionately to the individual potencies of the antibodies (synergistic action), as has been shown for combinations of two, three or four mAbs directed against different epitopes on the HIV-1 envelope glycoprotein (Li *et al.* 1998, Zwick *et al.* 2001, Xu *et al.* 2001). We have observed synergistic neutralization of SARS-CoV. Human mAb CR3014 isolated from a naïve antibody phage library and non-competing mAb CR022 isolated from an immune library were combined in equimolar ratios and the concentration required for 100% and 50% neutralization of SARS-CoV was determined. Applying a mathematical model based on the mass-action law principle (Chou and Talalay 1984) revealed that CR3014 and CR3022 neutralize SARS-CoV synergistically *in vitro*, with dose reduction indices of 3 and 20, respectively. Mechanistically, cooperative binding of mAbs may induce conformational changes in the antigen thereby altering affinities (allosteric effect), lead to intermolecular interaction of bivalent antibodies or result in interaction of the Fc regions of antibodies brought into close contact (Klonisch *et al.* 1996). As determined by ELISA, CR3014 and CR3022 bind simultaneously to the RBD of the S1 subunit of the glycoprotein spike of SARS-CoV. However, measuring separate, sequential and simultaneous binding of the mAbs in Biacore did not reveal enhanced affinity of the antibody mixture over the individual affinities, especially for C3022. Neutralization escape variants generated with one anti-SARS-CoV mAb were effectively neutralized by the other. When administered prophylactically to ferrets before intratracheal challenge with SARS-CoV, mAb CR3014 prevented lung pathology due to viral replication and abolished shedding of the virus in pharyngeal secretions at a dose of 10mg/kg (ter Meulen *et al.* 2004). The synergistic action of the mAbs may allow for a lower total combined antibody dose to reach the same potency, which increases the economic attractiveness of a potential product for immunoprophylaxis of SARS.

4.3.   Synergy of antibodies and antiviral drugs

Few experimental studies have addressed the potential synergy of neutralizing antibodies and antiviral drugs, however, data from clinical studies combining immune serum and antivirals may allow some conclusions.

Orthotopic liver transplantation for hepatitis B virus (HBV) infection was limited until recently by poor graft and patient outcomes caused by recurrent HBV. Long-term immunoprophylaxis with hepatitis B immune globulin (HBIG) dramatically improved post-transplantation survival, but recurrent HBV still occurred in up to 36% of the recipients due to the occurrence of immune escape mutants. More recently, combination HBIG and lamivudine has been shown to effectively prevent HBV recurrence in patients post-transplantation. However, optimum doses and duration for these drugs are not yet clear. The combination of lamivudine 300 mg/day and low-dose HBIG was reported to prevent post-transplantation recurrence of hepatitis B, even in the presence of viral replication in the pre-transplant period (Karasu *et al.* 2004). Furthermore, combination therapy resulted in a cost-effective strategy with an average cost-effectiveness ratio of $252,111/recurrence prevented compared with $362,570/recurrence prevented in the monotherapy strategy. It was concluded that combination prophylaxis with HBIG and lamivudine is highly effective in preventing recurrent HBV, may protect against the emergence of resistant mutants, and is significantly more cost-effective than HBIG monotherapy with its associated rate of recurrent HBV (Han *et al.* 2000). However, in low-risk liver transplant patients who were HBsAg-positive/HBV DNA-negative before transplantation and had received HBIg for at least 6 months without HBV recurrence, lamivudine alone was effective for prevention of HBV recurrence (Naoumov *et al.* 2001). Another study also concluded that lamivudine monotherapy after a short course of lamivudine and HBIg was equally as efficacious in preventing HBV recurrence as HBIg plus lamivudine during the first 18 months after liver transplantation (Buti *et al.* 2003).

In seronegative heart transplant recipients receiving an allograft from seropositive donors, double CMV prophylaxis consisting of CMV hyperimmune globulin and ganciclovir was shown to abolish CMV death and to prevent CMV disease (Bonaros *et al.* 2004). In contrast to these findings, a human monoclonal antibody directed against CMV gH for the treatment of newly diagnosed CMV retinitis in AIDS patients did not consistently produce clinical improvement (Borucki *et al.* 2004).

A synergistic effect of combining immune serum and the guanosine analogue ribavirin has been shown clinically for therapy of measles pneumonia (Barnard 2004, Stogner *et al.* 1993). In a mouse model of lymphocytic choriomeningits (LCMV) infection, monoclonal antibody expression in transgenic mice alone was unable to prevent the establishment of persistent infection due to generation of neutralization escape variants. Daily administration of ribavirin resulted in elimination of the virus after two weeks and virus isolated at earlier time points was shown to remain sensitive to mAb neutralization (Seiler *et al.* 2000). It was concluded that the main effect of ribavirin was reduction of viral replication efficiency, which rendered the appearance of antibody escape variants less frequent and lowered viral titers sufficiently so that CTLs were able to control the virus. In agreement with this study, virus-specific immune plasma was shown to enhance the beneficial effect of ribavirin on primate survival and virus replication after infection with another member of the arenavirus family, the Lassa virus (Jahrling *et al.* 1984).

5. CONCLUSION AND OUTLOOK

Many viral infections are currently prevented by passive prophylaxis with human immune serum, with only some being amenable to serum therapy. The state of monoclonal antibody technology now makes is possible to replace human sera by custom-tailored human monoclonal antibodies, thereby solving the safety, cost and supply problems inherent in blood products derived from human donors. However, only a few biotech companies have currently taken mAbs for HIV, hepatitis B, C, CMV and RSV infection in preclinical or clinical development. Economically, monoclonal antibodies are seen as less attractive antivirals than small molecule drugs due to the approximately two-fold increase in cost and time required to produce clinical-grade material. Scientifically, the possibly limited breadth of protection, generation of immune escape variants and uncertain clinical efficacy in case of established acute or chronic infection are perceived as major obstacles to the development of mAb products for control of viral infections.

We speculate that several of the antiviral mAbs which have produced unsatisfactory results in preclinical or clinical studies were not optimally selected. Based on our own data and published evidence we therefore propose a paradigm shift, predicting that a broad-scale screening for synergistic action of antibody cocktails as stand-alone or in combination with antivirals may solve many of the above outlined problems. Crucial to the discovery process of highly potent antiviral mAbs is a high-throughput screening based on *in vitro* or ideally *in vivo* neutralization, which should systematically investigate the possible neutralizing synergy of non-competing antibodies. Synergy of the mAbs with antivirals could also be included in an early screening step. If the potency of an individual antibody needs to be increased, selection for an improved k_{on} should be performed. Novel antibody formats based on multimerization of antigen-binding fragments may lead to molecules with high avidity and increased neutralizing potency.

Importantly, the state of recombinant antibody technology offers a unique opportunity to rapidly counteract epidemic threats of newly emerging diseases. It took less than a year to discover human mAbs which protect against SARS-coronavirus in different animal models and could be used for human ring vaccination in case of an outbreak or a bioterror attack. Avian influenza may represent another attractive target for passive immunization, if mAbs cross-reactive with the current Asian H5 strains can be found.

Economically, the increased production costs of mAbs and the longer production time compared to conventional small-molecules are set-off by a faster and more predictable clinical trajectory. The approval success rate for chimeric/humanized antibodies is currently 18-29% compared to 11% for new chemical entities (NEC) as a whole (Reichert 2005). Both academia and industry should therefore revisit the opportunity of using monoclonal antibodies for prophylaxis and therapy of viral infections.

ACKNOWLEDGEMENTS

We would like to thank all the colleagues from Crucell who have helped in many invaluable ways in the production of this chapter, in particular the Antibody Discovery group: J de Kruif, ABH Bakker, M Throsby, WE Marissen, RA Kramer, EN van den Brink, AQ Bakker, M Clijsters-van der Horst, TJ Visser, M Jongeneelen, S Thijsse, F Cox and M de Jong. We thank G. Turner for editing of the manuscript.

REFERENCES

Alfthan K, 1998, Surface plasmon resonance biosensors as a tool in antibody engineering. *Biosensors & Bioelectronics* **13**: 653-663.
Arakelov S, Gorny MK, Williams C, Riggin CH, Brady F, Collett MS, Zolla-Pazner S, 1993, Generation of neutralizing anti-B19 parvovirus human monoclonal antibodies from patients infected with human immunodeficiency virus. *J. Infect. Dis.* **168**: 580-585.
Baker M, 2005, Upping the ante on antibodies. *Nat. Biotechnol.* **23**: 1065-1072.
Bakker AB, Marissen WE, Kramer RA, Rice AB, Weldon WC, Niezgoda M, Hanlon CA, Thijsse S, Backus HH, de Kruif J, Dietzschold B, Rupprecht CE, Goudsmit J, 2005, Novel human monoclonal antibody combination effectively neutralizing natural rabies virus variants and individual *in vitro* escape mutants. *J.Virol.* **79**: 9062-9068.
Barnard DL, 2004, Inhibitors of measles virus. *Antivir. Chem. Chemother.* **15**: 111-119.
Bjorling E, von Garrelts E, Morner A, Ehnlund M, Persson MA, 1999, Human neutralizing human immunodeficiency virus type 2-specific Fab molecules generated by phage display. *J. Gen. Virol.* **80**: 1987-1993.
Boder ET, Midelfort KS, Wittrup KD, 2000, Directed evolution of antibody fragments with monovalent femtomolar antigen-binding affinity. *Proc. Natl. Acad. Sci. USA* **97**: 10701-10705.
Boeckh M, Bowden RA, Storer B, Chao NJ, Spielberger R, Tierney DK, Gallez-Hawkins G, Cunningham T, Blume KG, Levitt D, Zaia JA, 2001, Randomized, placebo-controlled, double-blind study of a cytomegalovirus-specific monoclonal antibody (MSL-109) for

prevention of cytomegalovirus infection after allogeneic hematopoietic stem cell transplantation. *Biol. Blood Marrow Transplant.* **7**: 343-51.

Bonaros NE, Kocher A, Dunkler D, Grimm M, Zuckermann A, Ankersmit J, Ehrlich M, Wolner E, Laufer G, 2004, Comparison of combined prophylaxis of cytomegalovirus hyperimmune globulin plus ganciclovir versus cytomegalovirus hyperimmune globulin alone in high-risk heart transplant recipients. *Transplantation* **77**: 890-7.

Borucki MJ, Spritzler J, Asmuth DM, Gnann J, Hirsch MS, Nokta M, Aweeka F, Nadler PI, Sattler F, Alston B, Nevin TT, Owens S, Waterman K, Hubbard L, Caliendo A, Pollard RB; AACTG 266 Team, 2004, A phase II, double-masked, randomized, placebo-controlled evaluation of a human monoclonal anti-Cytomegalovirus antibody (MSL-109) in combination with standard therapy versus standard therapy alone in the treatment of AIDS patients with Cytomegalovirus retinitis. *Antiviral Res.* **64**:103-11.

Buti M, Mas A, Prieto M, Casafont F, Gonzalez A, Miras M, Herrero JI, Jardi R, Cruz de Castro E, Garcia-Rey C, 2003, A randomized study comparing lamivudine monotherapy after a short course of hepatitis B immune globulin (HBIg) and lamivudine with long-term lamivudine plus HBIg in the prevention of hepatitis B virus recurrence after liver transplantation. *J. Hepatol.* **38**: 811-817.

Calarese DA, Lee HK, Huang CY, Best MD, Astronomo RD, Stanfield RL, Katinger H, Burton DR, Wong CH, Wilson IA, 2005, Dissection of the carbohydrate specificity of the broadly neutralizing anti-HIV-1 antibody 2G12. *Proc. Natl. Acad. Sci. USA* **102**: 13372-13377.

Casadevall A, Dadachova E, Pirofski LA, 2004, Passive antibody therapy for infectious diseases. *Nat. Rev. Microbiol.* **2**: 695-703.

Chou TC, Talalay P, 1984, Quantitative analysis of dose-effect relationships: The combined effects of multiple drugs or enzyme inhibitors. *Adv.Enzyme Regul.* **22**: 27-55.

Chowdhury PS, Pastan I, 1999, Improving antibody affinity by mimicking somatic hypermutation in vitro. *Nat. Biotechnol.* **17**: 568-572.

Daffis S, Kontermann RE, Korimbocus J, Zeller H, Klenk HD, ter Meulen J, 2005, Antibody responses against wild-type yellow fever virus and the 17D vaccine strain: characterization with human monoclonal antibody fragments and neutralization escape variants. *Virology* **337**: 262-272.

de Carvalho Nicacio C, Lundkvist A, Sjolander KB, Plyusnin A, Salonen EM, Bjorling E, 2000, A neutralizing recombinant human antibody Fab fragment against Puumala hantavirus. *J. Med. Virol.* **60**: 446-454.

de Carvalho Nicacio C, Williamson RA, Parren PW, Lundkvist A, Burton DR, Bjorling E, 2002, Neutralizing human Fab fragments against measles virus recovered by phage display. *J. Virol.* **76**: 251-258.

de Kruif J, Boel E, Logtenberg T, 1995, Selection and application of human single chain Fv antibody fragments from a semi-synthetic phage antibody display library with designed CDR3 regions. *J. Mol Biol.* **248**: 97-105.

Dezube BJ, Doweiko JP, Prope JA, Conway B, Hwang L, Terada M, Leece BA, Ohno T, Mastico RA, 2004, Monoclonal antibody hNM01 in HIV-infected patiets: a phase 1 study. *J. Clin. Virol.* **31S**: S45-S47.

Enssle K, Kurrle R, Kohler R, Muller H, Kanzy EJ, Hilfenhaus J, Seiler FR, 1991, A rabies-specific human monoclonal antibody that protects mice against lethal rabies. *Hybridoma* **10**: 547-556.

Eren R, Ilan E, Nussbaum O, Lubin I, Terkieltaub D, Arazi Y, Ben-Moshe O, Kitchinzky A, Berr S, Gopher J, Zauberman A, Galun E, Shouval D, Daudi N, Eid A, Jurim O, Magnius LO, Hammas B, Reisner Y, Dagan S, 2000, Preclinical evaluation of two human anti-hepatitis B virus (HBV) monoclonal antibodies in the HBV-trimera mouse model and in HBV chronic carrier chimpanzees. *Hepatology* 32: 588-596.

Eren R, Lubin I, Terkieltaub D, Ben-Moshe O, Zauberman A, Uhlmann R, Tzahor T, Moss S, Ilan E, Shouval D, Galun E, Daudi N, Marcus H, Reisner Y, Dagan S, 1998, Human monoclonal antibodies specific to hepatitis B virus generated in a human/mouse radiation chimera: the Trimera system. *Immunology* 93: 154-161.

Ferrantelli F, Ruprecht RM (2002). Neutralizing antibodies against HIV – back in the major leagues? *Curr Opin Immunol.* 14: 495-502.

Ferrantelli F, Rasmussen RA, Buckley KA, Li PL, Wang T, Montefiori DC, Katinger H, Stiegler G, Anderson DC, McClure HM, Ruprecht RM, 2004, Complete protection of neonatal rhesus macaques against oral exposure to pathogenic simian-human immuno-deficiency virus by human anti-HIV monoclonal antibodies. *J. Infect. Dis.* 189: 2167-2173.

Galun E, Eren R, Safadi R, Ashour Y, Terrault N, Keeffe EB, Matot E, Mizrachi S, Terkieltaub D, Zohar M, Lubin I, Gopher J, Shouval D, Dagan S, 2002, Clinical evaluation (phase I) of a combination of two human monoclonal antibodies to HBV: safety and antiviral properties. *Hepatology.* 35: 673-679.

Goncalvez AP, Men R, Wernly C, Purcell RH, Lai CJ, 2004, Chimpanzee Fab fragments and a derived humanized immunoglobulin G1 antibody that efficiently cross-neutralize dengue type 1 and type 2 viruses. *J.Virol.* 78:12910-12918.

Greenough TC, Babcock GJ, Roberts A, Hernandez HJ, Thomas WD Jr, Coccia JA, Graziano RF, Srinivasan M, Lowy I, Finberg RW, Subbarao K, Vogel L, Somasundaran M, Luzuriaga K, Sullivan JL, Ambrosino DM, 2005, Development and characterization of a severe acute respiratory syndrome-associated coronavirus-neutralizing human monoclonal antibody that provides effective immunoprophylaxis in mice. *J. Infect. Dis.* 191: 507-514.

Griffiths AD, Malmqvist M, Marks JD, Bye JM, Embleton MJ, McCafferty J, Baier M, Holliger KP, Gorick BD, Hughes-Jones NC, Hoogenboom HR, Winter G, 1993, Human anti-self antibodies with high specificity from phage display libraries. *EMBO* 12: 725-734.

Griffin D, Levine B, Tyor W, Ubol S, Despres P, 1997, The role of antibody in recovery from alphavirus encephalitis. *Immunol. Rev.* 159: 155-161.

Habersetzer F, Fournillier A, Dubuisson J, Rosa D, Abrignani S, Wychowski C, Nakano I, Trepo C, Desgranges C, Inchauspe G, 1998, Characterization of human monoclonal antibodies specific to the hepatitis C virus glycoprotein E2 with *in vitro* binding neutralization properties. *Virology* 249: 32-41.

Han SH, Ofman J, Holt C, King K, Kunder G, Chen P, Dawson S, Goldstein L, Yersiz H, Farmer DG, Ghobrial RM, Busuttil RW, Martin P, 2000, An efficacy and cost-effectiveness analysis of combination hepatitis B immuneglobulin and lamivudine to prevent recurrent hepatitis B after orthotopic liver transplantation compared with hepatitis B immune globulin monotherapy. *Liver Transpl.* 6: 741-8.

Hanlon CA, DeMattos CA, DeMattos CC, Niezgoda M, Hooper DC, Koprowski H, Notkins A, Rupprecht CE, 2001, Experimental utility of rabies virus-neutralizing human monoclonal antibodies in post-exposure prophylaxis. *Vaccine* 19: 3834-42.

Haynes BF, Fleming J, St Clair EW, Katinger H, Stiegler G, Kunert R, Robinson J, Scearce RM, Plonk K, Staats HF, Ortel TL, Liao HX, Alam SM, 2005, Cardiolipin polyspecific autoreactivity in two broadly neutralizing HIV-1 antibodies. *Science* 2308: 1906-1908.

He Y, Honnen WJ, Krachmarov CP, Burkhart M, Kayman SC, Corvalan J, Pinter A, 2002, Efficient isolation of novel human monoclonal antibodies with neutralizing activity against HIV-1 from transgenic mice expressing human Ig loci. *J. Immunol.* **169**: 595-605.

Higo-Moriguchi K, Akahori Y, Iba Y, Kurosawa Y, Taniguchi K, 2004, Isolation of human monoclonal antibodies that neutralize human rotavirus. *J. Virol.* **78**: 3325-3332.

Hollinger P, Hudson PJ, 2005, Engineered antibody fragments and the rise of single domains. *Nat. Biotechnol.* **9**: 1126-1136.

Hong HJ, Ryu CJ, Hur H, Kim S, Oh HK, Oh MS, Park SY, 2004, In vivo neutralization of hepatitis B virus infection by an anti-preS1 humanized antibody in chimpanzees. *Virology* **318**: 134-141.

Hoogenboom H, 2005, Selecting and screening recombinant antibody libraries. *Nature Biotechnology* **23**: 1105-1116.

Hoogenboom HR, de Bruine AP, Hufton SE, Hoet RM, Arends JW, Roovers RC, 1998, Antibody phage display technology and its applications. *Immunotechnology* **4**: 1-20.

Iacono-Connors LC, Smith JF, Ksiazek TG, Kelley CL, Schmaljohn CS, 1996, Characterization of Langat virus antigenic determinants defined by monoclonal antibodies to E, NS1 and preM and identification of a protective, non-neutralizing preM-specific monoclonal antibody. *Virus Res.* **43**: 125-136.

Jahrling PB, Peters CJ, Stephen EL, 1984, Enhanced treatment of Lassa fever by immune plasma combined with ribavirin in cynomolgous monkeys. *J. Infect. Dis.* **149**: 420-427.

Jakobovits A., 1998, Production and selection of antigen-specific fully human monoclonal antibodies from mice engineered with human Ig loci. Adv. *Drug. Deliv. Rev.* **31**: 33-42.

Johnson S, Griego SD, Pfarr DS, Doyle ML, Woods R, Carlin D, Prince GA, Koenig S, Young JF, Dillon SB, 1999, A direct comparison of the activities of two humanized respiratory syncytial virus monoclonal antibodies: MEDI-493 and RSHZl9. *J. Infect. Dis.* **180**: 35-40.

Johnson S, Oliver C, Prince GA, Hemming VG, Pfarr DS, Wang SC, Dormitzer M, O'Grady J, Koenig S, Tamura JK, Woods R, Bansal G, Couchenour D, Tsao E, Hall WC, Young JF, 1997, Development of a humanized monoclonal antibody (MEDI-493) with potent in vitro and in vivo activity against respiratory syncytial virus. *J.Infect.Dis.* **176**: 1215-1224.

Karasu Z, Ozacar T, Akyildiz M, Demirbas T, Arikan C, Kobat A, Akarca U, Ersoz G, Gunsar F, Batur Y, Kilic M, Tokat Y, 2004, Low-dose hepatitis B immune globulin and higher-dose lamivudine combination to prevent hepatitis B virus recurrence after liver transplantation. *Antivir Ther.* **9**: 921-7.

Kausmally L, Waalen K, Lobersli I, Hvattum E, Berntsen G, Michaelsen TE, Brekke OH, 2004, Neutralizing human antibodies to varicella-zoster virus (VZV) derived from a VZV patient recombinant antibody library. *J. Gen. Virol.* **85**: 3493-3500.

Keck ZY, Sung VM, Perkins S, Rowe J, Paul S, Liang TJ, Lai MM, Foung SK, 2004, Human monoclonal antibody to hepatitis C virus E1 glycoprotein that blocks virus attachment and viral infectivity. *J. Virol.* **78**: 7257-7263.

Kellermann SA, Green LL, 2002, Antibody discovery: the use of transgenic mice to generate human monoclonal antibodies for therapeutics. *Curr. Opin. Biotechnol.* **13**: 593-597.

Kim SJ, Jang MH, Stapleton JT, Yoon SO, Kim KS, Jeon ES, Hong HJ, 2004, Neutralizing human monoclonal antibodies to hepatitis A virus recovered by phage display. *Virology* **318**: 598-607.

Klasse PJ and Sattentau QJ, 2002, Occupancy and mechanism in antibody-mediated neutralization of animal viruses. *J. Gen. Virol.* **83**: 2091-2108.

Klonisch T, Panayotou G, Edwards P, Jackson AM, Berger P, Delves PJ, Lund T, Roitt IM,1996, Enhancement in antigen binding by a combination of synergy and antibody capture. *Immunology* **89**: 165-171.

Koch J, Liang M, Queitsch I, Kraus AA, Bautz EK, 2003, Human recombinant neutralizing antibodies against hantaan virus G2 protein. *Virology* **308**: 64-73.

Koefoed K, Farnaes L, Wang M, Svejgaard A, Burton DR, Ditzel HJ, 2005, Molecular characterization of the circulating anti-HIV-1 gp120-specific B cell repertoire using antibody phage display libraries generated from pre-selected HIV-1 gp120 binding PBLs. *J. Immunol. Methods.* **297**: 187-201.

Kontermann RE, Dübel S (Eds.), 2001, Antibody engineering, Springer, Heidelberg.

Kramer RA, Cox F, van der Horst M, van der Oudenrijn S, Res PC, Bia J, Logtenberg T, de Kruif J, 2003, A novel helper phage that improves phage display selection efficiency by preventing the amplification of phages without recombinant protein. *Nucleic Acids Res.* **31**: e59

Kramer RA, Marissen WE, Goudsmit J, Visser TJ, Clijsters-Van der Horst M, Bakker AQ, de Jong M, Jongeneelen M, Thijsse S, Backus HH, Rice AB, Weldon WC, Rupprecht CE, Dietzschold B, Bakker AB, de Kruif J, 2005, The human antibody repertoire specific for rabies virus glycoprotein as selected from immune libraries. *Eur.J.Immunol.* **35**: 2131-45.

Li A, Baba TW, Sodroski J, Zolla-Pazner S, Gorny MK, Chou TC, Baba TW, Ruprecht RM, 1998, Synergistic neutralization of simian-human immunodeficiency virus SHIV-vpu+ by triple and quadruple combinations of human monoclonal antibodies and high-titer anti-human immunodeficiency virus type 1 immuoglobulins. *J. Virol.* **72**: 3235-3240.

Liang M, Mahler M, Koch J, Ji Y, Li D, Schmaljohn C, Bautz EK, 2003, Generation of an HFRS patient-derived neutralizing recombinant antibody to Hantaan virus G1 protein and definition of the neutralizing domain. *J. Med. Virol.* **69**: 99-107.

Lipovsek D, Pluckthun A, 2004, In-vitro protein evolution by ribosome display and mRNA display. *J. Immunol. Methods* **290**: 51-67.

Lonberg N, 2005, Human antibodies from transgenic animals. *Nat. Biotechnol.* **23**: 1117-1125.

Marissen WE, Kramer RA, Rice A, Weldon WC, Niezgoda M, Faber M, Slootstra JW, Meloen RH, Clijsters-van der Horst M, Visser TJ, Jongeneelen M, Thijsse S, Throsby M, de Kruif J, Rupprecht CE, Dietzschold B, Goudsmit J, Bakker AB, 2005, Novel rabies virus-neutralizing epitope recognized by human monoclonal antibody: fine mapping and escape mutant analysis. *J.Virol.* **79**: 4672-8.

Moulard M, Phogat SK, Shu Y, Labrijn AF, Xiao X, Binley JM, Zhang MY, Sidorov IA, Broder CC, Robinson J, Parren PW, Burton DR, Dimitrov DS, 2002, Broadly cross-reactive HIV-1-neutralizing human monoclonal Fab selected for binding to gp120-CD4-CCR5 complexes. *Proc Natl Acad Sci USA* **99**: 6913-6918.

Mozdzanowska K, Furchner M, Washko G, Mozdzanowski J, Gerhard W, 1997, A pulmonary influenza virus infection in SCID mice can be cured by treatment with hemagglutinin-specific antibodies that display very low virus-neutralizing activity in vitro. *J. Virol.* **71**: 4347-4355.

Naoumov NV, Lopes AR, Burra P, Caccamo L, Iemmolo RM, de Man RA, Bassendine M, O'Grady JG, Portmann BC, Anschuetz G, Barrett CA, Williams R, Atkins M, 2001, Randomized trial of lamivudine versus hepatitis B immunoglobulin for long-term prophylaxis of hepatitis B recurrence after liver transplantation. *J. Hepatol.* **34**:888-894.

Nejatollahi F, Hodgetts SJ, Vallely PJ, Burnie JP, 2002, Neutralising human recombinant antibodies to human cytomegalovirus glycoproteins gB and gH. *FEMS Immunol. Med. Microbiol.* **34**: 237-244.

Parren PW, Burton DR., 2001, The antiviral activity of antibodies in vitro and in vivo. *Adv. Immunol.* **77**: 195-262.

Parren PW, Geisbert TW, Maruyama T, Jahrling PB, Burton DR., 2002, Pre- and postexposure prophylaxis of Ebola virus infection in an animal model by passive transfer of a neutralizing human antibody. *J. Virol.* **76**: 6408-6412.

Poignard P, Sabbe R, Picchio GR, Wang M, Gulizia RJ, Katinger H, Parren PW, Mosier DE, Burton DR, 1999, Neutralizing antibodies have limited effects on the control of established HIV-1 infection in vivo. *Immunity* **10**: 431-438.

Purtscher M, Trkola A, Gruber G, Buchacher A, Predl R, Steindl F, Tauer C, Berger R, Barrett N, Jungbauer A, 1994, A broadly neutralizing human monoclonal antibody against gp41 of human immunodeficiency virus type 1. *AIDS Res. Hum. Retroviruses* **10**:1651-1658.

Rehermann B, Nascimbeni M, 2005, Immunology of hepatitis B virus and hepatitis C virus infection. *Nat. Rev. Immunol.* **5**: 215-229.

Reichert JM, Rosenzweig CJ, Faden LB, Dewitz MC, 2005, Monoclonal antibody successes in the clinic. *Nat. Biotechnol.* **9**: 1073-1078.

Sawyer LA, 2000, Antibodies for the prevention and treatment of viral diseases. *Antiviral Res.* **47**: 57-77.

Schmaljohn C, Cui Y, Kerby S, Pennock D, Spik K. Production and characterization of human monoclonal antibody Fab fragments to vaccinia virus from a phage-display combinatorial library. *Virology* **258**: 189-200.

Schofield DJ, Glamann J, Emerson SU, Purcell RH, 2000, Identification by phage display and characterization of two neutralizing chimpanzee monoclonal antibodies to the hepatitis E virus capsid protein. *J. Virol.* **74**: 5548-5555.

Seiler P, Senn BM, Klenermann P, Kalinke U, Hengartner H, Zinkernagel RM, 2000, Additive effect of neutralizing antibody and antiviral drug treatment in preventing virus escape and persisence. *J. Virol.* **74**: 5896-5901.

Steinitz M, Klein G, Koskimies S, Makel O., 1977, EB virus-induced B lymphocyte cell lines producing specific antibody. *Nature* **269**: 420-2.

Stiegler G, Armbruster C, Vcelar B, Stoiber H, Kunert R, Michael NL, Jagodzinski LL, Ammann C, Jager W, Jacobson J, Vetter N, Katinger H, 2002, Antiviral activity of the neutralizing antibodies 2F5 and 2G12 in asymptomatic HIV-1-infected humans: a phase I evaluation. *AIDS* **16**: 2019-2025.

Stogner SW, King JW, Black-Payne C, Bocchini J., 1993, Ribavirin and intravenous immune globulin therapy for measles pneumonia in HIV infection. *South. Med. J.* **86**: 1415-8.

Sui J, Li W, Murakami A, Tamin A, Matthews LJ, Wong SK, Moore MJ, Tallarico AS, Olurinde M, Choe H, Anderson LJ, Bellini WJ, Farzan M, Marasco WA, 2004, Potent neutralization of severe acute respiratory syndrome (SARS) coronavirus by a human mAb to S1 protein that blocks receptor association. *Proc. Natl. Acad. Sci. USA* **101**: 2536-2541.

ter Meulen J, Bakker AB, van den Brink EN, Weverling GJ, Martina BE, Haagmans BL, Kuiken T, de Kruif J, Preiser W, Spaan W, Gelderblom HR, Goudsmit J, Osterhaus AD, 2004, Human monoclonal antibody as prophylaxis for SARS coronavirus infection in ferrets. *Lancet* **363**: 2139-41.

Thompson J, Pope T, Tung JS, Chan C, Hollis G, Mark G, Johnson KS, 1996, Affinity maturation of a high-affinity human monoclonal antibody against the third hypervariable loop of human immunodeficiency virus: use of phage display to improve affinity and broaden strain reactivity. *J. Mol. Biol.* **256**: 77-88.

Traggiai E, Becker S, Subbarao K, Kolesnikova L, Uematsu Y, Gismondo MR, Murphy BR, Rappuoli R, Lanzavecchia A., 2004, An efficient method to make human monoclonal antibodies from memory B cells: potent neutralization of SARS coronavirus. *Nat. Med.* **10**: 871-5.

van den Brink EN, ter Meulen J, Cox F, Jongeneelen MA, Thijsse A, Throsby M, Marissen WE, Rood PM, Bakker AB, Gelderblom HR, Martina BE, Osterhaus AD, Preiser W, Doerr HW, de Kruif J, Goudsmit J, 2005, Molecular and biological characterization of human monoclonal antibodies binding to the spike and nucleocapsid proteins of severe acute respiratory syndrome coronavirus. *J.Virol.* **79**: 1635-44.

Wang CY, Sawyer LS, Murthy KK, Fang X, Walfield AM, Ye J, Wang JJ, Chen PD, Li ML, Salas MT, Shen M, Gauduin MC, Boyle RW, Koup RA, Montefiori DC, Mascola JR, Koff WC, Hanson CV, 1999, Postexposure immunoprophylaxis of primary isolates by an antibody to HIV receptor complex. *Proc. Natl. Acad. Sci. USA.* **96**: 10367-10372.

Wei X, Decker JM, Wang S, Hui H, Kappes JC, Wu X, Salazar-Gonzalez JF, Salazar, MG, Kilby JM, Saag MS, Komarova NL, Nowak MA, Hahn BH, Kwong PD, Shaw GM, 2003, Antibody neutralization and escape by HIV-1. *Nature* **422**: 307-312.

Wolbank S, Kunert R, Stiegler G, Katinger H, 2003, Characterization of human class-switched polymeric (Immunoglobulin M [IgM] and IgA) anti-human immunodeficiency virus type 1 antibodies 2F5 and 2G12. *J. Virol.* **77**: 4095-4103.

Wu H, Pfarr DS, Tang Y, An LL, Patel NK, Watkins JD, Huse WD, Kiener PA, Young JF. Ultra-potent antibodies against respiratory syncytial virus: effects of binding kinetics and binding valence on viral neutralization. *J. Mol. Biol.* **350**: 126-144.

Xu W, Smith-Franklin BA, Li PL, Wood C, He J, Du Q, Bhat GJ, Kankasa C, Katinger H, Cavacini LA, Posner MR, Burton DR, Chou TC, Ruprecht RM, 2001, Potent neutralization of primary human immunodeficiency virus clade C isolates with a synergistic combination of human monoclonal antibodies raised against clade B. *J. Hum. Virol.* **4**: 55-61.

Yang W-P, Green K, Pinz-Sweeny S, Briones AT, Buron DR, Barbas CF, 1995, CDR walking mutagenesis for the affinity maturation of a potent human anti-HIV antibody into the picomolar range. *J. Mol. Biol.* **254**: 392-403.

Zwick MB, Parren PW, Saphire EO, Church S, Wang M, Scott JK, Dawson PE, Wilson IA, Burton DR, 2003, Molecular features of the broadly neutralizing immunoglobulin G1 b12 required for recognition of human immunodeficiency virus type 1 gp120. *J. Virol.* **77**: 5863-5876.

Zwick MB, Wang M, Poignard P, Stiegler G, Katinger H, Burton DR, Parren W, 2001, Neutralization synergy of human immunodeficiency virus type 1 primary isolates by cocktails of broadly neutralizing antibodies. *J. Virol.* **75**: 12198-12208.

Zwick MB, Jensen R, Church S, Wang M, Stiegler G, Kunert R, Katinger H, Burton DR, 2005, Anti-human immunodeficiency virus type 1 (HIV-1) antibodies 2F5 and 4E10 require surprisingly few crucial residues in the membrane-proximal external region of glycoprotein gp41 to neutralize HIV-1. *J Virol.* **79**: 1252-1261.

Chapter 4.3

VECTOR-BASED ANTIVIRAL THERAPY

A. ENSSER

Institut für Klinische und Molekulare Virologie, Friedrich-Alexander Universität Erlangen-Nürnberg, Schlossgarten 4, 91054 Erlangen, Germany

Abstract: Viral and non-viral vector-based gene transfer and RNA-based inhibition are not only essential for basic molecular research in virology but can be developed into a valuable addition of traditional drug therapy. The efficiency of RNAi has largely supplant brought other nucleic acid based drugs out of focus, although the central problem remains efficient and long-term delivery of these novel drugs. Expansion of virus-specific T-cells and (re)targeting of immune cells to virus infected cells are further promising ways to enhance antiviral defence. Vaccination strategies based on DNA as well as on vectors like recombinant modified vaccinia virus have reached early clinical stages.

1.   INTRODUCTION

> *It is only by Beelzebul, the prince of demons,*
> *that this man casts out demons.* Matth. 12:24

The feasibility of therapeutic gene transfer is widely accepted for cancer and genetic disease, and two thirds of human clinical trials were as yet performed in cancer patients. Herpesviruses, lentiviruses, papillomaviruses and hepatitis B and C viruses (HBV, HCV) all cause significant burden of disease in man. They establish persistent infections that are in many aspects similar to genetic disease or cancer, and several viruses are in fact cancerogenic. Similar attempts as in cancer gene therapy have thus been applied to viral infections, while the success of vaccination against infectious disease has on the other hand prompted studies to vaccinate against cancer. The majority of studies in humans that use gene transfer in infectious disease (about

E. Bogner and A. Holzenburg (eds.), New Concepts of Antiviral Therapy, 507–532.
© *2006 Springer. Printed in the Netherlands.*

7% of more than 1000 trials world wide) are HIV-related (68 of 72 regis-tered trials on infectious disease, as of June 2005); the few others are on im-munotherapy directed against Cytomegalovirus (CMV), Epstein Barr virus (EBV) and HCV (The Journal of Gene Medicine Clinical Trial Site, 2005; Office of Biotechnological Activities at NIH and the Recombinant DNA Advisory Committee, 2005). The ingenuity of researchers has come up with an abundance of antiviral targeting strategies, including promising recent developments such as RNA interference, strategies to specifically direct an antiviral immune response toward infected cells, and vaccination procedures that involve gene-modified antigen presenting cells. Many techniques are still in a preclinical or conceptual stage. The major hurdle of most vector-based approaches remains the delivery of the therapeutic principle to the (virus-infected) target cell. Since preemptive application of antiviral gene transfer will more than likely remain an absolute exception in the foreseeable future, the viral foe will usually have a head start; in addition, for reasons of biosafety, the carefully controlled therapeutic vector has a fundamental dis-advantage versus the spread of the infectious agent which is not hindered by such considerations.

2.   VECTORS FOR ANTIVIRAL THERAPY

"It'll be great to have polymers if they work, but I don't think they're efficient enough yet — viruses have learned to do this over millions of years" (Inder Verma, in: Clayton, 2004)

This review will focus on potential treatments that involve transfer of a plasmid or viral vector based expression cassette as the therapeutic principle. Some of the approaches discussed here are also being developed for direct short term treatment, such as synthetic nucleic acids and proteins.

The final goal of efficient *and* selective transduction of target cells with a therapeutic vector in the patient *in vivo* is not in sight. Current transfer sytems lack sufficient target cell selectivity, although various attempts are being made to adapt existing vector systems (Tab. 1). These include pseudotyping retro- or lentiviral particles, modifying retroviral envelope proteins (Sandrin *et al.*, 2003; Verhoeyen and Cosset, 2004), construction of tropism-modified adenoviruses (Mizuguchi and Hayakawa, 2004) or Adeno-associated virus (AAV) vectors (Buning *et al.*, 2003), or integration of modified glycopro-teins into herpesviruses (reviewed in Grandi *et al.*, 2004).

2.1   Integrating viral vectors

"With AAV and retrovirus vectors you're just rolling the dice"
(William M. Pardridge, in: Clayton, 2004)

The successful correction of X-linked severe combined immunodeficiency (SCID-X1) in young children by retroviral transfer of the cytokine receptor common gamma chain (γc) (Hacein-Bey-Abina *et al.*, 2002) had been greeted enthusiastically as the fulfillment of the promises gene therapy had not been able to keep for more than a decade. It was soon followed by a tragic setback when three children developed treatment-associated T-cell leukemia that was associated with LMO2 gene expression, very likely the consequence of insertional activation by the retroviral vector (Hacein-Bey-Abina *et al.*, 2003; Kaiser, 2005).

Insertion mutagenesis is the most significant problem of the otherwise convenient and widely used integrating vectors (Baum *et al.*, 2003). Retroviral insertion mutagenesis had been observed previously in animal models; rhesus monkey infected by a retroviral vector contaminated with replication competent retrovirus developed T-cell leukemia (Donahue *et al.*, 1992), and retroviral mutagenesis was later also recognized in murine models that allowed for longer observation times (Li *et al.*, 2002b). It was recently reported for the first time for a non-primate equine infectious anemia virus (EIAV) based lentiviral vector (Themis *et al.*, 2005). Why have those side effects not been observed before? It has been estimated that in the French trial the number of retroviral insertion events exceeded those of all previous studies together. High vector dose, usage of a conventional vector with complete LTR, and the growth stimulatory properties of the transduced γc gene, plus increased viability and engraftment of gene-modified cells may in summary have precipitated the occurrence of side effects. Compared to the prognosis of SCID-X1 patients in which matched related donors are unavailable, however, the current survival rates and outcome in the remaining patients of the French and related gene therapy trials represent a significant advance in the treatment of this severe disease (Baum *et al.*, 2004).

Dose-finding will be important, as is development/application of newer generations of retro- and lentiviral vectors that will improve the outcome. These are self-inactivating (SIN) due to deletion of the enhancer elements in the LTR U3 region and can therefore avoid the direct activation of genes in the vicinity of the integration site. Nevertheless, dose-dependent genotoxicity by insertional gene inactivation is likely associated with any integrating viral vector. Safety concern might be overcome by including suicide genes that allow ablation of vector-transduced cells.

Table 1. Viral Vectors

Advantage	Disadvantage	Possible solution
Adenovirus		
High capacity	Large and complex genome	
High Titers	Recombination with Ad5 sequences from HEK293 cell line	
		Packaging cell lines with minimal sequence overlap
Transduction of resting, differentiated as well as proliferating cells	No active replication of vector genome: dilution effect in proliferating cells	
No integration into host cell genome	No long-term persistence: repeated applications necessary	
Direct *in vivo* application possible	Toxicity at high doses (Gelsinger case)	
		Establish secure dose
	Inflammation after repeated application	
		Sequential use of alternative vector serotypes
	Preexistent immunity reduces efficiency: High seroprevalence of adenovirus infection, multiple serotypes	
		Vectors based on unrelated animal adenoviruses
	Recombination with wildtype Adenovirus:	
		Vectors based on unrelated animal adenoviruses
	Tissue specific and/or polarized CAR expression	
		Tropism/Receptor modified vectors
Established clinical vector	Established side effects	

Advantage	Disadvantage	Possible solution
Retrovirus		
Small genome, simple structure: Easy construction of complementing packaging lines	Recombination with complementing sequences in packaging cell lines: high risk of insertion mutagenesis by replication competent retroviruses (RCR):	
		Improved packaging cell lines with separate transgenes *Reduced homologies by codon wobbling* *Control for RCR*
In vivo gene transfer possible No mobilization by primate Lentiviruses or human endogenous retroviruses (low sequence homology)	Murine retrovirus vectors inactivated by primate complement	
		Pseudotyping
Titers between 10^4 and 10^7 cfu/ml (concentration possible, intermediate stability of viral particles)	Low titer (when compared to adenovirus or AAV)	
		Pseudotyping with VSV-G can enhance titer
Colinear integration (LTR-gene-LTR): intact structure of integrated expression cassette	Transgene copy number correlates with expression (and risk for insertion mutagenesis)	
		Double copy vectors
Transduction depends on proliferation: Targeted transduction of proliferating (tumor) cells	Transduction depends on proliferation: T-cell stimulation to proliferation increases HIV replication No integration in resting/terminally differentiated cells	
		Introduce lentiviral elements for active nuclear transport of preintegration complex
Stable integration into host cell genome: Transgene is preserved in further cell divisions	Insertion mutagenesis: Oncogene activation, Tumor suppressor inactivation	
		Selection against harmful integrants (?) *Targeted integration* *Self-inactivating vectors*
Established clinical vector	Established side effects	

Table 1. continued

Lentivirus	
As above but	As above but
Infection of non-dividing cells	Mobilization by superinfecting lentiviruses
Possibly lower risk of insertion mutagenesis ?	Parental viruses are primate pathogens
Herpesvirus	
High packaging capacity	Large and complex genome
	Minimal vectors with reduced set of viral genes *Amplicons*
Active episomal persistence	Many unknown functions Unknown pathogenicity factors Tumorigenic potential of parental virus (EBV, HVS)
	Further attenuation *Further research*
Broad tropism in primary infection	Phenotypic changes in transduced cells
Specific tropism of persistent infection Neurotropic (HSV-1) B-cells (EBV) T-cells (H. saimiri)	Recombination with persisting wildtype Herpesviruses *Vectors based on unrelated animal herpesviruses*
Transduction of resting, differentiated as well as proliferating cells	Preexisting Immunity: High seroprevalence of human infection
First clinical trials with oncolytic HSV-1 vectors	Other than HSV-1: Mostly experimental
Adeno-associated Virus	
Small genome, simple structure	Small capacity (to 4,5 kb)
Stable virus particles	Helper virus required (Adenovirus, HSV) Pathogenicity/Immunogenicity of helper virus
	Define helper functions: separate expression of *helper function in producer cells*
High Titers possible (10^{12}/ml)	Concentration and purification from helper virus by ultracentrifugation necessary
Transduction of resting, differentiated as well as proliferating cells	Rep-overexpression toxic: only transient packaging systems. Insufficient knowledge of Rep-function
	Packaging lines with conditional Rep-expression
Stable Integration into target cell genome	Insertion mutagenesis Stable integration in Chromosome 19 is infrequent: requires high MOI and Rep-protein
	Lack of integration in absence of Rep: Episomal persistence for limited time period, Dilution effect in proliferating cells
Apathogenic properties of parental AAV and Dependoviruses	Preexisting Immunity: High seroprevalence of natural parvovirus infection
Established clinical vector	

However, expression of prodrug-activating genes can cause undesired elimination of transduced cells when the suicide gene is recognized by the immune system (Riddell *et al.*, 1996). Furthermore, in the context of a transformation and replication competent rhadinoviral vector, the Herpes simplex virus thymidine kinase (HSV-TK) gene unexpectedly enhanced viral pathogenicity in primates (Hiller *et al.*, 2000a), despite having been able to eliminate transduced cells *in vitro* (Hiller *et al.*, 2000b).

Dependoviruses and Adeno-associated virus (AAV) vectors have not been commonly associated with side effects from integration (Bell *et al.*, 2005), although tumors have been observed after AAV-transduction in mice (Donsante *et al.*, 2001). Site specific integration of AAV-vectors at chromosome 19q13, which was previously considered the theoretical basis for long-term AAV gene transfer, is now recognized to be a rather infrequent event. Although it occurs at the high multiplicities of infection in nature, it requires approximately 100 to 1000 infectious particles per target cell after transduction with AAV vectors (Duan *et al.*, 1998; Schnepp *et al.*, 2003; Nakai *et al.*, 2001). Reports by the Kay group provided insights into the nature of rAAV2 vector integration into chromosomes in quiescent somatic cells in animals and human subjects (Nakai *et al.*, 2003), describing the host chromosomal effects of rAAV2 integration in mice. This result was greeted with mixed enthusiasm (Kohn and Gänsbacher, 2003), but complemented earlier data from tissue culture (Miller *et al.*, 2002). An extended analysis of AAV-integration sites in mice confirmed that AAV has the potential to act as an insertional mutagen (Nakai *et al.*, 2005; Miller *et al.*, 2005), although it does not seem to induce chromosome breaks by itself (Miller *et al.*, 2004).

2.2   Episomal vector development

Episomal persistence and replication in the infected cell is a shared characteristic of the Herpesvirus and Polyomavirus life cycle, though integration occurs and is associated with significant pathology in humans (cervical/genital/skin cancer by human papillomaviruses) or chicken (Marek's Disease Herpesvirus, MDV). EBV integration has been occasionally described in cultured cell lines but is considered a rare event *in vivo* (Gulley *et al.*, 1992; Delecluse *et al.*, 1993). Herpesviruses other than MDV do not regularly integrate and possess a number of attractive features such as large packaging capacity, a mechanism for active replication and segregation of the viral episome, a broad host range, and infection of non-dividing cells. This encouraged the development of a number of episomally persisting vectors based on Herpes simplex virus (HSV), CMV, EBV and the animal *Herpesvirus saimiri* (Cotter and Robertson, 1999; Mazda, 2002; Conese *et al.*, 2004; Oehmig *et al.*, 2004; Doody *et al.*, 2005; Wieser *et al.*, 2005).

Vectors build on RNA-virus replicons such as Alphavirus lack the capability for long-term persistence. Since they elicit a strong immune response, they will possibly find an application in immunotherapy and vaccination studies, where only transient expression is required and their immune stimulating properties may be advantageous.

2.3 Nonviral vectors

Recombinant or DNA-based vaccines hitherto have not made their way into advanced clinical testing, nor are recombinant viral vector-based vaccines licensed in Europe or the US. Nevertheless, transfer of naked DNA into skin or muscle has been successfully applied for DNA-vaccination studies in animals, also in combination with booster strategies that involve viral expression vectors (Yang *et al.*, 2003); this has been able to provide protection from *Filovirus* challenge, and similar to adenoviral vector expressed Ebola virus glycoprotein alone rapidly protects primates (Sullivan *et al.*, 2003).

Encapsulated forms of DNA have been delivered successfully *in vivo* to the liver or joints; transfer using cationic or receptor targeted liposomes is nonetheless inefficient, as the vehicle is frequently cleared from the circulation by phagocytic cells before the target is reached. *Ex vivo* transfer of DNA by lipids or electroporation, also in a flow-through format, may be possible and a practicable way of nonviral gene transfer (summarized in Rössig and Brenner, 2004).

3.   VIRUS-SPECIFIC DNA- OR RNA-TARGETING STRATEGIES

There are basically two lines of attack, (A) directed against the exogenous target such as RNA-virus genomes and viral transcripts and, (B) to interfere with transcripts of host cell cofactors that are important for the invasion, replication, or pathogenicity.

DNA or RNA-targeting drugs have the advantage that by their sequence-specific action they are much easier to design compared to drugs that need to interact with structured targets such as cellular or viral proteins (Tab. 4.3.2). Multiple sequence targets can be combined in one formulation, although efficiency might be decreased by saturation, e.g. of RNA-interference (RNAi) pathways. The initial assumption that the non-proteinaceous nature of nucleic acid-based drugs prevents host immune response that would otherwise limits efficacy, did not hold true. Innate immunity is also activated via Toll-like receptors (TLR) that recognize CpG motifs and dsRNA, resulting in stimulation of the interferon system.

3.1   Delivery of nucleic acids

The most important barrier for all nucleic acid-based treatments is their intrinsic low capacity to cross the cellular membrane, as they must be delivered into the inside of the cell. Before they can arrive there in sufficient amounts, they must remain stable and should have no toxicity in the host. Ribozymes have not been approved for clinical use so far, and the only licensed antisense DNA drug (Fomivirsen, which is directed against human CMV for treatment of retinitis) is not a great commercial success due to more reliable alternative treatments. After delivery across the cell membrane, the drug has to be released e.g. from the endosome and find its way to its target, which has to be structurally accessible. Compared to short-lived transfer of synthetic molecules, it seems therefore more attractive to perform vector-mediated transfer of expression cassettes for enzymatically active ribozymes, decoy molecules and specifically for siRNA-precursors (below), which also use physiological cellular mechanisms for transport and processing.

3.2   miRNA

Cellular micro-RNAs (miRNAs) are an ancient physiological mechanism for posttranscriptional gene silencing (PTGS). PTGS by miRNA or RNA-interference (RNAi) is a common eukaryotic defense pathway conserved all the way from plants over fungi to animals. It was first noted more than a decade ago in plants, then fungi, nematodes, flies and higher animals. The underlying molecular mechanism of miRNA and RNAi are highly similar and have been elucidated mostly by the work of Tuschl and colleagues (Tuschl *et al.*, 1999; Zamore *et al.*, 2000; Elbashir *et al.*, 2001a, 2001b).

Briefly, RNA polymerase II transcribes mammalian miRNA genes into long primary miRNAs (pri-miRNAs), which are then processed in the nucleus into 70–80-nucleotide hairpin-like precursor miRNAs (pre-miRNAs) by Drosha, a nuclear RNAse III which is in complex with DGCR8, an RNA binding protein. Exportin-5 together with its cofactor Ran-GTP transport pre-miRNAs to the cytoplasm, where they are processed by the RNAse III Dicer into mature 22-nucleotide duplex miRNAs. The Dicer containing complex processes long double-stranded RNA precursors into the small interfering RNAs (siRNAs) that mediate RNAi. Beginning at the duplex end with the lower thermodynamic stability, miRNA duplexes are unwound, and the miRNA strand that has its 5' terminus at this end is the future mature miRNA (also called guide RNA). Unwound mature miRNAs are then incorporated into a miRNA ribonucleoprotein complex analogous to the RNA-induced silencing complex (RISC). The miRNA/RISC then downregulate the target by translational inhibition or target mRNA cleavage. This is determined by the degree of complementarity between the miRNA and target gene, in combination with Argonaute proteins, which are present in RISC. All Argonaute proteins can bind miRNAs and siRNAs, and Argonaute2 (eIF2C2) is the catalytic component of the RISC that mediates targeted RNA cleavage. Near-perfect complementarity results in cleavage, followed by general RNA degradation of the targets, while partial complementarity causes translational inhibition, though the exact mechanism for translational inhibition is not known. Translational inhibition seems more common than miRNA-directed cleavage in animals (reviewed in Wienholds and Plasterk, 2005; see also Chapter 1.3, Figure 5). Given that the recognition of miRNA usually involves 1-2 mismatches to its target sequence, current bioinformatic prediction methods for miRNA and their potential targets are not dependable, and detection of miRNAs, their targets, and validation of target specificity so far mostly rely on experimentation.

As indicated, PTGS by RNAi is a conserved mechanism of antiviral defense, all the way from plants to animals This is indicated by the finding that several viruses have evolved proteins interfering with RNAi. Viral suppressors of PTGS were first recognized in plants (reviewed in Voinnet *et al.*, 1999; Roth *et al.*, 2004), and plants in response seem to have evolved secondary defense mechanisms targeting these viral proteins (Li *et al.*, 1999; Savenkov and Valkonen, 2002). Likewise, suppression of RNA interference by animal viral pathogens was recently described (Li *et al.*, 2002a; Lichner *et al.*, 2003; Bucher *et al.*, 2004; Delgadillo *et al.*, 2004; Li *et al.*, 2004; Soldan *et al.*, 2005).

Interestingly, it was recently found that some viruses, mostly large-genome DNA-viruses, have evolved or acquired expression of miRNAs (Pfeffer *et al.*, 2004, 2005; Samols *et al.*, 2005; Cai *et al.*, 2005). A first viral miRNA function has been revealed in that the simian polyomavirus 40 encoded miRNA regulates large T antigen expression and reduces susceptibility of infected cells against cytolysis by cytotoxic T-cells (Sullivan *et al.*, 2005). The cellular or viral targets of other virus encoded miRNAs are unknown.

3.3   RNA interference

*"RNAi is what antisense never was. It's robust and reproducible,
and it exists in nature"* (Inder Verma, in: Clayton, 2004)

Within a few years, the technique of RNAi has been widely and successfully employed in the dissection of cellular and viral functions, and the enormous potential for therapeutic application has been rapidly recognized (reviewed in Shankar *et al.*, 2005; Stevenson, 2004). Briefly, the concept has been successfully applied to therapeutic approaches directed against HIV (Berkhout, 2004; Boden *et al.*, 2004; Cullen, 2005), HCV (reviewed in Chapter 1.3 by Frese and Bartenschlager), SARS-CoV (Wu *et al.*, 2005), and using RNAi was even suggested as a way to breed non-permissive mosquitoes in order to interrupt Dengue virus transmission (Sanchez-Vargas *et al.*, 2004).

Table 2. Nuclei-Acid based Strategies

Strategy	Advantages	Disadvantages
Antisense Oligonucleotides	Straightforward design and production, inexpensive Can be modified for systemic delivery Introns and exons as possible targets	Protein binding possible (Aptamer activity) Exogenous delivery of synthetic molecules necessary Off-target effects possible Interferon-induction by CpG
Ribozymes	Discrimination of single nucleic acid polymorphisms (SNP) Introns and exons, subcellular compartments as targets *In vitro* and *in vivo* production possible Target specificity can be changed Simple catalytic domain Can correct defects	Limited target choice, GUC triplet required Protein binding possible (Aptamer activity) Dependent on *in vivo* folding and compartment
DNAzymes	Straightforward design and production, inexpensive Catalytic properties Can be modified for systemic delivery Introns and exons as possible targets	Protein binding possible (Aptamer activity) Dependent on *in vivo* folding and compartment Exogenous delivery of synthetic molecules necessary Off-target effects possible
RNA-interference	Based on natural mechanism Straightforward design and production, relatively inexpensive Effective at low concentration Exogenous and endogenous delivery possible Tissue specific expression possible Sequence specific: some discrimination of nucleic acid polymorphisms (SNP)	Only mRNA as target Induction of Interferon possible Off-target effects possible

Modified from (Scherer and Rossi, 2003)

In most cases RNA interference is efficient and highly specific. However, induction of the interferon system by the Toll-like receptor (TLR) pathway can result from recognition of byproducts of siRNA production or the specific siRNA itself. On the other hand, TLR pathway activation by a given siRNA can be experimentally ruled out *in vitro* and *in vivo* in various models, and has not been shown to interfere with the degree or specificity of RNA interference. More difficult to overcome, but possibly less severe, is the interference with off-target cellular RNAs that has been observed sporadically (Jackson *et al.*, 2003; Scacheri *et al.*, 2004). The potential off-target effects of the respective therapeutic RNA must therefore be studied extensively in the homologous species, e.g. by genome-wide expression analysis. A further aspect is the evaluation of such strategies *in vivo*: the high sequence specificity of RNA interference will limit the predictive value of any preclinical animal model, since it is not unlikely that a given siRNA will find different off-targets in other host genetic backgrounds.

The delivery of the therapeutic principle to the target cell faces similar problems as with other aforementioned techniques that rely on transfer of short nucleic acids or vector-mediated transfer. Synthetic precursor RNAs need to be protected from nucleases and can be delivered using techniques such as liposomes or ligand mediated delivery across the membrane of the target cell. Some cell types can also take up nucleic acids directly. Nevertheless, the effects of synthetic RNA will be transient and short-lived. Sustained as well as regulated intracellular expression of short-hairpin structured siRNA precursors (shRNA) has been achieved by a variety of integrating and non-integrating viral vectors, along with their particular advantages and disadvantages (Tab. 1). Widely used examples are retro- and lentiviral SIN vectors where the shRNA expression cassette is located in the 3'LTR region; two copies of the shRNA cassette are integrated into the target cell after reverse transcription has duplicated the 3'LTR (Barton and Medzihitov, 2002; An *et al.*, 2003; Wiznerowicz and Trono, 2003; Lee *et al.*, 2003); AAV have also been used (Tomar *et al.*, 2003).

4. TARGETING VIRUS-INFECTED T-CELLS

The targeting of infected cells by cytotoxic vectors is conceivable via viral or cellular proteins that are presented exclusively or preferentially at the cell surface. Productive infection with enveloped viruses is usually associated with expression of viral (glyco-)proteins at the surface of cells. Infected cells frequently also show upregulated expression of specific surface molecules such as the MHC-I-like ligands for NK-cell recognition via NKG2D.

Unsurprisingly, viruses have evolved evasion mechanisms (Alcami and Koszinowski, 2000a, 2000b, 2000c; Cerwenka and Lanier, 2003; Lodoen *et al.*, 2004; Krmpotic *et al.*, 2005). Glycoprotein-deficient rhabdoviruses modified to express CD4, or CD4 plus CXCR4 have been designed as "Magic bullets" for selective infection (Mebatsion *et al.*, 1997; Mebatsion and Conzelmann, 1996) or even killing (Schnell *et al.*, 1997) of cells displaying HIV envelope glycoprotein at their surface. Analogous targeting could be achieved by retroviruses pseudotyped with CD4 or CXCR4 (Endres *et al.*, 1997; Somia *et al.*, 2000), or anti-CCR5-scFv pseudotyped Sindbis vectors (Aires *et al.*, 2005). However, neither strategy has been translated to clinical applications so far.

5.   EXPRESSION OF ANTIVIRAL MOLECULES

5.1   Dominant negative proteins or receptors

Soluble forms of the main HIV-receptor CD4 can block infection and moreover CD4-TCRζ fusion proteins have been suggested, in which viral binding to the chimeric molecule will conditionally induce cytotoxicity before the virus can establish infection and replicate. Both approaches suffer from the fact that primary isolates are less dependent on CD4 and can infect via alternative receptors. An intracellular variant of CD4 (CD4-KDEL) that is retained in the endoplasmatic reticulum can block viral glycoprotein-transport (Buonocore and Rose, 1993). Intrakines are modified chemokines such as MIP-1α or RANTES designed to be retained in the ER; this results in ER-retention of the HIV coreceptor CCR5, reducing its expression at the cell surface.

Mutated forms of viral proteins that can exert trans-dominant negative effects on viral replication have been described for HIV Tat and Rev, and also for Gag, Env, and the protease. Clinical trials with transdominant-negative RevM10 showed increased survival of RevM10-expressing T-cells in AIDS-patients. A strong immune response to the Rev protein was observed, initiating further investigations into Rev-based therapeutic vaccines (Bevec *et al.*, 1992, 1996). Cellular transport factors such as eIF-5A are required for Rev-mediated nuclear export. Dominant-negative mutants of eIF-5A have been shown to inhibit the replication of HIV, and eIF-5A can also be a target for small molecule inhibitors of hypusine modification (Hauber *et al.*, 2005).

5.2 Intracellular expression of Antibodies or Intrabodies

Antibodies or single chain antibodies derived from the variable regions of the immunoglobulin heavy and light chains (intrabodies) have been derived against viral targets such as Env, Tat, Rev, reverse transcriptase, integrase or cellular targets required for efficient HIV-replication. Intrabodies targeting the viral reverse transcriptase and integrase can counteract early in the HIV lifecycle and theoretically can prevent HIV integration and as a result confer immunity from infection. Humanization of the primarily murine monoclonal antibodies will be necessary to decrease the immunogenic potential. Intracellular expression of LANA-specific intrabodies inhibited persistence of Kaposi sarcoma associated herpesvirus (KSHV, HHV8) in lymphoma cells (Corte-Real *et al.*, 2005), indicating that such an approach could eventually be used to overcome persistent herpesvirus infections.

5.3 Virus-specific expression of prodrug-activiting enzymes

Gene-directed enzyme prodrug therapy or gene-prodrug activation therapy relates to the transfer and expression of enzymes that metabolize nontoxic prodrugs and convert them into active cytotoxic substances. HSV thymidine kinase (HSV TK), Cytochrome P450, or Cytosinedeaminase from *E. coli* or other species are the most widely used genes, with many others under consideration.

The HIV-LTR is strongly activated by the HIV transactivator protein TAT. Therefore, it has been used in conjunction with marker genes as a sensor to detect replication of HIV (Mcans *et al.*, 1997), and for HIV infection-specific gene expression. Conditional expression of TK driven by the HIV-LTR has been shown to confer to infected cells a rather specific susceptibility to Ganciclovir mediated toxicity (Caruso and Bank, 1997; Miyake *et al.*, 2001). Transfer of TK via a HIV-based lentiviral vector was used to eliminate CD4 tumor cells from adult T-cell leukemia, which is caused by HTLV-I (Obaru *et al.*, 1996). A further possibility would be conditional expression of a toxin. This approach does not require a prodrug, but it is essential that undesired expression, in the absence of the inducing stimulus, does not occur under any circumstance.

5.4   HIV-directed preclinical and clinical studies

Due to the tropism of HIV for cells of the immune system, but also reflecting funding resources, most of T-cell directed vector- and nucleic acid-based antiviral therapies have been aimed at HIV (Gilboa and Smith, 1994). These have included transfer of antisense oligonucleotides, ribozymes or siRNA targeted at viral sequences or to host factors critical for virus replication, e.g. the HIV coreceptor CCR-5; expression of TAR or REV decoy RNAs; expression of intrabodies directed against TAT, REV; expression of transdominant Rev-mutants such as RevM10, and targeting of HIV-coreceptors (summarized in Rössig and Brenner, 2004).

Stimulation of T-cells for efficient transduction can lead to unwanted increases in HIV replication. Further problems with nucleic-acid based therapies (antisense, RNAi, ribozymes) arise from the high mutation rate and enormous sequence variability of HIV-Quasispecies, which contrasts with the high sequence specificity of e.g. RNAi and ribozymes. It is predictable that resistant HIV-1 variants will arise rapidly *in vivo*. Selection of conserved sequences in the LTR or within leader sequences may partially avoid this evasion mechanism, as will simultaneously targeting of multiple essential HIV-1 sequences or of targeting essential host genes that do not mutate. RNA decoy requires overexpression of TAR- or RRE-Sequences (e.g. by Pol-III-transcription) but is also less susceptible to sequence variation, since it targets viral proteins. Ribozymes have enzymatic activity and are therefore only needed in smaller amounts than classical antisense strategies, though the natural mechanism of RNAi seems to require even smaller amounts of the therapeutic RNA-molecules (reviewed by Peracchi, 2004).

The success of highly active antiretroviral therapy (HAART) with conventional drugs has largely decreased the demand for such novel therapies as a first line approach; however, there may be a place as a supplement, e.g. for transient salvage therapy to eliminate multi-resistant HIV by RNAi.

6.   ANTIVIRAL IMMUNOTHERAPY

In the recent years, a valuable addition to conventional tumor therapy has been brought about by specific monoclonal antibodies. These reagents are directed either against antigens preferentially or exclusively expressed on tumors, such as Her2, or in other cases serve to eliminate specific cell populations, that include the tumor cells but also normal cells, such as anti-CD20 (Rituximab). Although responses could not be achieved in all patients with tumors expressing high levels of the respective target antigens, a focus of attention has been created toward targeted therapeutics, to develop alternative and potentially more efficient reagents.

Cytotoxic T-cell immunity is central to the control of viral infection. T lymphocytes physiologically recognize antigens through TCR interaction with short peptides presented by MHC class I or II molecules. Activation of naïve T cells depends on professional antigen-presenting cells (APCs) that provide necessary co-stimulatory signals for initial and clonal expansion. Without co-stimulation, unresponsiveness and clonal anergy may be the outcome of TCR activation. Furthermore, virus-specific cytotoxic T-cells can not be derived from immunologically naïve individuals.

A variety of techniques to bypass the need for immunization for the generation of immune effector cells with desired recognition specificity have been worked out and will be described below. However, large scale cultivation of (transfected or transduced) T-cells is laborious and technically challenging, as is expansion of virus-specific T-cells when and starting from small number of CTL in the periphery; complex and time consuming protocols for enrichment are necessary to achieve the quantities (~10^7 cells) that are considered to be required for therapy.

6.1   (Vector-mediated) Expansion of antiviral CTL

Adoptive T-cell therapy has been successfully applied to tumors, including EBV-related post-transplant lymphoproliferative disorders (PTLD) (Papadopoulos et al., 1994; Rooney et al., 1995, 1998), and to Cytomegalovirus (CMV)-related disease in immune suppressed transplant recipients (Walter et al., 1995; Peggs et al., 2003; Einsele and Hebart, 2004). Adoptive transfer of gene-marked tumor and HIV-specific CTL has shown intact homing to sites of HIV-replication and local function, as well as good persistence, although therapeutic benefits were limited (Brodie et al., 1999, 2000).

While effective treatment options exist for EBV-related PTLD in the form of Yttrium-90-Labeled Ibritumomab Tiuxetan radioimmunotherapy or Rituximab Immunotherapy (Jaeger et al., 2005; Milpied et al., 2000), CMV-related disease is still highly problematic; CMV reactivation can cause severe morbidity and mortality in immune compromised patients and those with HIV infection, even with appropriate antiviral drug treatment. The constellation of CMV-seronegative donor/CMV-seropositive recipient in allogenic hematopoietic progenitor cell transplantation is at particular risk. In such patients, the recovery of CMV-specific cytotoxic T lymphocytes (CTL) seems to be vital in the reconstitution of CMV specific immunity. Remarkably, beneficial effects were shown for adoptive transfer of

CMV-specific T cell clones or polyclonal CTL from CMV seropositive donors that were expanded *in vitro* (reviewed in Einsele and Hebart, 2004).

6.2   Expression of Virus-specific TCRs

Cytolytic effector cells can be genetically modified to carry virus-specific TCR or chimeric receptors based on antibody recognition sites on the surface: Antibody-mediated recognition of viral antigens can thereby be linked with cytolytic effector cells, which then possibly have potential for improved localization, homing and enhanced efficacy.

Transfer of tumor-specific TCRα and β-chains (or their specificity determining regions) to effector T-cells has been done in murine models (Kessels *et al.*, 2001, 2005; Willemsen *et al.*, 2003; Tahara *et al.*, 2003) and can probably be also achieved for virus-specific TCRs. Such transferred receptors will recognize processed peptides that can be derived from diverse viral proteins, not restricted to molecules expressed at the cell surface. However, recognition of MHC-presented peptides will depend on the correct HLA background. The formation of heterodimeric TCR consisting of one original and one transferred TCR chain could in theory give rise to new, potentially autoreactive T-cell specificities.

6.3   Directing immune response to viral or infection-specific cellular proteins at the cell surface

Strategies have been developed to specifically target activated T-cells towards infected cells that circumvent MHC-restriction of antigen recognition. The immunoreceptor ("T-body") strategy relies on transfer of cytolytic T-cells that carry an antigen-specific, recombinant receptor with signaling properties at their surface. In principle, immunoreceptor grafted T-cells can be directed against every cell defined by the particular target antigen at the cell surface.

Construction of chimeric TCR can be done via gene fusion in which a specific extracellular domain derived from a monoclonal antibody is combined with signal transducing components from T-cells. The specific complementarity determining regions (CDR) of the antibodies variable heavy and light chain, i.e. a single-chain antibody fragment (scFv), mediates recognition of target cells. This is coupled via a transmembrane domain to specific signaling domains of T-cell signal transducing molecules such as the TCRζ-chain and the costimulatory molecule CD28 (Abken *et al.*, 2002). Similar chimeric receptors have been derived from binding domains of natural ligands or antibodies fused directly to the TCRα and TCRβ chains, although this may result in formation of heterodimeric TCR when expressed on T-lymphocytes. Upon antigen binding, the chimeric receptors generate activating signals in the effector cell similar to those initiated by the TCR complex.

Delivery of such recombinant molecules into NK- or T-cells with cytolytic capacity has been shown to direct them toward the target which is specifically recognized by the antibodies CDR. The major advantage of this approach compared to expression of a conventional TCR is that recognition is not restricted to HLA, which would allow preparation, expansion and storage of virus specific cytotoxic cells for use "of-the-shelf". Furthermore, presentation of a peptide by MHC is not required. This is specifically interesting as a number of viruses, including human CMV, are able to downmodulate MHC from the surface of infected cells (Mocarski, Jr., 2004; Hewitt, 2003). On the other hand, the target molecule has to be presented on the surface of the infected cell (or tumor cell) in a form that is recognizable for the antibody CDR. Processed peptides derived from target molecules that are presented on the MHC are usually not recognized.

In initial clinical studies, infusion of such cells into patients proved to be safe and transient therapeutic effects have been observed. Initial HIV-directed clinical trials lacked therapeutic effectiveness and persistence of modified cells *in vivo*. There are inherent limitations of the immunoreceptor technology: Impaired persistence may result from recognition of the chimeric receptor as foreign, resulting in elimination. Furthermore, T-helper cells are required for efficient CTL function. Impaired signaling capacity that does not achieve the appropriate signal strength in the right context, and lack corresponding costimulatory signals, could also lead to anergy or elimination. Enhanced receptor molecules have therefore been constructed that include functional domains from costimulatory molecules. The promising technique could also be improved by usage of humanized molecules, as well as transduction of cells with specificity against a strong antigen, such as EBV, to improve persistence (summarized in Rössig and Brenner, 2003, 2004).

6.4 Problems arising from performance enhanced immune cells

Graft versus host disease like effects resulting from autoaggressive behavior of transferred autologous CTL have been observed in other contexts. The altered specificity of T-bodies and heterodimeric TCR can possibly induce autoimmune side effects, though they have not been observed with the relatively small patient numbers treated to date. Lymphoproliferative syndrome can be caused by transduced cells that result from insertion mutagenesis by the vector; this danger might be partially overcome by self-inactivating vectors, by introduction of suicide genes to ablate the unwanted cells, or by gene transfer into mature cells that may be easier to eliminate than stem cells. Possible solutions could be preselection of transferred cells for integration sites, if suitable strategies can be developed. Immunogenicity or undesired behavior of suicide genes could be overcome by novel physiological suicide genes such as CD20 (for elimination by Rituximab, Introna *et al.*, 2000), or conditional proapoptotic molecules derived from endogenous genes (summarized in Rössig and Brenner, 2004).

7. VECTOR- AND NUCLEIC ACID-BASED VACCINATION STRATEGIES

Numerous vector-based vaccination and immunity enhancing strategies have been developed, also employing specific gene transfer into professional APC such as Langerhans or dentritic cells (DC) (Jenne *et al.*, 2001). RNA- and DNA-virus-based replicons that are under development for prophylactic and therapeutic vaccination are mostly targeted against HIV, viral hepatitis, HPV-associated cervical cancer (Chapter 2.5 by L. Gissmann), as well as toward more fashionable and recent threats such as SARS-CoV, Filovirus, and those relating to "Biodefense" (Lee *et al.*, 2005). The reader is referred to recent reviews focusing on Adenovirus- (Tatsis and Ertl, 2004) Alphavirus- (Lundstrom, 2001), and DNA-vaccine vectors (Srivastava and Liu, 2003; Duenas-Carrera, 2004; Giri *et al.*, 2004). Vaccines based on Adenoviruses or on gene-modified poxviruses (Drexler *et al.*, 2004; Moss, 1996), such as modified Vaccinia strain Ankara (MVA), may be specifically considered in situations where adequate medical supply chains and financial resources do not exist. Vaccine formulations based on (lyophilized) DNA are also highly stable under adverse circumstances. The success of Variola elimination did crucially depend on the enormous stability, effectiveness and low cost of the Vaccinia virus. MVA-based vaccination trials showed good safety and immunogenicity of MVA expressing HIV Nef in HIV patients (Harrer *et al.*, 2005), and there is promising data on regression of precancerous cervical lesions in a preclinical study of MVA expressing HPV E2 (Corona Gutierrez *et al.*, 2004).

8. CONCLUSIONS

The intensified use of genetically modified viruses as gene transfer vectors has to consider a number of potential risk factors and their implications for activities with viral vectors. Specific and careful risk assessment is necessary from the perspective of naturally occurring cross-species transfer of viruses, and of possible host-range mutants resulting either from cell culture (using non-host species cells) or tropism engineering (Louz *et al.*, 2005). It remains open if viral vector-based antiviral therapy can be curative in an already established persistent infection. Yet such strategies using gene modified cells may be useful to (transiently) support the host immune system in its struggle with the viral intruder until its own resources have sufficiently recovered. The challenge will be to maintain the balance.

ACKNOWLEDGEMENTS

The author thanks Elke Heck, Florian Full and Frank Wucherpfennig for critical reading of the manuscript and Helmut Fickenscher for a previous compilation of Table 4.3-1. This work was supported by the Deutsche Forschungsgemeinschaft (SFB466 and 643), the Interdisciplinary Center for Clinical Research (IZKF: Genesis, Diagnostics and Therapy of Inflammation Processes) at the University of Erlangen-Nuremberg and the Wilhelm Sander-Stiftung (2002.033.1).

REFERENCES

Abken, H., Hombach, A., Heuser, C., Kronfeld, K., and Seliger, B., 2002, Tuning tumor-specific T-cell activation: a matter of costimulation? *Trends Immunol.* **23**: 240-245

Aires, d.S., Costa, M.J., Corte-Real, S., and Goncalves, J., 2005, Cell type-specific targeting with sindbis pseudotyped lentiviral vectors displaying anti-CCR5 single-chain antibodies. *Hum. Gene Ther.* **16**: 223-234

Alcami, A., Koszinowski, U.H., 2000a, 2000b, 2000c, Viral mechanisms of immune evasion. *Mol. Med. Today.* **6**: 365-372; *Immunol. Today* **21**: 447-455; *Trends Microbiol.* **8**: 410-418

An, D.S., Xie, Y., Mao, S.H., Morizono, K., Kung, S.K., and Chen, I.S., 2003, Efficient lentiviral vectors for short hairpin RNA delivery into human cells. *Hum. Gene Ther.* **14**: 1207-1212

Barton, G.M., Medzhitov, R., 2002, Retroviral delivery of small interfering RNA into primary cells. *Proc. Natl. Acad. Sci. U.S.A.* **99**: 14943-14945

Baum, C., Dullmann, J., Li, Z., Fehse, B., Meyer, J., Williams, D.A. *et al.*, 2003, Side effects of retroviral gene transfer into hematopoietic stem cells. *Blood* **101**: 2099-2114

Baum, C., von Kalle, C., Staal, F.J., Li, Z., Fehse, B., Schmidt, M. *et al.*, 2004, Chance or necessity? Insertional mutagenesis in gene therapy and its consequences. *Mol. Ther.* **9**: 5-13

Bell, P., Wang, L., Lebherz, C., Flieder, D.B., Bove, M.S., Wu, D. *et al.*, 2005, No Evidence for Tumorigenesis of ΛΛV Vectors in a Large-Scale Study in Mice. *Mol. Ther.* **12**: 299-306

Berkhout, B., 2004, RNA interference as an antiviral approach: targeting HIV-1. *Curr Opin. Mol. Ther.* **6**: 141-145

Bevec, D., Dobrovnik, M., Hauber, J., and Bohnlein, E., 1992, Inhibition of human immunodeficiency virus type 1 replication in human T cells by retroviral-mediated gene transfer of a dominant-negative Rev trans-activator. *Proc. Natl. Acad. Sci. U.S.A.* **89**: 9870-9874

Bevec, D., Jaksche, H., Oft, M., Wohl, T., Himmelspach, M., Pacher, A. *et al.*, 1996, Inhibition of HIV-1 replication in lymphocytes by mutants of the Rev cofactor eIF-5A. *Science* **271**: 1858-1860

Boden, D., Pusch, O., and Ramratnam, B., 2004, HIV-1-specific RNA interference. *Curr. Opin. Mol. Ther.* **6**: 373-380

Brodie, S.J., Lewinsohn, D.A., Patterson, B.K., Jiyamapa, D., Krieger, J., Corey, L. *et al.*, 1999, In vivo migration and function of transferred HIV-1-specific cytotoxic T cells. *Nat Med.* **5**: 34-41

Brodie, S.J., Patterson, B.K., Lewinsohn, D.A., Diem, K., Spach, D., Greenberg, P.D. *et al.*, 2000, HIV-specific cytotoxic T lymphocytes traffic to lymph nodes and localize at sites of HIV replication and cell death. *J. Clin .Invest.* **105**: 1407-1417

Bucher, E., Hemmes, H., de Haan, P., Goldbach, R., and Prins, M., 2004, The influenza A virus NS1 protein binds small interfering RNAs and suppresses RNA silencing in plants. *J. Gen. Virol.* **85**: 983-991

Buning, H., Ried, M.U., Perabo, L., Gerner, F.M., Huttner, N.A., Enssle, J. *et al.*, 2003, Receptor targeting of adeno-associated virus vectors. *Gene. Ther.* **10**: 1142-1151

Buonocore, L., Rose, J.K., 1993, Blockade of human immunodeficiency virus type 1 production in CD4+ T cells by an intracellular CD4 expressed under control of the viral long terminal repeat. *Proc. Natl. Acad. Sci. U.S.A.* **90**: 2695-2699

Cai, H., Tian, X., Hu, X.D., Li, S.X., Yu, D.H., and Zhu, Y.X., 2005, Combined DNA vaccines formulated either in DDA or in saline protect cattle from Mycobacterium bovis infection. *Vaccine* **23**: 3887-3895

Caruso, M., Bank, A., 1997, Efficient retroviral gene transfer of a Tat-regulated herpes simplex virus thymidine kinase gene for HIV gene therapy. *Virus Res.* **52**: 133-143

Cerwenka, A., Lanier, L.L., 2003, NKG2D ligands: unconventional MHC class I-like molecules exploited by viruses and cancer. *Tissue Antigens* **61**: 335-343

Clayton, J., 2004, RNA interference: the silent treatment. *Nature* **431**: 599-605

Conese, M., Auriche, C., and Ascenzioni, F., 2004, Gene therapy progress and prospects: episomally maintained self-replicating systems. *Gene Ther.* **11**: 1735-1741

Corona Gutierrez, C.M., Tinoco, A., Navarro, T., Contreras, M.L., Cortes, R.R., Calzado, P. *et al.*, 2004, Therapeutic vaccination with MVA E2 can eliminate precancerous lesions (CIN 1, CIN 2, and CIN 3) associated with infection by oncogenic human papillomavirus. *Hum. Gene Ther.* **15**: 421-431

Corte-Real, S., Collins, C., Aires, d.S., Silmas, P., Barbas, I.C., Chang, Y. *et al.*, 2005, Intrabodies targeting the Kaposi's sarcoma-associated herpesvirus latency antigen inhibit viral persistence in lymphoma cells. *Blood*: DOI 10.1182/blood-2005-04-1627

Cotter, M.A., Robertson, E.S., 1999, Molecular genetic analysis of herpesviruses and their potential use as vectors for gene therapy applications. *Curr. Opin. Mol. Ther.* **1**: 633-644

Cullen, B.R., 2005, Does RNA interference have a future as a treatment for HIV-1 induced disease? *AIDS Rev.* **7**: 22-25

Delecluse, H.J., Bartnizke, S., Hammerschmidt, W., Bullerdiek, J., and Bornkamm, G.W., 1993, Episomal and integrated copies of Epstein-Barr virus coexist in Burkitt lymphoma cell lines. *J. Virol.* **67**: 1292-1299

Delgadillo, M.O., Saenz, P., Salvador, B., Garcia, J.A., and Simon-Mateo, C., 2004, Human influenza virus NS1 protein enhances viral pathogenicity and acts as an RNA silencing suppressor in plants. *J. Gen. Viro.* **85**: 993-999

Donahue, R.E., Kessler, S.W., Bodine, D., McDonagh, K., Dunbar, C., Goodman, S. *et al.*, 1992, Helper virus induced T cell lymphoma in nonhuman primates after retroviral mediated gene transfer. *J. Exp. Med.* **176**: 1125-1135

Donsante, A., Vogler, C., Muzyczka, N., Crawford, J.M., Barker, J., Flotte, T. *et al.*, 2001, Observed incidence of tumorigenesis in long-term rodent studies of rAAV vectors. *Gene Ther.* **8**: 1343-1346

Doody, G.M., Leek, J.P., Bali, A.K., Ensser, A., Markham, A.F., and de Wynter, E.A., 2005, Marker gene transfer into human haemopoietic cells using a herpesvirus saimiri-based vector. *Gene Ther.* **12**: 373-379

Drexler, I., Staib, C., and Sutter, G., 2004, Modified vaccinia virus Ankara as antigen delivery system: how can we best use its potential? *Curr. Opin. Biotechnol.* **15**: 506-512

Duan, D., Sharma, P., Yang, J., Yue, Y., Dudus, L., Zhang, Y. *et al.*, 1998, Circular intermediates of recombinant adeno-associated virus have defined structural characteristics responsible for long-term episomal persistence in muscle tissue. *J. Virol.* **72**: 8568-8577

Duenas-Carrera, S., 2004, DNA vaccination against hepatitis C. *Curr Opin.Mol Ther.* **6**: 146-150

Einsele, H., Hebart, H., 2004, CMV-specific immunotherapy. *Hum. Immunol* **65**: 558-564

Elbashir, S.M., Harborth, J., Lendeckel, W., Yalcin, A., Weber, K., and Tuschl, T., 2001a, Duplexes of 21-nucleotide RNAs mediate RNA interference in cultured mammalian cells. *Nature* **411**: 494-498

Elbashir, S.M., Lendeckel, W., and Tuschl, T., 2001b, RNA interference is mediated by 21- and 22-nucleotide RNAs. *Genes Dev.* **15**: 188-200

Endres, M.J., Jaffer, S., Haggarty, B., Turner, J.D., Doranz, B.J., O'Brien, P.J. *et al.*, 1997, Targeting of HIV- and SIV-infected cells by CD4-chemokine receptor pseudotypes. *Science* **278**: 1462-1464

Gilboa, E., Smith, C., 1994, Gene therapy for infectious diseases: the AIDS model. *Trends Genet.* **10**: 139-144

Giri, M., Ugen, K.E., and Weiner, D.B., 2004, DNA vaccines against human immunodeficiency virus type 1 in the past decade. *Clin.Microbiol. Rev.* **17**: 370-389

Grandi, P., Spear, M., Breakefield, X.O., and Wang, S., 2004, Targeting HSV amplicon vectors. *Methods* **33**: 179-186

Gulley, M.L., Raphael, M., Lutz, C.T., Ross, D.W., and Raab-Traub, N., 1992, Epstein-Barr virus integration in human lymphomas and lymphoid cell lines. *Cancer* **70**: 185-191

Hacein-Bey-Abina, S., Le Deist, F., Carlier, F., Bouneaud, C., Hue, C., de Villartay, J.P. *et al.*, 2002, Sustained correction of X-linked severe combined immunodeficiency by ex vivo gene therapy. *N. Engl. J. Med.* **346**: 1185-1193

Hacein-Bey-Abina, S., von Kalle, C., Schmidt, M., McCormack, M.P., Wulffraat, N., Leboulch, P. *et al.*, 2003, LMO2-associated clonal T cell proliferation in two patients after gene therapy for SCID-X1. *Science* **302**: 415-419

Harrer, E., Bauerle, M., Ferstl, B., Chaplin, P., Petzold, B., Mateo, L. *et al.*, 2005, Therapeutic vaccination of HIV-1-infected patients on HAART with a recombinant HIV-1 nef-expressing MVA: safety, immunogenicity and influence on viral load during treatment interruption. *Antivir. Ther.* **10**: 285-300

Hauber, I., Bevec, D., Heukeshoven, J., Kratzer, F., Horn, F., Choidas, A. *et al.*, 2005, Identification of cellular deoxyhypusine synthase as a novel target for antiretroviral therapy. *J. Clin. Invest* **115**: 76-85

Hewitt, E.W., 2003, The MHC class I antigen presentation pathway: strategies for viral immune evasion. *Immunology* **110**: 163-169

Hiller, C., Tamguney, G., Stolte, N., Matz-Rensing, K., Lorenzen, D., Hor, S. *et al.*, 2000a, Herpesvirus saimiri pathogenicity enhanced by thymidine kinase of herpes simplex virus. *Virology* **278**: 445-455

Hiller, C., Wittmann, S., Slavin, S., and Fickenscher, H., 2000b, Functional long-term thymidine kinase suicide gene expression in human T cells using a herpesvirus saimiri vector. *Gene Ther.* **7**: 664-674

Introna, M., Barbui, A.M., Bambacioni, F., Casati, C., Gaipa, G., Borleri, G. *et al.*, 2000, Genetic modification of human T cells with CD20: a strategy to purify and lyse transduced cells with anti-CD20 antibodies. *Hum. Gene Ther.* **11**: 611-620

Jackson, A.L., Bartz, S.R., Schelter, J., Kobayashi, S.V., Burchard, J., Mao, M. *et al.*, 2003, Expression profiling reveals off-target gene regulation by RNAi. *Nat. Biotechnol* **21**: 635-637

Jaeger, G., Linkesch, W., Temmel, W., and Neumeister, P., 2005, AntiCD20 monoclonal antibody-based radioimmunotherapy of relapsed chemoresistant aggressive post-transplantation B-lymphoproliferative disorder in heart-transplant recipient. *Lancet Oncol.* **6**: 629-631

Jenne, L., Schuler, G., and Steinkasserer, A., 2001, Viral vectors for dendritic cell-based immunotherapy. *Trends Immunol* **22**: 102-107

Kaiser, J., 2005, American Society of Gene Therapy meeting. Retroviral vectors: a double-edged sword. *Science* **308**: 1735-1736

Kessels, H.W., Wolkers, M.C., and Schumacher, T.N., 2005, Gene transfer of MHC-restricted receptors. *Methods Mol. Med.* **109**: 201-214

Kessels, H.W., Wolkers, M.C., van, d.B., van der Valk, M.A., and Schumacher, T.N., 2001, Immunotherapy through TCR gene transfer. *Nat. Immunol* **2**: 957-961

Kohn, D.B., Gänsbacher, B., 2003, Letter to the editors of Nature from the American Society of Gene Therapy (ASGT) and the European Society of Gene Therapy (ESGT). *J. Gene Med.* **5**: 641

Krmpotic, A., Hasan, M., Loewendorf, A., Saulig, T., Halenius, A., Lenac, T. *et al.*, 2005, NK cell activation through the NKG2D ligand MULT-1 is selectively prevented by the glycoprotein encoded by mouse cytomegalovirus gene m145. *J.Exp. Med.* **201**: 211-220

Lee, J.S., Hadjipanayis, A.G., and Parker, M.D., 2005, Viral vectors for use in the development of biodefense vaccines. *Adv. Drug Deliv. Rev.* **57**: 1293-1314

Lee, M.T., Coburn, G.A., McClure, M.O., and Cullen, B.R., 2003, Inhibition of human im-munodeficiency virus type 1 replication in primary macrophages by using Tat- or CCR5-

specific small interfering RNAs expressed from a lentivirus vector. *J. Virol.* **77**: 11964-11972

Li, H., Li, W.X., and Ding, S.W., 2002a, Induction and suppression of RNA silencing by an animal virus. *Science* **296**: 1319-1321

Li, H.W., Lucy, A.P., Guo, H.S., Li, W.X., Ji, L.H., Wong, S.M. *et al.*, 1999, Strong host resistance targeted against a viral suppressor of the plant gene silencing defence mechanism. *EMBO J.* **18**: 2683-2691

Li, W.X., Li, H., Lu, R., Li, F., Dus, M., Atkinson, P. *et al.*, 2004, Interferon antagonist proteins of influenza and vaccinia viruses are suppressors of RNA silencing. *Proc. Natl. Acad. Sci. U.S.A.* **101**: 1350-1355

Li, Z., Dullmann, J., Schiedlmeier, B., Schmidt, M., von Kalle, C., Meyer, J. *et al.*, 2002b, Murine leukemia induced by retroviral gene marking. *Science* **296**: 497

Lichner, Z., Silhavy, D., and Burgyan, J., 2003, Double-stranded RNA-binding proteins could suppress RNA interference-mediated antiviral defences. *J. Gen. Virol.* **84**: 975-980

Lodoen, M.B., Abenes, G., Umamoto, S., Houchins, J.P., Liu, F., and Lanier, L.L., 2004, The cytomegalovirus m155 gene product subverts natural killer cell antiviral protection by disruption of H60-NKG2D interactions. *J. Exp. Med.* **200**: 1075-1081

Louz, D., Bergmans, H.E., Loos, B.P., and Hoeben, R.C., 2005, Cross-species transfer of viruses: implications for the use of viral vectors in biomedical research, gene therapy and as live-virus vaccines. *J. Gene. Med*: doi 10.1002/jgm.794

Lundstrom, K., 2001, Alphavirus vectors for gene therapy applications. *Curr. Gene Ther.* **1**: 19-29

Mazda, O., 2002, Improvement of nonviral gene therapy by Epstein-Barr virus (EBV)-based plasmid vectors. *Curr. Gene Ther.* **2**: 379-392

Means, R.E., Greenough, T., and Desrosiers, R.C., 1997, Neutralization sensitivity of cell culture-passaged simian immunodeficiency virus. *J. Virol.* **71**: 7895-7902

Mebatsion, T., Conzelmann, K.K., 1996, Specific infection of CD4+ target cells by recombinant rabies virus pseudotypes carrying the HIV-1 envelope spike protein. *Proc. Natl. Acad. Sci. U.S.A.* **93**: 11366-11370

Mebatsion, T., Finke, S., Weiland, F., and Conzelmann, K.K., 1997, A CXCR4/CD4 pseudotype rhabdovirus that selectively infects HIV-1 envelope protein-expressing cells. *Cell* **90**: 841-847

Miller, D.G., Petek, L.M., and Russell, D.W., 2004, Adeno-associated virus vectors integrate at chromosome breakage sites. *Nat. Genet.* **36**: 767-773

Miller, D.G., Rutledge, E.A., and Russell, D.W., 2002, Chromosomal effects of adeno-associated virus vector integration. *Nat. Genet.* **30**: 147-148

Miller, D.G., Trobridge, G.D., Petek, L.M., Jacobs, M.A., Kaul, R., and Russell, D.W., 2005, Large-scale analysis of adeno-associated virus vector integration sites in normal human cells. *J. Virol.* **79**: 11434-11442

Milpied, N., Vasseur, B., Parquet, N., Garnier, J.L., Antoine, C., Quartier, P. *et al.*, 2000, Humanized anti-CD20 monoclonal antibody (Rituximab) in post transplant B-lympho-proliferative disorder: a retrospective analysis on 32 patients. *Ann. Oncol.* **11 Suppl 1**: 113-116

Miyake, K., Iijima, O., Suzuki, N., Matsukura, M., and Shimada, T., 2001, Selective killing of human immunodeficiency virus-infected cells by targeted gene transfer and inducible gene expression using a recombinant human immunodeficiency virus vector. *Hum. Gene Ther.* **12**: 227-233

Mizuguchi, H., Hayakawa, T., 2004, Targeted adenovirus vectors. *Hum. Gene Ther.* **15**: 1034-1044

Mocarski, E.S., Jr., 2004, Immune escape and exploitation strategies of cytomegaloviruses: impact on and imitation of the major histocompatibility system. *Cell Microbiol.* **6**: 707-717

Moss, B., 1996, Genetically engineered poxviruses for recombinant gene expression, vaccination, and safety. *Proc. Natl. Acad. Sci. U.S.A.* **93**: 11341-11348

Nakai, H., Montini, E., Fuess, S., Storm, T.A., Grompe, M., and Kay, M.A., 2003, AAV serotype 2 vectors preferentially integrate into active genes in mice. *Nat. Genet.* **34**: 297-302

Nakai, H., Wu, X., Fuess, S., Storm, T.A., Munroe, D., Montini, E. *et al.*, 2005, Large-scale molecular characterization of adeno-associated virus vector integration in mouse liver. *J. Virol.* **79**: 3606-3614

Nakai, H., Yant, S.R., Storm, T.A., Fuess, S., Meuse, L., and Kay, M.A., 2001, Extrachromosomal recombinant adeno-associated virus vector genomes are primarily responsible for stable liver transduction in vivo. *J. Virol.* **75**: 6969-6976

Obaru, K., Fujii, S., Matsushita, S., Shimada, T., and Takatsuki, K., 1996, Gene therapy for adult T cell leukemia using human immunodeficiency virus vector carrying the thymidine kinase gene of herpes simplex virus type 1. *Hum. Gene Ther.* **7**: 2203-2208

Oehmig, A., Fraefel, C., Breakefield, X.O., and Ackermann, M., 2004, Herpes simplex virus type 1 amplicons and their hybrid virus partners, EBV, AAV, and retrovirus. *Curr. Gene Ther.* **4**: 385-408

Office of Biotechnological Activities at NIH and the Recombinant DNA Advisory Committee, http://www4.od.nih.gov/oba/rac/clinicaltrial.htm

Papadopoulos, E.B., Ladanyi, M., Emanuel, D., Mackinnon, S., Boulad, F., Carabasi, M.H. *et al.*, 1994, Infusions of donor leukocytes to treat Epstein-Barr virus-associated lymphoproliferative disorders after allogeneic bone marrow transplantation. *N. Engl. J. Med.* **330**: 1185-1191

Peggs, K.S., Verfuerth, S., Pizzey, A., Khan, N., Guiver, M., Moss, P.A. *et al.*, 2003, Adoptive cellular therapy for early cytomegalovirus infection after allogeneic stem-cell transplantation with virus-specific T-cell lines. *Lancet* **362**: 1375-1377

Peracchi, A., 2004, Prospects for antiviral ribozymes and deoxyribozymes. *Rev. Med. Virol.* **14**: 47-64

Pfeffer, S., Sewer, A., Lagos-Quintana, M., Sheridan, R., Sander, C., Grasser, F.A. *et al.*, 2005, Identification of microRNAs of the herpesvirus family. *Nat. Med.* **2**: 269-276

Pfeffer, S., Zavolan, M., Grasser, F.A., Chien, M., Russo, J.J., Ju, J. *et al.*, 2004, Identification of virus-encoded microRNAs. *Science* **304**: 734-736

Riddell, S.R., Elliott, M., Lewinsohn, D.A., Gilbert, M.J., Wilson, L., Manley, S.A. *et al.*, 1996, T-cell mediated rejection of gene-modified HIV-specific cytotoxic T lymphocytes in HIV-infected patients. *Nat. Med.* **2**: 216-223

Rooney, C.M., Smith, C.A., Ng, C.Y., Loftin, S., Li, C., Krance, R.A. *et al.*, 1995, Use of gene-modified virus-specific T lymphocytes to control Epstein-Barr-virus-related lymphoproliferation. *Lancet* **345**: 9-13

Rooney, C.M., Smith, C.A., Ng, C.Y., Loftin, S.K., Sixbey, J.W., Gan, Y. *et al.*, 1998, Infusion of cytotoxic T cells for the prevention and treatment of Epstein-Barr virus-induced lymphoma in allogeneic transplant recipients. *Blood* **92**: 1549-1555

Rössig, C., Brenner, M.K., 2003, Chimeric T-cell receptors for the targeting of cancer cells. *Acta. Haematol.* **110**: 154-159

Rössig, C., Brenner, M.K., 2004, Genetic modification of T lymphocytes for adoptive immunotherapy. *Mol. Ther.* **10**: 5-18

Roth, B.M., Pruss, G.J., and Vance, V.B., 2004, Plant viral suppressors of RNA silencing. *Virus Research* **102**: 97-108

Samols, M.A., Hu, J., Skalsky, R.L., and Renne, R., 2005, Cloning and identification of a microRNA cluster within the latency-associated region of Kaposi's sarcoma-associated herpesvirus. *J. Virol.* **79**: 9301-9305

Sanchez-Vargas, I., Travanty, E.A., Keene, K.M., Franz, A.W., Beaty, B.J., Blair, C.D. *et al.*, 2004, RNA interference, arthropod-borne viruses, and mosquitoes. *Virus Res.* **102**: 65-74

Sandrin, V., Russell, S.J., and Cosset, F.L., 2003, Targeting retroviral and lentiviral vectors. *Curr. Top. Microbiol. Immunol.* **281**: 137-178

Savenkov, E.I., Valkonen, J.P., 2002, Silencing of a viral RNA silencing suppressor in transgenic plants. *J. Gen. Virol.* **83**: 2325-2335

Scacheri, P.C., Rozenblatt-Rosen, O., Caplen, N.J., Wolfsberg, T.G., Umayam, L., Lee, J.C. *et al.*, 2004, Short interfering RNAs can induce unexpected and divergent changes in the levels of untargeted proteins in mammalian cells. *Proc. Natl. Acad. Sci. U.S.A.* **101**: 1892-1897

Scherer, L.J., Rossi, J.J., 2003, Approaches for the sequence-specific knockdown of mRNA. *Nat.Biotechnol.* **21**: 1457-1465

Schnell, M.J., Johnson, J.E., Buonocore, L., and Rose, J.K., 1997, Construction of a novel virus that targets HIV-1-infected cells and controls HIV-1 infection. *Cell* **90**: 849-857

Schnepp, B.C., Clark, K.R., Klemanski, D.L., Pacak, C.A., and Johnson, P.R., 2003, Genetic fate of recombinant adeno-associated virus vector genomes in muscle. *J. Virol.* **77**: 3495-3504

Shankar, P., Manjunath, N., and Lieberman, J., 2005, The prospect of silencing disease using RNA interference. *JAMA* **293**: 1367-1373

Soldan, S.S., Plassmeyer, M.L., Matukonis, M.K., and Gonzalez-Scarano, F., 2005, La Crosse virus nonstructural protein NSs counteracts the effects of short interfering RNA. *J. Virol.* **79**: 234-244

Somia, N.V., Miyoshi, H., Schmitt, M.J., and Verma, I.M., 2000, Retroviral vector targeting to human immunodeficiency virus type 1-infected cells by receptor pseudotyping. *J. Virol.* **74**: 4420-4424

Srivastava, I.K., Liu, M.A., 2003, Gene vaccines. *Ann Intern. Med.* **138**: 550-559

Stevenson, M., 2004, Therapeutic potential of RNA interference. *N. Engl. J. Med.* **351**: 1772-1777

Sullivan, C.S., Grundhoff, A.T., Tevethia, S., Pipas, J.M., and Ganem, D., 2005, SV40-encoded microRNAs regulate viral gene expression and reduce susceptibility to cytotoxic T cells. *Nature* **435**: 682-686

Sullivan, N.J., Geisbert, T.W., Geisbert, J.B., Xu, L., Yang, Z.Y., Roederer, M. *et al.*, 2003, Accelerated vaccination for Ebola virus haemorrhagic fever in non-human primates. *Nature* **424**: 681-684

Tahara, H., Fujio, K., Araki, Y., Setoguchi, K., Misaki, Y., Kitamura, T. *et al.*, 2003, Reconstitution of CD8+ T cells by retroviral transfer of the TCR alpha beta-chain genes isolated from a clonally expanded P815-infiltrating lymphocyte. *J. Immunol* **171**: 2154-2160

Tatsis, N., Ertl, H.C., 2004, Adenoviruses as vaccine vectors. *Mol. Ther.* **10**: 616-629

The Journal of Gene Medicine Clinical Trial Site, http://82.182.180.141/trials/index.html

Themis, M., Waddington, S.N., Schmidt, M., von Kalle, C., Wang, Y., Al Allaf, F. *et al.*, 2005, Oncogenesis Following Delivery of a Nonprimate Lentiviral Gene Therapy Vector to Fetal and Neonatal Mice. *Mol. Ther.* **12**: 763-771

Tomar, R.S., Matta, H., and Chaudhary, P.M., 2003, Use of adeno-associated viral vector for delivery of small interfering RNA. *Oncogene* **22**: 5712-5715

Tuschl, T., Zamore, P.D., Lehmann, R., Bartel, D.P., and Sharp, P.A., 1999, Targeted mRNA degradation by double-stranded RNA in vitro. *Genes Dev.* **13**: 3191-3197

Verhoeyen, E., Cosset, F.L., 2004, Surface-engineering of lentiviral vectors. *J. Gene Med.* **6 Suppl 1**: S83-S94

Voinnet, O., Pinto, Y.M., and Baulcombe, D.C., 1999, Suppression of gene silencing: a general strategy used by diverse DNA and RNA viruses of plants. *Proc. Natl. Acad. Sci. U.S.A.* **96**: 14147-14152

Walter, E.A., Greenberg, P.D., Gilbert, M.J., Finch, R.J., Watanabe, K.S., Thomas, E.D. *et al.*, 1995, Reconstitution of cellular immunity against cytomegalovirus in recipients of allogeneic bone marrow by transfer of T-cell clones from the donor. *N. Engl. J. Med.* **333**: 1038-1044

Wienholds, E., Plasterk, R.H., 2005, MicroRNA function in animal development. *FEBS Lett.*: doi:10.1016/j.febslet.2005.07.070

Wieser, C., Stumpf, D., Grillhosl, C., Lengenfelder, D., Gay, S., Fleckenstein, B. *et al.*, 2005, Regulated and constitutive expression of anti-inflammatory cytokines by nontransforming herpesvirus saimiri vectors. *Gene Ther.* **12**: 395-406

Willemsen, R.A., Debets, R., Chames, P., and Bolhuis, R.L., 2003, Genetic engineering of T cell specificity for immunotherapy of cancer. *Hum. Immunol* **64**: 56-68

Wiznerowicz, M., Trono, D., 2003, Conditional suppression of cellular genes: lentivirus vector-mediated drug-inducible RNA interference. *J. Virol.* **77**: 8957-8961

Wu, C.J., Huang, H.W., Liu, C.Y., Hong, C.F., and Chan, Y.L., 2005, Inhibition of SARS-CoV replication by siRNA. *Antiviral Res.* **65**: 45-48

Yang, Z.Y., Wyatt, L.S., Kong, W.P., Moodie, Z., Moss, B., and Nabel, G.J., 2003, Overcoming immunity to a viral vaccine by DNA priming before vector boosting. *J. Virol.* **77**: 799-803

Zamore, P.D., Tuschl, T., Sharp, P.A., and Bartel, D.P., 2000, RNAi: double-stranded RNA directs the ATP-dependent cleavage of mRNA at 21 to 23 nucleotide intervals. *Cell* **101**: 25-33

INDEX